Microwaves in Chemistry Applications

Advances in Green and Sustainable Chemistry

Microwaves in Chemistry Applications

Fundamentals, Methods, and Future Trends

Aparna Das

Department of Mathematics and Natural Sciences, College of Sciences and Human Studies, Prince Mohammad Bin Fahd University, Al Khobar, Kingdom of Saudi Arabia

Bimal Krishna Banik

Department of Mathematics and Natural Sciences, College of Sciences and Human Studies, Prince Mohammad Bin Fahd University, Al Khobar, Kingdom of Saudi Arabia

ELSEVIER

Elsevier
Radarweg 29, PO Box 211, 1000 AE Amsterdam, Netherlands
The Boulevard, Langford Lane, Kidlington, Oxford OX5 1GB, United Kingdom
50 Hampshire Street, 5th Floor, Cambridge, MA 02139, United States

Library of Congress Cataloging-in-Publication Data
A catalog record for this book is available from the Library of Congress

British Library Cataloguing-in-Publication Data
A catalogue record for this book is available from the British Library

ISBN: 978-0-12-822895-1

For information on all Elsevier publications
visit our website at https://www.elsevier.com/books-and-journals

Publisher: Susan Dennis
Acquisitions Editor: Anneka Hess
Editorial Project Manager: Lena Sparks
Production Project Manager: Sruthi Satheesh
Cover Designer: Victoria Pearson

Typeset by SPi Global, India

Contents

Authors' biographies xi
Preface xv

Part I
Introduction to microwave chemistry

1. Foundational principles of microwave chemistry

1.1 Introduction 3
1.2 Electromagnetic spectrum 3
1.3 Microwave radiation 5
1.3.1 Properties of microwaves 6
1.3.2 Advantages and disadvantages of microwave radiation 6
1.3.3 Applications of microwave radiation 7
1.4 Microwave chemistry 10
1.4.1 Microwave heating 10
1.4.2 Principles of microwave heating 13
1.4.3 Microwave versus conventional heating 16
1.4.4 Advantages of microwave heating 17
1.4.5 Applications of microwave heating 19
1.5 Conclusion 21
Acknowledgments 21
References 21

2. Microwave equipment for chemistry

2.1 Introduction 27
2.2 The basic structure of the microwave device 27
2.2.1 Microwave generator 28
2.2.2 High-frequency power supplies 32
2.2.3 Isolators 32
2.2.4 Waveguide 32
2.2.5 Power monitors 33
2.2.6 Applicator (cavity) 33
2.3 Domestic microwave ovens in the laboratory 35
2.3.1 Modifications in domestic microwave ovens 35
2.4 Commercial microwave reactors 47

2.5 Conclusion 56
Acknowledgments 56
References 56

3. Modeling and interpreting microwave effects

3.1 Introduction 61
3.2 Modeling of microwave heat transfer in liquids 62
3.3 Modeling of microwave heat transfer in solids 80
3.4 Conclusion 98
Acknowledgments 98
References 98

Part II
Methods in microwave chemistry

4. Microwave-assisted synthesis of oxygen- and sulfur-containing organic compounds

4.1 Introduction 107
4.2 Oxygen-containing organic compounds 107
4.2.1 Xanthene derivatives 107
4.2.2 Furan derivatives 109
4.2.3 Cinnamic acid and coumarin derivatives 114
4.3 Organosulfur compounds 123
4.4 Other heterocyclic compounds 129
4.4.1 Benzoselenophenes 129
4.4.2 Phenanthrene 130
4.4.3 Arylphosphonates 131
4.4.4 Alcohols and acetones 133
4.5 Conclusion 137
Acknowledgments 138
References 138

5. Microwave-assisted synthesis of *N*-heterocycles

5.1 Introduction 143
5.2 Pyrimidines 144
5.3 Steroidal derivatives 147
5.4 Thiazoles 149
5.5 Imines 151
5.6 Oxadiazoles 153
5.7 Tetrazoles and triazoles 154
5.8 Quinolines 156
5.9 Indolizines 157
5.10 β-Lactams 159
5.11 Pyrroles 163

5.12 Aza cyclo-octanes 164
5.13 Schiff bases 165
5.14 Chalcone 168
5.15 Quinoxalines 170
5.16 Carbazoles 171
5.17 Ipsapirones 174
5.18 Thiazoles 176
5.19 Acridine-acetazolamide conjugates 184
5.20 Conclusion 188
Acknowledgments 188
References 188

6. Microwave-assisted oxidation and reduction reactions

6.1 Introduction 199
6.2 Oxidation reactions 199
6.2.1 Oxidation of polyphenolic molecules 199
6.2.2 Oxidation of alcohols 202
6.2.3 Oxidation of carbohydrates 208
6.3 Reduction reactions 216
6.3.1 Pd/C-catalyzed reduction of levulinic acid induced by microwave 216
6.3.2 Reduction of nitroarenes assisted by microwave using NiO nanoparticles 218
6.3.3 Wolff-Kishner reduction assisted by microwave 219
6.3.4 Reduction of trimethylsilyl carbonyl compounds mediated by microwave in the presence of sodium borohydride 219
6.3.5 Reduction of carbonyl compounds induced by microwave in glycerol 220
6.3.6 Crossed Cannizzaro reaction using barium hydroxide mediated by microwave 222
6.3.7 Microwave-induced reduction of levulinic acid with alcohols in the presence of Ru/C 223
6.3.8 Microwave-assisted reduction of acetophenones using Ni-Al alloy in water 224
6.3.9 Microwave-assisted one-pot selective azido reduction/tandem cyclization in condensed and solid phase with nickel boride 224
6.3.10 Reduction of β-carboline imines by chiral auxiliaries induced by microwave 225
6.3.11 Microwave-induced reduction of α, β-unsaturated carbonyl compounds using sodium borohydride supported on magnesium sulfate 226
6.3.12 Reduction of carbonyl compounds using sodium borohydride supported on alumina induced by microwave 228

6.3.13 Microwave-assisted bioreduction of aromatic aldehydes using aloe vera 228
6.3.14 Microwave-induced reduction of aromatic nitro compounds with ammonium chloride and zinc dust 228
6.3.15 Molybdenum hexacarbonyl and DBU-mediated reduction of nitro compounds induced by microwave 229
6.3.16 Microwave-induced solvent-free reduction of aromatic nitro compounds with alumina-supported hydrazine 230
6.3.17 Reduction of carbonyl compounds with sodium borohydride mediated by microwave 231
6.3.18 Reduction of alkene groups in β-lactams induced by microwave 233
6.3.19 Reductive cyclization induced by microwave 236
6.4 Conclusion 236
Acknowledgments 236
References 237

7. Microwave-assisted enzymatic reactions

7.1 Introduction 245
7.2 Lipases 245
7.3 Cellulase 255
7.4 Glycosidases 259
7.5 Protease 262
7.6 Baker's yeast 265
7.7 Conclusion 271
Acknowledgments 272
References 272

Part III
Applications of microwave chemistry

8. Microwave-assisted sterilization

8.1 Introduction 285
8.2 Microwave sterilization in medical industry 286
8.3 Microwave sterilization in food industry 288
8.4 Sterilization using microwave plasmas 304
8.5 Sterilization using combined system 313
8.6 Conclusion 320
Acknowledgments 320
References 320

9. Microwave-assisted CVD processes for diamond synthesis

9.1 Diamond: Structure, properties, and applications 329
9.1.1 Natural diamonds 330
9.1.2 Synthetic diamond 332
9.1.3 High-pressure high-temperature method 333
9.1.4 Chemical vapor deposition method 333
9.2 Synthesis of the single-crystal diamond 338
9.3 Synthesis of polycrystalline diamond film 341
9.4 Diamond doping 346
9.4.1 Co-deposition (composites) 352
9.5 Ultrananocrystalline diamond films 355
9.6 Conclusion 364
Acknowledgments 364
References 364

10. Future trends in microwave chemistry and biology

10.1 Current trends in microwave technology 375
10.2 Future trends in microwave technology 378
10.2.1 Organic and inorganic chemistry 378
10.2.2 Material science 379
10.2.3 Biochemistry 379
10.2.4 Flow chemistry 380
10.2.5 Communication 380
10.2.6 Earth science 380
10.2.7 Food industry 381
10.2.8 Medicinal chemistry 381
10.2.9 Space shuttles 381
10.2.10 Portable microwave 382
Acknowledgments 382
References 382

Index 385

Authors' biographies

Aparna Das, Ph.D.
Department of Mathematics and Natural Sciences, College of Sciences and Human Studies, Prince Mohammad Bin Fahd University, Al Khobar, Kingdom of Saudi Arabia.

Aparna Das obtained her Ph.D. in material science/nanophysics from the Joseph Fourier University based upon the work on "Semiconductor Quantum Dots for Optochemical Sensor Application" at French Alternative Energies and Atomic Energy Commission (CEA), in France. Her postdoctoral experiences include working as a research scientist at the California NanoSystems Institute and Electric Engineering Department, University of California, Los Angeles (UCLA), in the United States, and as an experienced researcher at the Nanowiring-Marie Curie Initial Training Network, Georg August Universität Göttingen, in Germany. Currently, she is working as an assistant professor at Prince Mohammad Bin Fahd University, in Saudi Arabia.

Dr. Das's research interests include computer-assisted physicochemical methods, quantum mechanical calculations, interdisciplinary science (biology, chemistry, and physics) for drug development, computer-aided drug design, microwave applications, photochemical reactions, III-nitride-based chemical sensors, solar cells, optoelectronic devices, and synthesis of thin layers and nanostructures including quantum wells, quantum dots, and nanowires. Dr. Das also has experience in molecular-beam epitaxy, plasma-enhanced chemical vapor deposition, sputtering, atomic force microscopy, scanning electron microscopy, X-ray diffraction and spectroscopy, UV and IR photoluminescence, transmission spectroscopy, time-resolved photoluminescence, Raman spectroscopy, hall effect measurements, spin coating, reactive-ion etching, electron beam evaporator, rapid thermal processing, electron-beam lithography, photolithography, wire bonding, profilometer and wafer saw, optical fibers (cleaving, cutting, splicing, launching light, loss measurements, and characterization of various parameters), and in the operation of various lasers, light-emitting diodes, spectrophotometers, fluorometers, interferometers, and photomultiplier tubes.

Dr. Das is the author of a book and she has written several book chapters published by internationally reputed publishers. She has published approximately 50 papers and has 52 contributions to international conferences. She has received several conference invitations to present her research works. Importantly, a significant portion of her research contributions is undergoing editorial reviews by many journals. Dr. Das has served as a guest editor of two international journals and she has acted as a reviewer. Dr. Das has obtained several competitive grants and has filed a patent application. She has received the young women researcher award, European Microscopy Society (EMS) outstanding paper awards, and several international fellowships such as the Marie-Curie fellowship, CEA-CNRS research fellowship, and Brain Korea 21 fellowship. Dr. Das's research works have been recognized by Bentham Publishers (Press release at the AAAS Site), Heidelberg University, and Prince Mohammad Bin Fahd University.

Bimal Krishna Banik, Ph.D., C.Chem., F.R. S.C., F.I.C.S., F.I.S.R.O.S.E.T., F.R.S.C.S.
Department of Mathematics and Natural Sciences, College of Sciences and Human Studies, Deanship of Research Development, Prince Mohammad Bin Fahd University, Al Khobar, Kingdom of Saudi Arabia.

Bimal Krishna Banik conducted his doctoral research at the Indian Association for the Cultivation of Science, Calcutta. Then, he pursued postdoctoral research at Case Western Reserve University (United States) and Stevens Institute of Technology (United States). Dr. Banik was a tenured Full Professor in chemistry and First President's Endowed Professor in Science & Engineering at the University of Texas-Pan American and the Vice President of Research & Education Development of the Community Health Systems of South Texas. At present, Dr. Banik is Professor and Senior Researcher of the Deanship of Research Development & College of Natural Sciences at the Prince Mohammed Bin Fahd University in Kingdom of Saudi Arabia.

Professor Banik has taught organic and medicinal chemistry to B.S., M.S., and Ph.D. students in United States universities for many years. His teaching skills are exceptionally strong and these are proved by several thousand students' and peer's evaluations. He has mentored approximately 400 students, 20 postdoctoral fellows, and 7 Ph.D. research scientists and has advised 22 university faculties. Professor Banik has acted as the advisor of two students' organizations and societies that have 1400 students.

Professor Banik has conducted synthetic chemistry and chemical biology research on ovary, colon, breast, blood, prostate, brain, pancreas, and skin cancers (also on NCI 60 cancer cell lines); antibiotics; hormones; catalysis; green chemistry; natural products; and microwave-induced reactions. As a principal investigator (PI), he has been awarded $7.25 million in grants from USA NIH and USA NCI. Importantly, he has 492 peer-reviewed publications along with 502 presentation abstracts. The number of citations of his publications is more than 7500. His research has been exposed in media approximately 200 times. Professor Banik has served as the PI of a joint green chemistry symposium between United States and India. He has presided over 20 symposiums at the American Chemical Society (ACS) National Meetings and over 2 dozen conferences at the state, national, and international levels, including 1 at the Nobel Prize celebration in Germany. In the capacity of chair, he has introduced more than 300 speakers. He is a reviewer of 93, editorial board member of 28, editor-in-chief of 12, founder of 4, associate editor of 4, and guest editor of 10 journals. As the editor-in-chief, he has recruited approximately 200 associate editors, regional editors, and editorial board members from different countries. He is an examiner of NSF, NCI, NRC, DOE, ACS, and international grant applications; reviewer of promotion and tenure of faculty of national and international universities; examiner of doctoral theses; and panel member of NSF and NCI/NIH grant sections. Over the years, he has served as the chair/member of more than 100 scientific committees. Professor Banik has served as the chair of the University of Texas MD Anderson Cancer Center's drug discovery symposiums and directed the NCI-funded analytical chemistry core research laboratory.

Professor Banik is ranked within the top 2% of researchers in the world. He has received the Indian Chemical Society's (ICS) Lifetime Achievement Award in 2018; Mahatma Gandhi Pravasi Honor Medal from the UK Parliament; ICS's Professor P.K. Bose Endowment Medal; Dr. M.N. Ghosh Gold Medal; University of Texas Board of Regents' Outstanding Teaching Award; 5 top-cited papers awards by Elsevier Journals; approximately 50 certificates of excellence in his profession; Indian Association Community Service Award; ACS Member Service Award; NCI webpage recognition; best researcher and mentor award by the UTPA; chosen as one of the World's Most Influential People on Earth in Year-2016 by US News Corporation; Burdwan University Eminent Alumnus recognition; First President's Endowed Professorship at the UTPA in its 87 years of history; UTPA's award for excellence in international studies. Some of his international research presentations are considered as keynote, plenary, inaugural, and award-winning lectures. Dr. Banik has received more than 200 invitations to deliver lectures in United States, India, United Kingdom, Germany, China, Hong Kong, Greece, Italy, France, Jamaica, Sweden, Japan, Singapore, Pakistan, Norway, Bangladesh, Canada, Mexico, Vietnam, South Korea, Thailand, Saudi Arabia, United Arab Emirates, Argentina, Portugal, Switzerland, Venezuela, Brazil, Spain, New Zealand, Egypt, Austria, Australia, and Turkey. He is also invited to write research and text books by major publishers, including Wiley, Elsevier, Springer, Springer Nature, Taylor & Francis, De Gruter, Bentham, Thompson, Linus, Nova, Pearson, Cengage, Houghton Mifflin, and PMU Press.

Preface

Microwaves are electromagnetic waves with wavelengths ranging from as short as one millimeter to as long as one meter (a frequency range of 300 MHz to 300 GHz). That is, microwaves occupy a place in the electromagnetic spectrum with frequency above ordinary radio waves, and below infrared light. Microwaves are widely used in modern science and engineering technology, for example, in point-to-point communication links, wireless networks, microwave radio relay networks, radar, satellite and spacecraft communication, chemical reaction, medical diathermy and cancer treatment, remote sensing, radio astronomy, particle accelerators, spectroscopy, industrial heating, collision avoidance systems, garage door openers and keyless entry systems, and for cooking food in microwave.

The main focus of our book is the application of microwave in chemistry, biology, and engineering. Microwave chemistry is the science of applying microwave radiation to chemical reactions. Applications of microwave chemistry in organic chemistry, inorganic chemistry, and engineering are described in this book. Microwave technology has certain benefits over conventional ovens: reaction rate acceleration, milder reaction conditions, higher chemical yield, lower energy usage, and different reaction selectivity. Selected benefits of the microwave technology will be discussed in our book with examples.

Although the use of microwave heating in chemical modification can be traced back to the 1950s, the real reaction kinetics and mechanism behind the microwave effect is still a mystery. As this book is covering several significant topics like modeling and interpreting microwave effects and future trends, this will be a promising acquisition in the field of microwave technology.

This book has 10 chapters written by Aparna Das and Bimal Krishna Banik, and it is divided into 3 parts.

Part I describes the introduction to microwave chemistry, and it has three chapters: foundational principles of microwave chemistry, microwave equipment for chemistry, and modeling and interpreting microwave effects. These three chapters summarize the origin of microwave chemistry and the basic principle behind the microwave effect; different equipment related to the microwave chemistry and their functions, and different theoretical studies and computerized simulation works for interpreting microwave effects.

Part II describes the methods in microwave chemistry, and it has four chapters: microwave-assisted synthesis of oxygen and sulfur heterocycles, microwave-assisted synthesis of nitrogen heterocycles, reductions and oxidations using microwave chemistry, and enzyme-mediated reactions using microwave chemistry. These four chapters describe the microwave-assisted synthesis, recent microwave-assisted catalytic reactions performed in different media, several examples of microwave-assisted organic reduction and oxidation reactions, the rate of enzyme-mediated reactions, and the reaction kinetics and different enzyme-mediated reactions.

Part III focuses on the applications of microwave chemistry, and it has three chapters: sterilization method in chemistry, biology, and medicine; microwave-enhanced CVD processes for diamond synthesis; and future trends in microwave chemistry and biology. These three chapters explore the applications of microwave chemistry for sterilization, diamond synthesis, and the future potential of the microwave.

Many innovative approaches are yet to come with likely with microwave in the future days. Because of the trend, it can be predicted that microwave would be used in new materials science, medicine, engineering, and enzyme-related science. The future application of microwave in flow and scale-up science may be affected by this technology. New ways of using microwave can also be demonstrated. The application of microwave irradiation on diverse research demonstrates a clear superiority over conventional heating. Despite this, it is not understood if microwaves will be used into the manufacturing processes.

The authors are highly grateful to all scientists who have contributed in microwave-induced various processes as described in this book. This book has been possible because of the development of many significant new areas of science which are unbelievable. The authors are highly appreciative to Elsevier, particularly to Ms. Susan Dennis (Publisher), Ms. Anneka Hess (Acquisitions Editor), Ms. Lena Sparks (Editorial Project Manager), Ms. Sruthi Satheesh (Production Project Manager), Ms. Victoria Pearson (Cover Designer) and Ms. Narmatha Mohan (copyrights coordinator). Ms. Hess and Ms. Sparks help the authors in many ways, including providing a constant encouragement. Our thanks go to Dr. Bela Torok and Dr. Tomothy Dransfield, the series editors of "Advances in Green and Sustainable Chemistry."

Aparna Das
Prince Mohammad Bin Fahd University, Khobar,
Kingdom of Saudi Arabia
Bimal Krishna Banik
Prince Mohammad Bin Fahd University, Khobar,
Kingdom of Saudi Arabia

Part I

Introduction to microwave chemistry

Chapter 1

Foundational principles of microwave chemistry

Aparna Das* and Bimal Krishna Banik*

Department of Mathematics and Natural Sciences, College of Sciences and Human Studies, Prince Mohammad Bin Fahd University, Al Khobar, Kingdom of Saudi Arabia

**Corresponding authors: E-mails: aparnadasam@gmail.com (Aparna Das); bimalbanik10@gmail.com; bbanik@pmu.edu.sa (Bimal Krishna Banik)*

1.1 Introduction

Many scholars have described the applications of microwaves in different disciplines. For example, microwave chemistry is applied in various fields such as pharmaceuticals, biotechnology, plastics, petroleum, and chemicals. In general, microwave-induced methods are economical, fast, high yielding, and selective in most of these cases. This chapter discusses the foundational principle of microwave chemistry and its history of origin, advantages, and disadvantages.

1.2 Electromagnetic spectrum

The entire range of electromagnetic (EM) radiation is known as the electromagnetic spectrum (Fig. 1). It covers electromagnetic waves with frequencies ranging from below 1 Hz to above 10^{25} Hz and consists of radio waves, microwaves, infrared, visible light, ultraviolet, X-rays, and gamma rays.

An electromagnetic wave comprises two constituents, an electric field and a magnetic field (Fig. 2). The electric field and magnetic field oscillate perpendicular to each other, and the wave moves in a direction perpendicular to these fields [1–6]. Thus, it forms a transverse wave. The electromagnetic waves do not need a medium for transmission, and it can move freely through a vacuum.

An electromagnetic wave can be described in terms of its frequency (f), wavelength (λ), or photon energy (E). The frequency and wavelength are related by the wave equation:

$$\lambda = \frac{c}{f}$$

Microwaves in Chemistry Applications. https://doi.org/10.1016/B978-0-12-822895-1.00005-9

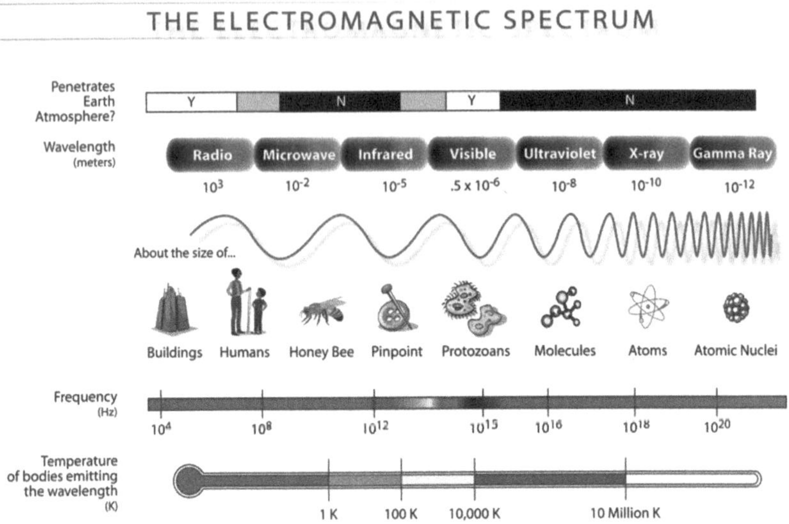

FIG. 1 Schematic of the electromagnetic spectrum that depicting various properties. *(Reproduced from Nasa (public domain).)*

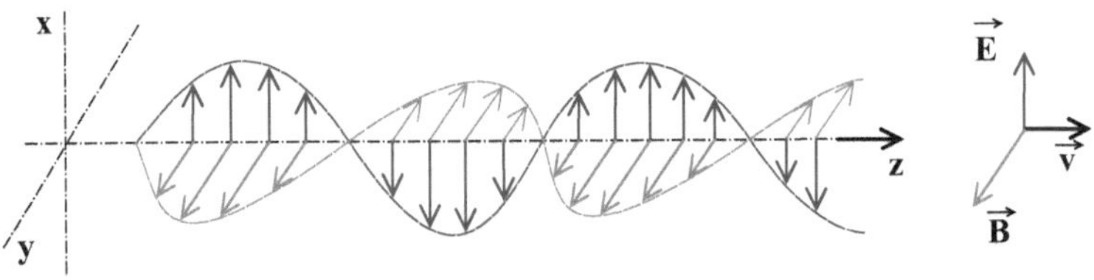

FIG. 2 Schematics of a sinusoidal electromagnetic wave propagating along the positive z-direction. *(Modified from Wikipedia.)*

$$f = \frac{E}{h}$$

$$E = \frac{hc}{\lambda}$$

where $h = 6.62607015 \times 10^{-34}$ J s, Planck's constant, and $c = 299{,}792{,}458$ m/s, the speed of light in vacuum [7].

Table 1 shows the approximate wavelengths, frequencies, and energies for selected regions of the electromagnetic spectrum. The arrangements are as follows: γ: gamma rays; HX: hard X-rays; SX: soft X-rays; EUV: extreme ultraviolet; NUV: near ultraviolet; visible; NIR: near-infrared; MIR: mid-infrared; FIR: far infrared; EHF: extremely high frequency; SHF: superhigh frequency; UHF: ultrahigh frequency; VHF: very high frequency; HF: high frequency; MF: medium frequency; LF: low frequency; VLF: very low frequency; ULF: ultralow frequency; SLF: superlow frequency; ELF: extremely low frequency.

TABLE 1 Electromagnetic spectrum wavelength/frequency/energy range.

CLASS	FREQUENCY	WAVELENGTH	ENERGY
Y	300 EHz	1 pm	1.24 MeV
HX	30 EHz	10 pm	124 keV
	3 EHz	100 pm	12.4 keV
SX	300 PHz	1 nm	1.24 keV
EUV	30 PHz	10 nm	124 eV
NUV	3 PHz	100 nm	12.4 eV
NIR	300 THz	1 μm	1.24 eV
MIR	30 THz	10 μm	124 meV
FIR	3 THz	100 μm	12.4 meV
EHF	300 GHz	1 mm	1.24 meV
SHF	30 GHz	1 cm	124 μeV
UHF	3 GHz	1 dm	12.4 μeV
VHF	300 MHz	1 m	1.24 μeV
HF	30 MHz	10 m	124 neV
MF	3 MHz	100 m	12.4 neV
LF	300 kHz	1 km	1.24 neV
VLF	30 kHz	10 km	124 peV
VF/ULF	3 kHz	100 km	12.4 peV
SLF	300 Hz	1 Mm	1.24 peV
ELF	30 Hz	10 Mm	124 feV
	3 Hz	100 Mm	12.4 feV

Modified and adapted from Wikipedia.

As this book is intended for microwave chemistry, the following sections only discuss the properties of microwave radiation. Some sources describe microwaves as radio waves or a subclass of the radio waveband. On the contrary, some classify them as radio waves and microwaves as distinct types of electromagnetic radiation.

1.3 Microwave radiation

Microwaves are lying between infrared radiation and radio waves in the electromagnetic spectrum. Their frequency ranges from 300 MHz to 300 GHz that corresponds to the wavelength of 1 m–1 mm [8–14]. The microwave electromagnetic spectrum is separated into subbands as shown in Table 2. The L-band is at lower microwave frequency ranges and is used mainly for communication. The higher frequency ranges, for instance, W band, are appropriate for analytical techniques like spectroscopic techniques [15].

TABLE 2 Microwave frequency bands.

Frequency range (GHz)	Band	Applications
1–2	L	Satellite, cellular phones
2–4	S	Satellite, cellular phones
4–8	C	Satellite, Wifi
8–12	X	Radar
12–18	Ku	Satellite TV, radar
18–26.5	K	Microwave backhaul
26.5–40	Ka	5G cellular, microwave backhaul
30–50	Q	5G cellular, microwave backhaul
40–60	U	Radar, experimental
50–75	V	WLAN
60–90	E	Microwave backhaul
75–110	W	Radar
90–140	F	Radar, experimental
110–170	D	Radar, experimental

1.3.1 Properties of microwaves

The important properties associated with microwave radiation are as follows.

- It radiates electromagnetic energy with a shorter wavelength.
- It passes through the ionosphere without being reflected.
- It can be easily attenuated within shorter distances.
- It travels in a straight line and is reflected by metal surfaces.
- It passes through the atmosphere.
- It can penetrate haze, light rain, and snow, smoke, and clouds.
- It passes through glass and plastics.
- It can be absorbed by water molecules.
- Microwave currents can flow even through a thin layer of a cable.

1.3.2 Advantages and disadvantages of microwave radiation

Microwaves have many advantages as well as a few disadvantages and are listed in Table 3.

1.3.3 Applications of microwave radiation

Microwaves are widely used in modern technology, some of which is not possible with other radiations. The main applications are listed in Table 4.

TABLE 3 Advantages and disadvantages of microwave radiation.

Advantages	Disadvantages
• Supports larger bandwidth and hence more information is transmitted. For this reason, microwaves are used for point-to-point communications • More antenna gain is possible • Higher data rates are transmitted as the bandwidth is more • Antenna size gets reduced, as the frequencies are higher • Low power consumption as the signals are of higher frequencies • The effect of fading gets reduced by using a line of sight propagation • Provides an effective reflection area in the radar systems • Satellite and terrestrial communications with high capacities are possible • Low-cost miniature microwave components can be developed • Effective spectrum usage with a wide variety of applications in all available frequency ranges of operation	• The cost of equipment or installation cost is high depending upon specifications • They are hefty and occupy more space depending upon the nature of the instrument • Electromagnetic interference may occur • Variations in dielectric properties with temperatures may occur • The inherent inefficiency of electric power

TABLE 4 Applications of microwaves.

Wireless communications	For long-distance telephone calls
	WiMAX operations
	Bluetooth
	Broadcast auxiliary services
	Outdoor broadcasting transmissions
	Remote pickup unit
	Direct broadcast satellite
	Studio/transmitter link
	Personal communication systems
	Cellular video systems
	Wireless local area networks
	Automobile collision avoidance system

Continued

TABLE 4 Applications of microwaves—cont'd

Commercial uses	Garage door openers
	Burglar alarms
	Police speed detectors
	Cell phones, pagers, wireless LANs
	Identification by noncontact methods
	Satellite television, XM radio
	Remote sensing
	Motion detectors
Electronics	Fast jitter-free switches
	HF generation
	Phase shifters
	Tuning elements
	Spread spectrum systems
	Electronic countermeasure systems
Military and radar	Radars to detect the range and speed of the target
	Air traffic control
	SONAR applications
	Weather forecasting
	Minesweeping applications
	Navigation of ships
	Speed limit enforcement
	Communications
Food industry	Microwave ovens used for cooking and reheating
	Preheating applications
	Food processing applications
	Precooking
	Drying potato chips
	Roasting food grains/beans
	Absorbing water molecules
	Moisture leveling
Industrial uses	Vulcanizing rubber
	Drying and reaction processes
	Analytical chemistry applications
	Processing ceramics
	Surface modification
	Polymer matrix

TABLE 4 Applications of microwaves—cont'd

	Chemical vapor processing
	Sterilizing pharmaceuticals
	Powder processing
	Chemical synthesis
	Power transmission
	Waste remediation
	Tunnel boring
	Breaking up coal seams
	Breaking rock/concrete
	Curing of cement
	Fusion reactors
	RF lighting
	Active denial systems
Medical applications	Monitoring heartbeat
	Tumor detection
	Lung water detection
	Regional hyperthermia
	Local heating
	Therapeutic applications
	Angioplasty
	Microwave Acoustic imaging
	Microwave tomography
Spectroscopy	Electron paramagnetic resonance spectroscopy
	To know the free radicals in materials
	To know about unpaired electrons in chemicals
	Electron chemistry
Radio astronomy	Mark cosmic microwave background radiation
	Detection of many radiations in the universe and earth's atmosphere
	Detection of powerful waves in the universe
Navigation	Global positioning system (GPS)
	Global navigation satellite systems
Semiconductor processing techniques	Chemical vapor deposition
	Reactive ion etching
Research applications	Nuclear resonances
	Atomic resonances

1.4 Microwave chemistry

The branch of science applying microwave radiation to chemical reactions is called microwave chemistry [16–20].

1.4.1 Microwave heating

In the early years, the Bunsen burner was the main heat source used for chemical synthesis, which was invented by Robert Bunsen in 1855. Bunsen burner makes a single-open gas flame and is utilized for heating, combustion, and sterilization [21–26]. A model of the Bunsen burner with a needle valve is shown in Fig. 3A. In this setup, the hose barb for the gas tube is on the left and the needle valve for gas flow regulation is on the right side. Different types of flames can be produced using Bunsen burner by changing the airflow in the throat holes (Fig. 3B). The flame number "1" will obtain if the air hole is closed, and this is the default one and safety flame mainly used for lighting; the flame number "2" will obtain if the air hole is slightly open; the flame number "3" will obtain if air hole is half-open; and flame number "4" will obtain if air hole is fully open. Other burners such as Teclu burner, Meker burner, and Tirrill burner are also existed, following the same principle and are the most important alternatives to the Bunsen burner.

Later, Bunsen burner was replaced by hot plates or oil baths (Fig. 4) [27]. These conventional heating methods have disadvantages such as long heating time, wall effect (temperature maximums occur on the reaction vessel/chamber wall surface), limited reaction control, difficulties in the cooling process,

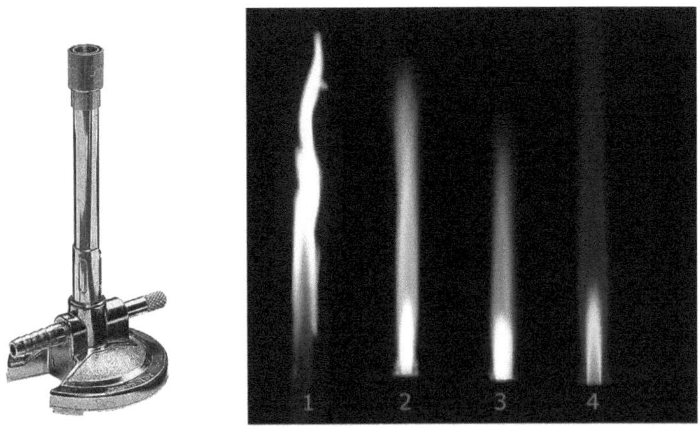

(A) (B)

FIG. 3 (A) Bunsen burner with a needle valve. (B) Different types of flames from the burner. *(Modified from Wikipedia.)*

(A) (B)

FIG. 4 (A) Oil bath technique. (B) Hot plate. *(Reproduced from Wikipedia.)*

and high heat losses. In the 21st century, synthesis by microwave energy has been an increasingly interesting subject in the scientific community [28–32].

Originally, microwave energy was used only for heating food. In the 1940s, Percy Spencer invented the microwave oven and used it for the first time [33]. Even though the application of microwave heating in chemical alterations can be traced back to the 1950s, the first reports on the application of microwave heating to accelerate chemical synthesis were published in the mid-1980s [34, 35]. Table 5 shows the evolution of microwave chemistry. Microwave

TABLE 5 Evolution of microwave chemistry.

Year	Developments
1946	As a method of heating, microwave radiation was discovered
1947	The commercial domestic microwave oven was first introduced
1978	CEM Corporation built up the first microwave laboratory instrument and the instrument is used to analyze moisture in solid materials
1980–82	Microwave irradiation used to dry organic materials
1983–85	Microwave irradiation applied for chemical analysis
1986	Published papers about microwave heating in chemical analysis and synthesis
1990s	Several developments in microwave chemistry field
1990	The first high-pressure vessel (HPV 80) was developed
1992–96	A microwave batch system reactor and a single-mode cavity system were developed by CEM
2000	The first commercial microwave synthesizer was developed

energy can be utilized in organic synthesis [36] as well as inorganic synthesis [37–42].

In the early days, experiments were mostly carried out in a domestic household microwave oven without any pressure or temperature measurement. The reactions often resulted in violent explosions owing to the rapid uncontrolled heating of solvents. However, in the first 15 years of microwave chemistry, domestic household microwave ovens were increasingly employed for chemical synthesis. After the introduction of dedicated microwave reactors in the market, a dramatic increase of interest in microwave chemistry was observed. Fig. 5 shows the number of scientific publications reported on the microwave from 2011 through 2020.

Most of the commercially available dedicated microwave reactors and domestic kitchen microwave ovens for chemical transformation operate at a frequency of 2.45 GHz that is corresponding to a wavelength of 12.25 cm. In that way, interference with telecommunication and cellular phonc frequencies will be avoided. Besides, close to this frequency (2.45 GHz), the microwave energy absorption by liquid water is maximum. The other approved band for industrial applications is 915 MHz. Studies showed that energy associated with microwave irradiation is insufficient to cleave molecular bonds [43]. In detail, the energy associated with microwave radiation is in the 1.24×10^{-6}–1.24×10^{-3} eV range; these energies are much lower than ionization energies of biological compounds (13.6 eV), of covalent bond energies such as hydrogen bonds (2 eV), OH-(5 eV), van der Waals intermolecular interactions (lower than 2 eV), and the energy of Brownian motion (2.7×10^{-3} eV at 37°C) [44–46]. However, microwave radiation provides unique thermal effects that can be very beneficial for chemical synthesis.

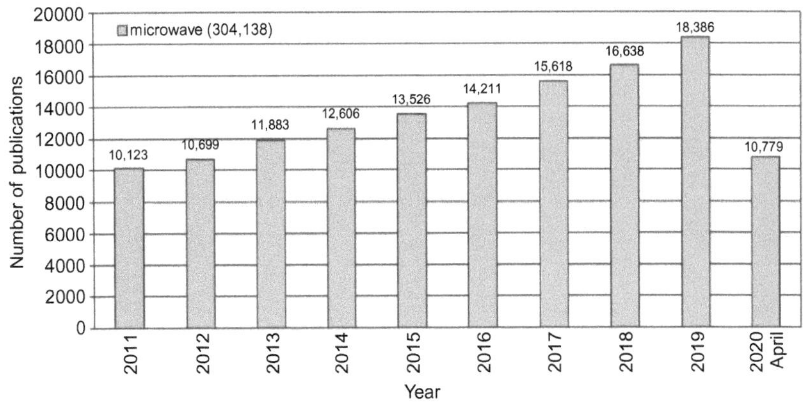

FIG. 5 Number of scientific publications reported on the microwave between 2011 and 2020. *(Reproduced with permission from J. Wojnarowicz, T. Chudoba, W. Lojkowski, A review of microwave synthesis of zinc oxide nanomaterials: reactants, process parameters and morphologies, Nanomaterials 10 (2020) 1086.)*

1.4.2 Principles of microwave heating

Microwave heating is based on the effective heating of materials by dielectric heating effects. The heating of materials by microwave radiation has been described thoroughly by several researchers [31, 47–75]. Microwave heating effects work by four major mechanisms, namely dipolar polarization (rotation of dipoles), ionic conduction (ionic polarization), electronic polarization (atomic polarization), and interfacial polarization (surface polarization) as shown in Fig. 6. When a polar molecule having an electrical dipole moment aligns with the oscillating electromagnetic field, the dipolar rotation takes place. In this process, due to the molecular friction and dielectric loss, energy is lost as heat. The ionic conduction mechanism results from the dissolved (dissociated) charged particles (usually ions). Under the influence of the electric component of microwave radiation, the charged particles oscillate back and forth. The back and forth oscillations induce the collisions of charged

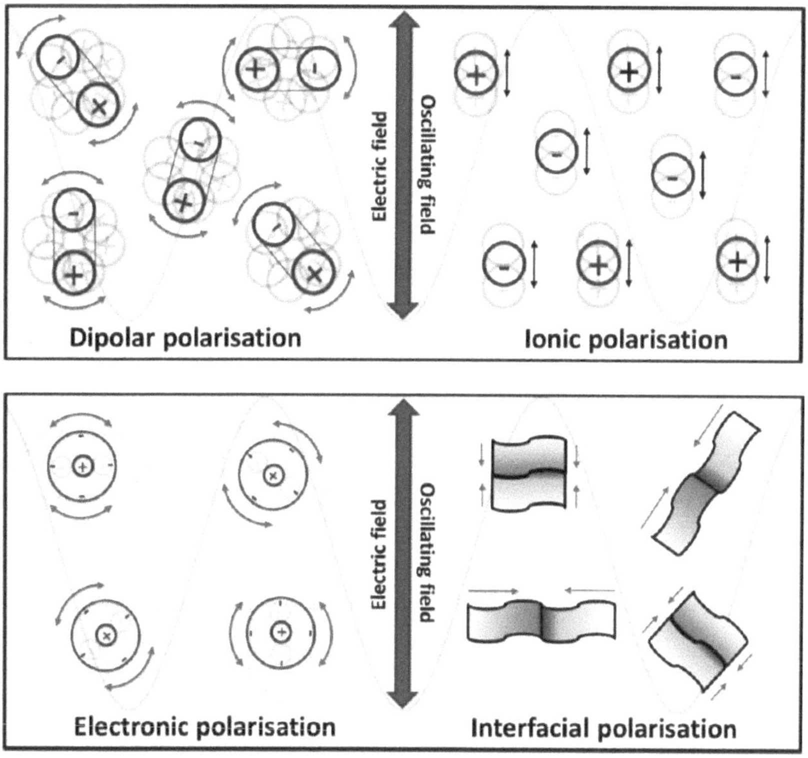

FIG. 6 Schematic representation of dipolar polarization, ionic conduction, electronic polarization, and interfacial polarization and under microwave conditions. *(Reproduced with permission from J. Wojnarowicz, T. Chudoba, W. Lojkowski, A review of microwave synthesis of zinc oxide nanomaterials: reactants, process parameters and morphologies. Nanomaterials 10 (2020) 1086.)*

particles with neighboring atoms or molecules, which are responsible for producing heat energy. Compared with the dipolar polarization mechanism, the absorption of microwave radiation by this mechanism is more effective and the transfer of energy is quicker. In the third mechanism, the induction of a dipole moment occurs because of the shifting of the center of the electron charge with respect to the nucleus. The fourth mechanism is based on the polarization of a material due to the accumulation of charges on the interface surface.

Thus, in order to be efficiently heated in the microwave, the material must possess certain dielectric properties. Under microwave treatment, the heating characteristics of a material depend on the ability of that material to convert electromagnetic energy into heat energy. The ability can be expressed as tan δ, the loss tangent. The tan δ values of some selected organic solvents are summarized in Table 6. For fast heating, a solvent with a high tan δ is required. At the same time, solvents with low tan δ values can also be useful for microwave synthesis. In that case, either substrates or reagents/catalysts must be polar and the overall dielectric properties of a reaction mixture allow sufficient heating by microwaves. Thus, microwave heating is also possible even with a nonpolar solvent. Besides, if the reaction mixture is nonpolar, passive heating elements can be added mixture to assist the heating process.

$$\tan\delta = \frac{\varepsilon''}{\varepsilon'}$$

ε'' = Dielectric loss (efficiency with which electromagnetic radiation is converted into heat).

ε' = Dielectric constant (polarizability of molecules in the electric field).

TABLE 6 Tan δ values of selected organic solvents in the microwave field at 2.45 GHz and 20°C [47, 76, 77].

Solvent	tan δ	Solvent	tan δ	Solvent	tan δ
Ethylene glycol	1.35	2-Butanol	0.44	Chloroform	0.09
Ethanol	0.94	Dichlorobenzene	0.28	Acetonitrile	0.06
DMSO	0.82	NMP	0.27	Ethyl acetate	0.05
2-Propanol	0.79	Acetic acid	0.17	Acetone	0.05
Formic acid	0.72	DMF	0.16	THF	0.04
Methanol	0.65	Dichloroethane	0.12	Dichloromethane	0.04
Nitrobenzene	0.58	Water	0.12	Toluene	0.04
1-Butanol	0.57	Chlorobenzene	0.10	Hexane	0.02

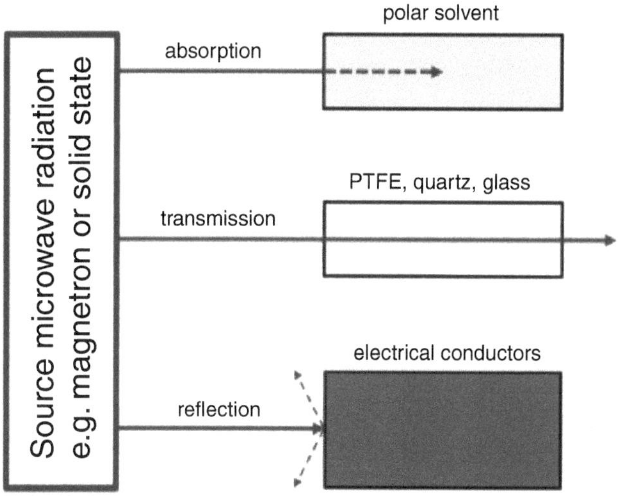

FIG. 7 Interaction of different substances with microwaves: Absorbing materials, insulation materials, and conductors. *(Reproduced with permission from J. Wojnarowicz, T. Chudoba, W. Lojkowski, A review of microwave synthesis of zinc oxide nanomaterials: reactants, process parameters and morphologies, Nanomaterials 10 (2020) 1086.)*

Materials can interact with microwaves in three ways (Fig. 7). Conducting materials (like metals) reflect microwaves and heating does not occur. Dielectric materials absorb microwaves, and heating takes place. For example, materials like water, acids, or polar organic solvents absorb microwaves. Insulating materials are transparent to microwaves and heating does not take place, for example, in the case of materials like polymers, quartz, or ceramics.

Microwave heating is based on electromagnetic induction characterized by quickness and economical. Studies described that heating of the material occurs directly in the structure of the material and not from the surface. The mechanism of changing microwave energy to heat can be expressed by the following formula:

$$P = 2\pi f \varepsilon' \varepsilon'' E^2$$

where P is the energy absorbed in a volume unit [W/m^3], f is the frequency of the microwave radiation [Hz], ε' is the permittivity of the material, ε'' is the dielectric material loss factor, E is the intensity of an electric field inside the material [V/m].

Microwave radiation acts on living organisms mainly in two ways. These are the nonthermal effects (reversible) and the thermal effects (nonrefundable). The interface between these two factors is called the threshold of the MW radiation, or power density. Thermal effects may lead to the weakening and killing of the microorganism. The safety conditions should be taken while staying in the area of higher intensity of electromagnetic radiation.

1.4.3 Microwave versus conventional heating

Fig. 8 compares microwave heating with conventional heating. During the conventional heating methods, the temperature of the surface of the vessel increases along with the internal materials. It is called wall heating. Due to this, a large amount of energy is lost to the environment through convection currents and conduction of materials. The higher surface temperatures making heat transfer from the outer surface to the internal materials. Nonuniform sample temperatures and higher thermal gradients are observed in this process [78, 79]. The effect of heating during the conventional method is not homogeneous and dependent on thermal conductivity, specific heat capacity, and density of materials.

As an example, consider 5 mL of ethanol boiled at 160°C in an open vessel oil bath heating conditions and a single-mode closed-vessel microwave irradiation. Fig. 9A compares the temperature profiles. It depicts that microwave heating allows for the rapid increase of solvent temperature and quick cooling as well. But in the case of an oil bath (conventional heating), the heating rate and cooling rate were very slow. Fig. 9B demonstrates the thermal behavior of microwave and oil bath heating methods. The microwave radiation raises the temperature of the whole volume evenly and simultaneously. On the contrary, in the oil bath heating method, the vessel wall is heated first and then the reaction mixture contacted with it. Inverted thermal gradient differences were reported between these two heating methods [80–82].

However, microwave radiation does not always heat the entire materials (full sample volume). When electromagnetic enters the surface of a given substance, a part of the radiation reflects from the surface and the remaining part penetrates inside the substance. The radiation that penetrated inside interacts

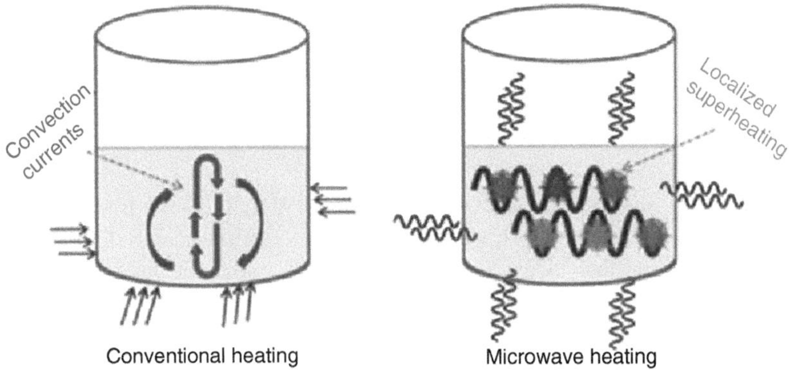

FIG. 8 Schematic representation of conventional and microwave heating mechanisms. *(Reproduced with permission from V.G. Gude, P. Patil, E. Martinez-Guerra, S. Deng, N. Khandan, Microwave energy potential for biodiesel production, Sustain. Chem. Process. 1 (2013) 5.)*

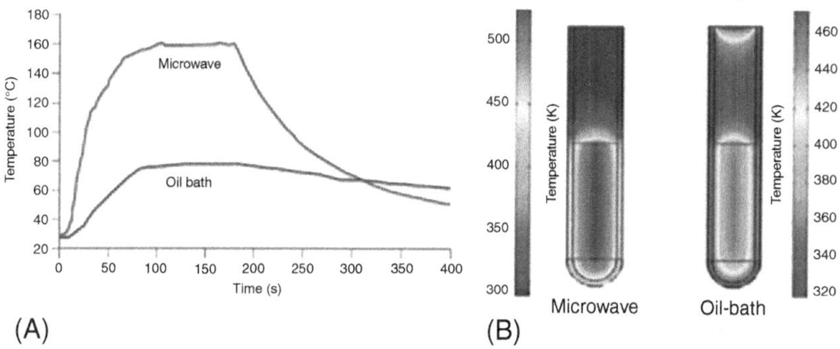

FIG. 9 (A) Temperature profile comparison. (B) Thermal behavior in the microwave and conventional (oil bath) heating methods. *(Reproduced with permission from V.G. Gude, P. Patil, E. Martinez-Guerra, S. Deng, N. Khandan, Microwave energy potential for biodiesel production, Sustain. Chem. Process. 1 (2013) 5.)*

with the molecules and ions of the substance. Depending on the substance's properties, the radiation penetrates the substance at various depths. In microwave heating, the penetration depth is an important parameter. The penetration depth of a field is the distance from the surface of the substance to a certain internal point where the magnitude of field strength decreases to $1/e$ (=36.8%) of the initial magnitude at the surface of the substance [83]. The penetration depth can be expressed by the formula [55, 73]:

$$d_p = \frac{\lambda_0}{2\pi}\left(\frac{1}{2\pi\varepsilon_0\varepsilon'}\right)^{1/2}\left[\left(1+(\tan\delta)^2\right)^{1/2}-1\right]^{-1/2}$$

where d_p is the penetration depth, λ_0 is the wavelength at vacuum conditions, μ_0 is the magnetic permeability of free space, μ' is the magnetic relative permeability, ε^0 is the permittivity in free space, ε' is the relative permeability, $\tan\delta$ is the loss angle.

Several researchers reported the penetration depth for different materials [73, 84, 85]. Penetration depths of the 2.45 GHz microwaves for selected solvents and materials are shown in Table 7. The microwave reactor's reaction vessels are mostly made of polytetrafluoroethylene (PTFE). PTFE has a low thermal conductivity (0.25 W/(m K)) and acts as a thermal insulator [86–88].

1.4.4 Advantages of microwave heating

Microwave heating has several important advantages compared to conventional heating, such as it is a contactless method (no direct contact of material with a heat source); minimized wall effect as the wall of the reaction chamber is not

TABLE 7 Penetration depths of the 2.45 GHz microwaves for selected solvents and materials.

Material	Penetration depth (cm)	Temperature (°C)
Water (distilled)	1.6	20
Water (distilled)	2.88	25
Water (distilled)	80	100
Water (ice)	1100	−12
0.125 M NaCl solution of salt water	0.88	25
0.5 M NaCl solution of salt water	0.45	25
2 M NaCl solution of salt water	0.14	25
Ethylene glycol	0.46	25
Methanol	0.68	25
Ethanol	0.93	25
1-Propanol	1.39	25
Acetone	7.07	25
Ethyl acetate	11.05	25
Xylene	28.32	25
Rubber, styrene-butadiene (SBR), vulc.	19	–
Nitrile rubber, natural	65	–
Aluminium oxide (Al_2O_3) ceram	3000	–
Polyethylene	4000	25
Polyethylene	5907.1	–
Polystyrene	7619.3	–
PTFE (Teflon)	9000	–
Quartz	20,000	–
Silver	0.33×10^{-4}	–
Zinc, pure (Zn)	1.24×10^{-4}	–
Copper (Cu)	1.3×10^{-4}	–
Aluminium 100% (Al)	0.86×10^{-4}	–
Aluminium (Al)	1.7×10^{-4}	–
Nickel (Ni)	2.7×10^{-4}	–

TABLE 7 Penetration depths of the 2.45 GHz microwaves for selected solvents and materials.—cont'd

Material	Penetration depth (cm)	Temperature (°C)
Iron (Fe)	3.2×10^{-4}	–
Titanium, pure (Ti)	3.3×10^{-4}	–
Stainless steel (304)	4.3×10^{-4}	–

Reproduced with permission from J. Wojnarowicz, T. Chudoba, W. Lojkowski, A review of microwave synthesis of zinc oxide nanomaterials: reactants, process parameters and morphologies, Nanomaterials 10 (2020) 1086.

heated directly; volumetric heating of the feedstock; precise and instantaneous electronic control; rapid energy transfer [89]; uniform heating, thus products with good quality will obtain [55–58, 60, 90, 91]; Shorter reaction time (shorter duration of processes) [49, 51, 92]; fewer side reactions [93, 94]; products with good selectivity and purity [52, 95]; operate under solvent-free conditions [68]; very high power densities developed in the processing zone [70]; superior moisture leveling [70]; high level of energy saving [87]; higher efficiency of production [70]; compact equipment [70]; shorter start-up time of apparatus; an important part of green chemistry approach [96–101]. Although it has several advantages, the main drawbacks are: this method needs expensive equipment and is unsuitable for scale-up; besides, in this method the reaction monitoring is unfeasible.

1.4.5 Applications of microwave heating

Microwave heating is used for various applications like food processing (pasteurization, cooking, and drying) [62, 66, 102]; industrial application (drying, wood curing, rubber curing, and vulcanization, disinfection, coal pretreatment, and processing, ceramic processing, polymer processing, polymeric composites, ceramic composites, melting of glasses, melting of metallic materials, roasting of tea/coffee beans, and plant extraction processes) [59, 65–69, 103–105]; waste treatment (medical waste, garbage, and sludge) [66]; medical applications (sterilization, drying, diagnosis, and novel methods for treating inoperable tumors) [66, 106]; analytical chemistry (laboratory sample processing, extraction, ashing, digestion, and moisture analysis) [52, 107]; and chemical synthesis: organic and inorganic synthesis (microwave-assisted synthesis of drugs, polymers, and nanomaterials) [31, 48–51, 53–61, 74, 80, 96–101, 108–119].

The application of microwave energy in organic and inorganic synthesis is an emerging and invaluable technique. In recent years, the use of microwave irradiation technique as an eco-friendly option for the synthesis of organic material has become very popular. The microwave-assisted organic synthetic method has several advantages compared to a conventional method such as uniform heating, shorter reaction time, improved yield, less by-product formation, selectivity, environmentally friendly, reduced pollution with ease in processing and handling, and also it promotes new reactions. The microwave-assisted synthetic method has accelerated reaction rates significantly of many chemical reactions, for example, hour's job in conventional methods can be done in a few seconds using the microwave method. Besides, some reactions that were difficult or unable to perform under conventional methods were made possible with the microwave method.

Several organic and inorganic synthetic methods employing microwave irradiation under both homogeneous and heterogeneous reaction conditions to prepare different heterocyclic compounds were reported by several researchers from all over the world. Also, to explore the probability of getting the biologically important moiety in higher yields and in shorter reaction times, the microwave irradiation methods become the first choice among researchers. Presently, the microwave-assisted synthesis approach has been used to conduct a diverse type of organic reactions including cycloaddition reaction; coupling; cyclization reactions; rearrangement; organocatalysis; condensation; cleavage; radical; hydrogenation; olefination reaction; protection and deprotection; oxidation and reduction reactions; heck reaction; Diels-alder reaction; Suzuki reaction; hydrolysis; Mannich reaction; dehydration; epoxidation; esterification. Different types of microwave ovens are currently used by researchers for the synthesis approach. As an automated microwave oven is extremely expensive and difficult to have this type of microwave oven in teaching undergraduate research courses, some research groups have employed domestic microwaves for synthesis purposes and showed comparable results.

The basic structure of the microwave power application apparatus and microwave device is discussed in Chapter 2 along with the current microwave features, costs, and benefits. The focus of the second part of the book (Chapters 4–7) is to provide some examples of microwave-assisted synthesis; especially, we discuss the application of microwave energy in organic synthesis. Examples from several families are considered. Mostly we explore the microwave-assisted (catalytic and noncatalytic) synthesis of pharmacologically significant compounds, such as pyrimidines, benzoselenophenes, xanthene, steroidal derivatives, thiazoles, imines, tetrazoles, triazoles, quinolines, indolizine, β-lactams, pyrroles, octane, Schiff bases, furans, quinoxaline, and coumarins. Selected applications of microwave heating such as the sterilization method in chemistry, biology, and medicine; microwave-enhanced CVD processes for diamond synthesis are discussed in part 3 along with future trends in microwave chemistry and biology.

1.5 Conclusion

In this chapter, we have described the properties, advantages, and importance of microwave radiation in several applications. Mainly the application of microwave heating in organic synthesis is described in detail, along with the basic principle of microwave heating.

Acknowledgments

AD and BKB are grateful to Prince Mohammad Bin Fahd University for support. BKB is also grateful to US NIH, US NCI, and Kleberg Foundation of Texas for financial support.

References

[1] J.C. Maxwell, A Dynamical Theory of the Electromagnetic Field, Philosophical Transactions of the Royal Society of London, 1865, pp. 459–512.
[2] M. Browne, Physics for Engineering and Science, second ed., McGraw Hill/Schaum, New York, 2013, p. 427.
[3] S. Cloude, An Introduction to Electromagnetic Wave Propagation and Antennas, Springer Science and Business Media, 1995, pp. 28–33.
[4] A. Bettini, A Course in Classical Physics, vol. 4—Waves and Light, Springer, 2016.
[5] P.A. Tipler, Physics for Scientists and Engineers: vol. 1: Mechanics, Oscillations and Waves, Thermodynamics, MacMillan, 1999.
[6] M. Purcell, Electricity and Magnetism, third ed., Cambridge University Press, New York, 2013.
[7] P.J. Mohr, B.N. Taylor, D.B. Newell, CODATA recommended values of the fundamental physical constants: 2006, Rev. Mod. Phys. 80 (2008) 633–730.
[8] S. Kumar, S. Shukla, Concepts and Applications of Microwave Engineering, PHI Learning Pvt. Ltd, 2014, p. 3.
[9] R.T. Hitchcock, Radio-Frequency and Microwave Radiation, American Industrial Hygiene Assn, 2004, p. 1.
[10] D.M. Pozar, Microwave Engineering, Addison–Wesley Publishing Company, 1993.
[11] G.A. Jones, D.H. Layer, T.G. Osenkowsky, National Association of Broadcasters Engineering Handbook, tenth ed., Taylor & Francis., 2013.
[12] R. Sorrentino, G. Bianchi, Microwave and RF Engineering, John Wiley & Sons, 2010.
[13] R.T. Hitchcock, Radio-Frequency and Microwave Radiation, third ed., American Industrial Hygiene Association, Fairfax, VA, 2004.
[14] M. Vollmer, Physics of the microwave oven, Phys. Educ. 39 (2004) 74–81.
[15] J.L. Krstenansky, I. Cotterill, Recent advances in microwave-assisted organic syntheses, Curr. Opin. Drug Discov. Devel. 4 (2000) 454–461.
[16] B.K. Banik, D. Bandyopadhyay, Advances in Microwave Chemistry, first ed., CRC Press, 2018.
[17] A. de la Hoz, A.D. Ortiz, A. Moreno, Microwaves in organic synthesis. Thermal and non-thermal microwave effects, Chem. Soc. Rev. 34 (2) (2005) 164–178.
[18] C.R. Strauss, R.W. Trainor, Developments in microwave-assisted organic chemistry, Aust. J. Chem. 48 (1995) 1665–1692.
[19] M. Kidwai, Dry media reactions, Pure Appl. Chem. 73 (2001) 147–151.

[20] C. Oliver Kappe, A. Stadler, D. Dallinger, Microwaves in Organic and Medicinal Chemistry, second ed., Completely Revised and Enlarged Edition, Wiley-VCH, Weinheim, 2012.
[21] W.B. Jensen, The origin of the Bunsen burner, J. Chem. Educ. 82 (2005) 518.
[22] G. Lockemann, The centenary of the Bunsen burner, J. Chem. Educ. 33 (1956) 20–21.
[23] A.J. Rocke, Bunsen burner, in: Oxford Companion to the History of Modern Science, 114, 2002.
[24] J.J. Griffith, Chemical Reactions–A Compendium of Experimental Chemistry, eighth ed., R Griffin and Co., Glasgow, 1838.
[25] E.R. Sanders, Aseptic laboratory techniques: volume transfers with serological pipettes and micropipettors, J. Vis. Exp. 63 (2012) 2754.
[26] M. Kohn, Remarks on the history of laboratory burners, J. Chem. Educ. 27 (1950) 514.
[27] A. Vogel, Vogel's Textbook of Practical Organic Chemistry, Prentice Hall, 1996.
[28] D. Adam, Out of the kitchen, Nature 421 (2003) 571.
[29] N. Leadbeater, Chem. World 1 (2004) 38–41.
[30] V. Marx, As microwave technology matures, it is catching on in drug discovery and development, Chem. Eng. News 82 (50) (2004) 14–16.
[31] C.O. Kappe, A. Stadler, D. Dallinger, Microwaves in Organic and Medicinal Chemistry, second ed., Wiley-VCH, Weinheim, 2012.
[32] A. Yarnell, HEATING TECHNOLOGY proves useful in proteomics and synthesis of peptides, proteins, and carbohydrates, Chem. Eng. News 85 (21) (2007) 32–33.
[33] J. Tang, Unlocking potentials of microwaves for food safety and quality, J. Food Sci. 80 (2015) E1776–E1793.
[34] R. Gedye, F. Smith, K. Westaway, H. Ali, L. Baldisera, L. Laberge, J. Rousell, The use of microwave ovens for rapid organic synthesis, Tetrahedron Lett. 27 (1986) 279–282.
[35] R.J. Giguere, T.L. Bray, S.M. Duncan, G. Majetich, Application of commercial microwave ovens to organic synthesis, Tetrahedron Lett. 27 (1986) 4945–4948.
[36] R. Cecilia, U. Kunz, T. Turek, Possibilities of process intensification using microwaves applied to catalytic microreactors, Chem. Eng. Process. 46 (2007) 870–881.
[37] J. Martín-Gil, F.J. Martín-Gil, M. José-Yacamán, L. Carapia-Morales, T. Falcón-Bárcenas, Microwave-assisted synthesis of hydrated sodium uranyl oxonium silicate, Pol. J. Chem. (2005) 1399–1403.
[38] J. Prado-Gonjal, M.E. Villafuerte-Castrejón, L. Fuentes, E. Morán, Microwave-hydrothermal synthesis of $BiFeO_3$, Mater. Res. Bull. 44 (2009) 1734–1737.
[39] K.J. Rao, B. Vaidhyanathan, M. Ganduli, P.A. Ramakrishnan, Synthesis of inorganic solids using microwaves, Chem. Mater. 11 (4) (1999) 882–895.
[40] J. Zhao, W. Yan, Microwave-assisted inorganic syntheses, in: Modern Inorganic Synthetic Chemistry, Elsevier, 2011, pp. 173–195.
[41] R.K. Sahu, M.L. Rao, S.S. Manoharan, Microwave synthesis of magnetoresistive $La_{0.7}Ba_{0.3}MnO_3$ using inorganic precursors, J. Mater. Sci. 36 (2001) 4099.
[42] D.M.P. Mingos, D. Baghurst, Tilden Lecture. Applications of microwave dielectric heating effects to synthetic problems in chemistry, Chem. Soc. Rev. 20 (1991) 1–47.
[43] E. Neas, M. Collins, in: H.M. Kingston, L.B. Jassie (Eds.), Introduction to Microwave Sample Preparation: Theory and Practice, American Chemical Society, Washington DC, 1988.
[44] A.C. Metaxas, R.J. Meredith, Industrial Microwave Heating, Peter Peregrinus Ltd, London, 1993.
[45] Z. Chemat-Djenni, B. Hamada, F. Chemat, Atmospheric pressure microwave assisted heterogeneous catalytic reactions, Molecules 12 (2007) 1399–1409.

[46] E.R. Peterson, Microwave chemical processing, Res. Chem. Intermed. 1 (1994) 93–96.

[47] K. Rana, S. Rana, Microwave reactors: a brief review on its fundamental aspects and applications, Open Access Lib. J. 1 (2014) 1–20.

[48] I. Bilecka, M. Niederberger, Microwave chemistry for inorganic nanomaterials synthesis, Nanoscale 2 (2010) 1358–1374.

[49] E. Mohammadi, M. Aliofkhazraei, M. Hasanpoor, M. Chipara, Hierarchical and complex ZnO nanostructures by microwave-assisted synthesis: morphologies, growth mechanism and classification, Crit. Rev. Solid State Mater. Sci. 43 (2018) 475–541.

[50] P.P. Falciglia, P. Roccaro, L. Bonanno, G. De Guidi, F.G.A. Vagliasi, S. Romano, A review on the microwave heating as a sustainable technique for environmental remediation/detoxification applications, Renew. Sust. Energ. Rev. 95 (2018) 147–170.

[51] N.E. Leadbeater, Microwave Heating as a Tool for Sustainable Chemistry, first ed., CRC Press, Boca Raton, FL, 2011.

[52] M. Gaba, N. Dhingra, Microwave chemistry: general features and applications, Indian J. Pharm. Educ. Res. 45 (2011) 175–183.

[53] N.E. Leadbeater, C.B. McGowan, Laboratory Experiments Using Microwave Heating, first ed., CRC Press, Boca Raton, FL, 2013.

[54] C.O. Kappe, Controlled microwave heating in modern organic synthesis, Angew. Chem. 43 (2004) 625–6284.

[55] G. Cravotto, D. Carnaroglio, Microwave Chemistry, first ed., De Gruyter, Boston, MA, 2017.

[56] G. Yang, S.-J. Park, Conventional and microwave hydrothermal synthesis and application of functional materials: a review, Materials 12 (2019) 1177.

[57] S. Horikoshi, N. Serpone, Microwaves in Nanoparticle Synthesis: Fundamentals and Applications, first ed., Wiley-VCH, Weinheim, Germany, 2013.

[58] G. Yang, S.-J. Park, Author response to comment on: conventional and microwave hydrothermal synthesis and application of functional materials: a review, Materials 12 (2019) 3640.

[59] S. Horikoshi, R.F. Schimann, J. Fukushima, N. Serpone, Microwave Chemical and Materials Processing, first ed., Springer, Singapore, 2018.

[60] B. Jalouli, A. Abbasi, S.M. Musavi Khoei, A comment on: "conventional and microwave hydrothermal synthesis and application of functional materials: a review", Materials 12 (2019) 3631.

[61] P. Lidstrom, J. Tierney, B. Wathey, J. Westman, Microwave assisted organic synthesis—a review, Tetrahedron 57 (2001) 9225–9283.

[62] S. Chandrasekaran, S. Ramanathan, T. Basak, Microwave food processing—a review, Food Res. Int. 52 (2013) 243–261.

[63] A. Feliczak-Guzik, M. Górzynska, A. Zielinska, A. Wawrzynczak, Promieniowanie mikrofalowe wlaboratorium, Laborant 7 (2014) 52–60.

[64] H. Feng, Y. Yin, J. Tang, Microwave drying of food and agricultural materials: basics and heat and mass transfer modeling, Food Eng. Rev. 4 (2012) 89–106.

[65] D. Agrawal, Microwave sintering of ceramics, composites and metallic materials, and melting of glasses, Trans. Indian Ceram. Soc. 65 (2006) 129–144.

[66] H. Vasudev, G. Singh, A. Bansal, S. Vardhan, L. Thakur, Microwave heating and its applications in surface engineering: a review, Mater. Res. Express 6 (2019) 102001.

[67] M. Oghbaei, O. Mirzaee, Microwave versus conventional sintering: a review of fundamentals, advantages and applications, J. Alloys Compd. 494 (2010) 175–189.

[68] S. Singh, D. Gupta, V. Jain, A.K. Sharma, Microwave processing of materials and applications in manufacturing industries: a review, Mater. Manuf. Process. 30 (2015) 1–29.

[69] R.R. Mishra, A.K. Sharma, A review of research trends in microwave processing of metal-based materials and opportunities in microwave metal casting, Crit. Rev. Solid State Mater. Sci. 41 (2016) 217–255.
[70] C. Leonelli, T.J. Mason, Microwave and ultrasonic processing: now a realistic option for industry, Chem. Eng. Process. 49 (2010) 885–900.
[71] R.S. Chavan, S.R. Chavan, Microwave baking in food industry: a review, Int. J. Dairy Sci. 5 (2010) 113–127.
[72] J. Sun, W. Wang, Q. Yue, Review on microwave-matter interaction fundamentals and efficient microwave-associated heating strategies, Materials 9 (2016) 231.
[73] T. Kim, J. Lee, K.-H. Lee, Microwave heating of carbon-based solid materials, Carbon Lett. 15 (2014) 15–24.
[74] D. Bogdal, Microwave-Assisted Organic Synthesis. One Hundred Reaction Procedures, first ed., Elsevier, Oxford, 2005.
[75] D.A. Jones, T.P. Lelyveld, S.D. Mavrofidis, S.W. Kingman, N.J. Mile, Microwave heating applications in environmental engineering—A review, Resour. Conserv. Recycl. 34 (2002) 75–90.
[76] J. Wojnarowicz, T. Chudoba, W. Lojkowski, A review of microwave synthesis of zinc oxide nanomaterials: reactants, process parameters and morphologies, Nanomaterials 10 (2020) 1086.
[77] B.L. Hayes, Microwave Synthesis: Chemistry at the Speed of Light, first ed., CEM Publishing, Durham, 2002.
[78] Y. Groisman, A. Gedanken, Continuous flow, circulating microwave system and its application in nanoparticle fabrication and biodiesel synthesis, J. Phys. Chem. C 112 (2008) 8802–8808.
[79] D. Bogdal, Microwave-Assisted Organic Synthesis, Elsevier, 2005.
[80] C.O. Kappe, D. Dallinger, The impact of microwave synthesis on drug discovery, Nat. Rev. Drug Discov. 5 (2006) 51–63.
[81] C.O. Kappe, Controlled microwave heating in modern organic synthesis, Angew. Chem. Int. Ed. 43 (2004) 6250–6284.
[82] Y. Li, W. Yang, Microwave synthesis of zeolite membranes: a review, J. Membr. Sci. 316 (2008) 3–17.
[83] A.C. Metaxas, Foundations of Electroheat: A Unified Approach, first ed., John Wiley and Sons, New York, NY, 1996.
[84] S. Horikoshi, N. Serpone, Microwave frequency effects in organic synthesis, in: A. de la Hoz, A. Loupy (Eds.), Microwaves in Organic Synthesis, third ed., Wiley-VCH, Weinheim, Germany, 2013.
[85] O.O.D. Afolabi, M. Sohail, Microwaving human faecal sludge as a viable sanitation technology option for treatment and value recovery—a critical review, J. Environ. Manag. 187 (2017) 401–415.
[86] Z. Fang, R.L. Smith Jr., X. Qi (Eds.), Production of Biofuels and Chemicals with Microwave, Biofuels and Biorefineries, Springer, Dodlek, Netherlands, 2015.
[87] W. Lojkowski, C. Leonelli, T. Chudoba, J. Wojnarowicz, A. Majcher, A. Mazurkiewicz, High-energy-low-temperature technologies for the synthesis of nanoparticles: microwaves and high pressure, Inorganics 2 (2014) 606–619.
[88] A. Majcher, J. Wiejak, J. Przybylski, T. Chudoba, J. Wojnarowicz, A novel reactor for microwave hydrothermal scale-up nanopowder synthesis, Int. J. Chem. React. Eng. 11 (2013) 361–368.
[89] J.S. Schanche, Microwave synthesis solutions from personal chemistry, Mol. Divers. 7 (2003) 291–298.

[90] C. Leonelli, W. Lojkowski, Main development directions in the application of microwave irradiation to the synthesis of nanopowders, Chem. Today 25 (2007) 34–38.

[91] J. Wojnarowicz, T. Chudoba, D. Smoleń, W. Łojkowski, A. Majcher, Mazurkiewicz, Examples of the nanoparticles produced by microwave solvothermal synthesis (MSS) route, Glass Ceram. 6 (2014) 8–11.

[92] J.J. Shah, K. Mohanraj, Comparison of conventional and microwave-assisted synthesis of benzotriazole derivatives, Indian J. Pharm. Sci. 76 (2014) 46–53.

[93] Y.J. Zhu, W.W. Wang, R.J. Qi, X.L. Hu, Microwave- assisted synthesis of single-crystalline tellurium nanorods and nanowires in ionic liquids, Angew. Chem. Int. Ed. 43 (2004) 1410–1414.

[94] Q. Li, W. Cao, J. Lei, X. Zhao, T. Hou, B. Fan, D. Chen, L. Zhang, H. Wang, H. Xu, et al., Synthesis and growth mechanism of ZnO rod-like nanostructures by a microwave-assisted low-temperature aqueous solution route, Cryst. Res. Technol. 49 (2014) 298–302.

[95] A. Rizzuti, C. Leonelli, Crystallization of aragonite particles from solution under microwave irradiation, Powder Technol. 186 (2008) 255–262.

[96] G. Stefanidis, A. Stankiewicz, Alternative Energy Sources for Green Chemistry, first ed., Royal Society of Chemistry, Cambridge, 2016.

[97] S.C. Ameta, P.B. Punjabi, R. Ameta, C. Ameta, Microwave-Assisted Organic Synthesis A Green Chemical Approach, first ed., CRC Press, Boca Raton, FL, 2015.

[98] O.G. Bhusnure, S.B. Gholve, P.S. Giram, T.A. Warad, V.S. Pangave, J.N. Sangshetti, Green approaches for the industrial production of active pharmaceutical ingredients, World J. Pharm. Res. 4 (2015) 629–648.

[99] A.K. Bose, M.S. Manhas, S.N. Ganguly, A.H. Sharma, B.K. Banik, More chemistry for less pollution: applications for process development, Synthesis 11 (2002) 1578–1591.

[100] A. Kumar, Y. Kuang, Z. Liang, X. Sun, Microwave chemistry, recent advancements and eco-friendly microwave-assisted synthesis of nanoarchitectures and their applications: a review, Mater. Today Nano 11 (2020) 100076.

[101] A. Pal, Microwave assisted synthesis: a green chemistry approach, Int. J. Eng. Sci. Invent. 7 (2018) 51–53.

[102] H. Feng, Y. Yin, J. Tang, Microwave drying of food and agricultural materials: basics and heat and mass transfer modelling, Food Eng. Rev. 4 (2012) 89–106.

[103] J. Huang, G. Xu, Y. Liang, G. Hu, P. Changa, Improving coal permeability using microwave heating technology—a review, Fuel 266 (2020) 117022.

[104] C. Leonelli, P. Veronesi, L. Denti, A. Gatto, L. Iuliano, Microwave assisted sintering of green metal parts, J. Mater. Process. Technol. 205 (2008) 489–496.

[105] S. Das, A.K. Mukhopadhyay, S. Datta, D. Basu, Prospects of microwave processing: an overview, Bull. Mater. Sci. 32 (2009) 1–13.

[106] J. Irving, M.F.J. Cepeda, F. Valdés, G. Guerrero, Microwave ablation: state-of-the-art review, Onco Targets. Ther. 8 (2015) 1627–1632.

[107] R.T. Giberson, R.S. Demaree Jr., Microwave Techniques and Protocols, first ed., Human Press INC., Totowa, NJ, 2001.

[108] M. Tsuji, Microwave-assisted synthesis of metallic nanomaterials in liquid phase, ChemistrySelect 2 (2017) 805–819.

[109] Y.J. Zhu, F. Chen, Microwave-assisted preparation of inorganic nanostructures in liquid phase, Chem. Rev. 114 (2014) 6462–6555.

[110] A. Feliczak-Guzik, M. Górzyńska, A. Zielińska, A. Wawrzyńczak, Promieniowanie mikrofalowe w laboratorium, Laborant 7 (2014) 52–60.

[111] S. Dworakowska, D. Bogdal, A. Prociak, Microwave-assisted synthesis of polyols from rapeseed oil and properties of flexible polyurethane foams, Polymers 4 (2012) 1462–1477.

[112] S. Dworakowska, D. Bogdał, F. Zaccheria, N. Ravasio, The role of catalysis in the synthesis of polyurethane foams based on renewable raw materials, Catal. Today 223 (2014) 148–156.
[113] D. Bogdał, A. Prociak, Microwave-Enhanced Polymer Chemistry and Technology, first ed., Blackwell Publishing Professional, Ames, IA, 2007.
[114] K. Kempe, C.R. Becer, U.S. Schubert, Microwave-assisted polymerizations: recent status and future perspectives, Macromolecules 44 (2011) 5825–5842.
[115] D. Adam, Microwave chemistry: out of the kitchen, Nature 421 (2003) 571–572.
[116] A.K. Nagariya, A.K. Meena, K. Kiran, A.K. Yadav, U.S. Niranjan, A.K. Pathak, B. Singh, M. M. Rao, Microwave assisted organic reaction as new tool in organic synthesis, J. Pharm. Res. 3 (2010) 575–580.
[117] K.S. Jignasa, T.S. Ketan, S.P. Bhumika, K.G. Anuradha, Microwave assisted organic synthesis: an alternative synthetic strategy, Der Pharma Chem. 2 (2010) 342–353.
[118] G. Kłosowski, D. Mikulski, A. Menka, Microwave-assisted one-step conversion of wood wastes into levulinic acid, Catalysts 9 (2019) 753.
[119] A. Díaz-Ortiz, P. Prieto, A. de la Hoz, A critical overview on the effect of microwave irradiation in organic synthesis, Chem. Rec. 19 (2019) 85–97.

Chapter 2

Microwave equipment for chemistry

Aparna Das* and Bimal Krishna Banik*
Department of Mathematics and Natural Sciences, College of Sciences and Human Studies, Prince Mohammad Bin Fahd University, Al Khobar, Kingdom of Saudi Arabia
Corresponding authors: E-mails: aparnadasam@gmail.com (Aparna Das); bimalbanik10@gmail.com, bbanik@pmu.edu.sa (Bimal Krishna Banik)

2.1 Introduction

In the early days, household microwave ovens were extensively used in chemical laboratories for synthesis. Nowadays, dedicated instrumentation with numerous features is available. However, scientists still use kitchen microwave ovens for scientific purposes. There are serious scientific as well as practical reasons to choose dedicated instrumentation. In this chapter, we describe the basic structure of a conventional microwave device and the modifications made in the domestic microwave ovens for different purposes. The last section of the chapter presents various commercial microwave reactors available in the market.

2.2 The basic structure of the microwave device

A schematic diagram of a household microwave is shown in Fig. 1. It consists of:

- Cavity magnetron
- High-voltage power source
- High-voltage capacitor
- Magnetron control circuit
- Short waveguide
- Metal cooking chamber
- Turntable and/or metal waveguide stirring fan
- Control panel.

The apparatus is subdivided into two parts: the microwave generation part and the applicator part. The microwave generation part consists of an

Microwaves in Chemistry Applications. https://doi.org/10.1016/B978-0-12-822895-1.00002-3

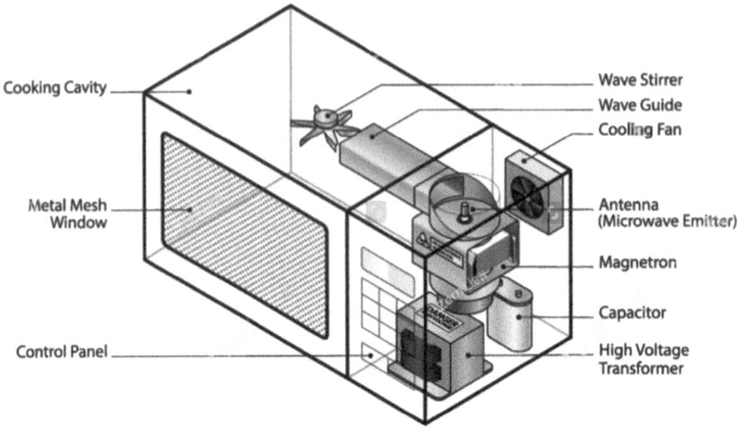

FIG. 1 Schematic diagram of a microwave oven showing the key internal components. *(Reproduced from Shutterstock.)*

independent microwave generator portion and a microwave transmission portion. The components in the generator parts are a microwave source, a power supply, and a cooling unit. The transmission portion consists of a waveguide, an isolator, a power monitor/power sensor, and a tuner. The samples to be irradiated should be placed in the applicator parts.

2.2.1 Microwave generator

Electric devices that generate microwaves are gyrotrons, magnetrons, and klystrons. Among them, the magnetron is comparatively inexpensive, produces a large power, and is commonly used in microwave ovens [1]. The magnetron is a high-powered vacuum tube oscillator and is the main part of most of the household microwave ovens (Fig. 2). In the 1930s, Albert W. Hull invented the magnetron tube [2, 3].

It consists of a cathode and an anode surrounded by an external magnet. The operation of the magnetron is based on the combined action of a magnetic field that is applied externally and an electric field that exists between the electrodes. The tube acts as a simple diode with the electrons flowing directly from the cathode to the anode, in the absence of the external magnetic field. During the process, the electrons emitted from the hot cathode are forcefully rotated by the magnetic field. In that way, a rotating electron cloud is created and this cloud excites a microwave field. Magnetrons are highly effective for converting input electrical power to output microwave power [4]. That is, it can generate microwave energies ranging from less than 1 kW to 1 MW or above.

Klystron is another microwave tube used for the generation and amplification of microwaves. American electrical engineers Russell and Sigurd Varian invented this tube and demonstrated it successfully in

FIG. 2 Magnetron with a portion removed to show the cavities. The antenna emitting microwave radiation is at the left. *(Reproduced from Wikipedia.)*

1937 [5–7]. Mainly two varieties of klystron are available. One is the klystron microwave oscillator also known as reflex klystron, and the second is the klystron microwave amplifier (two-cavity or multicavity klystron) [8]. The two-cavity klystron is the most popular type (Fig. 3).

Klystron amplifiers are also utilized for other applications in satellite systems, radar, television broadcasting, particle accelerators, and the medical industry [9]. It can be used in the ultrahigh-frequency region, 300 MHz to 3 GHz, up to 400 GHz. The two-cavity klystron consists of an electron source (heater), a cathode, and an anode like a conventional vacuum tube. It also has a collector region at the end of the electron stream. During the process, the electrons ejected from the cathode accelerated toward the anode due to the high dc potential between the electrodes. In that way, a focused beam of electrons is generated.

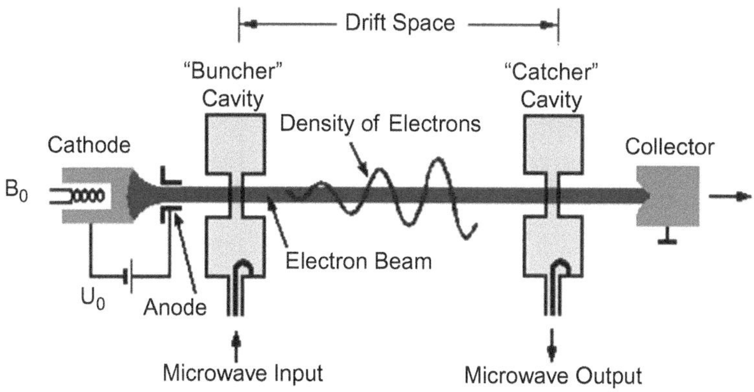

FIG. 3 Layout of the two-cavity klystron tube. *(Reproduced from Wikipedia.)*

Concerning the two-cavity-type klystron, the generated electron beam goes through a central hole in the first toroid-shaped cavity and then a similar second cavity and finally terminating at the collector region. The high levels of amplification of the beam are mainly due to the interaction of the cavities with the beam.

Reflex klystron has only one cavity and generates variable frequencies with low power and low efficiency (Fig. 4). It is widely used in applications in local oscillator in microwave receiver, in radar receiver, and pump oscillator in parametric amplifier [10]. It mainly consists of an electron gun and a focusing electrode at cathode potential. The emitted electrons from the cathode enter the cavity with a constant velocity where their velocity altered depends upon the cavity voltage. The magnetic field is used to focus the electron in the center of the tube.

Gyrotron is under the class of high-power linear-beam vacuum tubes and it produces millimeter-wave electromagnetic waves using the electron's cyclotron resonance in a very strong magnetic field. It can produce waves with output frequencies ranging from 20 GHz to 527 GHz [11, 12]. Thus, it can cover the electromagnetic spectrum from the microwave to the edge of the terahertz wave [13]. It can generate output powers from tens of kilowatts up to 2 MW. The gyrotron consists of an electron gun, an acceleration chamber, a resonance chamber immersed in a strong magnetic field, a mode converter, and a collector of electrons, all of the parts are situated inside a vacuum tube (Fig. 5). The electron gun generates the beam of electrons, and it is rotated by a strong magnetic field. Because of the changing electric and magnetic fields, the beam is accelerated and changed the form [14, 15].

Another type of vacuum tube, the traveling wave tube (TWT), is also used to amplify radio frequency signals in the microwave range [9]. In 1942, Rudolf

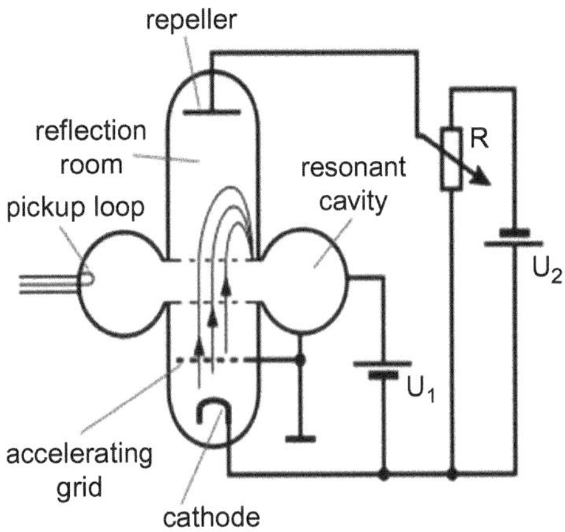

FIG. 4 Layout of the reflex klystron. *(Reproduced from Wikipedia.)*

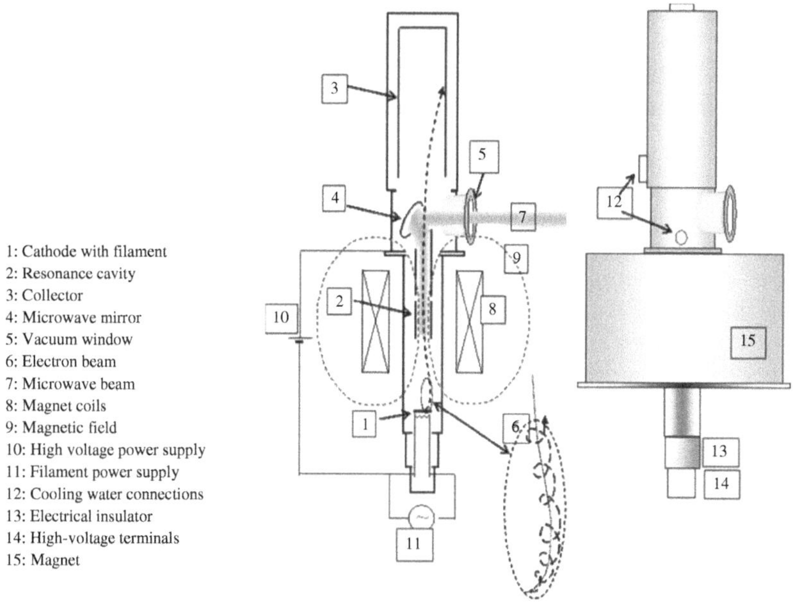

FIG. 5 Layout of gyrotron. *(Reproduced from Wikipedia.)*

Kompfner invented traveling wave tubes (TWTs). Same as in the klystron tube, the radio wave is amplified by absorbing power from a beam of electrons [9].

Various types of TWTs are available; among them, helix TWT and coupled-cavity TWT are the most popular ones. A cutaway view of a helix traveling wave tube is shown in Fig. 6. It normally operates in the frequency range of 300 MHz to 50 GHz and the output power ranges from a few watts to megawatts [9, 16].

Semiconductor-based microwave generators also were considered and showed remarkable results in communication applications. The use of solid-state electronics for microwave heating was first proposed already by McAvoy in the 1970s [17]. Later, other scholars also studied this in great detail

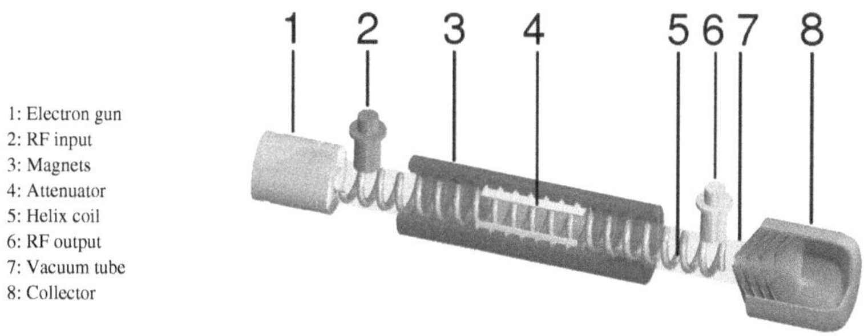

FIG. 6 Layout of traveling wave tubes. *(Reproduced from Wikipedia.)*

[18, 19]. Schwartz et al. reported the conceptual design and the experimental results of a 20-W transistor-based microwave heater for biological tests at 2.14 GHz [20]. But the application of semiconductor microwave generators in microwave heating was limited because of its low microwave power output and high cost.

2.2.2 High-frequency power supplies

Power supplies are needed to power the microwave generator. For example, some magnetrons require a high voltage of around 3 kV dc supply for its operation. Commonly a simple transformer or an electronic power converter passes energy to the magnetron.

2.2.3 Isolators

An isolator is a device that sends waves directly to the applicator and absorbs reflected waves using its built-in dummy load. In that way, it avoids the wave to return to the microwave generator and reduces the influence of reflected wave fluctuation. The excessive reflected power can sometimes damage generators. That is, the isolator functions to protect the magnetron. If the materials that are heated are not good absorbers of electromagnetic radiation, a significant amount of radiation will reflect back to the microwave source. A reflected wave also occurs due to the rotation of the turntable and the stirrer.

As the isolator is a nonreciprocal device, to achieve nonreciprocity in an isolator, it must needfully comprise a nonreciprocal material. At microwave frequencies, the most suitable nonreciprocal material is ferrite [21, 22]. The waves absorbed by the ferrite are released into the atmosphere in the form of heat energy. The ferrite-based isolators are classified into four groups: resonance isolators, Faraday rotation isolators, terminated circulators, and field-displacement isolators. Fig. 7 shows the resonance-absorption-type isolator.

2.2.4 Waveguide

The waveguide is defined as the hollow pipe used to carry waves from initial position to final position (Fig. 8). In radio frequency engineering and communications engineering, it carried radio waves. This type of waveguide is used as a transmission line generally at microwave frequencies, in equipment such as microwave ovens, satellite communications, radar sets, and microwave radio links. In the waveguide, the electromagnetic waves follow a zig-zag path, being repeatedly reflected between the sides of the guide. Waveguides can be made from either conductive or dielectric materials depending on the frequency. In the case of microwave transmission, a waveguide normally consists of a hollow metallic conductor.

FIG. 7 Resonance-absorption-type isolator. *(Reproduced from Wikipedia.)*

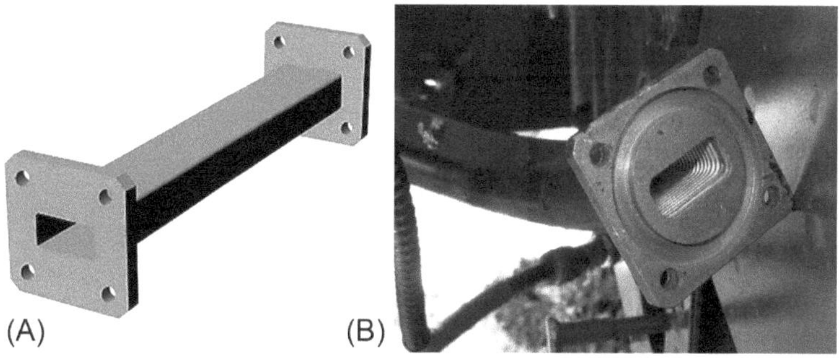

FIG. 8 (A) Drawing of a straight waveguide. (B) Image of a flexible waveguide from a J-band radar. *(Reproduced from Wikipedia.)*

2.2.5 Power monitors

This device monitors the power of the traveling and reflected waves propagating through the waveguide. It is important to measure the power of the wave during microwave heating.

2.2.6 Applicator (cavity)

The applicator is the heating tank and it heats the object placed inside it by using microwave irradiation. Depending on the application, different types of applicators such as batch type, conveyor type, waveguide type, traveling wave type, single (mono)-mode type, and multimode type are available.

In single-mode applicators, a relatively more stable single-standing wave is generated inside the cavity. The substrate to be heated is usually put in one of the maxima of the electromagnetic field as shown in Fig. 9. The disadvantage of single-mode applicators is that their volume and the material load are strictly limited in size. It is usually applied to process volumes up to 200 mL.

The multimode applicator is the most widely used one. It can be seen in domestic microwaves as well as in large-scale industrial microwaves. They normally have the shape of a rectangular closed metal box (cavity) as shown in Fig. 10. A great number of resonance modes exist inside the cavity, as the microwaves get reflected from the cavity walls. Due to the reflections from

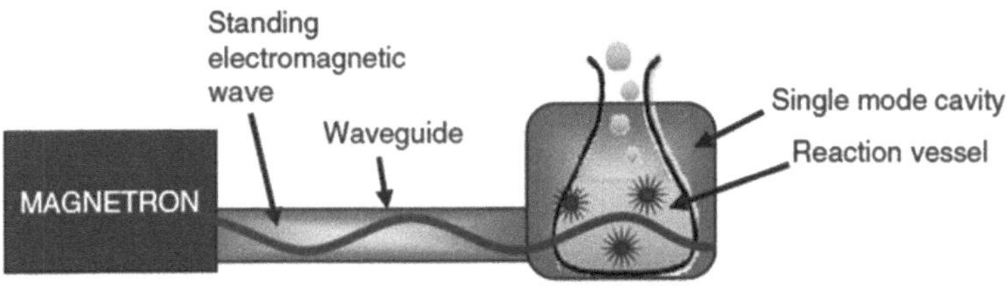

FIG. 9 Single-mode microwave applicator. *(Reproduced with permission from G. Stefanidis, A. Navarrete Muñoz, G. Strum, A. Stankiewcz, A helicopter view of microwave application to chemical processes: reactions, separations, and equipment concepts, Rev. Chem. Eng. 30 (2014) 233–259.)*

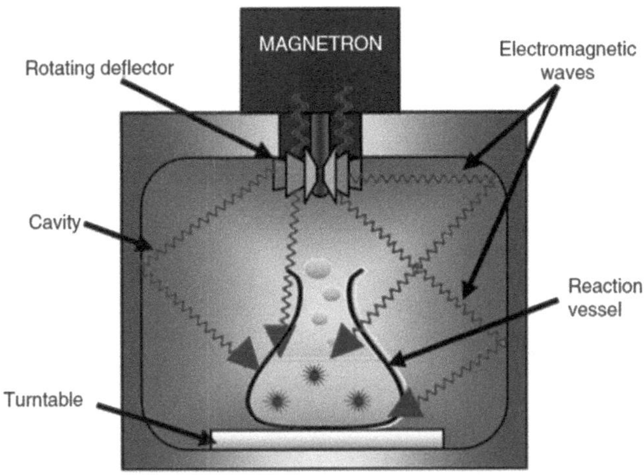

FIG. 10 Multimode microwave applicator. *(Reproduced with permission from G. Stefanidis, A. Navarrete Muñoz, G. Strum, A. Stankiewcz, A helicopter view of microwave application to chemical processes: reactions, separations, and equipment concepts, Rev. Chem. Eng. 30 (2014) 233–259.)*

the sidewalls, wave interference occurs inside the cavity. In that way, the microwave field that is applied to the material (substrate) is highly nonuniform irrespective of the use of a mode stirrer or a rotating disk to support the substrate. The main advantage of this type of applicator is that it can be made large in size to accommodate sizable volumes of substrate for processing.

Comparing single-mode and multimode cavities, the advantage of the single-mode cavities, compared to multimode ones, is that higher heating rates can be formed inside the cavity due to the higher electromagnetic field densities. Even though in a single-mode cavity the electromagnetic field pattern is well defined inside an empty cavity, it may vary drastically in the presence of the heated load inside it [23].

Traveling wave applicators support a microwave field passing in a single direction from the inlet to the outlet of a waveguide. Traveling wave applicators of different configurations are available.

2.3 Domestic microwave ovens in the laboratory

Still, domestic microwave ovens make useful results in the laboratory and it is easy to handle the instrument. The microwave-safe vessels used in these ovens are made from Teflon and polyetherimide, and they can withstand pressures up to 80 atm and temperatures up to 250°C [24]. At the same time, it has a lot of limitations. One of the main problems with these ovens is the chance of explosion during the heating of organic solvents in an open vessel. Several modifications have been used to avoid the explosion [25, 26]. Another limitation is the problem of temperature measurement. Most of the classical temperature sensors were failed to work at the conditions used for microwave heating. Scholars had tried to overcome this by using optical fiber thermometers even though the measurements were limited below 250°C.

2.3.1 Modifications in domestic microwave ovens

Several researchers have used a modified kitchen microwave oven to perform the chemical reactions. For instance, Halfadji et al. reported a procedure to determine the 26 polychlorinated biphenyls (PCBs) congeners in soil samples using microwave-assisted extraction with an open vessel (MAE-OV) [27]. A modified kitchen microwave oven (2450 MHz, M1933N Samsung, Korea) was used for this purpose. The schematic diagram of the setup is shown in Fig. 11.

Okmen et al. modified a conventional household microwave oven for recording temperature and weight measurements continuously during microwave drying of foods [28]. In that, modified instrument was furnished with an electronic balance and a set of standard thermocouples that were connected to a computer to collect data continuously (Fig. 12).

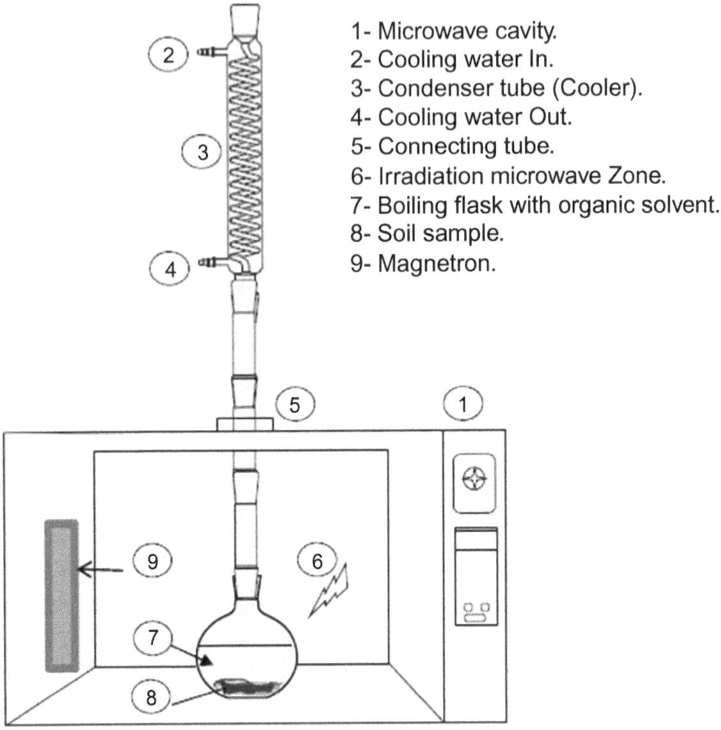

FIG. 11 Schematic diagram of a modified multimode domestic microwave oven open-vessel extraction method. *(Reproduced with permission from A. Halfadji, A. Touabet, Determination of 26 polychlorinated biphenyls congeners in soil samples using microwave-assisted extraction with open vessel and gas chromatography. Green Chem. Lett. Rev. 11 (3) (2018) 209–216.)*

Recently, a simple, fast, and low-cost method for the synthesis of metal-organic frameworks using a modified domestic microwave oven was reported [29]. Monitoring of temperature during the synthesis was also possible using this setup (Fig. 13).

Horikoshi et al. described a simple device based on a modified domestic microwave oven with a UV-Vis lamp encased in Teflon to photodegrade environmental pollutants in aqueous media (Fig. 14) [30]. The efficiency of the modified device was analyzed using the photodegradation of the agrochemical pollutant 2,4-dichlorophenoxyacetic acid. The coupled microwave-UV-Vis irradiation led to an enhancement in the decomposition rate of the target material in the modified microwave oven compared to the photocatalytic method alone. For instance, the degradation rates were 2×10^{-3} mM/min and 1.1×10^{-3} mM/min for photocatalytic/microwave method and photocatalytic method (with sixfold greater light irradiance), respectively.

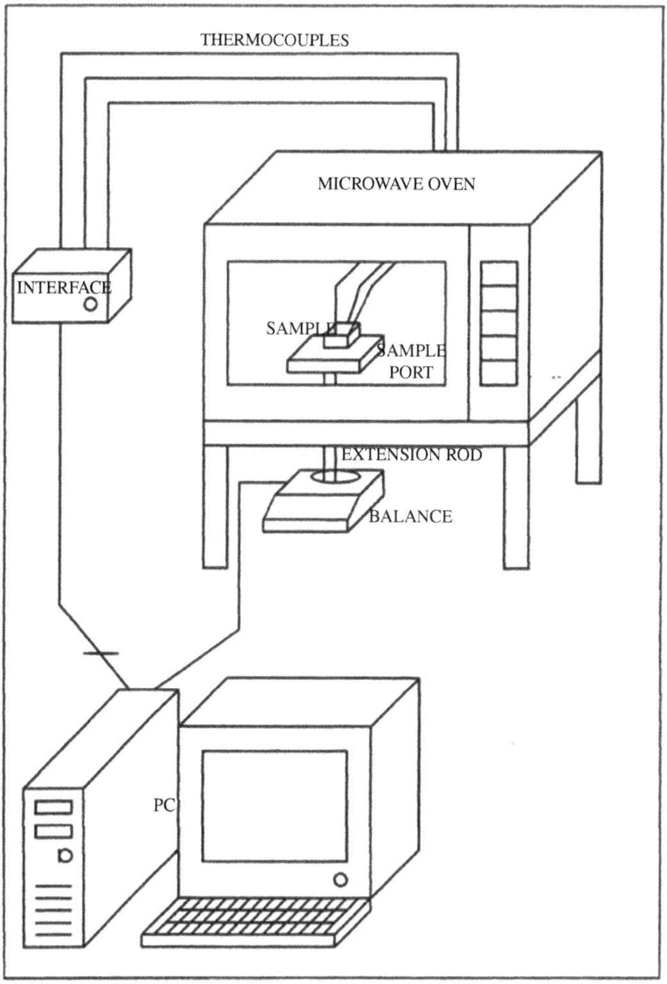

FIG. 12 Schematic diagram of a modified multimode domestic microwave oven open-vessel extraction method. *(Reproduced with permission from Z. Okmen, A. Bayindirli, Modification of a household microwave oven for continuous temperature and weight measurements during drying of foods. J. Microw. Power Electromag. Energ. 35 (4) (2000) 225–231.)*

Using a modified domestic microwave oven, a novel and economic method to synthesize and fabricate bioactive diopside ceramics were reported [31]. The schematic diagram of the setup is shown in Fig. 15.

Microwave-assisted pyrolysis was successfully demonstrated recently using a specially modified domestic microwave oven [32]. A programmable temperature controller and microwave output power level of 1000 W were used during the synthesis time (Fig. 16).

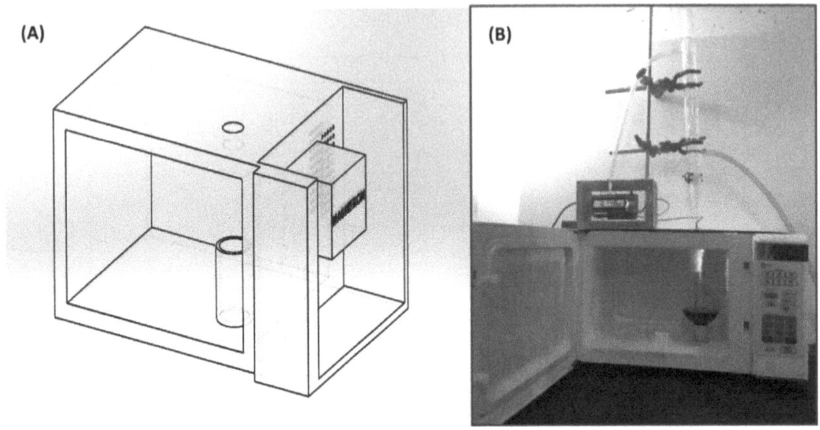

FIG. 13 (A) Positioning of the reactor inside the microwave oven. (B) Picture of the modified domestic microwave oven used for the synthesis of MOFs. *(Reproduced with permission from L. W. Aguiar, G.P. Otto, V.L. Kupfer, S.L. Fávaro, C.T.P. Silva, M.P. Moisés, L. de Almeida, M.R. Guilherme, E. Radovanovic, E.M. Girotto, A.W. Rinaldi, Simple, fast, and low-cost synthesis of MIL-100 and MIL-88B in a modified domestic microwave oven, Mater. Lett. 276 (2020) 128–127.)*

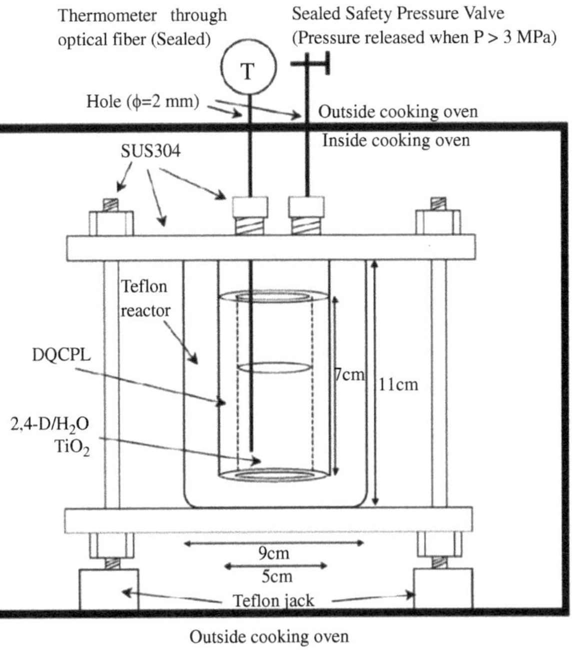

FIG. 14 Schematic diagram of the experimental setup of a Teflon batch-type reactor. *(Reproduced with permission from S. Horikoshi, H. Hidaka, N. Serpone, Environmental remediation by an integrated microwave/UV illumination technique: VI. A simple modified domestic microwave oven integrating an electrodeless UV-Vis lamp to photodegrade environmental pollutants in aqueous media. J. Photochem. Photobiol. A 161 (2-3) (2004) 221–225.)*

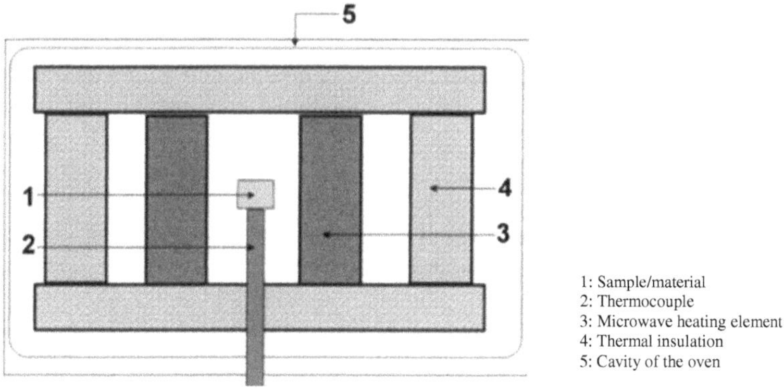

FIG. 15 Schematic diagram of the modified domestic microwave oven. *(Reproduced with permission from A. Harabi, S. Zouai, A new and economic approach to synthesize and fabricate bioactive diopside ceramics using a modified domestic microwave oven. Part 1: study of sintering and bioactivity. Int. J. Appl. Ceram. Technol. 11 (2014) 31–46.)*

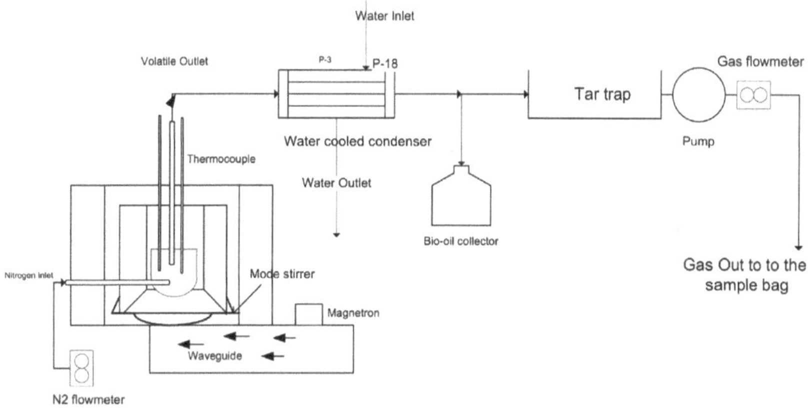

FIG. 16 Schematic diagram of the modified microwave oven. *(Reproduced with permission from S.A. Halim, J. Swithenbank, Simulation study of parameters influencing microwave heating of biomass. J. Energy Inst. 92 (4) (2019) 1191–1212.)*

Hussain et al. reported the production of highly upgraded biooil by pyrolysis of water hyacinth in a modified domestic microwave oven [33]. For this reaction, a domestic microwave oven of the Sharp Corporation Japan (frequency: 24.5 GHz, power: 0.8 kW) was slightly altered (Fig. 17).

A simple microwave photochemical reactor was also possible with a modified domestic microwave oven [34]. The experimental setup consists of an electrodeless discharge lamp placed into the reaction vessel as shown in Fig. 18. The microwave field produces ultraviolet radiation by the lamp at the same instant when it acts with the reaction mixture.

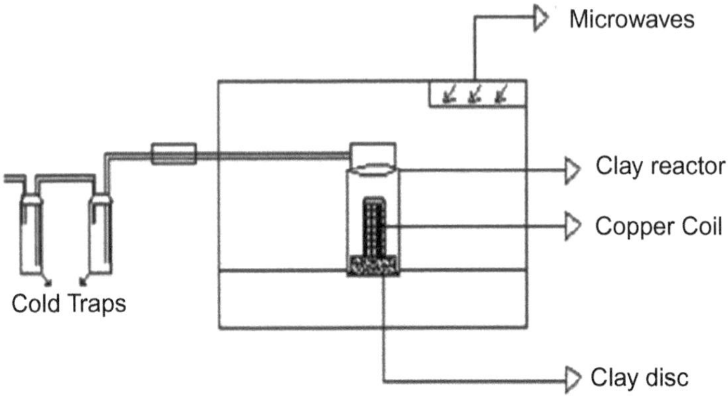

FIG. 17 Schematic diagram of the modified microwave oven used for the pyrolysis of biomass. *(Reproduced with permission from K. Hussain, N. Bashir, Z. Hussain, S.A. Sulaiman, Cement catalyzed conversion of biomass into upgraded bio-oil through microwave metal interaction pyrolysis in aluminum coil reactor, J. Anal. Appl. Pyrol. 129 (2018) 37–42.)*

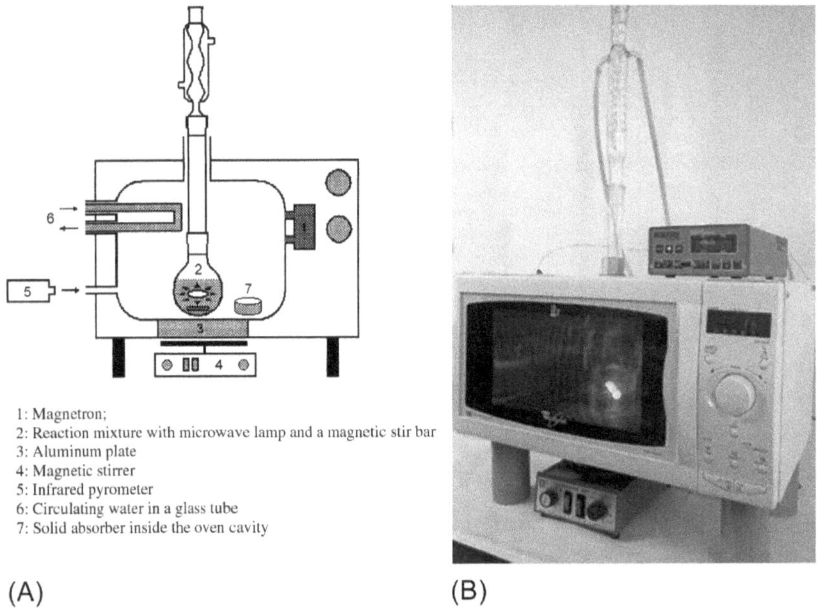

FIG. 18 (a) Photograph of the experimental setup. (b) A modified microwave oven for photochemistry. *(Reproduced with permission from P. Klán, M. Hájek, V. Církva, The electrodeless discharge lamp: a prospective tool for photochemistry: Part 3. The microwave photochemistry reactor. J. Photochem. Photobiol. A 140 (3) (2001) 185–189.)*

The same group reported the possibility of the microwave with two reflux apparatuses next to each other (Fig. 19).

To analyze the interactions of microwaves with dielectric substance and their conversion to thermal energy, the activities of ionic concentration

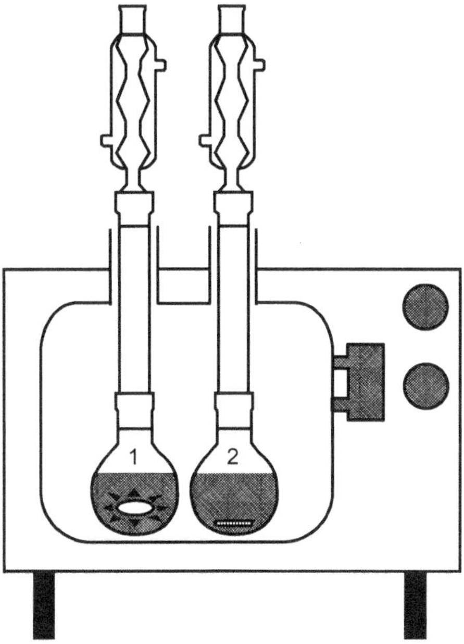

FIG. 19 An arrangement with two reflux apparatuses: (1) UV source and (2) vessel with a reaction mixture. *(Reproduced with permission from P. Klán, M. Hájek, V. Církva, The electrodeless discharge lamp: a prospective tool for photochemistry: Part 3. The microwave photochemistry reactor. J. Photochem. Photobiol. A 140 (3) (2001) 185–189.)*

have been analyzed and reported [35]. During the analysis, inorganic ions in aqueous solutions were treated to microwaves (2.45 GHz) in a modified oven (Fig. 20) under identical conditions. The changes in the solution temperatures with respect to deionized water were supervised in each case and a decrease in the temperature was found with an increase in the number of ions.

Numerous Diels-Alder and 1,3-dipolar cycloadditions to C_{60} were reported using a modified domestic microwave oven and a focused microwave reactor [36]. The top portion of the domestic microwave oven was perforated to accommodate a reflux condenser, as shown in Fig. 21. Besides, a pipe (10 cm long) was utilized to avoid microwave leakage and the turntable dish was replaced with a magnetic stirrer.

Recently, an application of microwave and ultrasound for degradation of sodium dodecyl sulfate was investigated in detail [37]. A modified domestic microwave oven (Whirlpool make, 700 W, 2.45 GHz) was used for this study (Fig. 22).

The dichlorination and decomposition of polychlorinated biphenyls in real waste transformer oil via a modified domestic microwave oven were reported (Fig. 23) [38]. The results showed that this modified microwave method is a fast, effective, and cheap method for polychlorinated biphenyls decomposition.

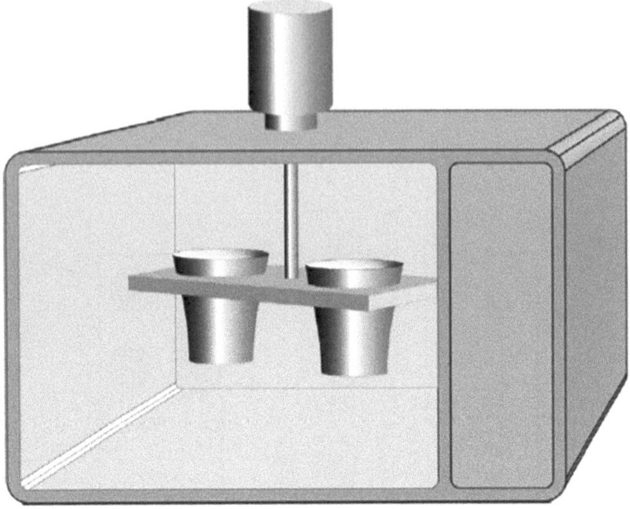

FIG. 20 Modified domestic microwave oven having rotating turntable with two identical arms to contain polystyrene cups. *(Reproduced with permission from J. Anwar, U. Shafique, W.-Uz Zaman, R. Rehman, M. Salman, A. Dar, J.M. Anzano, U. Ashraf, S. Ashraf, Microwave chemistry: effect of ions on dielectric heating in microwave ovens, Arab. J. Chem. 8 (1) (2015) 100–104.)*

Drying characteristics of wheat (Canada Western Red Spring wheat) were investigated using a modified domestic microwave convective oven. The effects of grain bed thickness, initial grain moisture, and microwave power level on the drying kinetics of the wheat were analyzed [39]. A commercially available microwave convection oven (2.45 GHz, 800 W, Panasonic) was used for this analysis. A schematic diagram of the setup is shown in Fig. 24. A Teflon plate was suspended using a fishing line nylon wire attached via the top of the oven to an electronic balance to take the data of moisture loss.

A domestic microwave oven was modified to operate as a microwave vacuum dryer with a turntable [40]. The performance of the dryer was assessed with different fruits and vegetables dried under vacuum. The schematic diagram of the microwave vacuum drying setup and the details of the rotary system are shown in Figs. 25 and 26, respectively. Different microwave powers, for example, 400 W, 700 W, and 1000 W were used to analyze the influence of power on the drying. The analysis depicted that it is possible to make dried vegetables and fruits with similar characteristics to those raised from a freeze-drying process. Besides, this method takes much shorter processing times compared to typical freeze-drying processes.

Effects of activated charcoal on dewaxing time in microwave hybrid heating are also analyzed using a modified microwave [41]. Fig. 27 shows a picture of the modified microwave oven.

The electrical energy consumption during melting of a known quantity of metallic material such as tin, zinc, aluminum, and brass was analyzed using

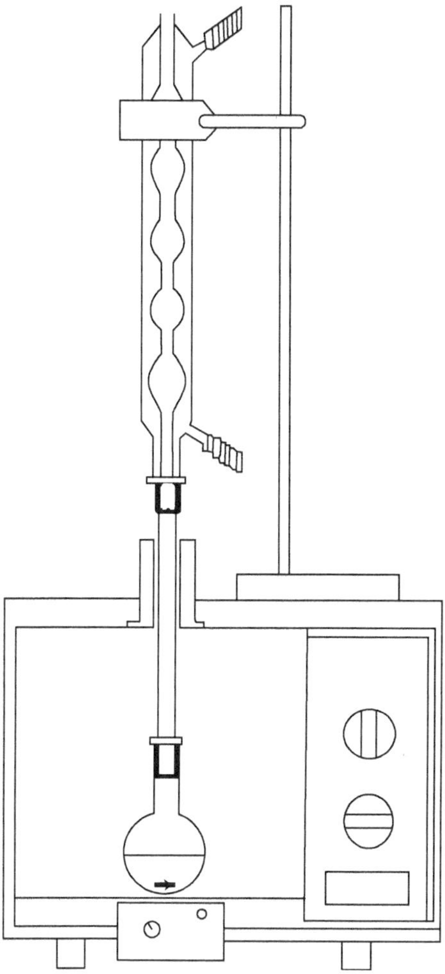

FIG. 21 A modified microwave oven for cycloaddition. *(Reproduced with permission from P. de la Cruz, A. de la Hoz, F. Langa, B. Illescas, N. Martin, Cycloadditions to [60] fullerene using microwave irradiation: a convenient and expeditious procedure. Tetrahedron 53 (7) (1997) 2599–2608.)*

a microwave oven [42]. A modified domestic microwave oven (900 W) operating at 2.45 GHz was used for the study (Fig. 28).

Horuz et al. demonstrated a programmable air-circulating hybrid domestic microwave oven (Arçelik, KMF 833 I, Turkey) [43]. Schematic representation of microwave-convective hybrid oven is shown in Fig. 29. Using that system, the drying behavior and kinetics of apricot halves were investigated at 120, 150, and 180 W microwave powers and 50°C, 60°C, and 70°C air temperature.

From a domestic microwave oven, a rotatable apparatus was developed to ensure uniform exposure of the material. Using this setup, a uniform dried ginger product with EO content of more than 0.33% and moisture content less than

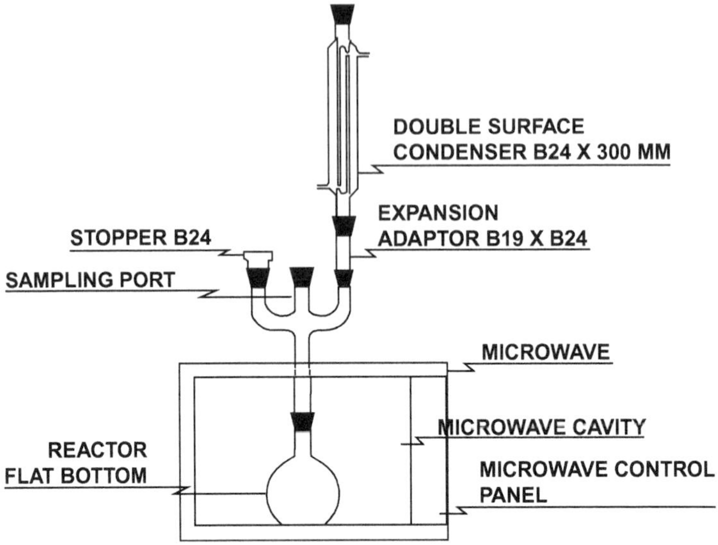

FIG. 22 A modified microwave oven for degradation of sodium dodecyl sulfate. *(Reproduced with permission from P.S. Bhandari, B.P. Makwana, P.R. Gogate, Microwave and ultrasound assisted dual oxidant based degradation of sodium dodecyl sulfate: efficacy of irradiation approaches and oxidants. J. Water Process. Eng. 36 (2020) 101316.)*

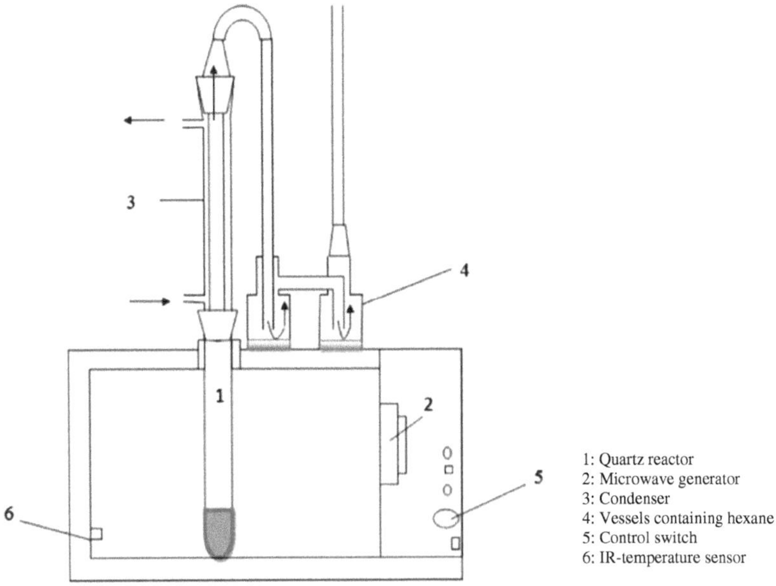

FIG. 23 A modified microwave oven for dechlorination and decomposition of polychlorinated biphenyls. *(Reproduced with permission from B. Kamarehie, A.J. Jafari, H.A. Mahabadi, Dechlorination and decomposition of Aroclor 1242 in real waste transformer oil using a nucleophilic material with a modified domestic microwave oven. J. Mater. Cycles Waste Manag. 16 (2014) 711-720.)*

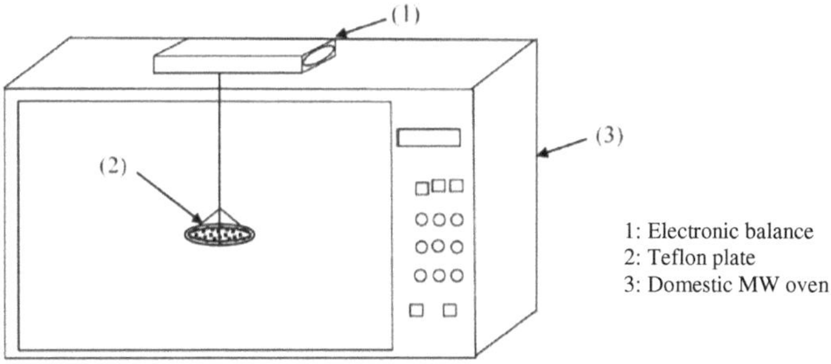

FIG. 24 A modified microwave oven for thin-layer drying of wheat. *(Reproduced with permission from M. Hemis, C.B. Singh, D.S. Jayas Microwave-assisted thin layer drying of wheat, Dry. Technol. 29 (10) (2011) 1240–1247.)*

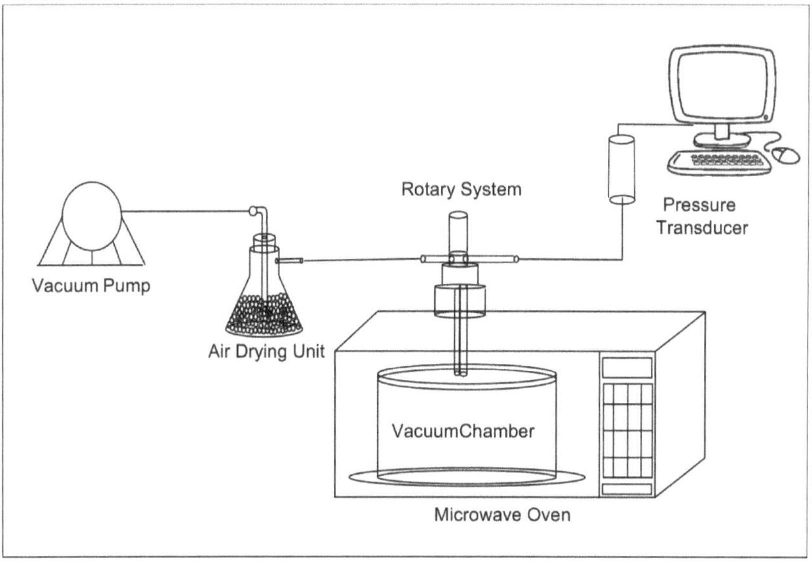

FIG. 25 Schematic diagram of the microwave vacuum drying with turntable. *(Reproduced with permission from R.L. Monteiro, B.A.M. Carciofi, A. Marsaioli, J.B. Laurindo, How to make a microwave vacuum dryer with turntable. J. Food Eng. 166 (2015) 276–284.)*

10% was demonstrated [44]. The modified microwave extraction combining hydrodiffusion and gravity setup is shown in Fig. 30. It consists of a domestic microwave oven with the maximum power of 800 W at 2450 MHz and is placed on a rotatable tray.

Based on a commercial microwave oven (700 W at 2450 MHz, Magnasonic MMMW5730, LG Electronics Inc., Korea), a microwave drying system was demonstrated by Cheng et al. [45]. The schematic diagram of the microwave drying setup is shown in Fig. 31.

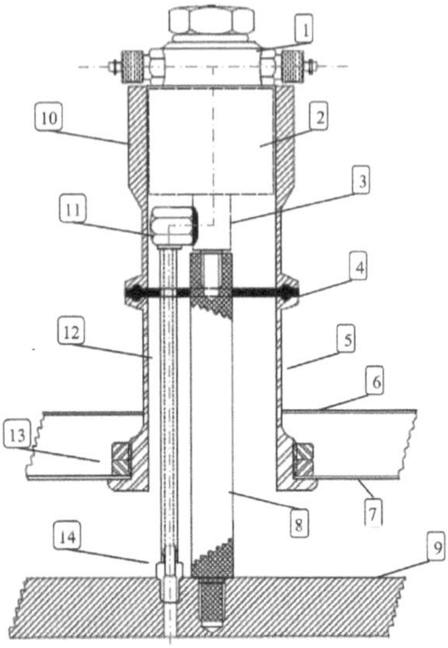

FIG. 26 Details of the rotary system. *(Reproduced with permission from R.L. Monteiro, B.A.M. Carciofi, A. Marsaioli, J.B. Laurindo, How to make a microwave vacuum dryer with turntable. J. Food Eng. 166 (2015) 276–284.)*

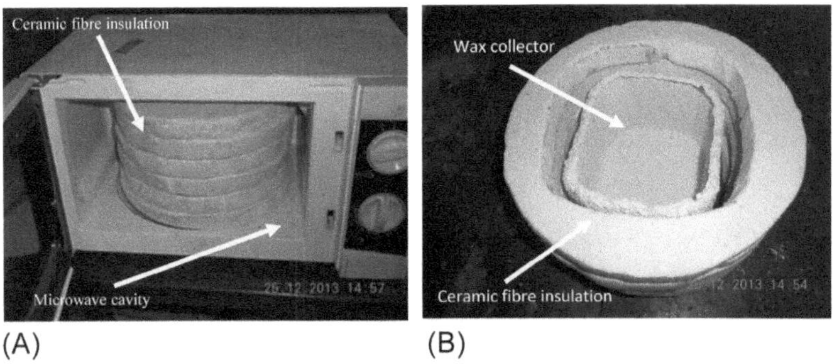

FIG. 27 (A) Modified microwave oven. (B) Wax collector and insulation arrangement in the modified microwave oven. *(Reproduced with permission from B. Yahaya, S. Izman, M.H. Idris, M.S. Dambatta, Effects of activated charcoal on dewaxing time in microwave hybrid heating. Procedia CIRP 26 (2015) 467–472.)*

By using a modified domestic microwave oven, a heater/dryer system was manufactured [46]. The schematic diagram of the system is shown in Fig. 32. Using the system, the dependency of the absorbed power on the local moisture content is measured by heating a wet material without evaporating.

The use of microwave technology in fixation was demonstrated using a modified domestic microwave oven [47]. The schematic view of the experimental setup is shown in Fig. 33. The objective of the study was to achieve quicker

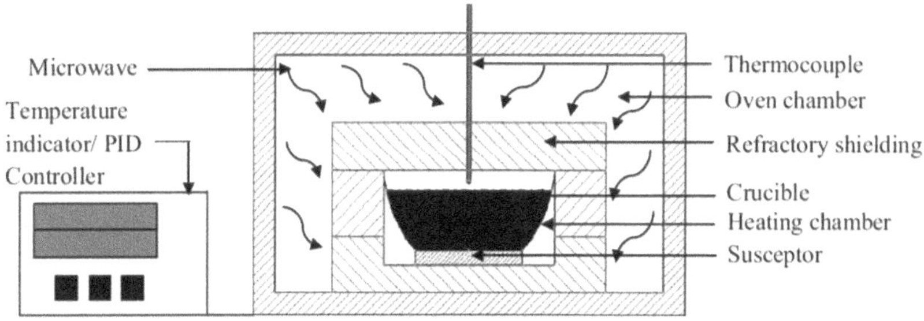

FIG. 28 Experimental setup of microwave melting. *(Reproduced with permission from S.M. Lingappa, M.S. Srinath, H.J. Amarendra, Melting of bulk non-ferrous metallic materials by microwave hybrid heating (MHH) and conventional heating: a comparative study on energy consumption. J Braz. Soc. Mech. Sci. Eng. 40 (2018) 1.)*

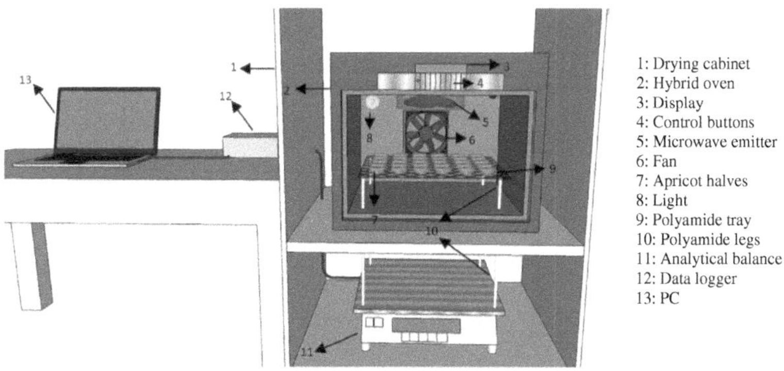

FIG. 29 Schematic representation of microwave-convective hybrid oven. *(Reproduced with permission from E. Horuz, H. Bozkurt, H. Karataş, M. Maskan, Drying kinetics of apricot halves in a microwave-hot air hybrid oven. Heat Mass Transfer 53 (2017) 2117-2127.)*

fixation regimes, lower concentrations of toxic and volatile reagents, and enhanced antigen detection on plant materials.

A very safe, fast, and practical Fries rearrangement with conventional $AlCl_3$ catalyst using a modified domestic microwave oven at atmospheric pressure was demonstrated [48]. The schematic diagram of the modified microwave is shown in Fig. 34.

2.4 Commercial microwave reactors

The development of controllable, safe, and efficient microwave reactors has gained attention because of the advent and popularization of green chemistry. Slowly and steadily, the domestic microwaves in the laboratory are being replaced with specialized microwave reactors. The dynamic development of

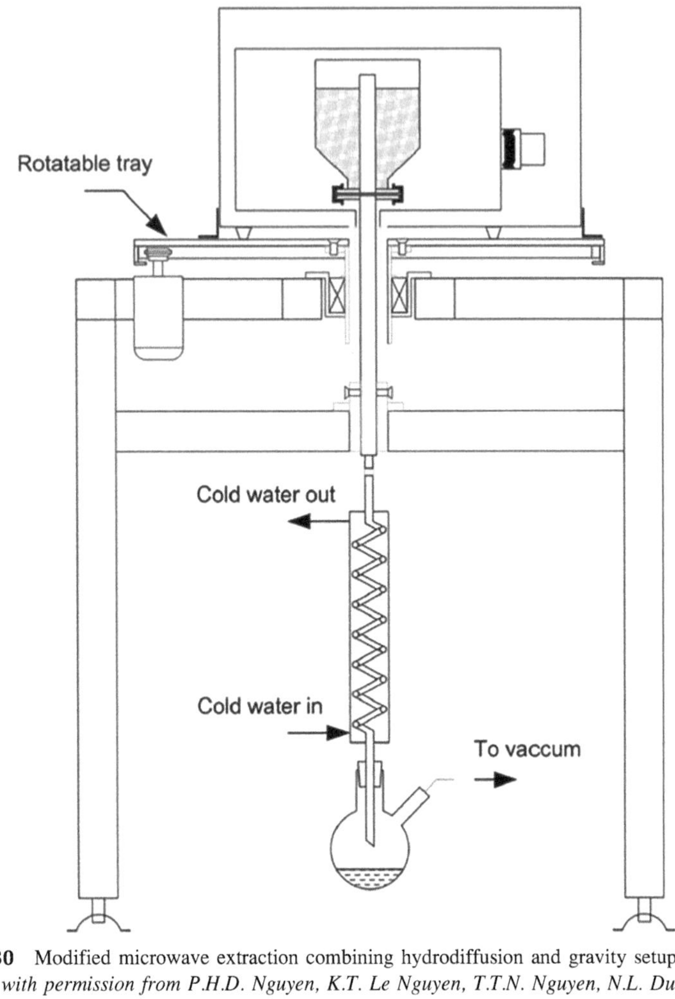

FIG. 30 Modified microwave extraction combining hydrodiffusion and gravity setup. *(Reproduced with permission from P.H.D. Nguyen, K.T. Le Nguyen, T.T.N. Nguyen, N.L. Duong, T.C. Hoang, T.T.P. Pham, D.-V.N. Vo, Application of microwave-assisted technology: a green process to produce ginger products without waste. J. Food Process Eng. 42 (2019) 12996.)*

microwave apparatus, in particular chemical reactors [49–55], and the decrease in apparatus prices heightened interest in microwave technologies furthermore. Over the past 30 years, control and measurement technologies have attained a very high development level. The current microwave reactors are equipped with advanced control systems and are safe to use. Besides, the current microwave reactors are characterized by high repeatability of processes, the optimization of reactor structure in terms of applied materials, size, and shape of the reaction chamber, recess structure, classes of microwave applicators, and automation. The new-generation reactor designs enable the rapid cooling of products [49, 56] and

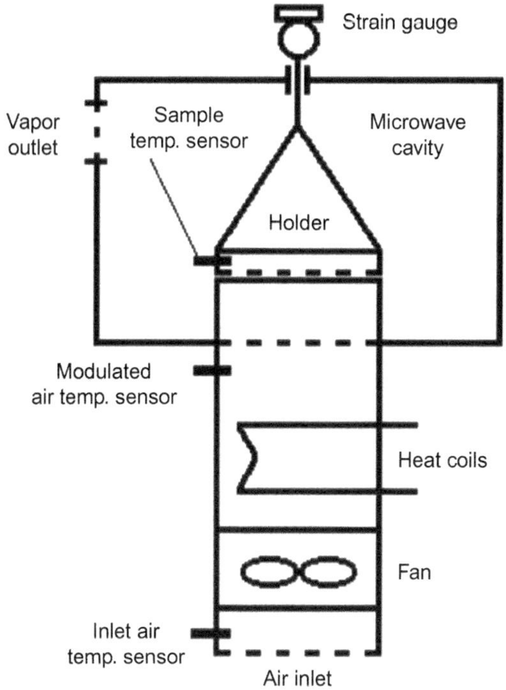

FIG. 31 Schematic diagram of the microwave drying setup. *(Reproduced with permission from W. M. Cheng, G.S.V. Raghavan, M. Ngadi, N. Wang, Microwave power control strategies on the drying process I. development and evaluation of new microwave drying system, J. Food Eng. 76 (2) (2006) 188–194.)*

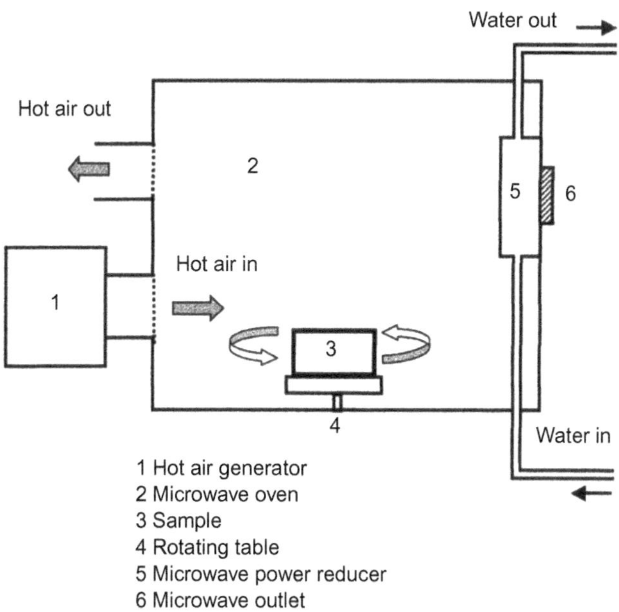

FIG. 32 Schematic diagram of the heater/dryer that is manufactured by modifying a domestic microwave oven. *(Reproduced with permission from H. Imakoma, D. Katsura, K. Kawamura, H. Iyota, N. Nishimura, Measurement of absorbed power of glass particle layer on moisture content and drying rate by combined convective and microwave drying. Dry. Technol. 23 (6) (2005) 1289–1301.)*

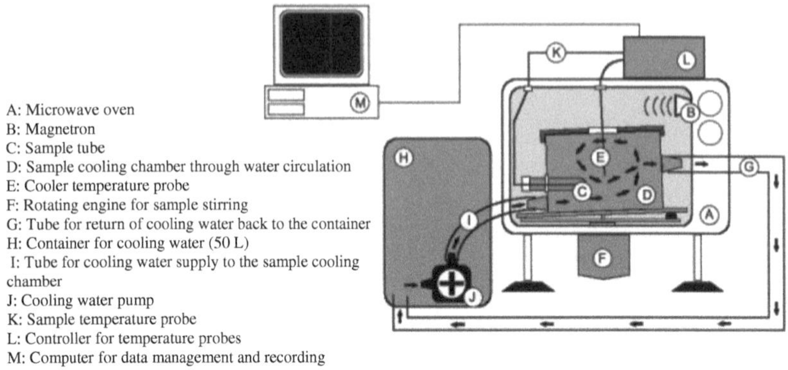

FIG. 33 Schematic view of the modified microwave oven. *(Reproduced with permission from F. Lería, R. Marco, F.J. Medina, Structural and antigenic preservation of plant samples by microwave-enhanced fixation, using dedicated hardware, minimizing heat-related effects. Microsc. Res. Tech. 65 (2004) 86–100.)*

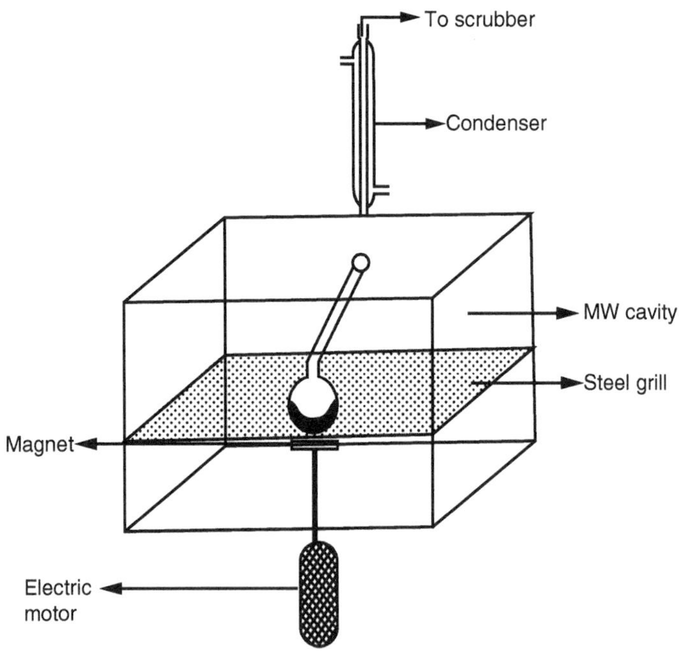

FIG. 34 Modified domestic microwave for Fries rearrangement. *(Reproduced with permission from B.M. Khadilkar, V.R. Madyar, Fries rearrangement at atmospheric pressure using microwave irradiation, Synth. Commun. 29 (7) (1999) 1195–1200.)*

real-time monitoring of temperature and pressure. However, microwave reactors can also experience failure [57]. In the following part, selected examples of commercial microwave reactors are included (Table 1).

The main producers of microwave reactors include (selected ones):

- www.berghof-instruments.com
- www.ertec.pl
- www.nade17.com
- www.cem.com
- www.sineomicrowave.com
- www.milestonesci.com
- www.anton-paar.com
- www.sairem.com
- www.chemspeed.com
- www.enbiogroup.pl
- www.cambrex.com
- www.labnano.pl
- www.weisstechnik.be

TABLE 1 **Selected examples of commercial microwave reactors used for chemical reactions.**

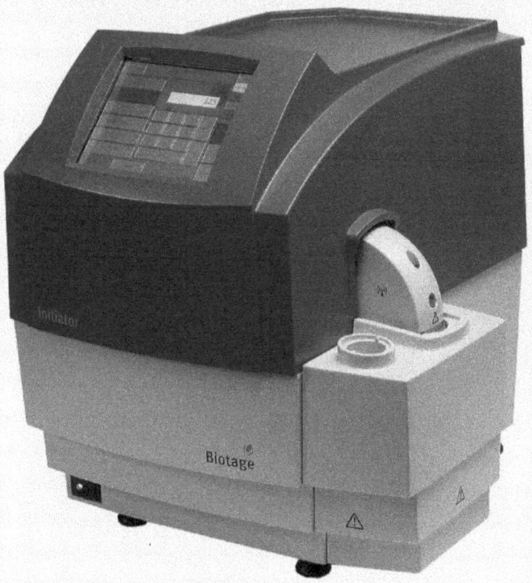	**Chemspeed's microwave reactor** (www.chemspeed.com)

Continued

TABLE 1 Selected examples of commercial microwave reactors used for chemical reactions—cont'd

	**Chemspeed's Swave platform with microwave** (www.chemspeed.com) It can perform microwave-assisted reactions but also other processes like dispensing of solid and liquid reagents, capping or crimping, vial gripping and transfer, sampling, and decapping
	**UWave-2000 Multifunctional Microwave Chemistry Reaction Workstation** (www.sineomicrowave.com) It has three energy sources—microwave, ultrasonic, and ultraviolet irradiation sources. It can be used separately or combined and can achieve the synergistic effect by multienergies Microwave source: 2450 MHz, 0–1000 W

TABLE 1 Selected examples of commercial microwave reactors used for chemical reactions—cont'd

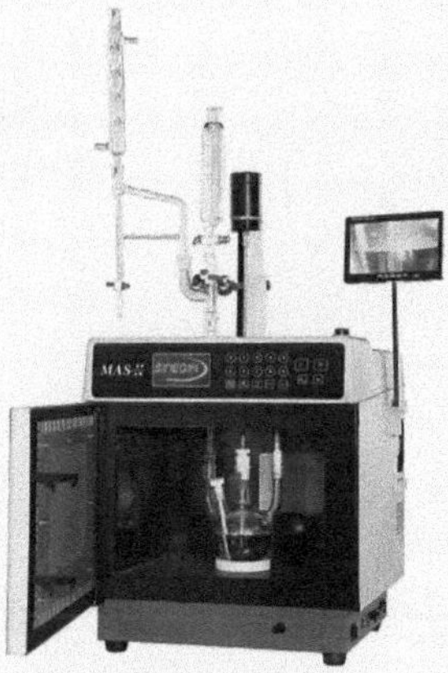	**MAS-II Plus Microwave Synthesis Workstation** (www.sineomicrowave.com) MAS-II Plus has a unique patented technology, dual-channel temperature detection technology; it can detect temperature through a PT thermocouple sensor and also using a noncontact infrared ray. The highest theoretical operation temperature of the equipment can be 900°C Microwave source: 2450 MHz, 0–1000 W
	**TANK Microwave Sample Preparation Workstation** (www.sineomicrowave.com) TANK has an advanced high-precision semiconductor pressure sensor. It also has two magnetrons, and its inverter microwave heating makes real-time adjustment of microwave output power according to the temperature and pressure feedback. In that way, the microwave field is more uniform and has more precise control

Continued

TABLE 1 Selected examples of commercial microwave reactors used for chemical reactions—cont'd

	**MDS-15 High-throughput Microwave Sample Preparation Workstation** (www.sineomicrowave.com) Microwave source: 2450 MHz, dual-magnetron design Installed power: 3600 W Maximum output power: 2200 W, nonpulse continuous automatic variable frequency control
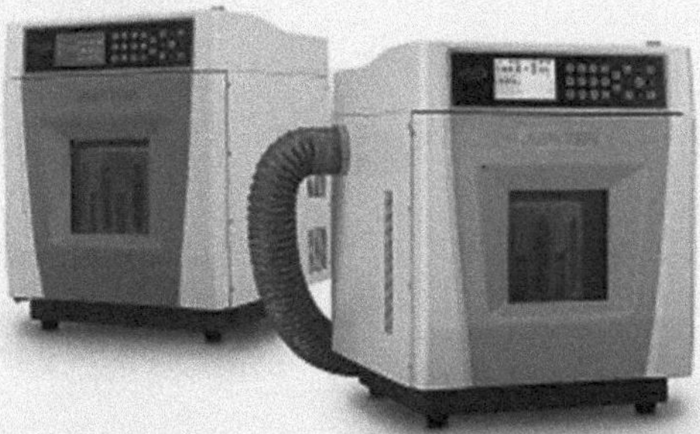	**Jupiter Series High-throughput Closed Microwave Digestion/Extraction Workstation** (www.sineomicrowave.com) Microwave frequency: 2450 MHz Installed power: 1800 W Maximum output power: 1300 W, nonpulse continuous automatic variable frequency control Sensor temperature range: 0–300°C Maximum pressure: 15 MPa (2250 psi)

TABLE 1 Selected examples of commercial microwave reactors used for chemical reactions—cont'd

	**Ultrahigh-throughput Closed Microwave Digestion/Extraction Workstation** (www.sineomicrowave.com) It has advanced and reliable pressure measuring technology by a piezoelectric crystal. The temperature inside the digestion vessel can be controlled by the combination of precise measurement of temperature inside the master vessel and IR temperature scanning
	**MDS-6G (SMART) Closed Microwave Digestion/Extraction System** (www.sineomicrowave.com) Microwave frequency: 2450 MHz Maximum output power: 1000 W High-precision platinum resistor temperature sensor Temperature range: 0–300°C Piezoelectric crystal pressure sensor control range: 0–10 MPa (1500 psi)

2.5 Conclusion

In this chapter, we have described the basic structure of the microwave device. The functions of microwave generators, high-frequency power supplies, isolators, waveguide, power monitors, and applicator are also described in detail. Besides, the modifications made in the domestic microwave ovens for different purposes are also included. The last section of the chapter displays various commercial microwave reactors available in the market.

Acknowledgments

AD and BKB are grateful to Prince Mohammad Bin Fahd University for support. BKB is also grateful to US NIH, US NCI, and Kleberg Foundation of Texas for financial support.

References

[1] E. Brookner, From $10,000 magee to $7 magee and $10 transmitter and receiver (T/R) on single chip, in: International Conference on the Origins and Evolution of the Cavity Magnetron, Bournemouth, 19-20 April, 2010, pp. 1–2.
[2] A.W. Hull, The effect of a uniform magnetic field on the motion of electrons between coaxial cylinders, Phys. Rev. 18 (1) (1921) 31–57.
[3] A.W. Hull, The magnetron, J. Am. Inst. Elect. Eng. 40 (9) (1921) 715–723.
[4] J. Brittain, The magnetron and the beginnings of the microwave age, Phys. Today 38 (7) (1985) 60–67.
[5] R.H. Varian, S.F. Varian, A high frequency oscillator and amplifier, J. Appl. Phys. 10 (5) (1939) 321.
[6] D. Varian, The Inventor and the Pilot, Pacific Books, 1983, p. 187.
[7] N.H. Pond, The Tube Guys, Russ Cochran, 2008, pp. 31–40.
[8] S.Y. Liao, Microwave Devices and Circuits, Dorling Kindersley, September 1990, p. 380.
[9] A.S. Gilmour, Klystrons, Traveling Wave Tubes, Magnetrons, Cross-Field Amplifiers, and Gyrotrons, Artech House, 2011, pp. 3–4.
[10] S.Y. Liao, Reflex Klystron, Dorling Kindersley, September 1990, pp. 391–392.
[11] M.A. Richards, J.A. Scheer, W.A. Holm, Power sources and amplifiers, in: Principles of Modern Radar: Basic Principles, SciTech Pub, 2010, p. 360.
[12] M. Blank, P. Borchard, S. Cauffman, K. Felch, M. Rosay, L. Tometich, Experimental demonstration of a 527 GHz gyrotron for dynamic nuclear polarization, in: Abstracts IEEE International Conference on Plasma Science (ICOPS), 1 June, 2013, p. 1.
[13] M. Thumm, State-of-the-art of high-power gyro-devices and free electron masers, J. Infrared Millim. Terahertz Waves 41 (1) (2020) 1.
[14] E. Borie, Review of gyrotron theory, EPJ Web Conf. 149 (2017), 04018.
[15] A. Kupiszewski, The Gyrotron: A High Frequency Microwave Amplifier, The Deep Space Network Progress Report 42 (52), 1979, pp. 8–12.
[16] J.C. Whitaker, The RF Transmission Systems Handbook, CRC Press, 2002, pp. 8.14–8.16.
[17] B.R. McAvoy, Solid State Microwave Oven, US Patent 3557333, 1971.
[18] A.B. Mackay, W.R. Tinga, W.A.G. Voss, Frequency agile sources for microwave ovens, J. Microwave Power 14 (1) (1979) 63–76.
[19] W.A.G. Voss, Solis state microwave oven development, J. Microw. Power 21 (1986) 188–189.

[20] E. Schwartz, A. Anaton, D. Huppert, E. Jerby, Transistor-based miniature microwave heater, in: 40th Annual Microwave Symposium Proceedings, IMPI International Microwave Power Institute, Boston, August, 2006.

[21] D.M. Pozar, Microwave Engineering, fourth ed., Wiley, 2011.

[22] A. Saib, M. Darques, L. Piraux, D. Vanhoenacker-Janvier, I. Huynen, An unbiased integrated microstrip circulator based on magnetic nanowired substrate, IEEE Trans. Microw. Theory Techn. 53 (6) (2005) 2043–2049.

[23] B. Ondruschka, W. Bonrath, D. Stuerga, Development and design of reactors in microwave-assisted chemistry, in: A. de la Hoz, A. Loupy (Eds.), Microwaves in Organic Synthesis, Wiley-VCH Verlag & Co. KGaA, Weinheim, Germany, 2012.

[24] A. Herzog, R. Klingner, U. Vogt, T. Graule, Wood-derived porous SiC ceramics by sol infiltration and carbothermal reduction, J. Am. Ceram. Soc. 87 (2004) 784.

[25] H. Sieber, C. Hoffmann, A. Kaindl, P. Greil, Biomorphic cellular ceramics, Adv. Eng. Mater. 2 (2000) 105.

[26] G.C.-T. Wei, Beta Sic powders produced by carbothermic reduction of silica in a high-temperature rotary furnace, J. Am. Ceram. Soc. 66 (1983) C-111.

[27] A. Halfadji, A. Touabet, Determination of 26 polychlorinated biphenyls congeners in soil samples using microwave-assisted extraction with open vessel and gas chromatography, Green Chem. Lett. Rev. 11 (3) (2018) 209–216.

[28] Z. Okmen, A. Bayindirli, Modification of a household microwave oven for continuous temperature and weight measurements during drying of foods, J. Microw. Power Electromag. Energ. 35 (4) (2000) 225–231.

[29] L.W. Aguiar, G.P. Otto, V.L. Kupfer, S.L. Fávaro, C.T.P. Silva, M.P. Moisés, L. de Almeida, M.R. Guilherme, E. Radovanovic, E.M. Girotto, A.W. Rinaldi, Simple, fast, and low-cost synthesis of MIL-100 and MIL-88B in a modified domestic microwave oven, Mater. Lett. 276 (2020) 128127.

[30] S. Horikoshi, H. Hidaka, N. Serpone, Environmental remediation by an integrated microwave/UV illumination technique: VI. A simple modified domestic microwave oven integrating an electrodeless UV-Vis lamp to photodegrade environmental pollutants in aqueous media, J. Photochem. Photobiol. A 161 (2-3) (2004) 221–225.

[31] A. Harabi, S. Zouai, A new and economic approach to synthesize and fabricate bioactive diopside ceramics using a modified domestic microwave oven. Part 1: study of sintering and bioactivity, Int. J. Appl. Ceram. Technol. 11 (2014) 31–46.

[32] S.A. Halim, J. Swithenbank, Simulation study of parameters influencing microwave heating of biomass, J. Energy Inst. 92 (4) (2019) 1191–1212.

[33] K. Hussain, N. Bashir, Z. Hussain, S.A. Sulaiman, Cement catalyzed conversion of biomass into upgraded bio-oil through microwave metal interaction pyrolysis in aluminum coil reactor, J. Anal. Appl. Pyrol. 129 (2018) 37–42.

[34] P. Klán, M. Hájek, V. Církva, The electrodeless discharge lamp: a prospective tool for photochemistry: Part 3. The microwave photochemistry reactor, J. Photochem. Photobiol. A 140 (3) (2001) 185–189.

[35] J. Anwar, U. Shafique, W.-U. Zaman, R. Rehman, M. Salman, A. Dar, J.M. Anzano, U. Ashraf, S. Ashraf, Microwave chemistry: effect of ions on dielectric heating in microwave ovens, Arab. J. Chem. 8 (1) (2015) 100–104.

[36] P. de la Cruz, A. de la Hoz, F. Langa, B. Illescas, N. Martin, Cycloadditions to [60] fullerene using microwave irradiation: a convenient and expeditious procedure, Tetrahedron 53 (7) (1997) 2599–2608.

[37] P.S. Bhandari, B.P. Makwana, P.R. Gogate, Microwave and ultrasound assisted dual oxidant based degradation of sodium dodecyl sulfate: efficacy of irradiation approaches and oxidants, J. Water Process. Eng. 36 (2020) 101316.

[38] B. Kamarehie, A.J. Jafari, H.A. Mahabadi, Dechlorination and decomposition of Aroclor 1242 in real waste transformer oil using a nucleophilic material with a modified domestic microwave oven, J. Mater. Cycles Waste Manag. 16 (2014) 711–720.

[39] M. Hemis, C.B. Singh, D.S. Jayas, Microwave-assisted thin layer drying of wheat, Dry. Technol. 29 (10) (2011) 1240–1247.

[40] R.L. Monteiro, B.A.M. Carciofi, A. Marsaioli, J.B. Laurindo, How to make a microwave vacuum dryer with turntable, J. Food Eng. 166 (2015) 276–284.

[41] B. Yahaya, S. Izman, M.H. Idris, M.S. Dambatta, Effects of activated charcoal on dewaxing time in microwave hybrid heating, Procedia CIRP 26 (2015) 467–472.

[42] S.M. Lingappa, M.S. Srinath, H.J. Amarendra, Melting of bulk non-ferrous metallic materials by microwave hybrid heating (MHH) and conventional heating: a comparative study on energy consumption, J Braz. Soc. Mech. Sci. Eng. 40 (2018) 1.

[43] E. Horuz, H. Bozkurt, H. Karataş, M. Maskan, Drying kinetics of apricot halves in a microwave-hot air hybrid oven, Heat Mass Transfer 53 (2017) 2117–2127.

[44] P.H.D. Nguyen, K.T. Le Nguyen, T.T.N. Nguyen, N.L. Duong, T.C. Hoang, T.T.P. Pham, D.-.V.N. Vo, Application of microwave-assisted technology: a green process to produce ginger products without waste, J. Food Process Eng. 42 (2019) 12996.

[45] W.M. Cheng, G.S.V. Raghavan, M. Ngadi, N. Wang, Microwave power control strategies on the drying process I. development and evaluation of new microwave drying system, J. Food Eng. 76 (2) (2006) 188–194.

[46] H. Imakoma, D. Katsura, K. Kawamura, H. Iyota, N. Nishimura, Measurement of absorbed power of glass particle layer on moisture content and drying rate by combined convective and microwave drying, Dry. Technol. 23 (6) (2005) 1289–1301.

[47] F. Lería, R. Marco, F.J. Medina, Structural and antigenic preservation of plant samples by microwave-enhanced fixation, using dedicated hardware, minimizing heat-related effects, Microsc. Res. Tech. 65 (2004) 86–100.

[48] B.M. Khadilkar, V.R. Madyar, Fries rearrangement at atmospheric pressure using microwave irradiation, Synth. Commun. 29 (7) (1999) 1195–1200.

[49] A. Majcher, J. Wiejak, J. Przybylski, T. Chudoba, J. Wojnarowicz, A novel reactor for microwave hydrothermal scale-up nanopowder synthesis, Int. J. Chem. React. Eng. 11 (2013) 361–368.

[50] S. Dabrowska, T. Chudoba, J. Wojnarowicz, W. Łojkowski, Current trends in the development of microwave reactors for the synthesis of nanomaterials in laboratories and industries: a review, Crystals 8 (2018) 379.

[51] P. Priecel, J.A. Lopez-Sanchez, Advantages and limitations of microwave reactors: from chemical synthesis to the catalytic valorization of biobased chemicals, ACS Sustain. Chem. Eng. 7 (2019) 3–21.

[52] A.J. Buttress, G. Hargreaves, A. Ilchev, T. Monti, A. Sklavounou, J. Katrib, P. Martin-Tanchereau, M.G. Unthank, D.J. Irvine, C.D. Dodds, Design and optimisation of a microwave reactor for kilo-scale polymer synthesis, Chem. Eng. Sci. X 2 (2019) 100022.

[53] T. Mitani, N. Hasegawa, R. Nakajima, N. Shinohara, Y. Nozaki, T. Chikata, T. Watanabe, Development of a wideband microwave reactor with a coaxial cable structure, Chem. Eng. J. 299 (2016) 209–216.

[54] C.O. Kappe, My twenty years in microwave chemistry: From kitchen ovens to microwaves that aren't microwaves, Chem. Rec. 19 (2019) 15–39.

[55] C. Leonelli, P. Veronesi, Microwave reactors for chemical synthesis and biofuels preparation, in: Z. Fang, R.L. Smith Jr., X. Qi (Eds.), Production of Biofuels and Chemicals with Microwave, first ed., Springer, Dordrecht, The Netherlands, 2015, pp. 17–40.
[56] J. Wojnarowicz, T. Chudoba, S. Gierlotka, W. Lojkowski, Effect of microwave radiation power on the size of aggregates of ZnO NPs prepared using microwave solvothermal synthesis, Nanomaterials 8 (2018) 343.
[57] S. Dabrowska, T. Chudoba, J. Wojnarowicz, W. Łojkowski, Problems of exploitations of microwave reactors for nanoparticles synthesis, J. Mach. Constr. Maint. 110 (2018) 41–47.

Chapter 3

Modeling and interpreting microwave effects

Aparna Das* and Bimal Krishna Banik*
Department of Mathematics and Natural Sciences, College of Sciences and Human Studies, Prince Mohammad Bin Fahd University, Al Khobar, Kingdom of Saudi Arabia
**Corresponding authors: e-mail: aparnadasam@gmail.com (Aparna Das); bimalbanik10@gmail.com, bbanik@pmu.edu.sa (Bimal Krishna Banik)*

3.1 Introduction

Investigations based on mathematical simulations were reported for microwave heating in solids as well as in liquid samples. For example, scientists reported the microwave heat transfer in solid foods [1–7] and liquids [8–11]. In 1986, it was demonstrated a model for the continuous-flow microwave system using the plug flow approach with the assumption that the microwave energy dissipates homogenously [12]. In certain studies, the microwave power absorbed was considered to decay exponentially into the sample (follows Lambert's law) [13, 14]. This was valid only for the large samples [15]. In the case of small or low loss dielectric materials, the spatial variations of electromagnetic fields and microwave power absorbed within samples were described with other mathematical expressions such as Maxwell's equations. Normally, during microwave heating of a liquid layer, the change of the microwave power level and the variation in an electric conductivity change the penetration degree and heat generation rate within the liquid layer. The rate of reflection of the microwave is related to the electric conductivity of the heating layer so that the effects of dielectric properties and the change of microwave power level are considered adequately. Besides, the local heating on the liquid surface by microwave induces the difference of surface tension on the surface layer and results in the flow due to the difference in surface tension. Thus, the influence of the surface tension effect is considered during analysis. Considering all, the main factors that affect microwave heating include volume, shape, and dielectric properties of the material as well as geometric parameters and design of the microwave unit [16]. All these factors cause it difficult to properly control the heating effect to get the wanted temperature distribution in the material.

Microwaves in Chemistry Applications. https://doi.org/10.1016/B978-0-12-822895-1.00007-2

3.2 Modeling of microwave heat transfer in liquids

Due to the ability of the microwave to penetrate the materials and create volumetric heat from within materials, this heating method is a much quicker and more energy-efficient method for liquid sample heating in comparison with conventional heating [17]. Many theories are reported to explain the intensification by microwave, and most of the studies have shown that the intensification is due to the superheating effect of the microwave. During the superheating, the liquid is heated greatly farther than its conventional boiling point [18, 19]. Consequently, a higher rate in reaction or mass transfer is possible owing to the higher reaction temperature [20, 21]. There are also many experimental works to explain the superheating effect of microwaves in liquid. The studies have shown that many solvents can be superheated up to 4–25°C by microwave [22]. The superheating effect is because of the limited bubble nucleation at the vessel wall during the heating, which makes excessive energy to be trapped in the liquid [22–24]. Chemat et al. demonstrated that the superheating temperature in organic solvents can be as high as 40°C [24]. Microwave heating can lead to an extended superboiling effect where the liquid stays superheated at the same time the bubble nucleation occurs [25].

One of the major problems related to the superheating effect is the inconsistency in the value of superheat temperatures reported for the same liquids by different scientists. For example, the superheating temperature reported for methanol was ranged from 6°C to 19°C [22, 24, 25]. Thus, the origin and behavior of the microwave superheating phenomenon are very complicated and temperamental, and very sensitive to various operating conditions. Therefore, the prediction of the superheating phenomenon and condition in real applications is challenging.

Ratanadecho et al. reported the numerical as well as experimental investigations of the modeling of microwave heating for liquid layers using a rectangular waveguide [26]. The study was executed with two different liquid layers such as water layer and NaCl-water solution layer. The proposed mathematical model used comprehensive two-dimensional heat and momentum equations together with a complete solution of the unsteady Maxwell's equations in the time domain to elaborate unsteady temperature and fluid flow fields and to examine many related aspects of the heating effects.

The modeling results depicted that the heating kinetic was strongly related to the dielectric properties of the material. The dielectric properties considered for the calculations are shown in Fig. 1. It is observed that the dielectric properties depend strongly on the temperature and electric conductivity of the liquid sample layer.

The analytical model for microwave heating of a liquid layer sample by a rectangular waveguide is shown in Fig. 2. The sample surface was exposed to the microwave in the form of a plane wave. Other surfaces of the sample were insulated, and the mass and heat fluxes were set to zero.

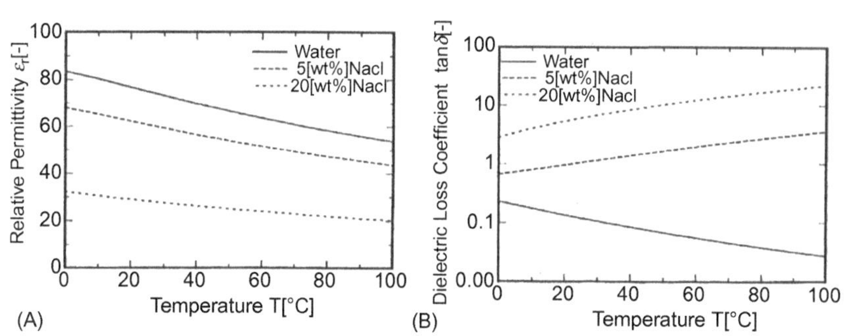

FIG. 1 Dielectric properties at different conditions used in the calculations: (A) relative permittivity and (B) dielectric loss coefficient. *(Reproduced with permission from P. Ratanadecho, K. Aoki, M. Akahori, A numerical and experimental investigation of the modeling of microwave heating for liquid layers using a rectangular wave guide (effects of natural convection and dielectric properties). Appl. Math. Model. 26 (2002) 449–472.)*

The calculations show that the change of microwave power level and electric conductivity value varies the penetration degree and heat generation rate within the liquid layer. Besides, the rate of reflection of the microwave is strongly related to the electric conductivity value within the liquid layer. The temperature and velocity field's distributions in the liquid layer are related to the electromagnetic fields. The theoretical results were in agreement with the experimental data.

Zhu et al. presented a numerical model to explain heat transfer in food liquids such as in apple sauce, skim milk, and tomato sauce [27]. To explain the electromagnetic field in the microwave cavity and the waveguide, the transient Maxwell's equations were evaluated by using the finite difference time domain method. Using the solution of the energy, momentum, and Maxwell's equations, the temperature field inside the applicator duct was calculated. The study helped to understand the effects of the geometry of the microwave cavity, the applicator diameter, applicator's location, and the dielectric properties of the

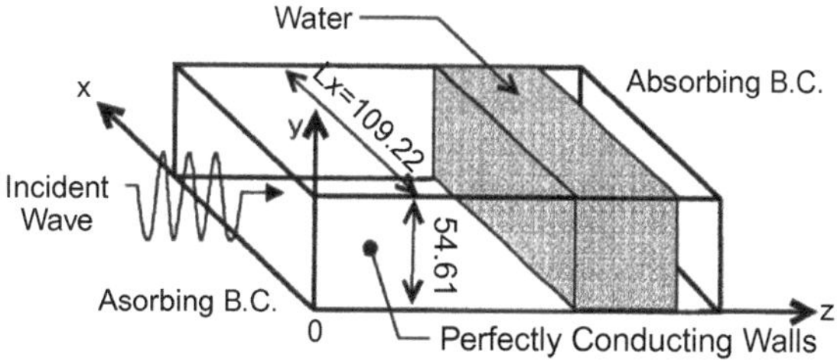

FIG. 2 The analytical model for microwave heating. *(Reproduced with permission from P. Ratanadecho, K. Aoki, M. Akahori, A numerical and experimental investigation of the modeling of microwave heating for liquid layers using a rectangular wave guide (effects of natural convection and dielectric properties). Appl. Math. Model. 26 (2002) 449–472.)*

fluid on the heating process. The data showed that the microwave heating pattern is strongly dependent on the dielectric properties of the liquid in the duct and the geometry of the microwave heating unit.

Expanding the diameter of the applicator increases the effective surface available to absorb the microwave energy and generally enhances the absorption of power in the liquid. At the same time, above the critical diameter of the applicator, an opposite effect was found. The critical diameter of the applicator is related to the geometry of the resonant cavity and the dielectric properties of the used liquid. The absorption of the microwave power depends on the location of the applicator and the shape of the resonant cavity.

Sabliov et al. numerically predicted the temperature of a liquid product heated in a continuous-flow focused microwave system [28].

The study was performed by coupling high-frequency electromagnetism, heat transfer, and fluid flow in ANSYS Multiphysics simulator. The prepared model was employed to analyze the temperature change in water processed in a 915-MHz microwave unit, under steady-state conditions. Continuous-flow focused microwave system geometry after meshing is shown in Fig. 3.

During the analysis, for two flow rates (1 and 2 L/min), the average temperature in the water as a function of distance traveled in the applicator tube was calculated as shown in Fig. 4. The effect of flow rate on the heating profile was computed using this. The data demonstrated that the average temperature of water increased from 25°C to 42°C at the flow rate of 1 L/min and to 34°C at the flow rate of 2 L/min.

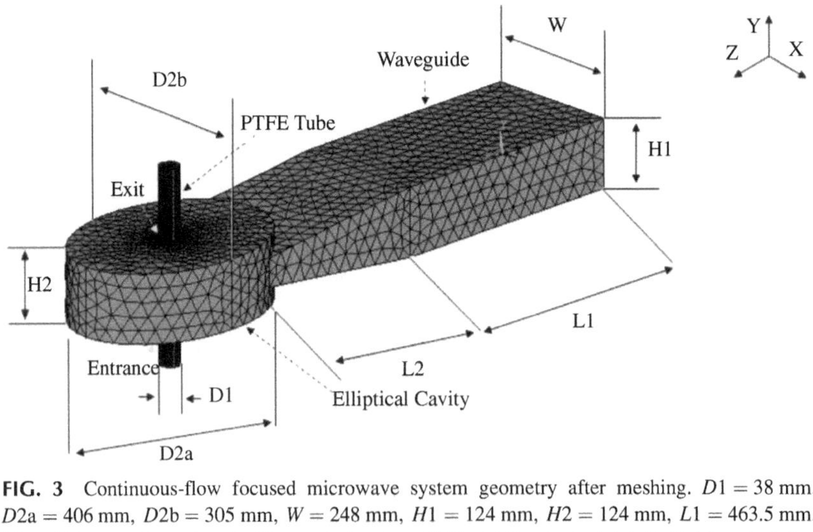

FIG. 3 Continuous-flow focused microwave system geometry after meshing. $D1 = 38$ mm, $D2a = 406$ mm, $D2b = 305$ mm, $W = 248$ mm, $H1 = 124$ mm, $H2 = 124$ mm, $L1 = 463.5$ mm, $L2 = 343$ mm. *(Reproduced with permission from C.M. Sabliov, D.A. Salvi, D. Boldor, High frequency electromagnetism, heat transfer and fluid flow coupling in ANSYS multiphysics. J. Microw. Power Electromagn. Energy. 41 (2006) 5–17.)*

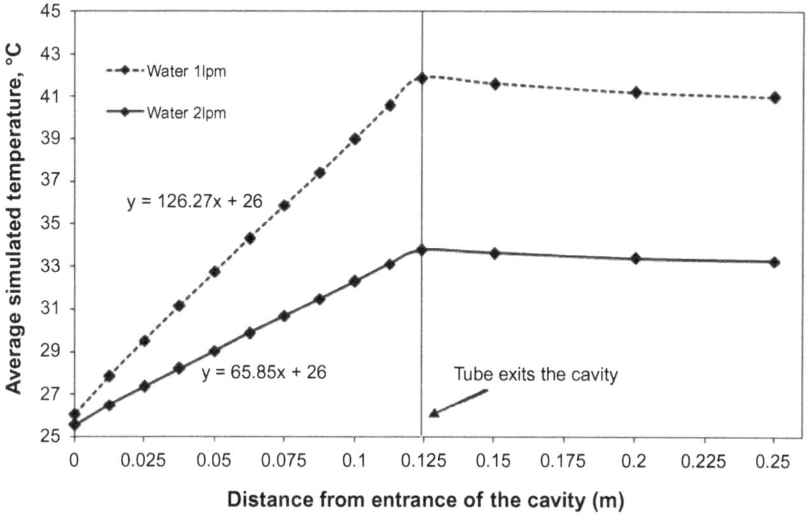

FIG. 4 Average simulated temperature in the water as the function of distance from the entrance of the cavity. *(Reproduced with permission from C.M. Sabliov, D.A. Salvi, D. Boldor, High frequency electromagnetism, heat transfer and fluid flow coupling in ANSYS multiphysics. J. Microw. Power Electromagn. Energy. 41 (2006) 5–17.)*

The radial temperature distribution in the heated fluid is illustrated in Fig. 5. In the liquid, at the center of the tube, the highest temperature regions were observed. Lower temperature regions were observed when the radial distance from the center increased. Finally, a slightly higher temperature region near the tube's wall was found corresponding to the energy distribution (Mathieu function). The energy distribution analysis also led to a similar temperature pattern.

The effects of turntable rotation, natural convection, power sources, and aspect ratio of the container on the temperature profiles of containerized liquid heated using microwave radiation were reported [29]. A detailed analysis of stream functions, temperature profiles, and time evolution of flow field was delivered. The study indicated that the rotation of the turntable does not help in attaining uniform heating in the case of the asymmetric heat source.

Ballast water release from oceangoing vessels exhibits a worldwide ecological problem in inland and coastal waters all over the world due to the formation of nonindigenous invasive species. Various technologies and methods have been reported in the past decade to overcome this aspect. Among those, continuous microwave heating is one of the appropriate methods to eliminate invasive species. Salvi et al. reported the first attempt to numerically model the flow distribution and temperature in ballast water heated using a continuous focused microwave system [30]. The numerical data were compared with experimental data at different salinities using model ballast water (0% and 1.5%) heated in a

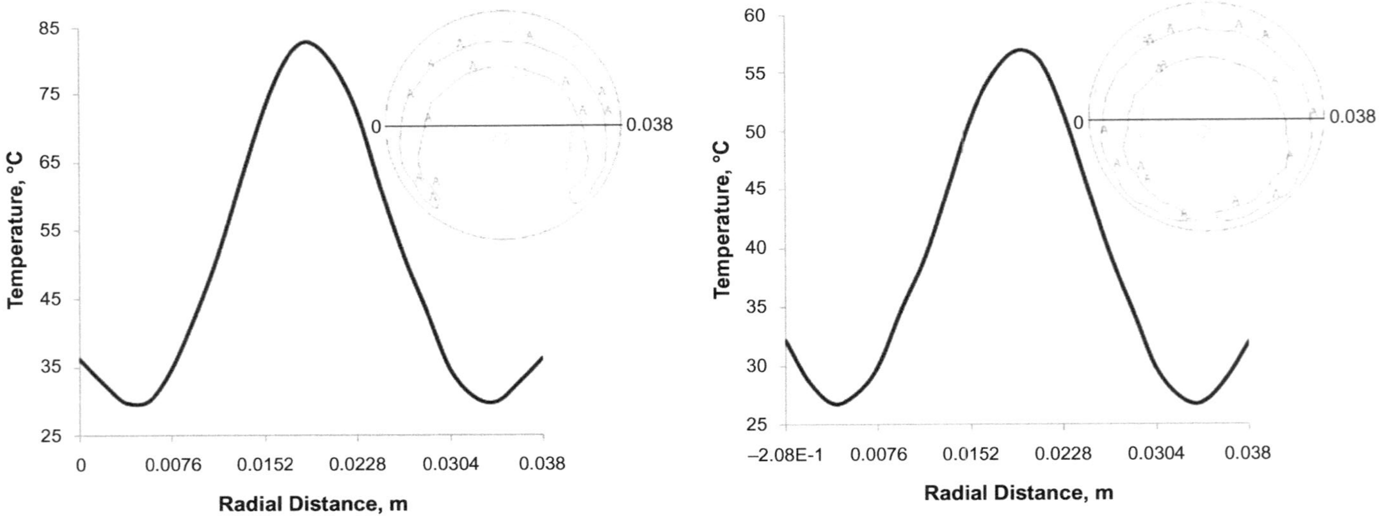

FIG. 5 Temperature variation along with the radial distance for water flowing (A) at the rate of 1 L/min and (B) at the rate of 2 L/min. *(Reproduced with permission from C.M. Sabliov, D.A. Salvi, D. Boldor, High frequency electromagnetism, heat transfer and fluid flow coupling in ANSYS multiphysics. J. Microw. Power Electromagn. Energy. 41 (2006) 5–17.)*

focused microwave unit (915 MHz, 4.5 kW). Various parameters were considered to check their influence on the ballast water temperature, such as flow rate (1 and 1.6 L/min) and water salinity (0% and 1.5%).

The data showed that the average cross-axial temperature increased linearly with distance from the entrance of the cavity and the temperature increase was influenced by the flow rates and salt concentrations. The highest temperature increase (40°C) was observed for the 1.5 % salt concentration at 1 L/min flow rate, and the lowest temperature increase (6°C) was obtained for the 0% salt concentration at 1.6 L/min flow rate.

Jian et al. reported the numerical analysis of the influence of stir during microwave heating on water [31]. To analyze this, the fluid field equations, coupled Maxwell's equations, and heat transport equations were figured out using the finite element method. The analysis was done with different dynamic viscosities, with high power and relative complex permittivities. The study depicted that the highest temperature occurs on the interface of the air and water. The accelerating stir could improve the uniformity of temperature when the water is heated under high power. However, if the rotation speed is fast enough, going on speeding up the stir cannot diminish the temperature difference.

Cha-um et al. investigated the heating process of water and oil using a microwave oven with a rectangular waveguide [32]. The numerical data were validated with the experimental data. The analytical model used for the study is shown in Fig. 6.

In this study, to describe the electromagnetic field in the sample and waveguide, the transient Maxwell's equations were examined using the finite difference time domain method. The velocity field and temperature profiles inside the sample were computed by the solution of the energy, momentum, and Maxwell's equations. Besides, the effects of various physical parameters such as microwave power, the position of the sample in a waveguide, size, and thickness of the sample were considered. Fig. 7A and B compares the temperature distribution within the water layer between the predicted and experimental results at $t = 60$ s, at various microwave power, along with the horizontal axis and vertical axis of a rectangular waveguide, respectively. It is found that the rate of temperature rise is significantly influenced by microwave power. The mathematical models are in good agreement with the experimental data.

Scientists investigated experimentally and numerically the microwave-induced natural convection in water [33]. The results depicted that during microwave heating of water in a resonant microwave field, hot spots (local over heatings) were formed, and they were majorly placed in the axis of the heated liquid. Consequently, natural convection arises, which is induced mainly in the strongest hot spot. The modeling and experimental analysis evidenced that in the time-dependent temperature profiles of hot spots, two characteristic periods could be distinguished. In the first period of microwave heating, the microwave-induced natural convection developed, and in the second period, circulation of

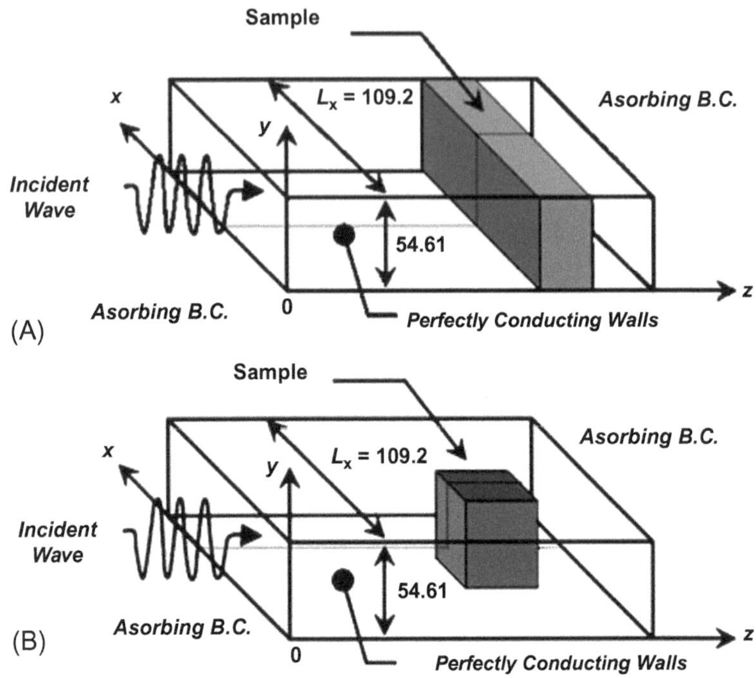

FIG. 6 The analytical model: (A) full load case, (B) partial load case. *(Reproduced with permission from W. Cha-um, P. Rattanadecho, W. Pakdee, Experimental and numerical analysis of microwave heating of water and oil using a rectangular wave guide: influence of sample sizes, positions, and microwave power. Food Bioprocess Technol. 4 (2011) 544–558.)*

the water began. Due to the circulation of liquid in the second period, the rates of temperature increase in the second period were lower than in the first one. Fig. 8 compares the simulated and experimental temperature differences for microwave heating without stirring at six axial positions of the fiber-optic probe.

Figs. 9 and 10 show the comparison of the simulated temperature differences at the fiber-optic position h20 for microwave without stirring and different incident microwave powers and frequencies, respectively. The results showed the increasing temperature of the water with the increasing incident power. Furthermore, the study was demonstrated that the variation of just 5 MHz from the standard frequency (2.45 GHz) alters the absorbed microwave power in water by 20%. The study also presented that natural convection is not capable of equalizing the temperature across the full volume of the liquid.

Studies showed that a uniform temperature distribution could be accomplished by continuous-flow microwave heating of low-viscosity fluid products like orange juice [34, 35], apple juice [36, 37], and milk [38–41]. However, nonuniform heating still played an issue for highly viscous liquids/foods. For instance, nonuniform temperature distribution was reported for green pea/carrot

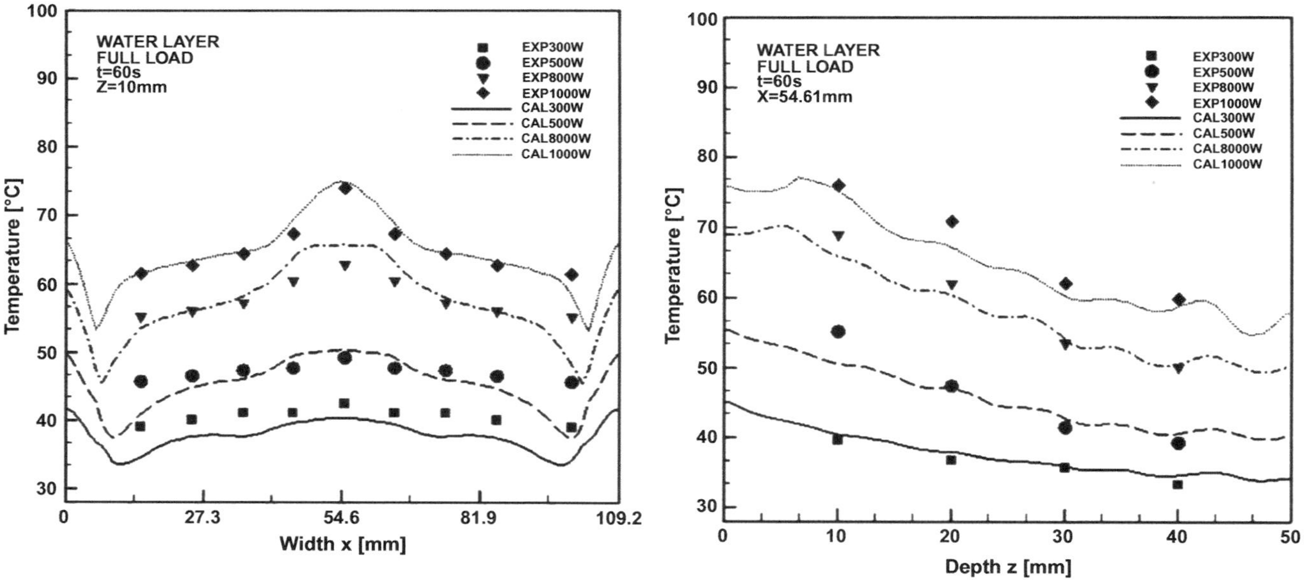

FIG. 7 Comparison of the temperature distribution within the water layer between the predicted and experimental results at various microwave powers. (A) Along with the horizontal axis ($z = 10$ mm) and (B) along with the vertical axis ($x = 54.61$ mm) of a rectangular waveguide. *(Reproduced with permission from W. Cha-um, P. Rattanadecho, W. Pakdee, Experimental and numerical analysis of microwave heating of water and oil using a rectangular wave guide: influence of sample sizes, positions, and microwave power. Food Bioprocess Technol. 4 (2011) 544–558.)*

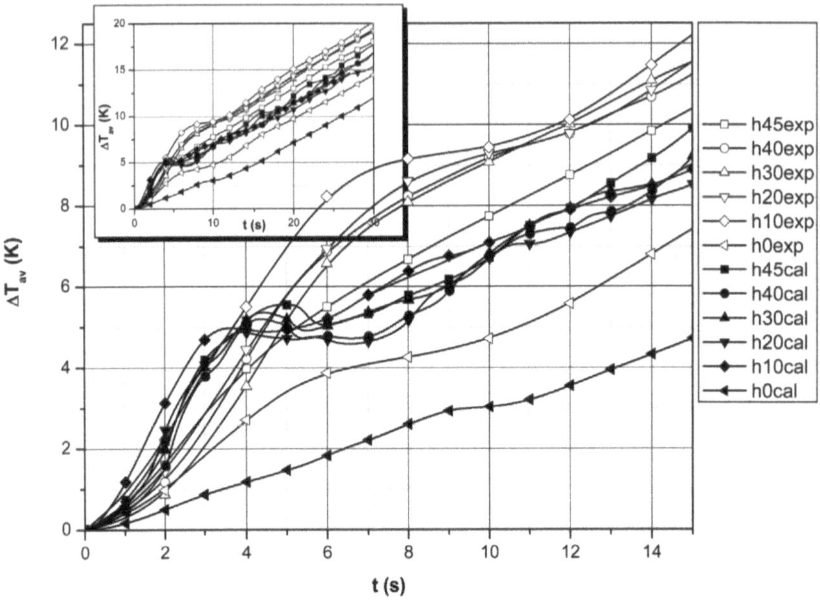

FIG. 8 Comparison of the experimental as well as calculated temperature profiles of water during microwave treatment without stirring at various axial positions of the fiber-optic probe. The data over the total time is shown in the inset figure. *(Reproduced with permission from R. Cherbański, L. Rudniak, Modelling of microwave heating of water in a monomode applicator–influence of operating conditions. Int. J. Therm. Sci. 74 (2013) 214–229.)*

puree [42], sweet potato puree [43, 44], and 0.5% CMC solution [45] in a continuous-flow microwave system. The nonuniform temperature distribution in highly viscous fluid foods was mainly because of the nonuniform electric field distribution inside the cavity. The configuration and diameter of the applicator tube, dielectric properties of the fluid, microwave power, flow rate, and temperature were the other important parameters affecting the temperature distribution in the product [42, 46]. To overcome the nonuniform temperature distribution issue, approaches like turbulent flow generation, installing static mixers, or using curved tubes instead of straight tubes may be utilized [42, 43, 47]. However, these approaches have limitations, for instance, difficult and time-consuming cleaning because of the complex geometry of static mixers that induces contamination risk and limits the usage of static mixers in food processing systems [48].

Tuta and Palazoğlu et al. investigated experimentally and numerically the heating of fluid foods in a continuous-flow microwave heating unit [49]. The microwave unit was specially designed with 915 MHz frequency and 5 kW maximum power, and the simulations were performed by using the finite element method with COMSOL Multiphysics.

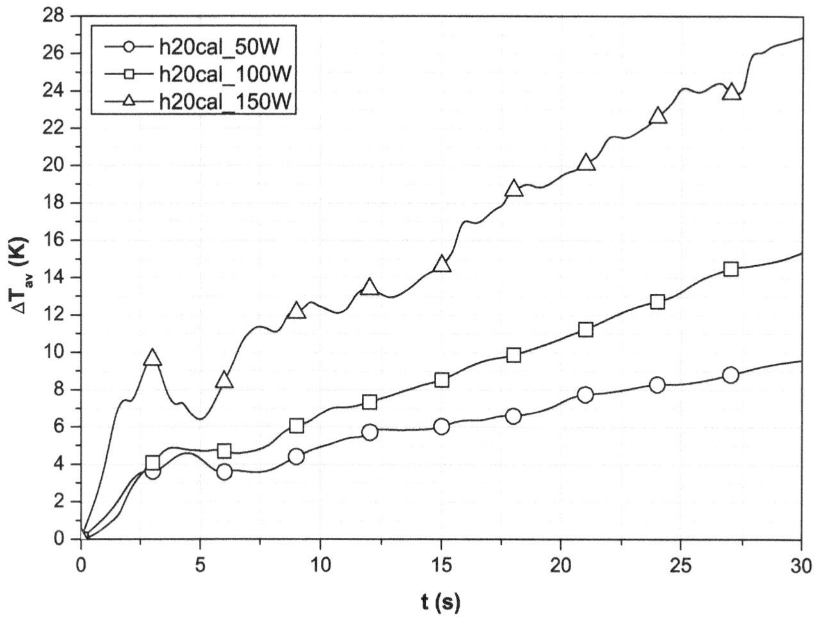

FIG. 9 Simulated temperature differences at the fiber-optic position h20 for microwave without stirring with different incident microwave powers. *(Reproduced with permission from R. Cherbański, L. Rudniak, Modelling of microwave heating of water in a monomode applicator–influence of operating conditions. Int. J. Therm. Sci. 74 (2013) 214–229.)*

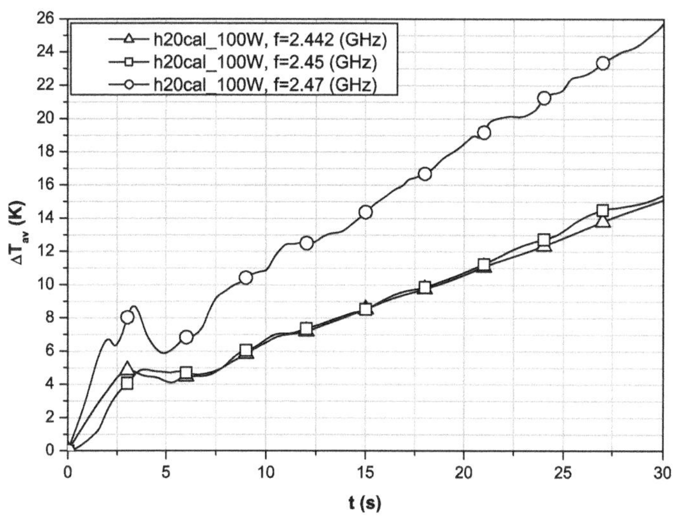

FIG. 10 Simulated temperature differences at the fiber-optic position h20 for microwave without stirring with various microwave frequencies. *(Reproduced with permission from R. Cherbański, L. Rudniak, Modelling of microwave heating of water in a monomode applicator–influence of operating conditions, Int. J. Therm. Sci. 74 (2013) 214–229.)*

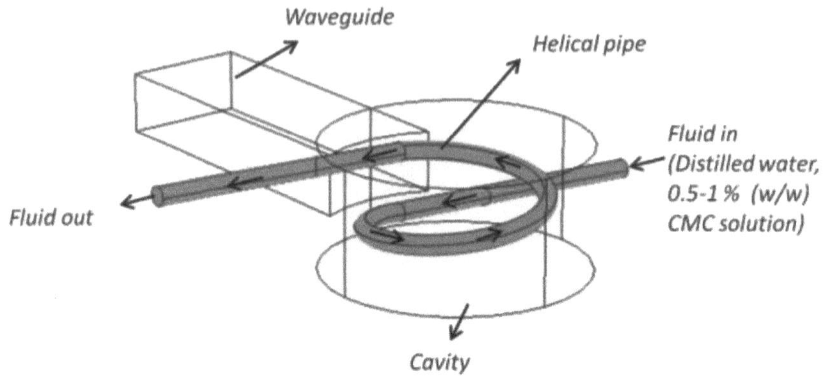

FIG. 11 Schematic diagram of a continuous-flow microwave system (COMSOL model geometry). *(Reproduced with permission from S. Tuta, T.K. Palazoğlu, Finite element modeling of continuous-flow microwave heating of fluid foods and experimental validation. J. Food Eng. 192 (2017) 79–92.)*

To verify the model, the experiments were carried with various fluids such as distilled water, 0.5% and 1% CMC solutions, and at various flow rates. Fig. 11 shows the COMSOL model geometry of a continuous-flow microwave unit. The unit was comprised of a PTFE tubing, and the tube was helically configured inside the cylindrical cavity. At the exit of the tube, the cross-sectional temperature distribution was calculated by using a specially designed thermocouple unit. Temperature values at seven points along the helical tube surface is shown in Fig. 12. Experimental and numerical data depicted that heating was more uniform for distilled water compared with CMC solutions at all flow rates (1, 2, and 3 L/min). The result demonstrated that more uniform heating could be attained for high-viscosity fluid food products by utilizing a helical tube in a continuous-flow microwave heating unit.

Yeong et al. reported another numerical model of the microwave heating of distilled water using COMSOL Multiphysics software [50]. They investigated the microwave effects on the heating rate by applying three frequencies (0.915, 2, and 2.45 GHz). Fig. 13 shows the COMSOL model used for the investigation.

Fig. 14 shows the variation in the temperature of the distilled water at three different frequencies. It was observed that the temperature increases with time at 2 and 2.45 GHz. In the case of 915-MHz frequency, even after 60 s, there was no significant temperature rise.

Thermal processing of fruit juices is normally founded on the inactivation of heat-resistant unwanted enzymes. Since peroxidase is recognized as a high heat-resistant enzyme and its intensity changes are related to certain conditions, the inactivation of the peroxidase enzyme is often considered as an index of processing adequacy [51–53]. Furthermore, the peroxidase enzyme is involved in the oxidation of a broad range of inorganic and organic substrates, leading to the degradation of sensory and off-flavors. Studies showed that microwave heating can be used conveniently to inactivate several enzymes, including peroxidase [54–56].

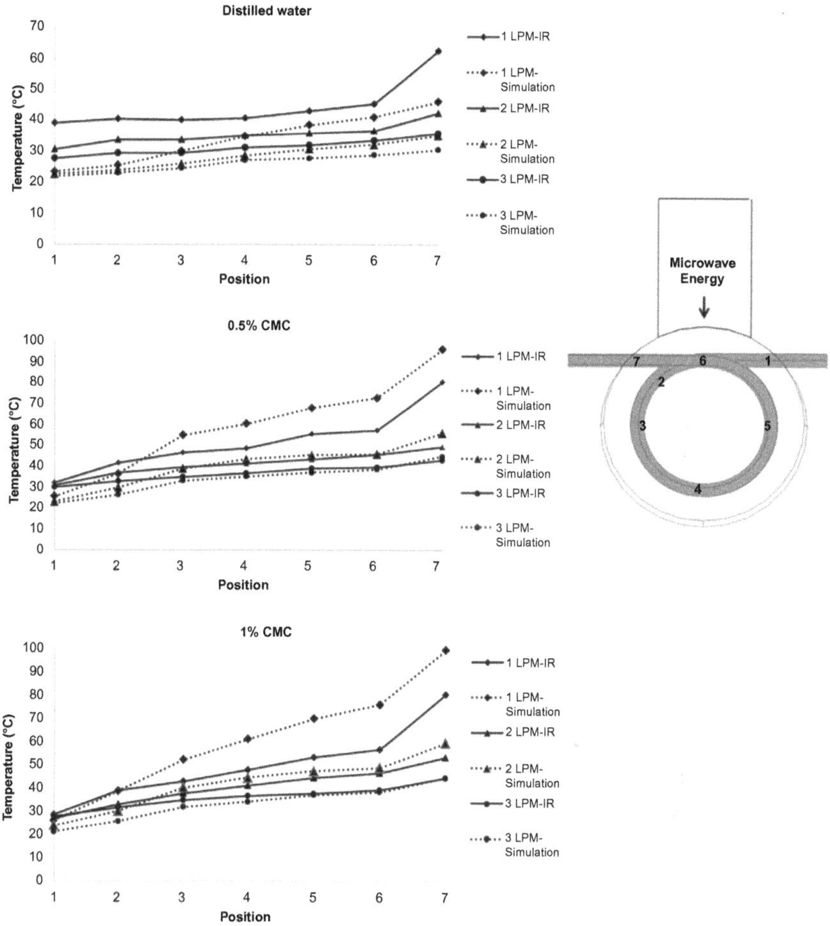

FIG. 12 Temperature values at seven points along the helical tube surface. *(Reproduced with permission from S. Tuta, T.K. Palazoğlu, Finite element modeling of continuous-flow microwave heating of fluid foods and experimental validation. J. Food Eng. 192 (2017) 79–92.)*

Kubo et al. demonstrated a multiphysics modeling of microwave processing in fruit juices for enzyme inactivation [57]. Microwave processing of fruit juice model solution in a container kept in a cylinder was analyzed by using the numerical model to find the peroxidase inactivation and temperature profiles along with processing. Fig. 15 shows the schematic diagram of the applicator and experimental setup. The numerical model was formulated by iteratively coupling electromagnetism, fluid flow, enzyme inactivation, and heat transfer.

Fig. 16 shows the obtained density and dielectric properties of the model juice solution at 2.45 GHz. The temperature dependence of the dielectric and

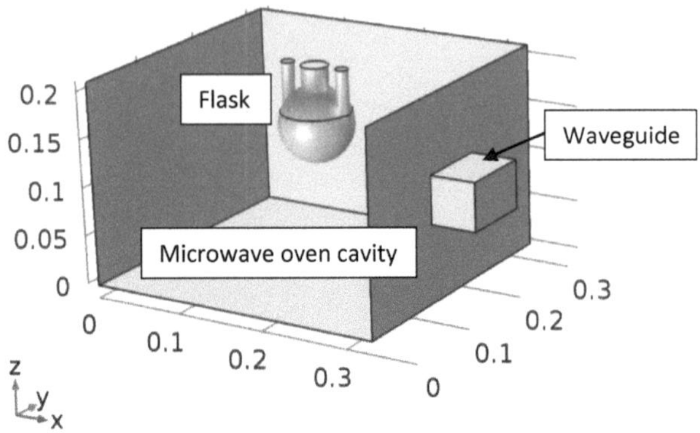

FIG. 13 COMSOL model used to investigate the microwave effects. *(Reproduced with permission from S.P. Yeong, M.C. Law, C.C.V. Lee, Y.S. Chan, Modelling batch microwave heating of water, IOP Conf. Ser.: Mater. Sci. Eng. 217 (2017) 012035. https://doi.org/10.1088/1757-899X/217/1/012035.)*

thermophysical properties has an important role in governing the numerical model equations.

Fig. 17A compares the predicted and experimental residual peroxidase activities after microwave processing. Because of the different temperature distribution, the global enzyme inactivation was also different (Fig. 17B). Microwave treatments at several combinations of time and temperature were also

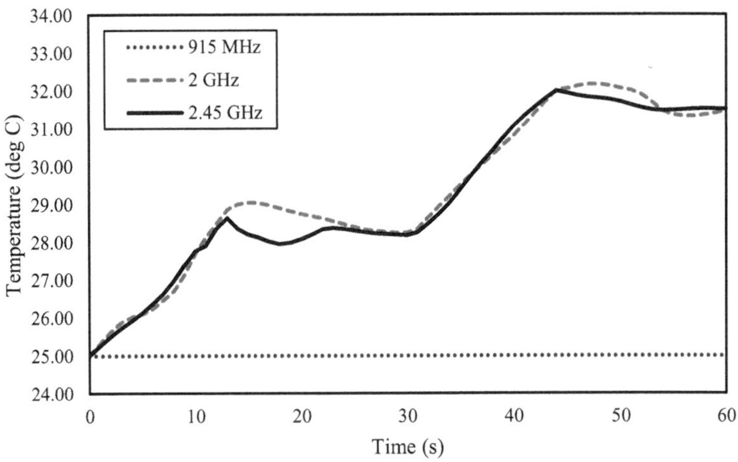

FIG. 14 Temperature of distilled water at several frequencies. *(Reproduced with permission from S.P. Yeong, M.C. Law, C.C.V. Lee, Y.S. Chan, Modelling batch microwave heating of water, IOP Conf. Ser.: Mater. Sci. Eng. 217 (2017) 012035. https://doi.org/10.1088/1757-899X/217/1/012035.)*

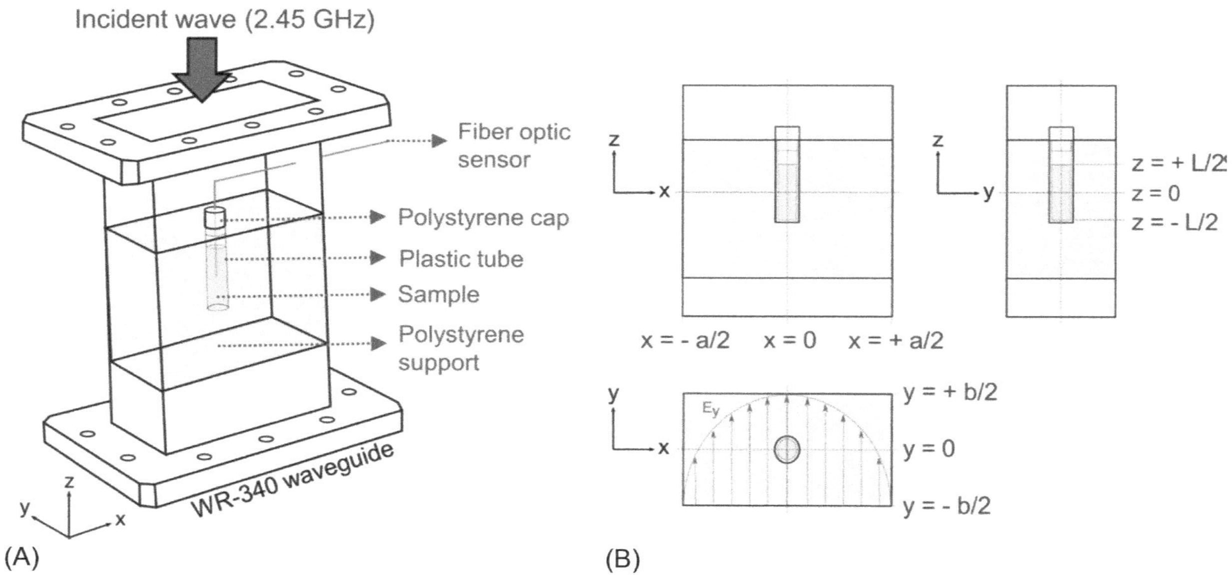

FIG. 15 (A) Schematic representation of the experimental setup for the microwave processing of fruit juice. (B) Arrangement of the sample, plastic tube, polystyrene cap, and support inside the rectangular waveguide. *(Reproduced with permission from M.T.K. Kubo, S. Curet, P.E.D. Augusto, L. Boillereaux, Multiphysics modeling of microwave processing for enzyme inactivation in fruit juices. J. Food Eng. 263 (2019) 366–379. https://doi.org/10.1016/j.jfoodeng.2019.07.011.)*

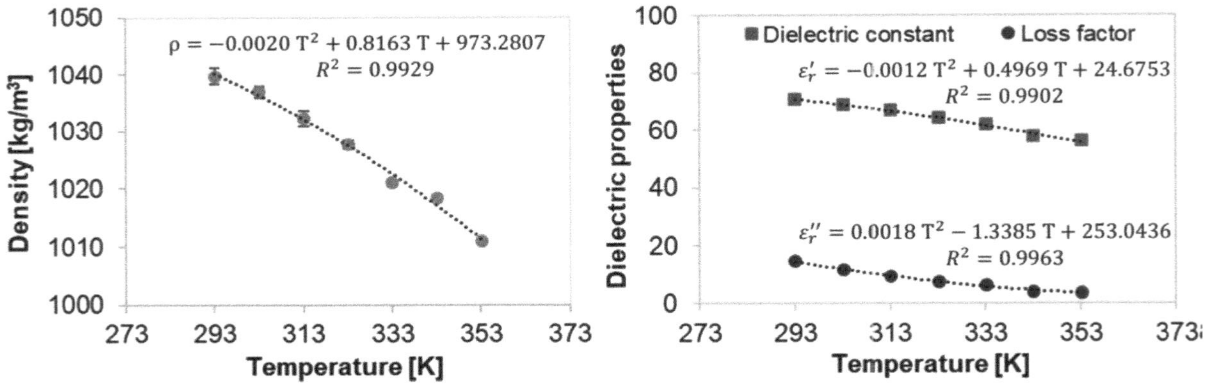

FIG. 16 (A) Density and (B) dielectric properties of the fruit juice model solution at 2.45 GHz as a function of temperature. *(Reproduced with permission from M.T. K. Kubo, S. Curet, P.E.D. Augusto, L. Boillereaux, Multiphysics modeling of microwave processing for enzyme inactivation in fruit juices, J. Food Eng. 263 (2019) 366–379. https://doi.org/10.1016/j.jfoodeng.2019.07.011.)*

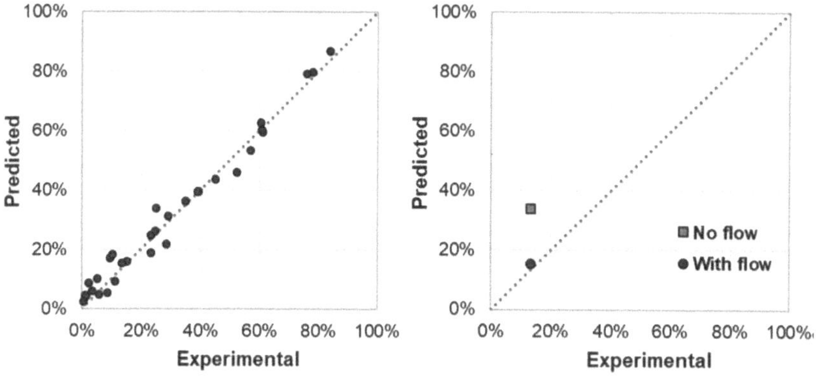

FIG. 17 (A) The experimental and predicted residual peroxidase activities in fruit juice model solution after microwave treatments. (B) The predicted and experimental residual peroxidase activities (with or without fluid flow) after the microwave treatment (66°C, 5 min). *(Reproduced with permission from M.T.K. Kubo, S. Curet, P.E.D. Augusto, L. Boillereaux, Multiphysics modeling of microwave processing for enzyme inactivation in fruit juices. J. Food Eng. 263 (2019) 366–379. https://doi.org/10.1016/j.jfoodeng.2019.07.011.)*

executed during the study. Fig. 18 compares the predicted and experimental values of the temperature profile at the geometric center of the sample.

Recently, Lee et al. demonstrated the numerical modeling of liquid heating and boiling under a microwave irradiation [58]. The successive multiphysics model was developed in OpenFOAM and compared with experimental data. Fig. 19 shows the computational domain used in this study.

In the study, the effect of antibumping granules was considered. The study depicted that during the microwave heating stage, the temperature distribution in the water sample is stratified, with a low temperature in the lower part and high temperatures in the upper part. With the heating time, the temperature uniformity in the sample deteriorates greatly (Fig. 20). It was also shown that there was no significant variation in the total power absorption with heating time (Fig. 21).

During the initial stage of the boiling process, the water was considerably superheated and no vapor bubble was generated. Evaporation takes place only at the free surface. After a period, the vapor bubble nucleation takes place when the temperature of the water at the antibumping granules reaches the saturation temperature. The vapor bubbles prevent the accumulation of superheat by removing most of the superheat in the water. The study also demonstrated that under microwave irradiation, the overall evaporation process was dominated by free surface evaporation. The study is useful to understand microwave-assisted procedures that require boiling, such as the microwave distillation process.

The same research group demonstrated the microwave heating and boiling of binary liquid mixtures using a new multiphysics model [59].

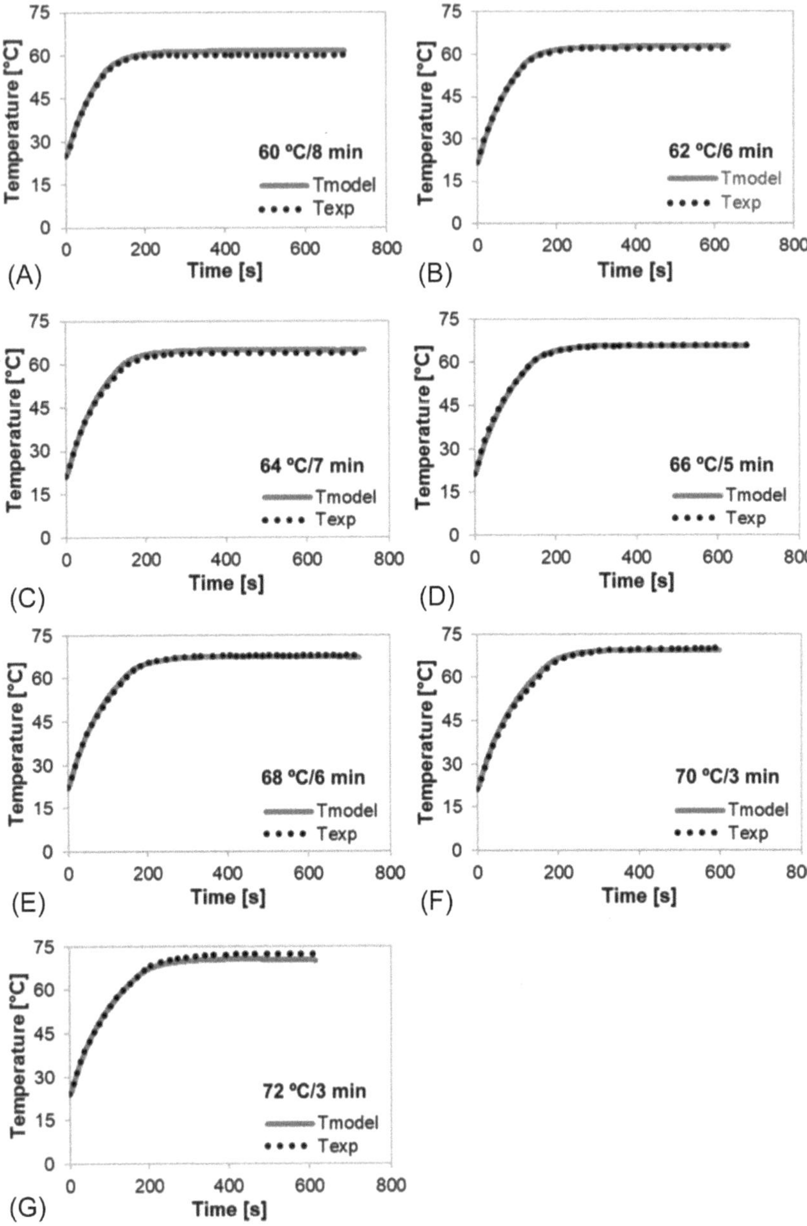

FIG. 18 Predicted and experimental temperature profiles at the center of the sample: (A) at 60°C/8 min, (B) at 62°C/6 min, (C) at 64°C/7 min, (D) at 66°C/5 min, (E) at 68°C/6 min, (F) at 70°C/3 min, and (G) at 72°C/3 min. *(Reproduced with permission from M.T.K. Kubo, S. Curet, P.E.D. Augusto, L. Boillereaux, Multiphysics modeling of microwave processing for enzyme inactivation in fruit juices, J. Food Eng. 263 (2019) 366–379. https://doi.org/10.1016/j.jfoodeng.2019.07.011.)*

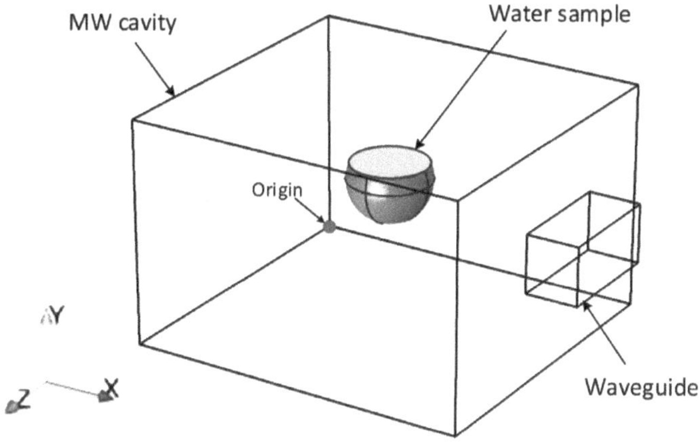

FIG. 19 Computational domain of modeling of single-phase microwave liquid heating. *(Reproduced with permission from G.L. Lee, M.C. Law, V.C.-C. Lee, Numerical modelling of liquid heating and boiling phenomena under microwave irradiation using OpenFOAM. Int. J. Heat Mass Transf. 148 (2020) 119096. https://doi.org/10.1016/j.ijheatmasstransfer.2019.119096.)*

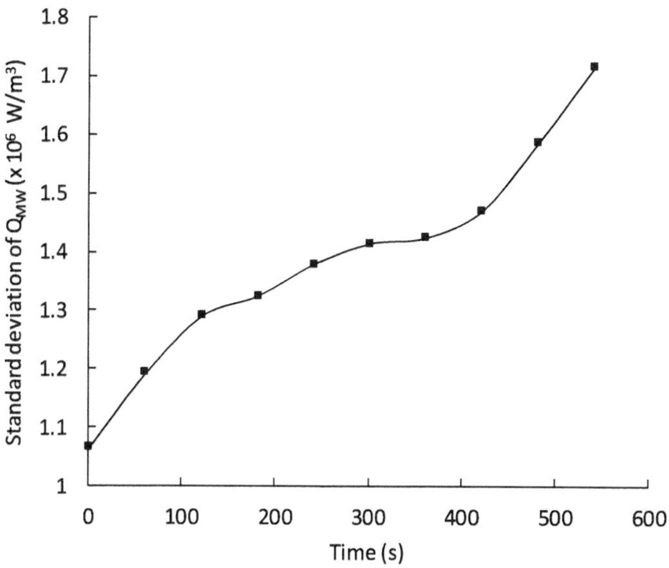

FIG. 20 Deviation of microwave power absorption with time. *(Reproduced with permission from G.L. Lee, M.C. Law, V.C.-C. Lee, Numerical modelling of liquid heating and boiling phenomena under microwave irradiation using OpenFOAM. Int. J. Heat Mass Transf. 148 (2020) 119096. https://doi.org/10.1016/j.ijheatmasstransfer.2019.119096.)*

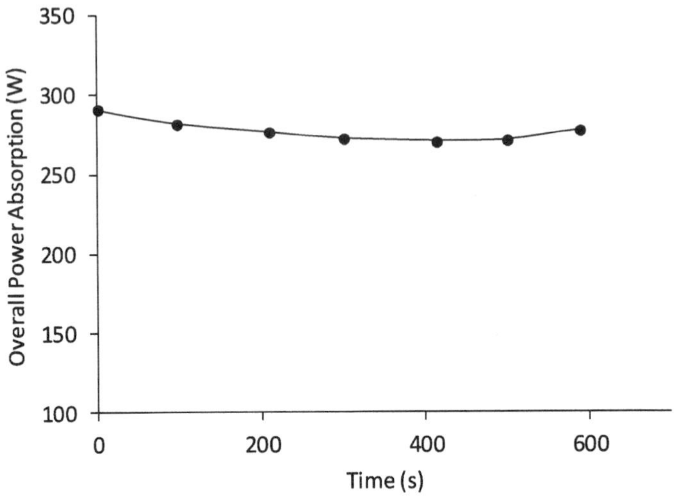

FIG. 21 Overall microwave power absorption with time. *(Reproduced with permission from G.L. Lee, M.C. Law, V.C.-C. Lee, Numerical modelling of liquid heating and boiling phenomena under microwave irradiation using OpenFOAM. Int. J. Heat Mass Transf. 148 (2020) 119096. https://doi.org/10.1016/j.ijheatmasstransfer.2019.119096.)*

3.3 Modeling of microwave heat transfer in solids

In the case of microwave heating of solid foodstuff, the energy equation is normally written as a conductive heat transfer with a generation term. To analyze the effects of the electromagnetic field distribution, the heat generation term has been modeled by scholars, using two different approaches: (a) by applying Lambert's law [2, 7] and (b) by solving Maxwell's equations [3, 60, 61]. Lambert's law is a simple power formulation, and it can be applied to simulate the temperature profiles for the thick food samples [60].

During microwave irradiation, the temperature distribution in the object is controlled by the absorption and interaction of radiation by the medium and the accompanying transport processes. So, modeling of microwave heating consists of coupling the models for microwave absorption and temperature distribution inside the object. Oliveira and Franca presented a model in which the absorbed microwave power was calculated by using Maxwell's equations and subsequently integrated as a source term in the transient heat equation [3]. For discretization of the governing equations, the finite element method was applied. The demonstrated method was applicable to simulate microwave heating of foodstuffs. The comparison analysis of the power distribution found by solving Lambert's equation to that by using Maxwell's showed that the sample size above which Lambert's law limit is valid is lower for slabs compared with cylinders. The computational analysis demonstrated that microwave heating is greatly dependent on the sample shape and size. The results also showed that heating is mostly dependent on the radiation frequency. The radiation penetration and power absorption are more effective at lower frequencies compared with higher frequencies. The simulation domain considered in this study is presented in Fig. 22.

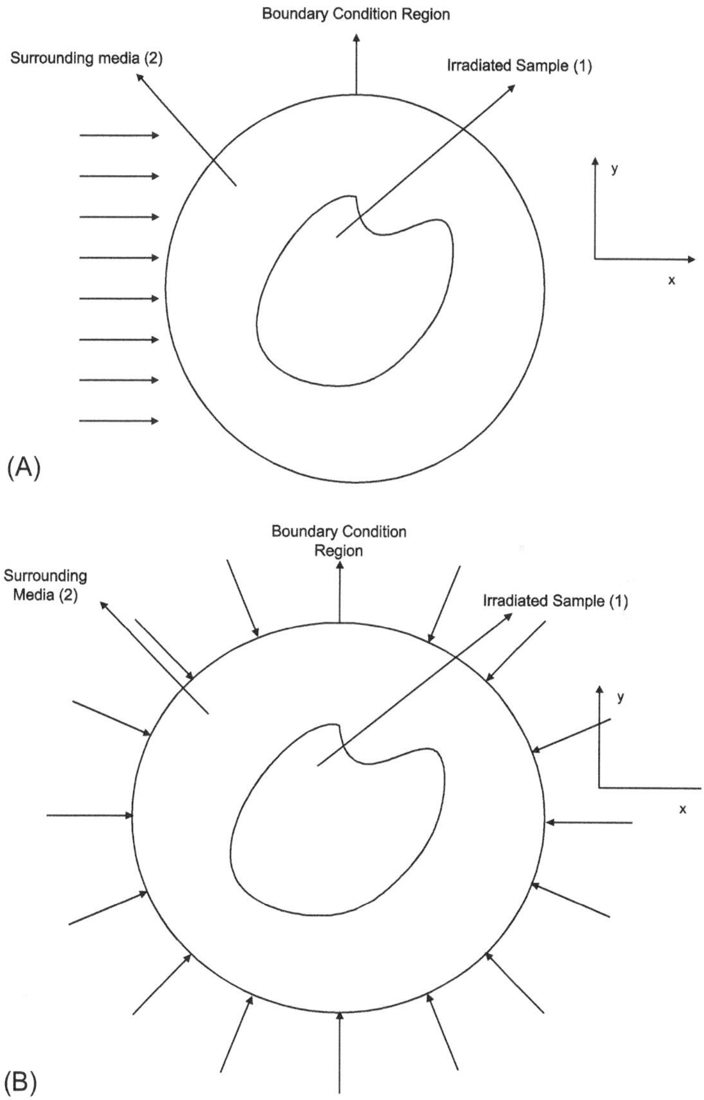

FIG. 22 Simulation domain: (A) lateral irradiation; (B) radial irradiation. *(Reproduced with permission from M.E.C. Oliveira, A.S. Franca, Microwave heating of foodstuffs. J. Food Eng. 53 (2002) 347–359. https://doi.org/10.1016/S0260-8774(01)00176-5.)*

The critical radius for cylindrical samples was also calculated during the study, and the relationship between the critical radius and the wave penetration depth is shown in Fig. 23.

Microwave power distributions for mashed potato samples irradiated at 2800 MHz are shown in Fig. 24. The power distributions predicted by Maxwell's equations are in agreement with Lambert's law predictions when the sample radius increases. But in smaller samples, most of the absorbed power was concentrated at the center, so Lambert's approximation resulted in a significant error. When the sample size increases, the amount of absorbed power increases at the surface and decreases in the center.

Several numerical techniques were reported for in simulation studies; among them, the two widely applied methods to model the heat transfers are the finite element method (FEM) and the finite difference method. Because of the flexibility in handling irregular geometrical configurations and matcrial properties, the finite element method has been used in a widespread way to solve the microwave heating in food [2, 7]. Romano et al. analyzed the effect of dimensions of microwave heating of cylindrical samples [5]. The numerical model was formulated by using a commercial package FEMLAB. The microwave absorption relationship was correctly formulated for cylindrical geometry with Lambert's law. The axis-symmetric model of the food system used in the study is shown in Fig. 25.

The absorbed power for cylindrical samples of four different foodstuffs such as potato, shrimp, marinated shrimp, and pea puree with three different

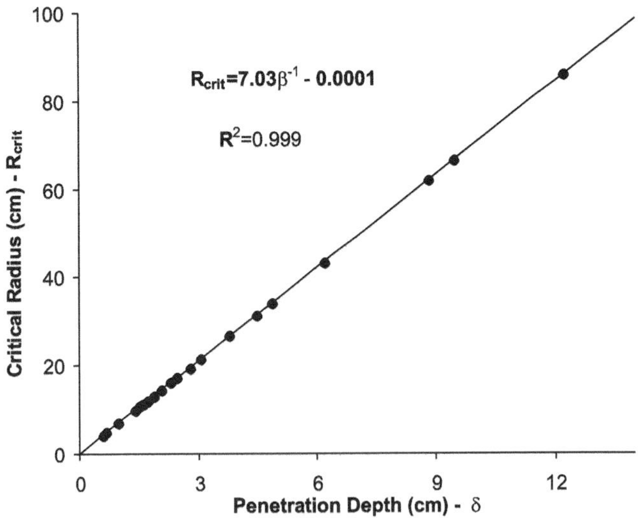

FIG. 23 The relationship between the critical radius and the wave penetration depth for cylindrical samples. *(Reproduced with permission from M.E.C. Oliveira, A.S. Franca, Microwave heating of foodstuffs. J. Food Eng. 53 (2002) 347–359. https://doi.org/10.1016/S0260-8774(01)00176-5.)*

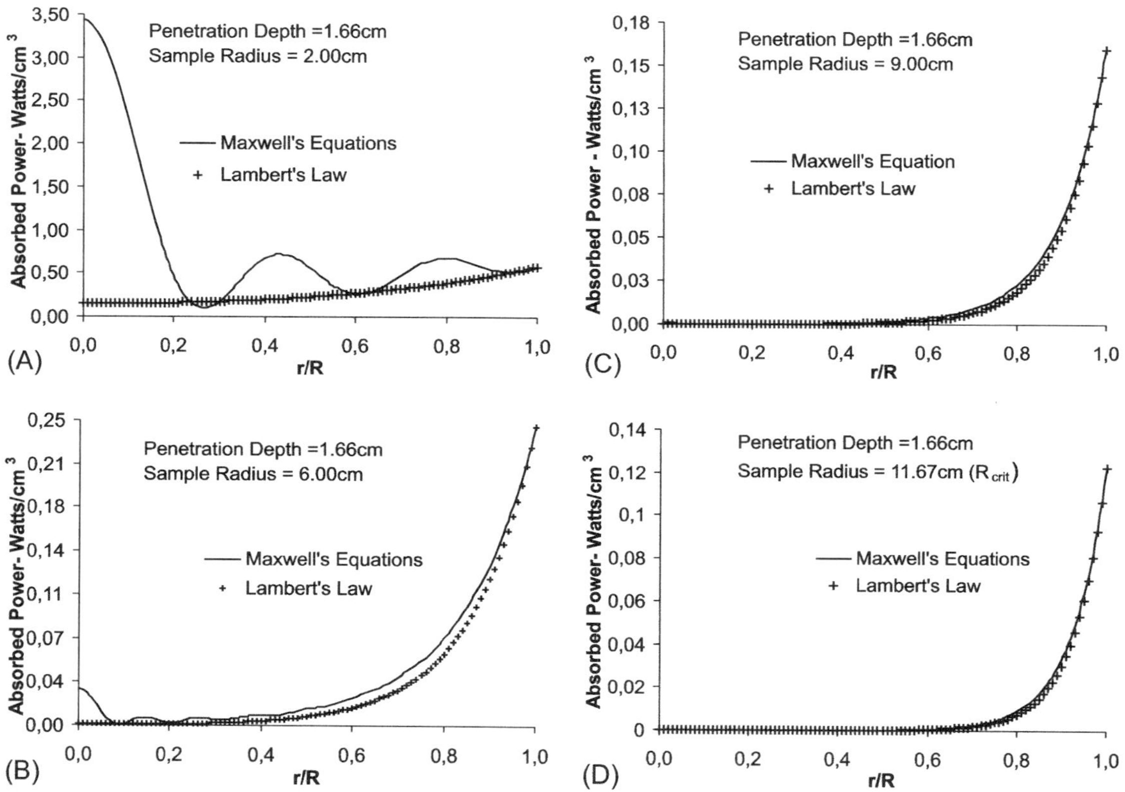

FIG. 24 Microwave power distribution for mashed potato samples with radius (A) 2 cm, (B) 6 cm, (C) 9 cm, and (D) 11.67 cm. *(Reproduced with permission from M.E.C. Oliveira, A.S. Franca, Microwave heating of foodstuffs. J. Food Eng. 53 (2002) 347–359. https://doi.org/10.1016/S0260-8774(01)00176-5.)*

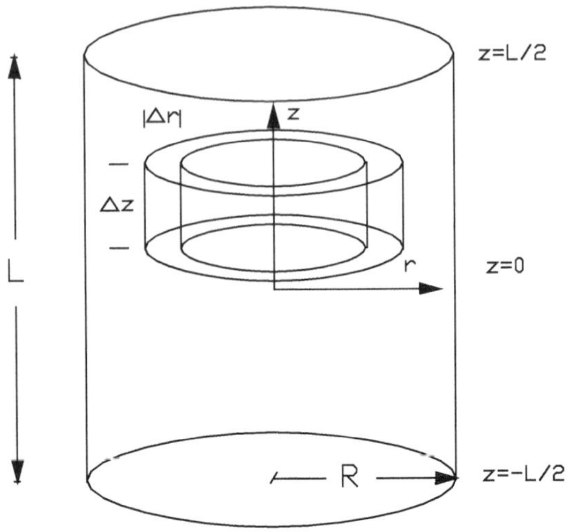

FIG. 25 Axis-symmetric model used for the analysis of the food system. *(Reproduced with permission from V.R. Romano, F. Marra, U. Tammaro, Modelling of microwave heating of foodstuff: study on the influence of sample dimensions with a FEM approach, J. Food Eng. 71 (2005) 233–241. https://doi.org/10.1016/j.jfoodeng.2004.11.036.)*

dimensions is shown in Fig. 26. The data showed that as the radius decreases, most of the absorbed power is concentrated at the center of the sample, which was in agreement with the experimental analysis by Zhou et al. [7]. When the sample radius increases, the absorbed power increases at the surface and decreases in the center.

Microwave drying or heating of hygroscopic porous media such as wood using a waveguide is a relatively interesting area of research. To gain insight into the phenomena that happen in the waveguide along with the distribution of temperature in the wood, an elaborate knowledge of absorbed power distribution is essential. Rattanadecho demonstrated a two-dimensional numerical model to predict the distribution of electromagnetic fields, power, and temperature distributions within the wood [62]. A three-dimensional finite difference time domain scheme was applied to find electromagnetic fields and absorbed power by solving Maxwell's equations. To determine unsteady temperature profiles, the finite difference method was employed.

The physical model used during the study for the microwave heating of wood using a rectangular waveguide is shown in Fig. 27. The assumption considered during this study is that a two-dimensional heat transfer model in the x- and z-directions is enough to describe the microwave heating processes in a rectangular waveguide [26, 63].

The experimental and predicted results of temperature distribution within the wood sample along with the horizontal axis of the rectangular waveguide are shown in Fig. 28. The data showed that the greatest temperature displays

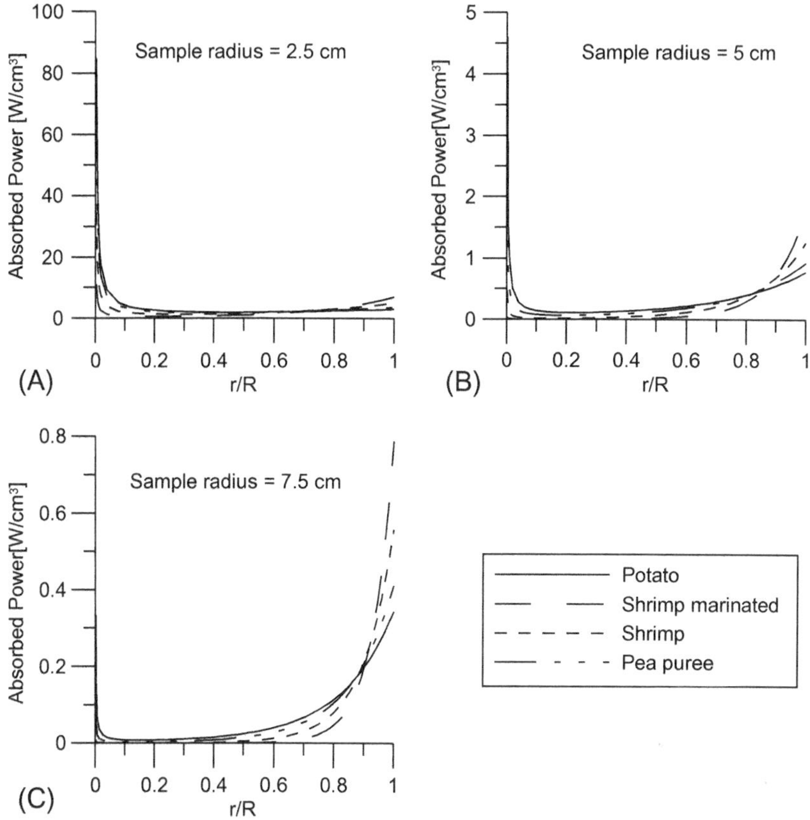

FIG. 26 Absorbed microwave power for samples of 2.5 cm radius and 4 cm length (A), 5 cm radius and 8 cm length (B), and 7.5 cm radius and 12 cm length (C). *(Reproduced with permission from V.R. Romano, F. Marra, U. Tammaro, Modelling of microwave heating of foodstuff: study on the influence of sample dimensions with a FEM approach. J. Food Eng. 71 (2005) 233–241. https://doi.org/10.1016/j.jfoodeng.2004.11.036.)*

in the center of the sample. Through an iterative process, the temperature dependence of wood dielectric properties was simulated. The simulations were executed illustrating the influence of working frequencies, irradiation times, and sample size.

Geedipalli et al. demonstrated numerically heat transfer in a combination-jet impingement oven [64]. It is simulated by coupling Maxwell's equations of electromagnetics with energy equation and applying experimentally evaluated heat transfer coefficients. A schematic diagram of the oven and waveguide used in this computational study is shown in Fig. 29. The contour plots of temperature depicted that during combination heating, the surface can be heated faster and overheating of the edge could be avoided partially. A 20%–30% increase in uniformity was observed during microwave-jet impingement combination

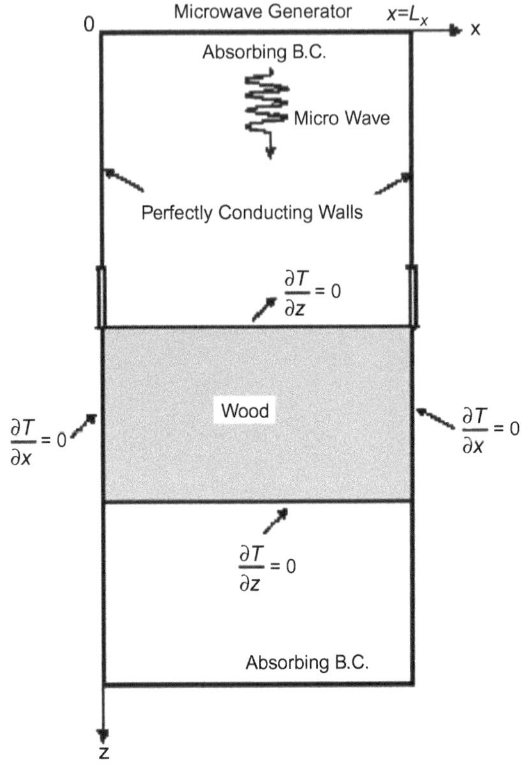

FIG. 27 Physical model. *(Reproduced with permission from P. Rattanadecho, The simulation of microwave heating of wood using a rectangular wave guide: influence of frequency and sample size. Chem. Eng. Sci. 61 (2006) 4798–4811. https://doi.org/10.1016/j.ces.2006.03.001.)*

heating over microwave-only heating. During the initial period of heating, the jet impingement heating-dominated over microwave heating near the surface and in the interior, the microwave heating was more significant. At later times, the roles were switched with microwave heating became more dominant on the surface and jet impingement took a more significant role in the interior heating of the food.

Fig. 30 compares the experimentally measured and numerically calculated temperatures in microwave heating, jet impingement heating, and combination heating.

The parameter complex dielectric permittivity expresses the behavior of the material in a microwave field. The imaginary part of permittivity also known as loss factor is used as a criterion for estimating the suitability of heating in a microwave oven. In the microwave heating process, both components of dielectric permittivity have a significant role.

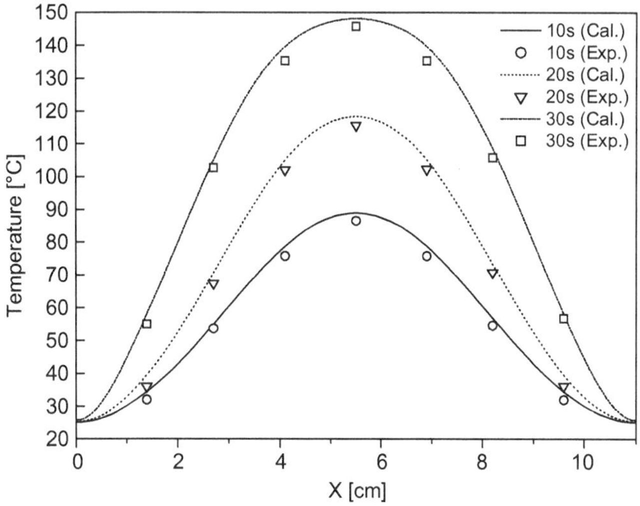

FIG. 28 Temperature distribution in the wood sample along the horizontal axis. *(Reproduced with permission from P. Rattanadecho, The simulation of microwave heating of wood using a rectangular wave guide: influence of frequency and sample size. Chem. Eng. Sci. 61 (2006) 4798–4811. https://doi.org/10.1016/j.ces.2006.03.001.)*

Lovás et al. presented the dependence of the real and imaginary permittivity on temperature as well as the temperature distributions at microwave frequencies for different minerals such as andesite, sulfide, and carbonate [65]. Applying the resonant cavity method, the real and imaginary permittivities, ε' and ε'', of minerals were measured at the frequency of 2216 MHz from room temperature to up to 1200°C (Fig. 31). The effect of temperature variation on the complex permittivity was most significant for the sulfides and siderite. To predict and visualize the temperature rise in the samples which are irradiated with microwave, the heat and Maxwell's equations were analyzed using the COMSOL MULTIPHYSICS program.

Fig. 32 shows the dependence of the penetration depth with temperature for pyrite, chalcopyrite, and galenite at 2216 MHz.

Microwave-assisted heating is also widely used for the pretreatment of biomass and the production of bioenergy and biofuels [66]. During the microwave treatment to get products with good quality and uniformity, complete decomposition of biomass inside the oven is important. Uniform temperature distribution inside the cavity is an important step to establish a good product quality, reduce material consumption, and save energy. To improve the efficiency of biomass heating by microwaves, recently, Fia et al. studied numerically the heating of sugar cane bagasse, wood, orange peel, and palm oil in domestic microwave oven [67].

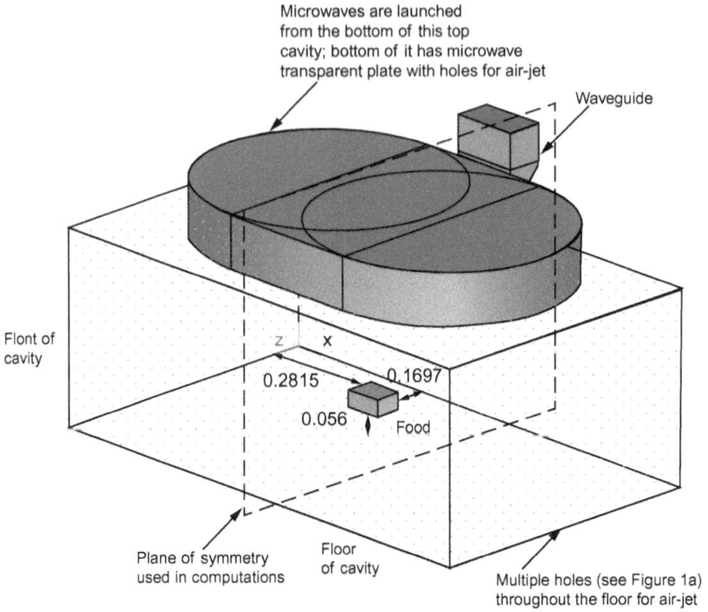

FIG. 29 Schematic of the waveguide and oven utilized in computational studies. *(Reproduced with permission from S. Geedipalli, A.K. Datta, V. Rakesh, Heat transfer in a combination microwave–jet impingement oven. Food Bioprod. Process. 86 (2008) 53–63. https://doi.org/10.1016/j.fbp.2007.10.016.)*

The study was mainly focused on the analysis of the power absorption capability of the biomasses in the sphere as well as in cylindrical shapes. The Maxwell equations coupled to the heat equation were used to compute the microwave electromagnetic field distribution in the samples as well as in the microwave cavity. The heating efficiency rate was calculated using the degree of thermal runaway and evolution of average temperature and their relationship with the energy absorbed by the samples and coefficient of variation, respectively. Energy consumption is an important matter in the production process. To reduce cost and to make the process energy efficient, the energy consumption needs to be minimalized before the phase transition temperature. The absorbed energy as a function of power, for cylinder and sphere, is shown in Fig. 33. The evolution of average temperature and energy absorbed as a function of power is shown in Fig. 34. The energy absorbed and average temperature followed the same trend when power is varied.

A slurry reactor is used in many catalytic reactions. It is a multiphase system in which solid particles are dispersed in a liquid. Selective and effective heating takes place when the liquid is microwave transparent (for example, nonpolar organic solvents), and the solid is a strong microwave absorber (for example, activated carbon). During the process, the microwave energy is dissipated in

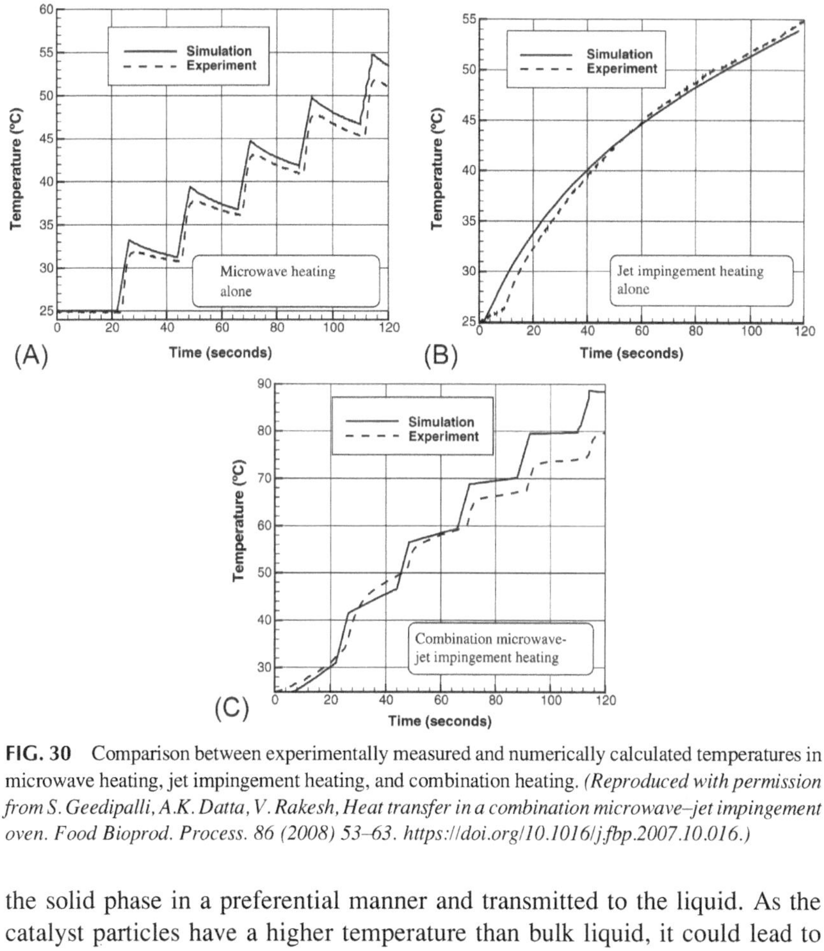

FIG. 30 Comparison between experimentally measured and numerically calculated temperatures in microwave heating, jet impingement heating, and combination heating. *(Reproduced with permission from S. Geedipalli, A.K. Datta, V. Rakesh, Heat transfer in a combination microwave–jet impingement oven. Food Bioprod. Process. 86 (2008) 53–63. https://doi.org/10.1016/j.fbp.2007.10.016.)*

the solid phase in a preferential manner and transmitted to the liquid. As the catalyst particles have a higher temperature than bulk liquid, it could lead to improve energy efficiency and cut downside reactions in the liquid phase. The efficiency of the microwave heating of slurries is related to the ability to control the temperature of the liquid phase and the solid particles.

Computational analysis using multiphysics software can give insights into the controlling processes of microwave heating of slurries. For instance, catalyst and operating conditions of slurry reactors can be optimized to improve the selective heating of catalyst particles. However, the large separation of scales brings down a stringent criterion on the computational mesh size, heading to excessively high computational requirements. Because of these, several scholars use simplistic models to simulate complex multiphase problems [68–75]. The lack of tools to predict and measure the temperature difference between the dispersed (solid particle) and continuum (liquid) phases obstructs the understanding and optimization of microwave-heated slurry reactors. Recently, Goyal et al. reported the simulations for microwave heating of slurries

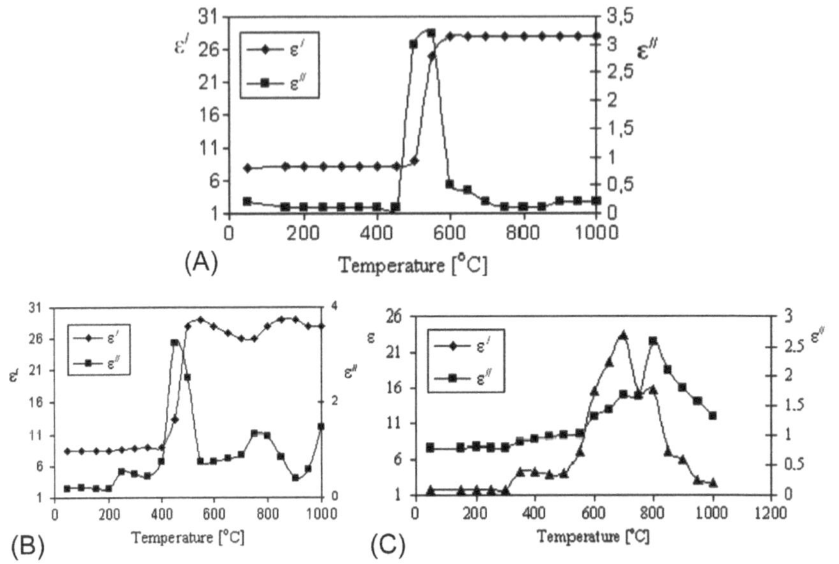

FIG. 31 The influence of temperature on the complex permittivity of (A) pyrite, (B) chalcopyrite, and (C) galenite. *(Reproduced with permission from M. Lovás, M. Kováčová, G. Dimitrakis, S. Čuvanová, I. Znamenáčková, Š. Jakabský, Modeling of microwave heating of andesite and minerals, Int. J. Heat Mass Transf. 53 (2010) 3387–3393. https://doi.org/10.1016/j.ijheatmasstransfer.2010.03.012.)*

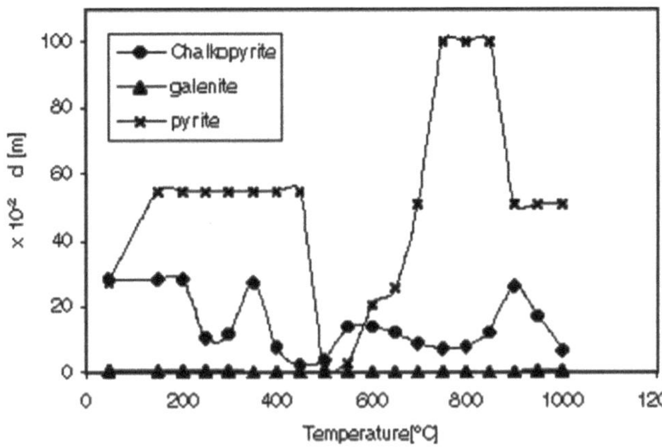

FIG. 32 The effect of the temperature on the penetration depth of pyrite, chalcopyrite, and galenite at 2216 MHz. *(Reproduced with permission from M. Lovás, M. Kováčová, G. Dimitrakis, S. Čuvanová, I. Znamenáčková, Š. Jakabský, Modeling of microwave heating of andesite and minerals, Int. J. Heat Mass Transf. 53 (2010) 3387–3393. https://doi.org/10.1016/j.ijheatmasstransfer.2010.03.012.)*

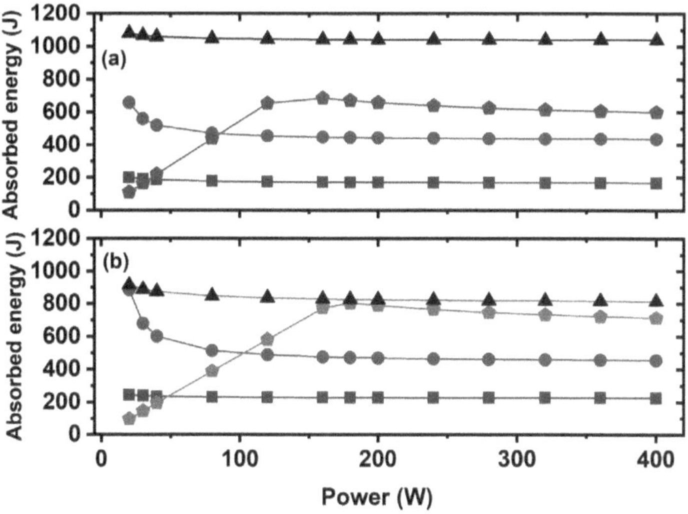

FIG. 33 Absorbed energy as the function of power, for (A) cylinder and (B) sphere. Wood (⬟), bagasse (■), orange peel (●), palm oil 18% (▲). *(Reproduced with permission from A.Z. Fia, J. Amorim, Heating of biomass in microwave household oven-A numerical study, Energy. 218 (2020) 119472.)*

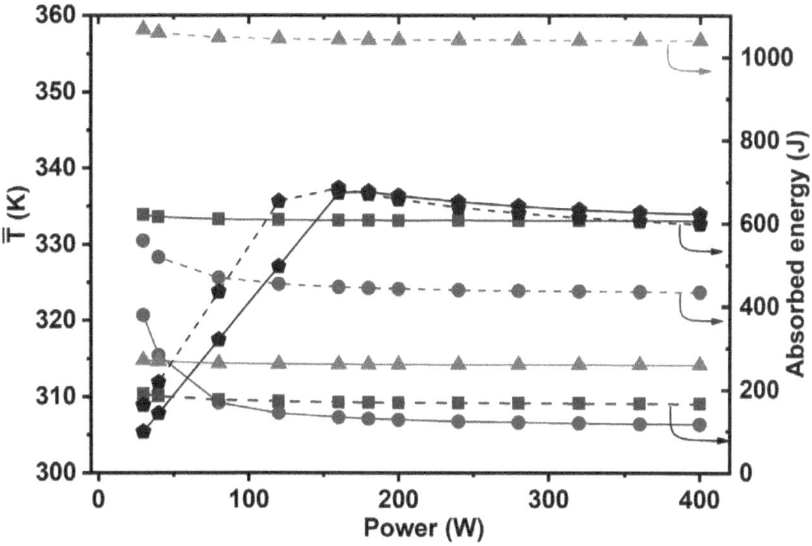

FIG. 34 Evolution of average temperature *(continuous lines)* and energy absorbed *(dashed lines)* as a function of power. Wood (⬟), bagasse (■), orange peel (●), palm oil 18% (▲). *(Reproduced with permission from A.Z. Fia, J. Amorim, Heating of biomass in microwave household oven-A numerical study, Energy. 218 (2020) 119472.)*

in a laboratory-scale microwave cavity and compared it with experimental analysis [76].

They used experiments, theory, and multiscale simulations to examine microwave heating of slurries comprising microwave-absorbing solid particles dispersed in a nonpolar liquid. During the study, using mathematical modeling and controlled experiments, an ideal slurry reactor was demonstrated and analytical expressions were developed to predict the temperature evolution of the solid and liquid phases during heating with the microwave. The experimental analysis showed a strong impact of the solid material and concentration on the rate of heating of slurries, compared with conventionally heated slurries. CEM Discover SP monomode microwave cavity was modeled for numerical analysis (Fig. 35). During this study, a multiscale simulation of microwave-heated slurries using a coarse-graining methodology to handle the large separation of length scales in slurries was also performed and compared with experimental analysis. The numerical analysis predicted that the temperature difference could be minimized by using micron-sized particles and low microwave power.

To achieve an ideal slurry, a uniform distribution of the solid particles is important. Mathematical expressions indicate that the required stirring speed

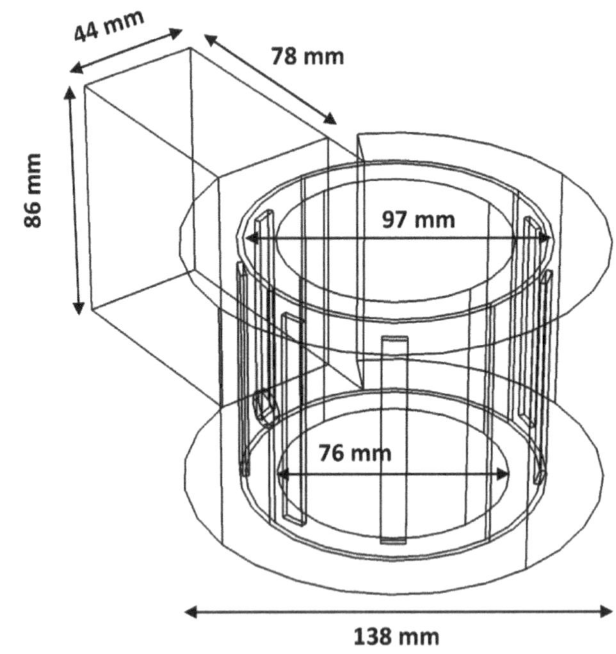

FIG. 35 Computational domain of the microwave cavity. *(Reproduced with permission from H. Goyal, S. Sadula, D.G. Vlachos, Microwave heating of slurries. Chem. Eng. J. (2020) 127892. https://doi.org/10.1016/j.cej.2020.127892.)*

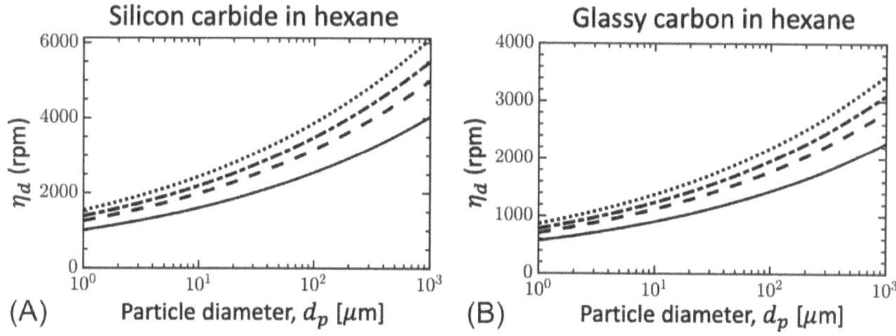

FIG. 36 Minimum stirring rate (η_d) required to suspend all the solid particles in a slurry for various particle volume fractions: *solid line* (1%), *dashed line* (5%), *dash-dot line* (10%), and *dotted line* (20%). *(Reproduced with permission from H. Goyal, S. Sadula, D.G. Vlachos, Microwave heating of slurries. Chem. Eng. J. (2020) 127892. https://doi.org/10.1016/j.cej.2020.127892.)*

increases with particle density, size, and concentration in a slurry. Thus, it is challenging to obtain uniformly dispersed concentrated slurries of denser and larger particles. The variation of minimum stirring rate (η_d) for silicon carbide and glassy carbon as a function of particle diameter for various particle volume fractions is shown in Fig. 36. The required stirring speed increases with particle size and particle concentration.

Microwave heating is playing an important role in coal processing such as grinding [77], drying [78], desulphurization [79], coking [80, 81], flotation [82] liquefaction [83], and pyrolysis [84]. Numerical analysis is a promising method to visualize and quantize microwave-coal interactions. However, numerical simulations are challenging for coal, as it is always assumed as a solid continuum in the current models and cannot accurately anticipate the thermodynamic behavior. Thus, only a few studies are reported in this area. Hong et al. predicted the heating pattern of coal using the finite element method [85]. To predict the thermal effects of microwaves on coal, Huang et al. demonstrated a coupled electromagnetic, heat, and mass transfer model [86, 87]. Li et al. demonstrated a coupled electromagnetic, heat transfer, and multiphase porous media model to analyze microwave heating of coal [88]. Heat transfer in porous media, thermal-dependent dielectric permittivity, multiphase transport, water evaporation, and mechanical damage were included in this study.

Fig. 37 shows the experimental apparatus and the geometrical model used for microwave heating of coal. The analysis showed that microwave absorption by coal makes a significant redistribution of the electromagnetic field in the cavity, setting up low-energy and high-energy regions. The thermal evolution of coal during microwave heating is shown in Fig. 38. The figure is also comparing the data from models developed by Hong et al. [85] and Huang et al. [86]. The nonuniform electromagnetic distribution together with microwave selective heating created hot and cold spots. Besides, surface heat convection and water evaporation exerted profound impacts on the thermal evolution of coal.

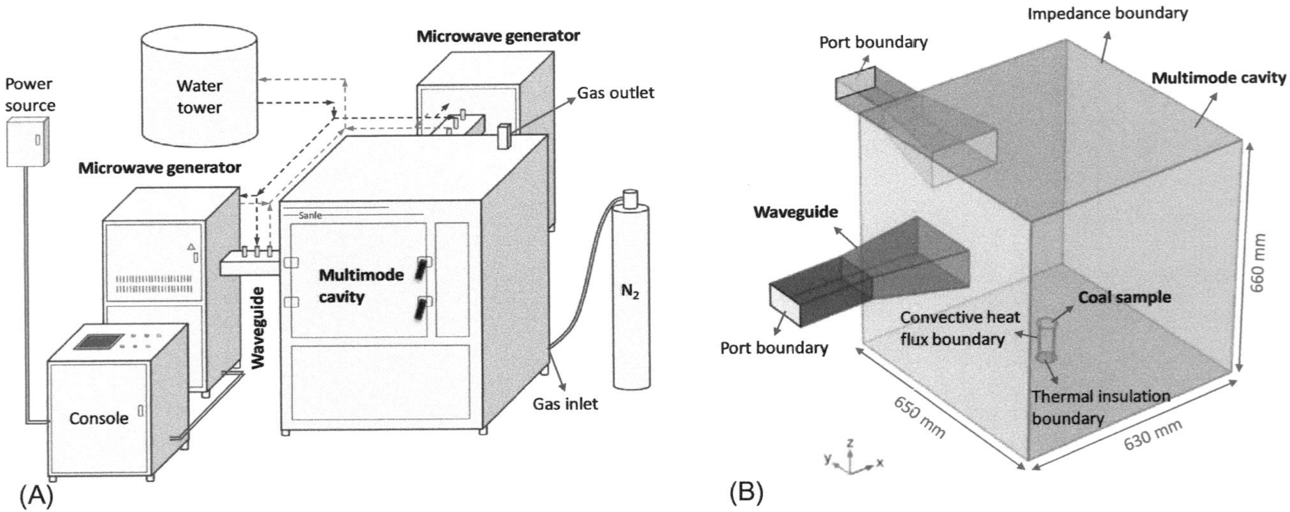

FIG. 37 (A) Experimental apparatus for microwave heating of coal and (B) the geometrical model. *(Reproduced with permission from H. Li, S. Shi, B. Lin, J. Lu, Y. Lu, Q. Ye, Z. Wang, Y. Hong, X. Zhu, A fully coupled electromagnetic, heat transfer and multiphase porous media model for microwave heating of coal. Fuel Process. Technol. 189 (2019) 49–61. https://doi.org/10.1016/j.fuproc.2019.03.002.)*

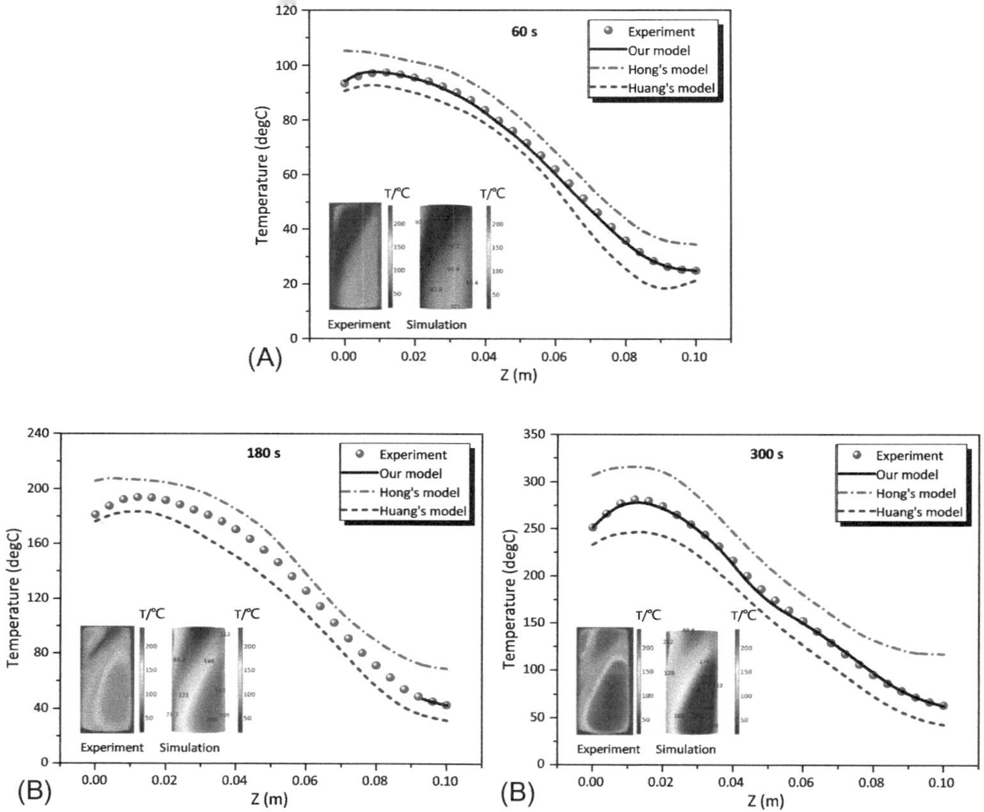

FIG. 38 Comparison of the experimental and simulated temperature of coal. *(Reproduced with permission from H. Li, S. Shi, B. Lin, J. Lu, Y. Lu, Q. Ye, Z. Wang, Y. Hong, X. Zhu, A fully coupled electromagnetic, heat transfer and multiphase porous media model for microwave heating of coal. Fuel Process. Technol. 189 (2019) 49–61. https://doi.org/10.1016/j.fuproc.2019.03.002.)*

Because the dielectric property of coal was constantly varying during microwave heating, when water evaporation is included in the model, the temperature increased nonlinearly.

To improve microwave heating performances, sometimes strongly absorbing materials are added to weak absorbing media. For example, studies showed that the addition of carbon particles could shorten the reaction time and increase the reaction rate for microwave pyrolysis of oil shale [89]. It is found that with the addition of coke as an enhancing agent, the pure sludge could be heated to 900°C and pyrolyzed efficiently [90–94]. Several other reports showed that the addition of strong microwave-absorbing materials can enhance microwave heating of waste oil [95], biomass [96], electronic waste [97], and degradation of organic compounds [98]. This temperature gradient is often known as the hot-spot effect. The quantitative study on microwave hot-spot effects is not easy mainly because of the difficulty of the direct measurement in temporal and spatial scales. Besides, the macroscopic temperature measurements can provide only an average and lagged temperature profile of the heating media, and also because of the various heat transfer conditions, it is difficult to capture the instantaneous temperature differences or to focus on a specific isolated region. Thus, numerical simulation is necessary to analyze the hot-spot effects with instantaneity or in microscales.

Wang et al. studied experimentally and numerically the hot spots occurring in microwave heating using SiC particles dispersed in paraffin oil as the sample [99]. Their study showed that hot spots occurred in such a mixture, and the SiC particles could have a higher temperature profile than the paraffin oil even at equilibrium. Hot-spot profiles were varied with different SiC particle geometry, heat generation, or heat transfer. Under certain conditions, the difference in temperature could exceed several hundred degrees Celsius. The efficiency of the effect is related to how quickly the heat generated in the strongly absorbing medium is transferred to the weakly absorbing media. Materials with strong microwave absorption, like Co, Fe, or Ni crystals, or novel carbon materials, can be used to enhance microwave heating.

Zeolites are hydrated aluminosilicates having a three-dimensional pore structure with cations in exchange positions. Zeolites have high porosity and elevated surface area, and are very well known as adsorbents, catalysts, catalytic supports, and ion exchangers [100]. During the catalytic processes or in the regeneration of sorbents, it is important to supply heat to these materials, and microwave heating is one of the possible methods to heat zeolite.

Nigar et al. reported experimentally and numerically microwave heating of a dry porous NaY zeolite fixed-bed in a quartz tube in a monomode rectangular resonant cavity [71]. For numerical analysis, the energy or heat, momentum equations were figured out together with Maxwell's electromagnetic field equations. The dielectric properties of NaY zeolite were calculated as a function of temperature for this analysis (Fig. 39), and considering these properties, the microwave heating of this material was simulated. Axial and radial temperature

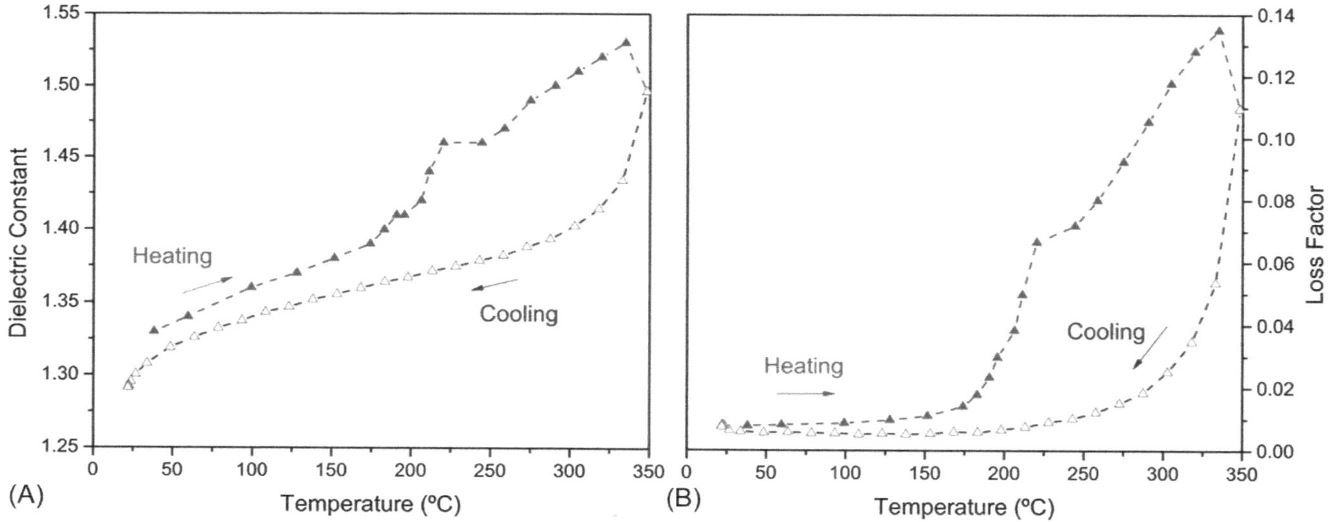

FIG. 39 Dielectric properties of NaY zeolite: (A) Dielectric constant (B) Loss factor as a function of temperature. *(Reproduced with permission from H. Nigar, G.S. J. Sturm, B. Garcia-Baños, F.L. Peñaranda-Foix, J.M. Catalá-Civera, R. Mallada, A. Stankiewicz, J. Santamaría, Numerical analysis of microwave heating cavity: combining electromagnetic energy, heat transfer and fluid dynamics for a NaY zeolite fixed-bed. Appl. Therm. Eng. 155 (2019) 226–238. https://doi.org/10.1016/j.applthermaleng.2019.03.117.)*

profiles, electric field distribution, and temperature evolution with time were demonstrated. Moreover, the developed model was able to predict thermal run-away for zeolites.

3.4 Conclusion

Over the past few decades, numerical simulation has become a promising tool to visualize and quantize the microwave heating process in several applications. In this chapter, we discussed numerical studies that are important for the microwave heating of liquids as well as solids. We considered the microwave heating of different samples such as water, oil, food, wood, minerals, and coal.

Acknowledgments

AD and BKB are grateful to Prince Mohammad Bin Fahd University for support. BKB is also grateful to US NIH, US NCI, and Kleberg Foundation of Texas for financial support.

References

[1] L.A. Campanone, N.E. Zaritzky, Mathematical analysis of microwave heating process, J. Food Eng. 69 (2005) 359–368.
[2] Y.E. Lin, R.C. Anantheswaran, V. Puri, Finite element analysis of microwave heating of solid foods, J. Food Eng. 25 (1995) 85–112, https://doi.org/10.1016/0260-8774(94)00008-W.
[3] M.E.C. Oliveira, A.S. Franca, Microwave heating of foodstuffs, J. Food Eng. 53 (2002) 347–359, https://doi.org/10.1016/S0260-8774(01)00176-5.
[4] R.B. Pandit, S. Prasad, Finite element analysis of microwave heating of potato—transient temperature profiles, J. Food Eng. 60 (2003) 193–202.
[5] V.R. Romano, F. Marra, U. Tammaro, Modelling of microwave heating of foodstuff: study on the influence of sample dimensions with a FEM approach, J. Food Eng. 71 (2005) 233–241, https://doi.org/10.1016/j.jfoodeng.2004.11.036.
[6] J. Zhang, A.K. Datta, Mathematical modeling of bread baking process, J. Food Eng. 75 (2006) 78–89.
[7] L. Zhou, V.M. Puri, R.C. Anantheswaran, G. Yeh, Finite element modeling of heat and mass transfer in food materials during microwave heating—model development and validation, J. Food Eng. 25 (1995) 509–529, https://doi.org/10.1016/0260-8774(94)00032-5.
[8] A.L. Bail, T. Koutchma, H.S. Ramaswamy, Modeling of temperature profiles under continuous tube-flow microwave and steam heating conditions, J. Food Process Eng. 23 (2000) 1–24.
[9] A. Datta, H. Prosetya, W. Hu, Mathematical modeling of batch heating of liquids in a microwave cavity, J. Microw. Power Electromagn. Energy. 27 (1992) 38–48.
[10] T. Ohlsson, In-flow microwave heating of pumpable foods, in: Developments in Food Engineering, Springer, 1994, pp. 322–324.
[11] C.M. Sabliov, K.P. Sandeep, J. Simunovic, High frequency electromagnetism coupled with conductive heat transfer–A method to predict temperature profiles in materials heated in a focused microwave system, in: Microwave and Radio Frequency Applications, Proc. 4th World Congress on Microwave and Radio Frequency Applications, Microwave Working Group, 2004, pp. 469–476.

[12] R.E. Mudgett, Microwave Properties and Heating Characteristics of Foods, Food Technology (USA), 1986.
[13] C. Saltiel, A.K. Datta, Heat and mass transfer in microwave processing, in: Advances in Heat Transfer, Elsevier, 1999, pp. 1–94.
[14] K.G. Ayappa, S. Brandon, J.J. Derby, H.T. Davis, E.A. Davis, Microwave driven convection in a square cavity, AICHE J. 40 (1994) 1268–1272.
[15] Q. Zhang, T.H. Jackson, A. Ungan, Numerical modeling of microwave induced natural convection, Int. J. Heat Mass Transf. 43 (2000) 2141–2154.
[16] R.C. Anantheswaran, L. Liu, Effect of viscosity and salt concentration on microwave heating of model non-Newtonian liquid foods in a cylindrical container, J. Microw. Power Electromagn. Energy. 29 (1994) 119–126.
[17] P. Pesheck, M. Lorence, Development of Packaging and Products for Use in Microwave Ovens, Elsevier, 2009.
[18] C.O. Kappe, B. Pieber, D. Dallinger, Microwave effects in organic synthesis: myth or reality? Angew. Chem. Int. Ed. 52 (2013) 1088–1094.
[19] A. Harrison, A.G. Whittaker, Microwave heating, in: J.A. McCleverty, T.J. Meyer (Eds.), Comprehensive Coordination Chemistry II, Pergamon, Oxford, 2003, pp. 741–745, https://doi.org/10.1016/B0-08-043748-6/01162-2.
[20] K. Werth, P. Lutze, A.A. Kiss, A.I. Stankiewicz, G.D. Stefanidis, A. Górak, A systematic investigation of microwave-assisted reactive distillation: influence of microwaves on separation and reaction, Chem. Eng. Process. Process Intensif. 93 (2015) 87–97.
[21] J.A. McCleverty, T.J. Meyer, Comprehensive Coordination Chemistry II: From Biology to Nanotechnology, Elsevier Pergamon, Amsterdam, Boston, 2004.
[22] D.M. PáMingos, Superheating effects associated with microwave dielectric heating, J. Chem. Soc. Chem. Commun. (1992) 674–677.
[23] R. Saillard, M. Poux, J. Berlan, M. Audhuy-Peaudecerf, Microwave heating of organic solvents: thermal effects and field modelling, Tetrahedron 51 (1995) 4033–4042.
[24] F. Chemat, E. Esveld, Microwave super-heated boiling of organic liquids: origin, effect and application, Chem. Eng. Technol. 24 (2001) 735–744.
[25] A. Ferrari, J. Hunt, A. Stiegman, G.B. Dudley, Microwave-assisted superheating and/or microwave-specific superboiling (nucleation-limited boiling) of liquids occurs under certain conditions but is mitigated by stirring, Molecules 20 (2015) 21672–21680.
[26] P. Ratanadecho, K. Aoki, M. Akahori, A numerical and experimental investigation of the modeling of microwave heating for liquid layers using a rectangular wave guide (effects of natural convection and dielectric properties), Appl. Math. Model. 26 (2002) 449–472.
[27] J. Zhu, A.V. Kuznetsov, K.P. Sandeep, Mathematical modeling of continuous flow microwave heating of liquids (effects of dielectric properties and design parameters), Int. J. Therm. Sci. 46 (2007) 328–341.
[28] C.M. Sabliov, D.A. Salvi, D. Boldor, High frequency electromagnetism, heat transfer and fluid flow coupling in ANSYS multiphysics, J. Microw. Power Electromagn. Energy. 41 (2006) 5–17.
[29] S. Chatterjee, T. Basak, S.K. Das, Microwave driven convection in a rotating cylindrical cavity: a numerical study, J. Food Eng. 79 (2007) 1269–1279.
[30] D.A. Salvi, D. Boldor, C.M. Sabliov, K.A. Rusch, Numerical and experimental analysis of continuous microwave heating of ballast water as preventive treatment for introduction of invasive species, J. Mar. Environ. Eng. 9 (2007).
[31] Y. Jian, X.Q. Yang, K.M. Huang, Numerical analysis of the influence of stir on water during microwave heating, Prog. Electromagn. Res. C 17 (2010) 105–119.

[32] W. Cha-um, P. Rattanadecho, W. Pakdee, Experimental and numerical analysis of microwave heating of water and oil using a rectangular wave guide: influence of sample sizes, positions, and microwave power, Food Bioprocess Technol. 4 (2011) 544–558.

[33] R. Cherbański, L. Rudniak, Modelling of microwave heating of water in a monomode applicator–influence of operating conditions, Int. J. Therm. Sci. 74 (2013) 214–229.

[34] S. Nikdel, C.S. Chen, M.E. Parish, D.G. MacKellar, L.M. Friedrich, Pasteurization of citrus juice with microwave energy in a continuous-flow unit, J. Agric. Food Chem. 41 (1993) 2116–2119, https://doi.org/10.1021/jf00035a055.

[35] S. Tajchakavit, H.S. Ramaswamy, Continuous-flow microwave heating of orange juice: evidence of nonthermal effects, J. Microw. Power Electromagn. Energy. 30 (1995) 141–148, https://doi.org/10.1080/08327823.1995.11688270.

[36] T.S. Gentry, J.S. Roberts, Design and evaluation of a continuous flow microwave pasteurization system for apple cider, LWT Food Sci. Technol. 38 (2005) 227–238, https://doi.org/10.1016/j.lwt.2004.05.016.

[37] S. Tajchakavit, H.S. Ramaswamy, P. Fustier, Enhanced destruction of spoilage microorganisms in apple juice during continuous flow microwave heating, Food Res. Int. 31 (1998) 713–722, https://doi.org/10.1016/S0963-9969(99)00050-2.

[38] P. Coronel, J. Simunovic, K.P. Sandeep, Temperature profiles within milk after heating in a continuous-flow tubular microwave system operating at 915 MHz, J. Food Sci. 68 (2003) 1976–1981, https://doi.org/10.1111/j.1365-2621.2003.tb07004.x.

[39] I. Sierra, C. Vidal-Valverde, A. Olano, The effects of continuous flow microwave treatment and conventional heating on the nutritional value of milk as shown by influence on vitamin B1 retention, Eur. Food Res. Technol. 209 (1999) 352–354, https://doi.org/10.1007/s002170050508.

[40] E. Valero, M. Villamiel, J. Sanz, I. Martinez-Castro, Chemical and sensorial changes in milk pasteurised by microwave and conventional systems during cold storage, Food Chem. 70 (2000) 77–81, https://doi.org/10.1016/S0308-8146(00)00074-1.

[41] M. Villamiel, R. López-Fandiño, N. Corzo, I. Martínez-Castro, A. Olano, Effects of continuous flow microwave treatment on chemical and microbiological characteristics of milk, Z. Lebensm. Unters. Forch. 202 (1996) 15–18, https://doi.org/10.1007/BF01229677.

[42] P. Kumar, P. Coronel, V.D. Truong, J. Simunovic, K.R. Swartzel, K.P. Sandeep, G. Cartwright, Overcoming issues associated with the scale-up of a continuous flow microwave system for aseptic processing of vegetable purees, Food Res. Int. 41 (2008) 454–461, https://doi.org/10.1016/j.foodres.2007.11.004.

[43] P. Coronel, V.-D. Truong, J. Simunovic, K.P. Sandeep, G.D. Cartwright, Aseptic processing of sweetpotato purees using a continuous flow microwave system, J. Food Sci. 70 (2005) E531–E536, https://doi.org/10.1111/j.1365-2621.2005.tb08315.x.

[44] L.E. Steed, V.-D. Truong, J. Simunovic, K.P. Sandeep, P. Kumar, G.D. Cartwright, K.R. Swartzel, Continuous flow microwave-assisted processing and aseptic packaging of purple-fleshed sweetpotato purees, J. Food Sci. 73 (2008) E455–E462, https://doi.org/10.1111/j.1750-3841.2008.00950.x.

[45] D. Salvi, J. Ortego, C. Arauz, C.M. Sabliov, D. Boldor, Experimental study of the effect of dielectric and physical properties on temperature distribution in fluids during continuous flow microwave heating, J. Food Eng. 93 (2009) 149–157, https://doi.org/10.1016/j.jfoodeng.2009.01.009.

[46] V. Romano, R. Apicella, Microwave heating of liquid foods, Engineering 7 (2015) 297–306, https://doi.org/10.4236/eng.2015.76026.

[47] J.A. Koutsky, R.J. Adler, Minimisation of axial dispersion by use of secondary flow in helical tubes, Can. J. Chem. Eng. 42 (1964) 239–246, https://doi.org/10.1002/cjce.5450420602.

[48] M. Eesa, CFD Studies of Complex Fluid flows in Pipes, d_ph, University of Birmingham, 2009. https://etheses.bham.ac.uk/id/eprint/413/. (Accessed 16 December 2020).

[49] S. Tuta, T.K. Palazoğlu, Finite element modeling of continuous-flow microwave heating of fluid foods and experimental validation, J. Food Eng. 192 (2017) 79–92.

[50] S.P. Yeong, M.C. Law, C.C.V. Lee, Y.S. Chan, Modelling batch microwave heating of water, IOP Conf. Ser.: Mater. Sci. Eng. 217 (2017) 012035, https://doi.org/10.1088/1757-899X/217/1/012035.

[51] M. Anese, S. Sovrano, Kinetics of thermal inactivation of tomato lipoxygenase, Food Chem. 95 (2006) 131–137, https://doi.org/10.1016/j.foodchem.2004.12.026.

[52] L. Dorantes-Alvarez, L. Parada-Dorantes, Blanching using microwave processing, in: H. Schubert, M. Regier (Eds.), The Microwave Processing of Foods, Woodhead Publishing, 2005, pp. 153–173, https://doi.org/10.1533/9781845690212.2.153.

[53] N. Stănciuc, I. Aprodu, E. Ioniță, G. Bahrim, G. Râpeanu, Exploring the process–structure–function relationship of horseradish peroxidase through investigation of pH- and heat induced conformational changes, Spectrochim. Acta A Mol. Biomol. Spectrosc. 147 (2015) 43–50, https://doi.org/10.1016/j.saa.2015.03.023.

[54] M. Benlloch-Tinoco, M. Igual, D. Rodrigo, N. Martínez-Navarrete, Comparison of microwaves and conventional thermal treatment on enzymes activity and antioxidant capacity of kiwifruit puree, Innovative Food Sci. Emerg. Technol. 19 (2013) 166–172, https://doi.org/10.1016/j.ifset.2013.05.007.

[55] M.E. Latorre, P.R. Bonelli, A.M. Rojas, L.N. Gerschenson, Microwave inactivation of red beet (Beta vulgaris L. var. conditiva) peroxidase and polyphenoloxidase and the effect of radiation on vegetable tissue quality, J. Food Eng. 109 (2012) 676–684, https://doi.org/10.1016/j.jfoodeng.2011.11.026.

[56] K.N. Matsui, J.A.W. Gut, P.V. de Oliveira, C.C. Tadini, Inactivation kinetics of polyphenol oxidase and peroxidase in green coconut water by microwave processing, J. Food Eng. 88 (2008) 169–176, https://doi.org/10.1016/j.jfoodeng.2008.02.003.

[57] M.T.K. Kubo, S. Curet, P.E.D. Augusto, L. Boillereaux, Multiphysics modeling of microwave processing for enzyme inactivation in fruit juices, J. Food Eng. 263 (2019) 366–379, https://doi.org/10.1016/j.jfoodeng.2019.07.011.

[58] G.L. Lee, M.C. Law, V.C.-C. Lee, Numerical modelling of liquid heating and boiling phenomena under microwave irradiation using OpenFOAM, Int. J. Heat Mass Transf. 148 (2020) 119096, https://doi.org/10.1016/j.ijheatmasstransfer.2019.119096.

[59] G.L. Lee, M.C. Law, V.C.-C. Lee, Numerical modelling and investigation of microwave heating and boiling phenomena in binary liquid mixtures using OpenFOAM, Int. J. Therm. Sci. 159 (2021) 106538, https://doi.org/10.1016/j.ijthermalsci.2020.106538.

[60] K.G. Ayappa, H.T. Davis, E.A. Davis, J. Gordon, Analysis of microwave heating of materials with temperature-dependent properties, AICHE J. 37 (1991) 313–322, https://doi.org/10.1002/aic.690370302.

[61] K.G. Ayappa, H.T. Davis, E.A. Davis, J. Gordon, Two-dimensional finite element analysis of microwave heating, AICHE J. 38 (1992) 1577–1592, https://doi.org/10.1002/aic.690381009.

[62] P. Rattanadecho, The simulation of microwave heating of wood using a rectangular wave guide: influence of frequency and sample size, Chem. Eng. Sci. 61 (2006) 4798–4811, https://doi.org/10.1016/j.ces.2006.03.001.

[63] P. Ratanadecho, K. Aoki, M. Akahori, A numerical and experimental study of microwave drying using a rectangular wave guide, Dry. Technol. 19 (2001) 2209–2234, https://doi.org/10.1081/DRT-100107495.

[64] S. Geedipalli, A.K. Datta, V. Rakesh, Heat transfer in a combination microwave–jet impingement oven, Food Bioprod. Process. 86 (2008) 53–63, https://doi.org/10.1016/j.fbp.2007.10.016.

[65] M. Lovás, M. Kováčová, G. Dimitrakis, S. Čuvanová, I. Znamenáčková, Š. Jakabský, Modeling of microwave heating of andesite and minerals, Int. J. Heat Mass Transf. 53 (2010) 3387–3393, https://doi.org/10.1016/j.ijheatmasstransfer.2010.03.012.

[66] A.A. Arpia, W.-H. Chen, S.S. Lam, P. Rousset, M.D.G. De Luna, Sustainable biofuel and bioenergy production from biomass waste residues using microwave-assisted heating: a comprehensive review, Chem. Eng. J. 403 (2020) 126233.

[67] A.Z. Fia, J. Amorim, Heating of biomass in microwave household oven-A numerical study, Energy 218 (2020) 119472.

[68] G. Wang, B. Liang, B.-C. Juang, A. Das, M.C. Debnath, D.L. Huffaker, Y.I. Mazur, M.E. Ware, G.J. Salamo, Comparative study of photoluminescence from $In_{0.3}Ga_{0.7}As/GaAs$ surface and buried quantum dots, Nanotechnology 27 (2016) 465701.

[69] N. Haneishi, S. Tsubaki, M.M. Maitani, E. Suzuki, S. Fujii, Y. Wada, Electromagnetic and heat-transfer simulation of the catalytic dehydrogenation of ethylbenzene under microwave irradiation, Ind. Eng. Chem. Res. 56 (2017) 7685–7692.

[70] L.M. Sanz-Moral, A. Navarrete, G. Sturm, G. Link, M. Rueda, G. Stefanidis, Á. Martín, Release of hydrogen from nanoconfined hydrides by application of microwaves, J. Power Sources 353 (2017) 131–137, https://doi.org/10.1016/j.jpowsour.2017.03.110.

[71] H. Nigar, G.S.J. Sturm, B. Garcia-Baños, F.L. Peñaranda-Foix, J.M. Catalá-Civera, R. Mallada, A. Stankiewicz, J. Santamaría, Numerical analysis of microwave heating cavity: combining electromagnetic energy, heat transfer and fluid dynamics for a NaY zeolite fixed-bed, Appl. Therm. Eng. 155 (2019) 226–238, https://doi.org/10.1016/j.applthermaleng.2019.03.117.

[72] H. Goyal, A. Mehdad, R.F. Lobo, G.D. Stefanidis, D.G. Vlachos, Scaleup of a single-mode microwave reactor, Ind. Eng. Chem. Res. 59 (2019) 2516–2523.

[73] J. Robinson, S. Kingman, D. Irvine, P. Licence, A. Smith, G. Dimitrakis, D. Obermayer, C.O. Kappe, Electromagnetic simulations of microwave heating experiments using reaction vessels made out of silicon carbide, Phys. Chem. Chem. Phys. 12 (2010) 10793–10800.

[74] G.S.J. Sturm, M.D. Verweij, T. van Gerven, A.I. Stankiewicz, G.D. Stefanidis, On the effect of resonant microwave fields on temperature distribution in time and space, Int. J. Heat Mass Transf. 55 (2012) 3800–3811, https://doi.org/10.1016/j.ijheatmasstransfer.2012.02.065.

[75] G.S.J. Sturm, M.D. Verweij, T. van Gerven, A.I. Stankiewicz, G.D. Stefanidis, On the parametric sensitivity of heat generation by resonant microwave fields in process fluids, Int. J. Heat Mass Transf. 57 (2013) 375–388, https://doi.org/10.1016/j.ijheatmasstransfer.2012.09.037.

[76] H. Goyal, S. Sadula, D.G. Vlachos, Microwave heating of slurries, Chem. Eng. J. (2020) 127892, https://doi.org/10.1016/j.cej.2020.127892.

[77] B.K. Sahoo, S. De, B.C. Meikap, Improvement of grinding characteristics of Indian coal by microwave pre-treatment, Fuel Process. Technol. 92 (2011) 1920–1928, https://doi.org/10.1016/j.fuproc.2011.05.012.

[78] C.A. Pickles, F. Gao, S. Kelebek, Microwave drying of a low-rank sub-bituminous coal, Miner. Eng. 62 (2014) 31–42, https://doi.org/10.1016/j.mineng.2013.10.011.

[79] S. Mesroghli, J. Yperman, E. Jorjani, R. Carleer, M. Noaparast, Evaluation of microwave treatment on coal structure and sulfur species by reductive pyrolysis-mass spectrometry method, Fuel Process. Technol. 131 (2015) 193–202, https://doi.org/10.1016/j.fuproc.2014.11.005.

[80] X. Guo, J. Ren, C. Xie, J. Lin, Z. Li, A comparison study on the deoxygenation of coal mine methane over coal gangue and coke under microwave heating conditions, Energy Convers. Manag. 100 (2015) 45–55, https://doi.org/10.1016/j.enconman.2015.04.079.

[81] E. Binner, M. Mediero-Munoyerro, T. Huddle, S. Kingman, C. Dodds, G. Dimitrakis, J. Robinson, E. Lester, Factors affecting the microwave coking of coals and the implications on microwave cavity design, Fuel Process. Technol. 125 (2014) 8–17, https://doi.org/10.1016/j.fuproc.2014.03.006.

[82] B.K. Sahoo, S. De, B.C. Meikap, Artificial neural network approach for rheological characteristics of coal-water slurry using microwave pre-treatment, Int. J. Min. Sci. Technol. 27 (2017) 379–386, https://doi.org/10.1016/j.ijmst.2017.01.022.

[83] H.S. Uhm, Y.H. Na, Y.C. Hong, D.H. Shin, C.H. Cho, Y.K. Park, High-efficiency gasification of low-grade coal by microwave steam plasma, Energy Fuel 28 (2014) 4402–4408.

[84] V. Abdelsayed, D. Shekhawat, M.W. Smith, D. Link, A.E. Stiegman, Microwave-assisted pyrolysis of Mississippi coal: a comparative study with conventional pyrolysis, Fuel 217 (2018) 656–667, https://doi.org/10.1016/j.fuel.2017.12.099.

[85] Y. Hong, B. Lin, H. Li, H. Dai, C. Zhu, H. Yao, Three-dimensional simulation of microwave heating coal sample with varying parameters, Appl. Therm. Eng. 93 (2016) 1145–1154, https://doi.org/10.1016/j.applthermaleng.2015.10.041.

[86] J. Huang, G. Xu, G. Hu, M. Kizil, Z. Chen, A coupled electromagnetic irradiation, heat and mass transfer model for microwave heating and its numerical simulation on coal, Fuel Process. Technol. 177 (2018) 237–245, https://doi.org/10.1016/j.fuproc.2018.04.034.

[87] J. Huang, G. Xu, Y. Chen, Z. Chen, Simulation of microwave's heating effect on coal seam permeability enhancement, Int. J. Min. Sci. Technol. 29 (2019) 785–789.

[88] H. Li, S. Shi, B. Lin, J. Lu, Y. Lu, Q. Ye, Z. Wang, Y. Hong, X. Zhu, A fully coupled electromagnetic, heat transfer and multiphase porous media model for microwave heating of coal, Fuel Process. Technol. 189 (2019) 49–61, https://doi.org/10.1016/j.fuproc.2019.03.002.

[89] K. El Harfi, A. Mokhlisse, M.B. Chanâa, A. Outzourhit, Pyrolysis of the Moroccan (Tarfaya) oil shales under microwave irradiation, Fuel 79 (2000) 733–742, https://doi.org/10.1016/S0016-2361(99)00209-4.

[90] J.A. Menéndez, M. Inguanzo, J.J. Pis, Microwave-induced pyrolysis of sewage sludge, Water Res. 36 (2002) 3261–3264, https://doi.org/10.1016/S0043-1354(02)00017-9.

[91] A. Domínguez, J.A. Menéndez, M. Inguanzo, P.L. Bernad, J.J. Pis, Gas chromatographic–mass spectrometric study of the oil fractions produced by microwave-assisted pyrolysis of different sewage sludges, J. Chromatogr. A 1012 (2003) 193–206, https://doi.org/10.1016/S0021-9673(03)01176-2.

[92] J.A. Menéndez, A. Domínguez, M. Inguanzo, J.J. Pis, Microwave-induced drying, pyrolysis and gasification (MWDPG) of sewage sludge: vitrification of the solid residue, J. Anal. Appl. Pyrolysis 74 (2005) 406–412, https://doi.org/10.1016/j.jaap.2004.10.013.

[93] A. Domínguez, J.A. Menéndez, M. Inguanzo, J.J. Pis, Investigations into the characteristics of oils produced from microwave pyrolysis of sewage sludge, Fuel Process. Technol. 86 (2005) 1007–1020, https://doi.org/10.1016/j.fuproc.2004.11.009.

[94] A. Domínguez, J.A. Menéndez, M. Inguanzo, J.J. Pís, Production of bio-fuels by high temperature pyrolysis of sewage sludge using conventional and microwave heating, Bioresour. Technol. 97 (2006) 1185–1193, https://doi.org/10.1016/j.biortech.2005.05.011.

[95] S.S. Lam, R.K. Liew, C.K. Cheng, H.A. Chase, Catalytic microwave pyrolysis of waste engine oil using metallic pyrolysis char, Appl. Catal. B Environ. 176–177 (2015) 601–617, https://doi.org/10.1016/j.apcatb.2015.04.014.

[96] P. Lahijani, M. Mohammadi, Z.A. Zainal, A.R. Mohamed, Improvement of biomass char-CO_2 gasification reactivity using microwave irradiation and natural catalyst, Thermochim. Acta 604 (2015) 61–66, https://doi.org/10.1016/j.tca.2015.01.016.

[97] H. Tu, T. Duan, Y. Ding, X. Lu, Y. Tang, Y. Li, Preparation of zircon-matrix material for dealing with high-level radioactive waste with microwave, Mater. Lett. 131 (2014) 171–173, https://doi.org/10.1016/j.matlet.2014.05.195.
[98] Y. Mao, Z. Xi, W. Wang, C. Ma, Q. Yue, Kinetics of solvent blue and reactive yellow removal using microwave radiation in combination with nanoscale zero-valent iron, J. Environ. Sci. 30 (2015) 164–172, https://doi.org/10.1016/j.jes.2014.09.030.
[99] W. Wang, B. Wang, J. Sun, Y. Mao, X. Zhao, Z. Song, Numerical simulation of hot-spot effects in microwave heating due to the existence of strong microwave-absorbing media, RSC Adv. 6 (2016) 52974–52981, https://doi.org/10.1039/C6RA05191J.
[100] H. Awala, J.-P. Gilson, R. Retoux, P. Boullay, J.-M. Goupil, V. Valtchev, S. Mintova, Template-free nanosized faujasite-type zeolites, Nat. Mater. 14 (2015) 447–451.

Part II

Methods in microwave chemistry

Chapter 4

Microwave-assisted synthesis of oxygen- and sulfur-containing organic compounds

Aparna Das* and Bimal Krishna Banik*

Department of Mathematics and Natural Sciences, College of Sciences and Human Studies, Prince Mohammad Bin Fahd University, Al Khobar, Kingdom of Saudi Arabia

**Corresponding authors: e-mails: aparnadasam@gmail.com (Aparna Das); bimalbanik10@gmail.com, bbanik@pmu.edu.sa (Bimal Krishna Banik)*

4.1 Introduction

The oxygen-containing heterocyclic (*O*-heterocycles) and sulfur-containing heterocyclic (*S*-heterocycles) compounds are the important classes of compounds in chemistry due to their wide range of applications. In this chapter, we discuss the microwave-assisted synthesis of *O*-heterocycles, *S*-heterocycles, and related compounds.

4.2 Oxygen-containing organic compounds

4.2.1 Xanthene derivatives

Xanthene derivatives are found in a large number of natural and synthetic bioactive molecules [1]. For example, benzoxanthenes are the important biologically active oxygen-containing heterocycles possessing anticancer [2], antimycobacterial [3], and potent modulators of intestinal *P*-glycoprotein [4]. Moreover, due to the useful spectroscopic properties, these xanthene derivatives were also used in laser technologies [5], in fluorescent materials for the visualization of biomolecule [6], and as dyes [7].

Ashok et al. described a simple, efficient, and eco-friendly method for the synthesis of pyranoxanthenes via a one-pot three-component reaction [8]. The reaction among aromatic aldehyde, 2,2-dimethylchroman-7-ol, and 1,3-cyclohexanedione was conducted under the microwave irradiation using molecular iodine in AcOH as a reaction medium. A variety of xanthene derivatives were synthesized in excellent yields using this protocol.

Microwaves in Chemistry Applications. https://doi.org/10.1016/B978-0-12-822895-1.00010-2

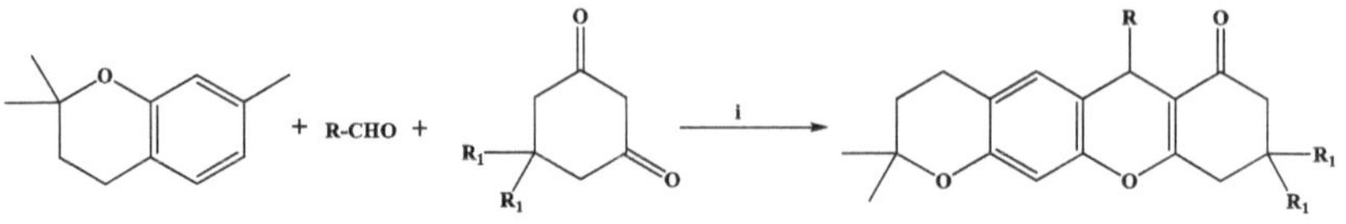

1 2 3 4

R= Aryl, Hetero aryl; R_1=H, CH_3

SCHEME 1 Synthesis of 2,2-dimethyl-6-aryl-3,4,6,8,9,10-hexahydropyrano[3,2-b]xanthen-7(2*H*)-ones and 6-aryl-2,2,9,9-tetramethyl-3,4,6,8,9,10-hexahydropyrano[3,2-b]xanthen-7(2*H*)-ones. Reagents and conditions: (i) I_2/AcOH, MW.

Scheme 1 shows the one-pot method for the three-component condensation of 2,2-dimethylchroman-7-ol (**1**), aryl/heteroaryl aromatic aldehydes (**2**), and 1,3-cyclohexadione/5,5-dimethyl-1,3-cyclohexanedione (**3**) for the synthesis of pyranoxanthenes (**4**) in good yields. To optimize the reaction conditions, the same reaction was investigated under different reaction conditions. The results demonstrated that the microwave irradiation led to an enhancement in the yield of the product and also the reaction time has also been reduced over the conventional heating method.

All the synthesized compounds were tested for in vitro antimicrobial and antioxidant activities. The results demonstrated that some compounds exhibited a significant antibacterial activity against the bacterial strains. For instance, the compounds containing NO_2 functional group on phenyl ring and five-membered thiophene ring showed very good antibacterial activity compared with other compounds.

An efficient and environmentally benign procedure for the synthesis of 1,8-dioxo-octahydroxanthene (**8**) by the condensation reaction between 5,5-dimethyl-1,3-cyclohexanedione (dimedone, **6** and **7**) and aldehydes (**5**) using the carboxy-functionalized ionic liquid under the microwave irradiation was described by Dadhania et al. [9].

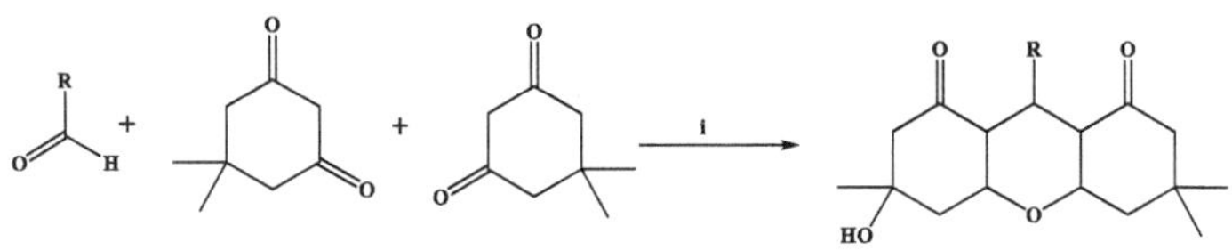

5 6 7 8

R= C_6H_5, 4-$O_2NC_6H_4$, 2-ClC_6H_4, 3-ClC_6H_4, 4-ClC_6H_4, 2-$O_2NC_6H_4$, 3-$O_2NC_6H_4$, 4-$H_3CC_6H_4$, 4-$MeOC_6H_4$, 4-HOC_6H_4, 4-HO, 3-$MeOC_6H_3$, 2-BrC_6H_4, 3-BrC_6H_4, 4-FC_6H_4, 2,5-$(MeO)_2C_6H_3$, 3,4-$(MeO)_2C_6H_3$, 3,4,5-$(MeO)_3C_6H_2$, 2-HOC_6H_4, 2-Furyl, 2-Thienyl, 2-Pyridyl

SCHEME 2 Synthesis of 1,8-dioxo-octahydroxanthene derivatives. Reagents and conditions: (i) [cmmim][BF4], MW.

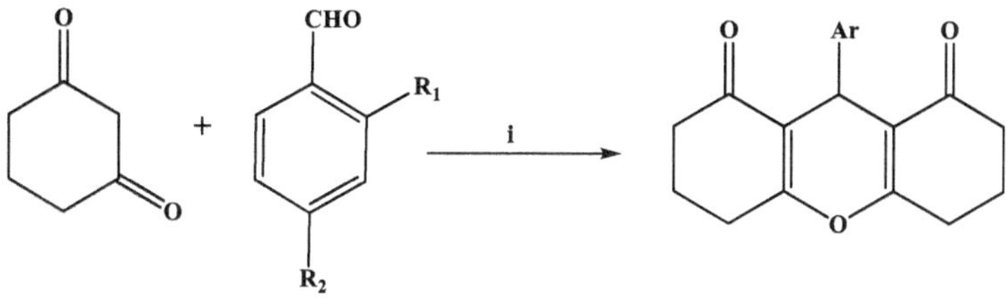

9 **10** **11**

R_1= H, CH_3, OCH_3, Cl; R_2= H, CH_3, OCH_3, Cl.

Ar=2,4-disubstitued benzaldehyde

SCHEME 3 Microwave-induced synthesis of 9-aryl-1,8-dioxo-octahyroxanthenes. Reagents and conditions: (i) BiI_3 (5–10 mol%), EtOH, MW, 2–6 min, 60°C.

The reaction of benzaldehyde and dimedone in the presence of 200 mg [cmmim][BF4] under the microwave irradiation is shown in Scheme 2. A series of experiments were carried out to investigate the influence of microwave irradiation on a reaction mixture and high purity, and the best yield was obtained at the power of 280 and 350 W.

Chavez et al. described the microwave-induced bismuth triiodide-catalyzed facile synthesis of octahydroxanthenes [10]. Microwave-induced reaction of 1,3-cyclohexanedione (**9**) with various aromatic aldehydes (**10**) in the presence of bismuth (III) salts as a catalyst was performed.

To optimize the reaction conditions and to identify the best catalyst system in this reaction, several bismuth (III) salts, for instance, bismuth nitrate pentahydrate, bismuth chloride, bismuth bromide, bismuth iodide, bismuth oxide, and bismuth subnitrate, were considered. Among the catalysts, bismuth iodide in ethanol provided good results and was included in this protocol. In this method, a mixture of the diketone (2 mmol) and aldehyde (1 mmol) was added to bismuth iodide (10 mol%) in ethanol (1 mL). Then, the reaction mixture was irradiated in a microwave oven at 60°C at a power of 300 W for 2–6 min (Scheme 3). At the final step, dichloromethane (10 mL) and water (2 mL) were added to the reaction mixture, and the organic layer was collected and dried with sodium sulfate.

4.2.2 Furan derivatives

Furan derivatives possess a wide range of biological activities such as antitumor, antibacterial, antifungal, antidepressant, analgesic, and hypoglycemic activities. Ashok et al. synthesized a series of chroman-4-one-fused benzofurans by cyclization of the corresponding chalcones under the microwave irradiation [11].

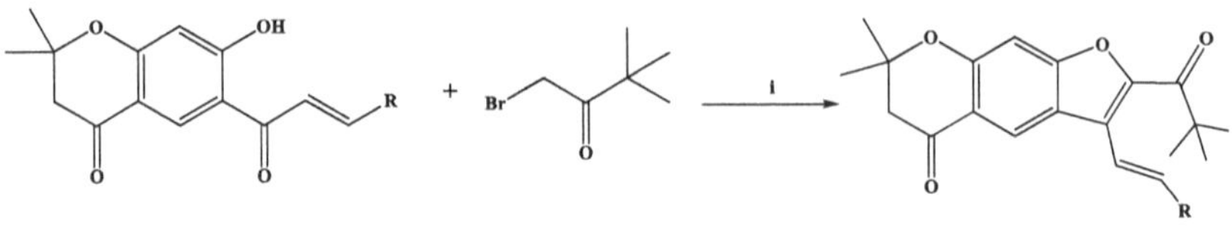

12 13 14

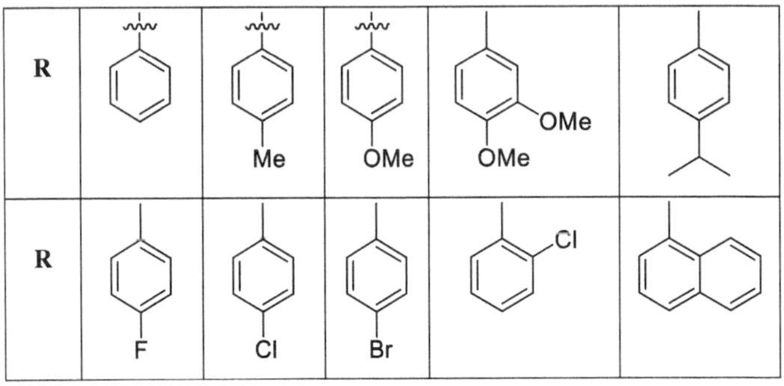

SCHEME 4 Synthesis of (*E*)-7,7-dimethyl-2-pivaloyl-3-aryl-6,7-dihydro-5*H*-furo-[3,2-g]chromen-5-ones. Reagents and conditions: (i) K_2CO_3, 2–4 min, MW.

During this process, chalcones (**12**) underwent cyclization with 1-bromo-3,3-dimethylbutan-2-one (**13**) to give the 2,2-dimethylchroman-4-one-based benzofurans (**14**) in good yields, as shown in Scheme 4.

The synthesized compounds were screened for their in vitro antibacterial activity by the disk diffusion method against two Gram-positive bacterial strains (*Staphylococcus aureus* and *Bacillus faecalis*) and two Gram-negative bacterial strains (*Escherichia coli* and *Klebsiella pneumonia*) at the concentrations of 20 and 40 μg/mL. The results were compared with that for the standard drug, ciprofloxacin. Two compounds showed the promising antibacterial activities compared with the standard drug. All other compounds were also tested for their antifungal activities against two pathogenic fungi (*Fusarium oxysporum* and *Aspergillus flavus*) by the poison plate technique. The results of the activity were compared with those of the activities of the standard antifungal drugs amphotericin-B and hymexazol. Interestingly, all compounds had a moderate activity against the tested fungal strains.

Microwave-assisted methods were used for the efficient solvent-free one-pot synthesis of isobenzofuran-1(3*H*)-ones using sulfamic acid as catalyst [12]. Using this method, isobenzofuran-1(3*H*)-ones (**17**) were synthesized in a good yield from 2-carboxybenzaldehyde (**15**) and substituted cyclic as well as non-cyclic ketones (**16**). Sulfamic acid was proved to be an excellent catalyst for this microwave-assisted condensation reaction. The general procedure for the microwave-assisted synthesis of isobenzofuran-1(3*H*)-ones (**17**) is described in Scheme 5.

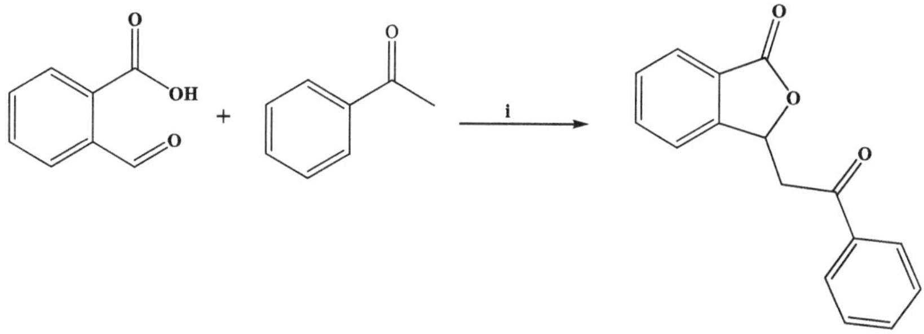

15 **16** **17**

SCHEME 5 Synthesis of isobenzofuran-1(3*H*)-ones using sulfamic acid as a novel catalyst under solvent-free condition. Reagents and conditions: (i) Sulfamic acid catalyst (20 mol%), MW, 2 min, 80°C, 1000 W, solvent-free.

During the process, a mixture of 2-carboxybenzaldehyde (10 mmol), ketone (12 mmol), and sulfamic acid (2 mmol) was irradiated using the microwave (1000 W) at 80°C under the solvent-free condition for 2 min, and the pure products were obtained in 94%–99% yields. Using similar conditions, different derivatives were synthesized by employing various substituted cyclic as well as noncyclic ketones as listed in Table 1.

The microwave-assisted one-pot linkage between furan and 3(2*H*)-furanone rings was demonstrated via the Et_3N-catalyzed domino condensation of the furan and benzofuran carboxylic acids with the available cyanopropargylic alcohols [13]. The desired products, 5-(2-furyl)-3(2*H*)-furanones (**20**), were formed in 59%–96% yields.

TABLE 1 Various substituted cyclic as well as noncyclic ketones used for the synthesis of isobenzofuran-1(3*H*)-ones.

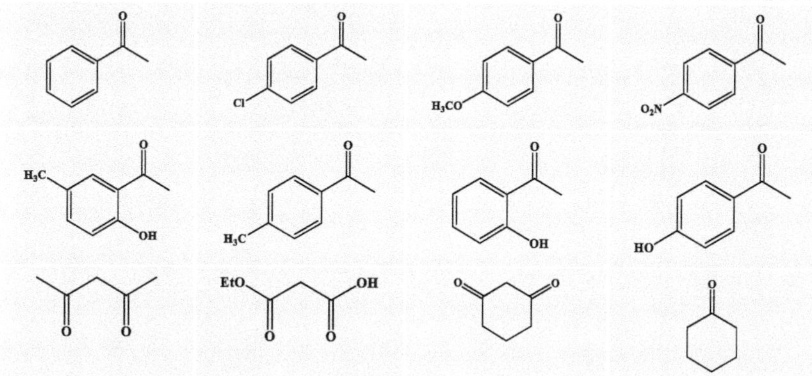

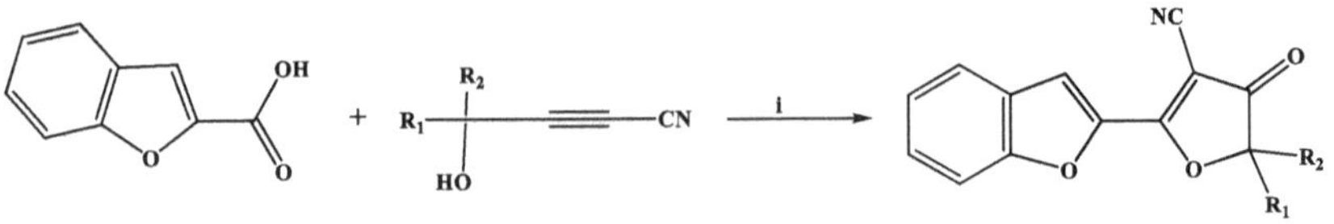

18 **19** **20**

SCHEME 6 Microwave-assisted synthesis of 5-(2-furyl)-3(2*H*)-furanones. Reagents and conditions: (i) Et_3N, MeCN, MW, 100°C, 2–17 h, $-H_2O$.

The synthesis of 5-(2-furyl)-3(2*H*)-furanones (**20**) is shown in Scheme 6. The reactions were carried out in MeCN in the presence of Et_3N under a microwave irradiation at 100°C and 1.2 atm for 2–17 h. Triethylamine acted as a catalyst in this synthesis. In the absence of Et_3N or other tertiary amines, the reaction did not occur. When a neat Et_3N was used as the only solvent, the reaction proceeded more slowly and the yield decreased from 85% to 18%, mainly due to the polymerization of the starting propargylic alcohol. The synthesized compounds (Table 2) are the promising building blocks for the synthesis of furan-based heterocyclic compounds.

An aurone is a heterocyclic compound that contains a benzofuran element associated with a benzylidene linked in position 2. It is a type of flavonoid, and a chalcone-like group is closed into a 5-membered ring instead of the 6-membered ring more typical of flavonoids. Natural and synthetic aurones have been shown to possess a wide spectrum of bioactivities such as antioxidant [14, 15], anticancer [16, 17], antiparasitic [18, 19], antiviral, antifungal,

TABLE 2 Synthesized 5-(2-furyl)-3(2*H*)-furanones (20).

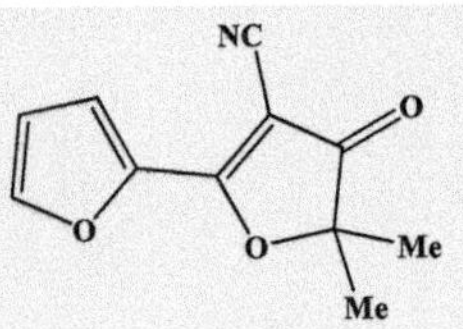 **20a**	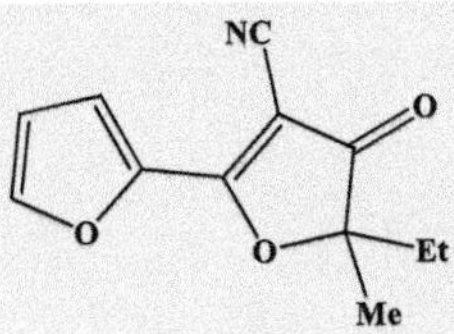 **20b**	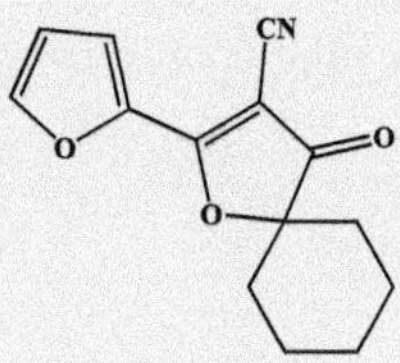 **20c**
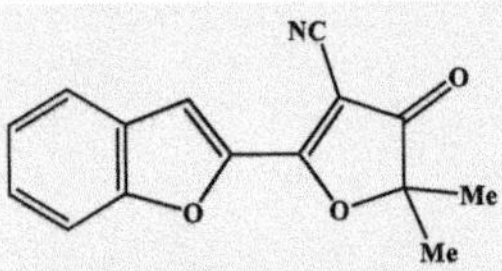 **20d**	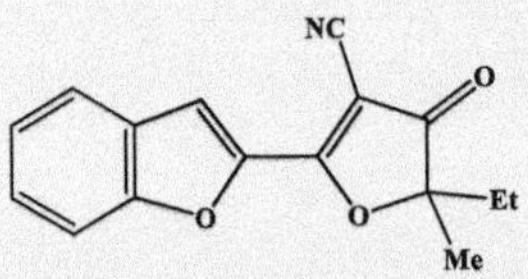 **20e**	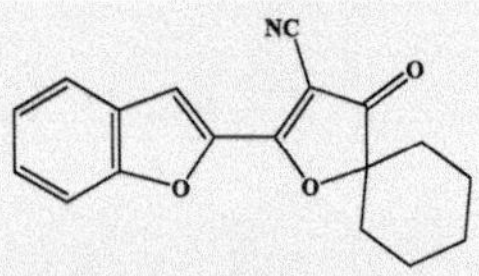 **20f**

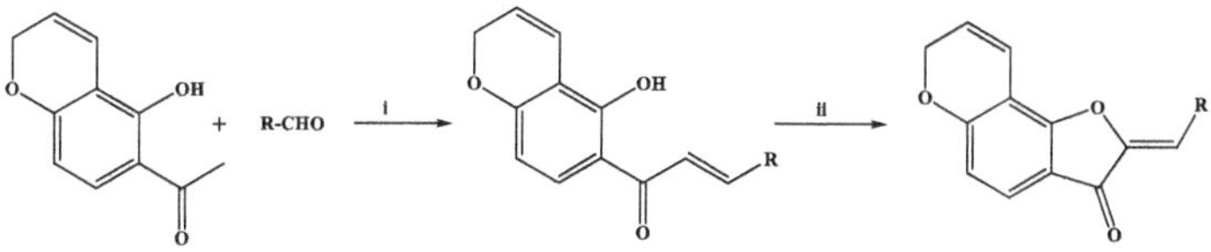

21 **22** **23**

R= phenyl, 4-bromophenyl, 2-chlorophenyl, 4-isopropylphenyl, 2,4-dichlorophenyl, 4-methylphenyl, 4-methoxyphenyl, 3,4-dimethoxyphenyl, 2-thiophenyl, 3-indolyl

SCHEME 7 Synthesis of 2-arylidene-2H-furo[2,3-f]chromen-3(7*H*)-ones (**23**). Reagents and conditions: (i) KOH, MW, 5–7 min, 180 W; (ii) $Hg(OAc)_2$, pyridine, MW, 180 W, 3–5 min.

neuroprotective [20], CDK1 inhibitor [21], antinociceptive [22], antimutagenic [23], and antidiabetic [24–26].

The microwave-assisted method of synthesis of some novel 2-arylidene-2*H*-furo[2,3-f]chromen-3(7*H*)-ones (**23**) in excellent yields was reported [27]. The synthetic route for the 2-arylidene-2*H*-furo[2,3-f]chromen-3(7*H*)-ones is depicted in Scheme 7. Microwave-assisted Claisen-Schmidt condensation between 1-(5-hydroxy-2*H*chromen-6-yl)ethanone (**21**) (0.1 g, 0.52 mmol) and substituted aromatic aldehydes (0.52 mmol) in the presence of ethanol (5 mL) and powdered KOH (1.04 mmol) produced (*E*)-3-(aryl)-1-(5-hydroxy-2*H*-chromen-6-yl)prop-2-en-1-ones (**22**). The microwave irradiation was done for 5–7 min. The produced chalcones (0.5 mmol) were then oxidatively cyclized using mercury (II) acetate (0.5 mmol), as the oxidative agent in pyridine (5 mL) under the microwave irradiation to provide the aurone derivatives in excellent yields.

All the synthesized compounds were screened for an antioxidant activity and an antimicrobial activity. The antioxidant activity was tested using 2,2-diphenyl-1-picrylhydrazyl and hydrogen peroxide radicals. The results demonstrated that some compounds exhibited a better radical scavenging ability and some showed a promising antimicrobial potency.

A new method for the synthesis of benzofuran-3(2H)-ones (**25**) under microwave conditions was reported by Hu et al [28]. The schematics of the reaction is shown in Scheme 8. The product, dihydrobenzofuranones, was obtained in 43%–58% yields.

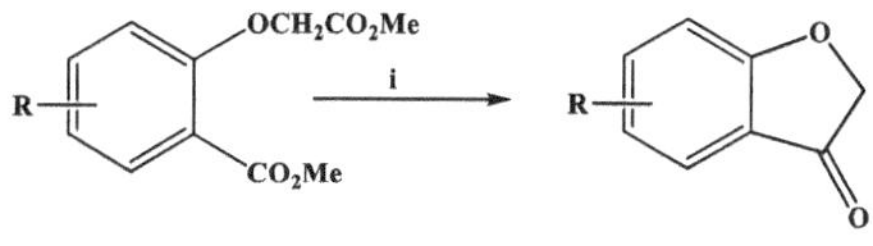

24 **25**

R= H, CH_3O, Alkyl,

SCHEME 8 Microwave-assisted synthesis of benzofuranones. Reagents and conditions: (i) CH_3OH/DMF, catalysts, MW, 30 min, 150°C.

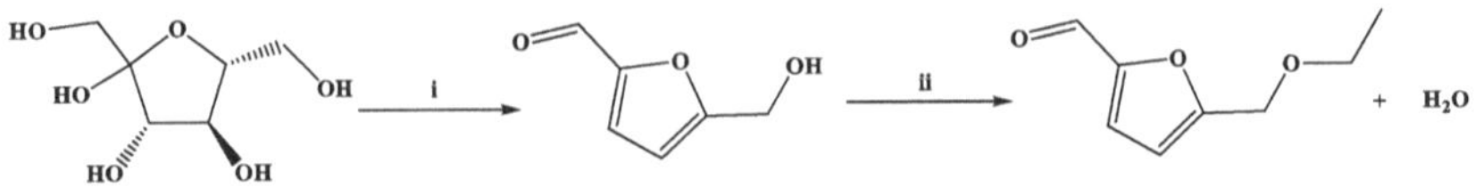

26 27 28

SCHEME 9 One-pot synthesis of 5-ethoxymethylfurfural from *D*-fructose. Reagents and conditions: (i), $-3H_2O$, ethanol, DESs, MW; (ii) ethanol, DESs, MW.

5-Ethoxymethylfurfural (EMF) (**28**) was synthesized in the one-pot from *D*-fructose using cheaper and environmentally friendly deep eutectic solvents (DESs) as catalysts at mild conditions under microwave conditions [29]. A variety of DESs was considered, and among them, the combination of choline chloride and oxalic acid gave the highest conversion of *D*-fructose (92%) with a yield of 5-ethoxymethylfurfural (74%) for 3 h at 343 K.

For the reaction, *D*-fructose (2×10^{-3} mol), ethanol (85×10^{-3} mol), and ChCl-oxalic acid (6×10^{-3} mol) were taken together and stirred at room temperature until a homogeneous solution was obtained. Then, the reaction was commenced by heating it in a microwave. The reaction of *D*-fructose (**26**) with ethanol leads to 5-hydroxymethyl furfural (HMF) (**27**), which further reacts with ethanol to produce 5-ethoxymethylfurfural (EMF) (**28**) as shown in Scheme 9.

4.2.3 Cinnamic acid and coumarin derivatives

Coumarin is also known as benzopyran-2-one, which is found naturally in some plants. Coumarin was first isolated by Vogel in 1820 [30, 31] from Tonka bean. In 1868, the English chemist William Henry Perkin first synthesized coumarin [32]. Later, the application of coumarin derivatives as bioactive molecules against different diseases gained a great interest from medicinal chemists for the synthesis of coumarin derivatives. Coumarin derivatives have manifested a wide spectrum of biological activities [33–38]. Fig. 1 shows the molecular structures of coumarin-containing drugs.

The one-pot microwave-assisted method was applied for the synthesis of cinnamic acid and coumarin derivatives [39]. This microwave-assisted synthesis started with an aldehyde synthon, and the reaction conditions were capable to produce cinnamic acid (**43**) or coumarin derivative (**49**).

The synthesis of cinnamic acid derivatives is shown in Scheme 10. During the synthetic process, the reaction between aldehyde (5.0 mmol) and ylide (5.5 mmol) was performed under solvent-free conditions in toluene (1.0 M, 5.0 mL). The reaction was accomplished at 150°C for 10 min. Six selected cinnamic acid derivatives are shown in Table 3.

Scheme 11 shows the microwave-assisted synthesis of coumarin. Aldehyde (5.0 mmol) and ylide (5.5 mmol) were placed into a microwave reaction vessel, and toluene (5.0 mL, 1.0 M) was added. Then, the reaction system was sealed

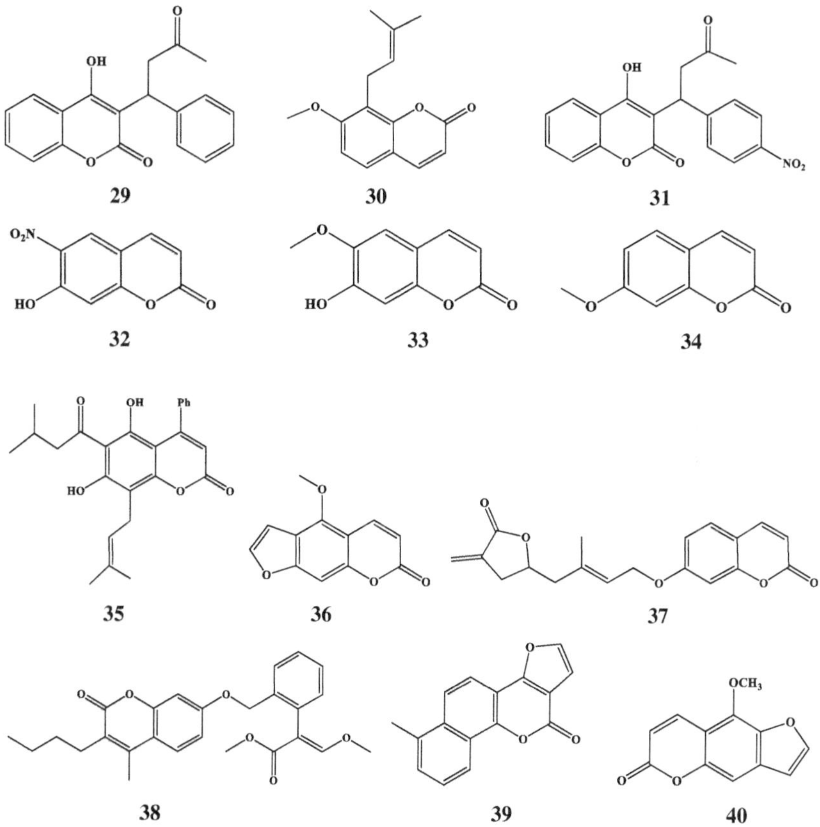

FIG. 1 Molecular structures of coumarin-containing drugs. Warfarin (**29**), osthole (**30**), acenocoumarol (**31**), Cou-NO_2 (**32**), scopoletin (**33**), herniarin (**34**), mammeisin (**35**), xanthotoxin (**36**), excavarin-A (**37**), coumoxystrobin (**38**), neotanshinlactone (**39**), bergapten (**40**).

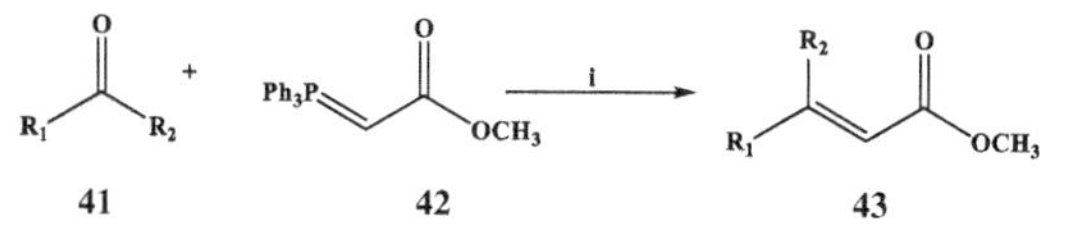

SCHEME 10 Microwave-assisted synthesis of cinnamic acid derivatives (**43**). Reagents and conditions: (i) Toluene (1 M), MW, 300 W, 150°C, 10 min.

with a silicone/polytetrafluoroethylene cap and placed in a microwave reactor, and the reaction was carried out at 150°C for 10 min. Four selected aldehyde (ketone) and corresponding coumarins are shown in Table 4.

The microwave-assisted synthesis of various 3-aryl-furo[3,2-c]coumarins was carried out by using two different methodologies [40]. In the first method, different 3-aryl-furo[3,2-c]coumarins (**49**) were synthesized by reacting various

TABLE 3 Six selected aldehyde (ketone) and corresponding cinnamic acid derivatives.

Aldehyde	Cinnamic acid derivatives	Aldehyde	Cinnamic acid derivatives
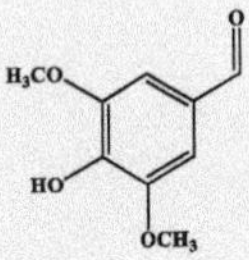 41a	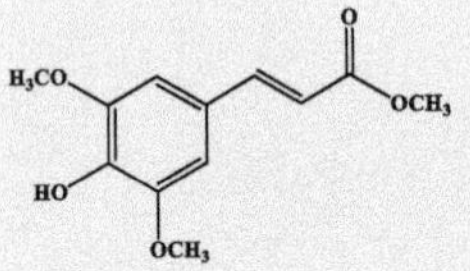 43a	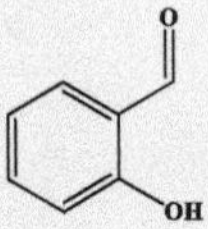 41b	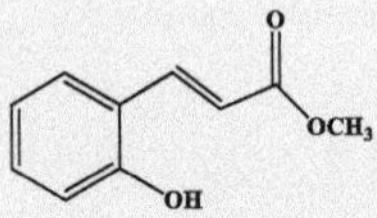 43b
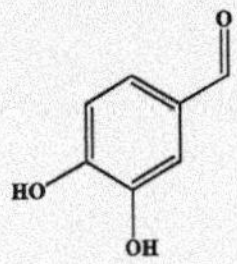 41c	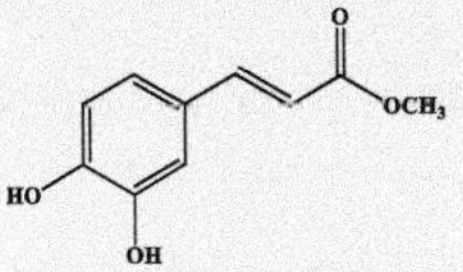 43c	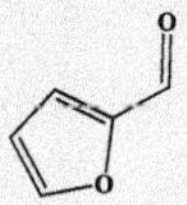 41d	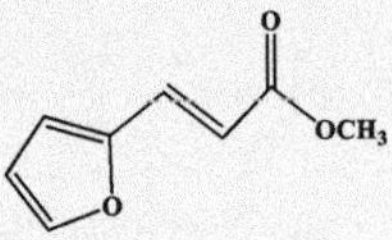 43d
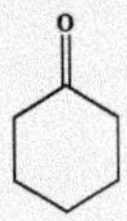 41e	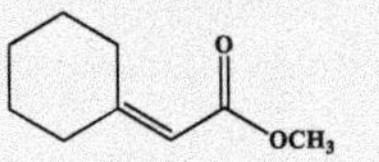 43e	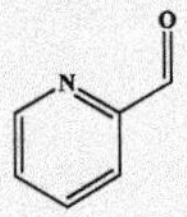 41f	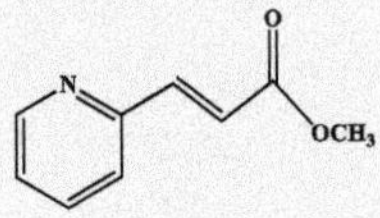 43f

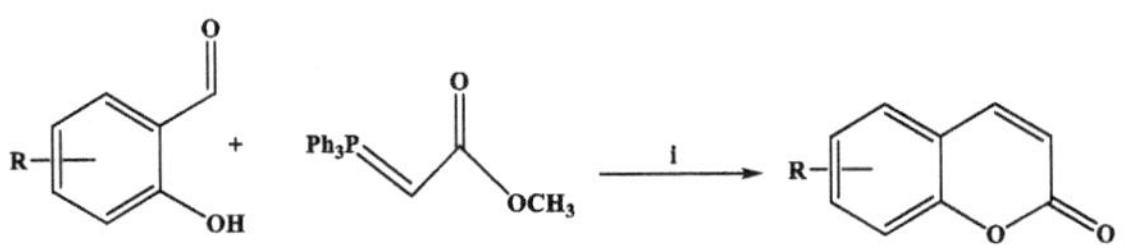

44 45 46

SCHEME 11 Microwave-assisted synthesis of coumarin. Reagents and conditions: (i) Toluene (1 M), MW, 300 W, 220°C, 60 min.

4-hydroxy coumarins (**47**) with appropriate 2-aryl-1-nitro ethenes (**48**) under Nef reaction condition. The reaction is shown in Scheme 12. During the synthesis, first, a solution of appropriate 4-hydroxy coumarin (0.002 mol) and 2-aryl-1-nitro-ethene (0.002 mol) in methanol (5 mL) containing piperidine in a catalytic amount was stirred at room temperature for 10 min, and then, the solution was irradiated for 5–7 min in the microwave at 240 W power.

In the second method, different 3-aryl-furo[3,2-c]coumarins (**52**) were synthesized by reacting various 4-hydroxy coumarins (**50**) with appropriate aroylmethyl bromides (**51**) under Feist-Benary reaction condition. The reaction is shown in Scheme 13. During this process, a solution of appropriate 4-hydroxy coumarin

TABLE 4 Four selected aldehyde (ketone) and corresponding coumarins.

Aldehyde	Coumarins	Aldehyde	Coumarins
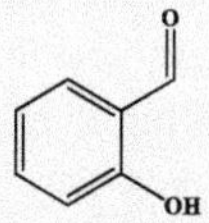 44a	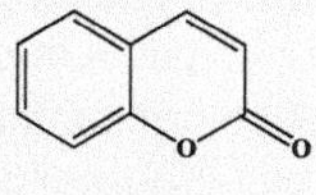 46a	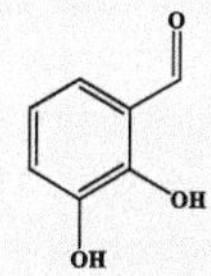 44b	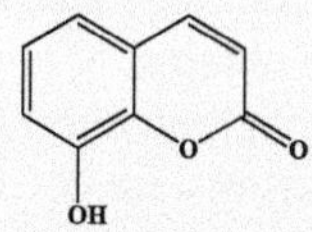 46b
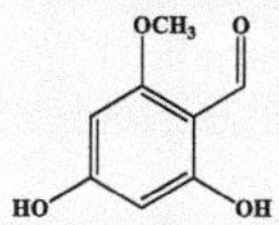 44c	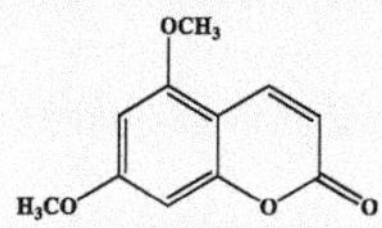 46c	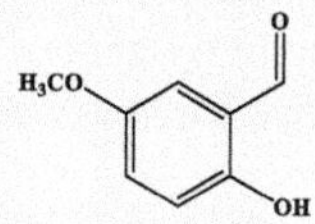 44d	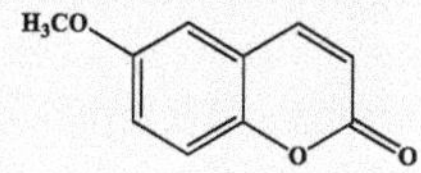 46d

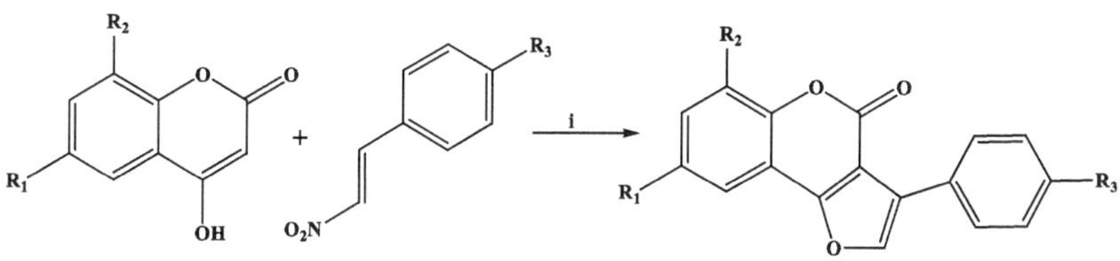

47 48 49

R_1=H,CH_3, Cl; R_2=H, CH_3; R_3=H, CH_3, OCH_3

SCHEME 12 Synthesis of 3-aryl-furo[3,2-c]coumarins (method 1). Reagents and conditions: (i) Piperidine, methanol, MW, 5–7 min, 240 W.

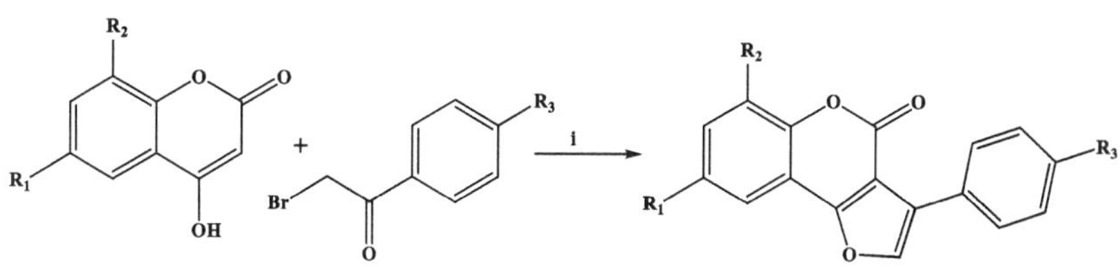

50 51 52

R_1=H,CH_3, Cl; R_2=H, CH_3; R_3=H, CH_3, OCH_3

SCHEME 13 Synthesis of 3-aryl-furo[3,2-c]coumarins (method 2). Reagents and conditions: (i) NH_4OAc, AcOH, MW, 2–3 min, 240 W.

(0.002 mol) and aroylmethyl bromide (0.002 mol) in acetic acid (4 mL) and ammonium acetate (0.01 mol) was stirred at the room temperature for 10 min and then irradiated for 2–3 min in the microwave at the power of 240 W.

All the synthesized compounds were screened for their antimicrobial activities against Gram-negative bacteria (*E. coli*) and Gram-positive bacteria (*Bacillus subtilis*). The compounds were also screened for their antifungal activity against fungi *Candida albicans*. The agar cup diffusion method was used for the evaluation of the antimicrobial activity. The activity data revealed that all the compounds exhibited a promising antibacterial activity against Gram-negative bacteria *E. coli* compared with streptomycin, and none of the compounds showed a better activity against Gram-positive bacterial strain *Bacillus subtilis* and fungi *Candida albicans* compared with standard drugs streptomycin and clotrimazole.

Osthole is a naturally occurring coumarin derivative, which is chemically known as an O-methylated coumarin. Osthole is found in many plants and possesses a wide spectrum of pharmacological and biochemical activities such as antiarrhythmia, antifungal, antidiabetic, anticancer, antiinflammatory, antiosteoporosis, hepatoprotection, and neuroprotection [41–49].

Microwave-assisted synthesis of a series of novel furo[3,2-c]coumarins as fused osthole derivatives, via the reaction of 4-hydroxycoumarins and b-ketoesters catalyzed by DMAP, was reported [50]. The synthetic paths are shown in Schemes 14–17. The compounds were prepared in moderate yields with toluene

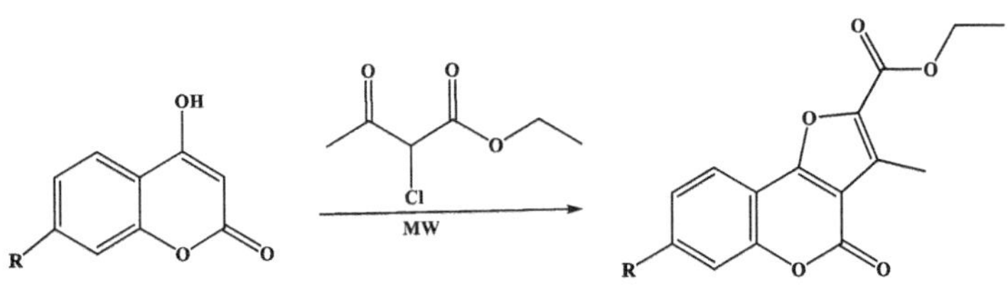

53 54

SCHEME 14 Microwave-assisted synthetic procedure for the preparation of osthole derivatives (**54**). Reagents and conditions: (i) Toluene, MW, 640 W.

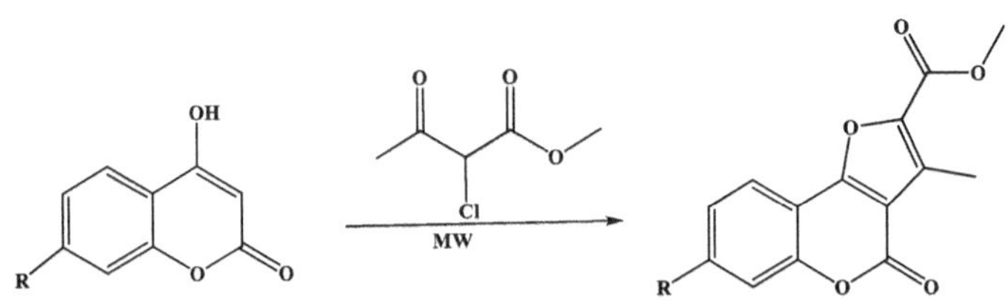

55 56

SCHEME 15 Microwave-assisted synthetic procedure for the preparation of osthole derivatives (**56**). Reagents and conditions: (i) Toluene, MW, 640 W.

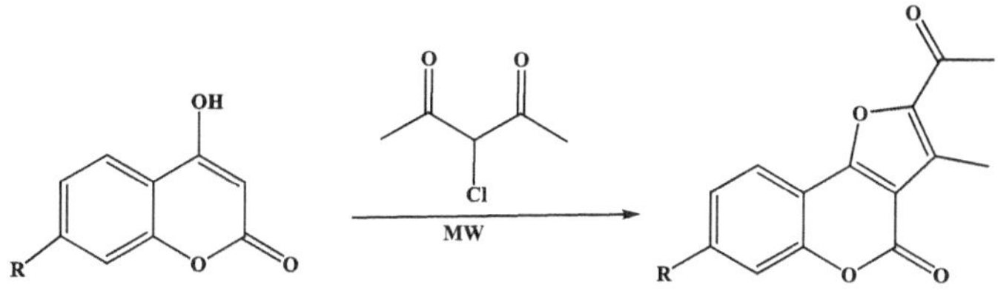

SCHEME 16 Microwave-assisted synthetic procedure for the preparation of osthole derivatives (**58**). Reagents and conditions: (i) Toluene, MW, 640 W.

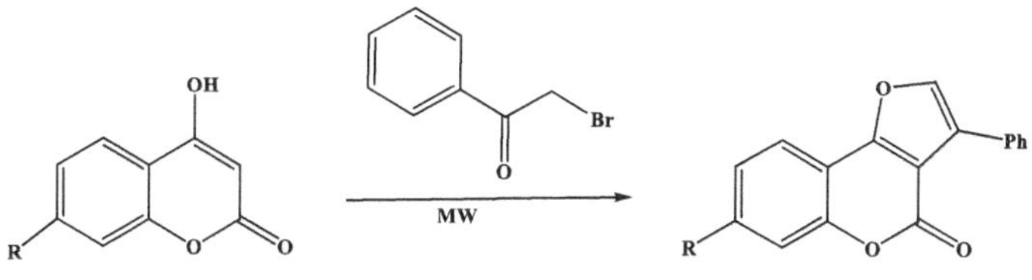

SCHEME 17 Microwave-assisted synthetic procedure for the preparation of osthole derivatives (**60**). Reagents and conditions: (i) Toluene, MW, 640 W.

as the solvent, using 640-W microwave irradiation, and DMAP as the catalyst. Depending upon the irradiative time, the yields varied from 61% to 77%. β-Ketoesters such as 3-chloropentane-2,4-dione, ethyl 2-chloroacetoacetate, methyl 2-chloro-3-oxobutanoate, and 2-bromoacetophenone were employed to synthesize the variety of furocoumarin derivatives.

All the target compounds were evaluated in vitro for their antifungal activity against six phytopathogenic fungi (*Botrytis cinerea*, *Colletotrichum capsici*, *Alternaria solani*, *Gibberella zeae*, *Cucumber anthrax*, and *Rhizoctonia solani*). Some compounds exhibited a potential activity in the primary assays.

A series of fused coumarin analogues, pyrano[3,2-c]chromene-2,5-diones (**63** and **66**), were also synthesized under microwave conditions [51].

The synthetic protocol is shown in Scheme 18. The optimal conditions were obtained using toluene as the solvent in the presence of DMAP under microwave irradiation (640 W) for 15 min. Compared with conventional heating, the reaction times were reduced from about 6 h to 15 min, and the isolated yields were improved from 30% to 76%.

All synthesized compounds were screened for their antifungal activity against *Botrytis cinerea*, *Alternaria solani*, *Colletotrichum capsici*, *Rhizoctonia solani*, and *Gibberella zeae*. The results depicted that some of the coumarin derivatives exhibited potent antifungal activities even at a concentration of less than 50 ppm.

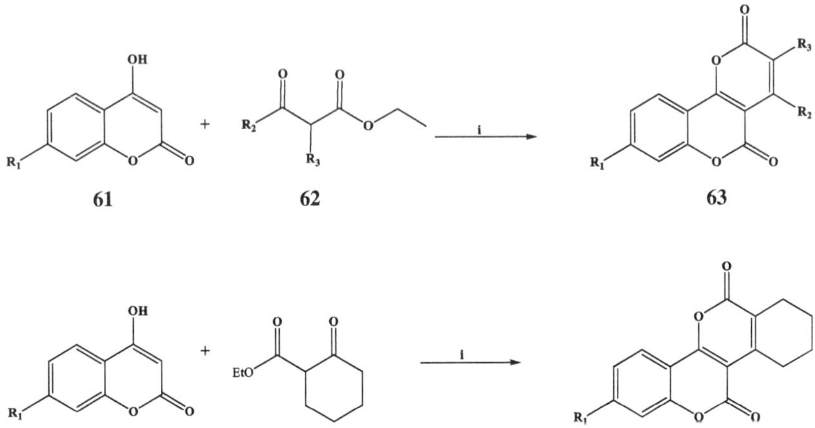

61 62 63

64 65 66

R1=H, OMe, OCH2CHCH2, O(CH2)3 CH3; R2=Me, CF3; R3=H, Me, Et, F

SCHEME 18 Microwave-assisted synthetic route for pyrano[3,2-c]chromene-2,5-diones. Reagents and conditions: (i) Toluene, DMAP, MW, 640 W, 15 min.

A series of hydroxycoumarin derivatives were also synthesized by microwave-induced reactions [52].

The compounds such as 4-(cyanomethoxy) chromen-2-one, 5-(cyanomethoxy)-4,7-dimethylchromen-2-one, and 6-acetyl-5-(cyanomethoxy)-4, 7-dimethylchromen-2-one were synthesized from 4-hydroxycoumarin, 5-hydroxy-4,7-dimethylcoumarin, and 6-acetyl-5-hydroxy-4,7-dimethylcoumarin, respectively, as shown in Scheme 19. The reaction of coumarins (**67**) with the alkylating agent chloroacetonitrile in acetone, using anhydrous potassium carbonate as a base, resulted in three novel O-substituted cyanomethoxy coumarins (**68**). Heating the 1-methyl-2-pyrrolidone solutions of the same substrates with the alkylating agent in the presence of anhydrous potassium carbonate at the temperature 130–140°C yielded the same products. The reaction using conventional methods took 16, 18, and 24 h to obtain the product in acetone, and for 6, 10, and 12 h in 1-methyl-2-pyrrolidone at 130–140°C, respectively. The *O*-alkylation reaction

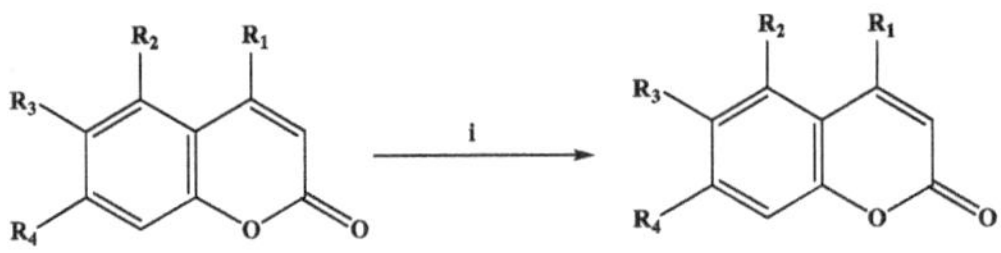

67 68

R1=OH, CH3, OCH2CN; R2= H, OH, OCH2CN, R3= H, OH, COCH3, OCH2CN; R4= H, CH3.

SCHEME 19 Microwave-assisted synthetic route for hydroxycoumarin derivatives. Reagents and conditions: (i) $ClCH_2CN$, K_2CO_3 (anh.), 1-methyl-2-pyrrolidone, MW, 12–15 min.

under microwave irradiation with the alkylating agent chloroacetonitrile afforded the products within a few minutes. For the microwave synthesis, a mixture of appropriate coumarin (7.5 mmol), chloroacetonitrile (8.25 mmol), and K_2CO_3 (22.5 mmol) were placed in a microwave flask, and then acetone or 1-methyl-2-pyrrolidone was added. The mixture was heated at the temperature of 130–140°C in the monomode microwave oven (300 W). Four cycles were needed to obtain 4-(cyanomethoxy)chromen-2-one, 5-(cyanomethoxy)-4,7-dimethylchromen-2-one, and 6-acetyl-5-(cyanomethoxy)-4,7-dimethylchromen-2-one, and the heating time for each cycle was 3 min.

Chromene and coumarin scaffolds are biologically important compounds. The hybrid compounds containing both coumarin and chromene moieties, called pyranocoumarins, may exhibit a better biological activity. Few examples of pyranocoumarin derivatives, which have shown an anti-HIV activity, are depicted in Fig. 2.

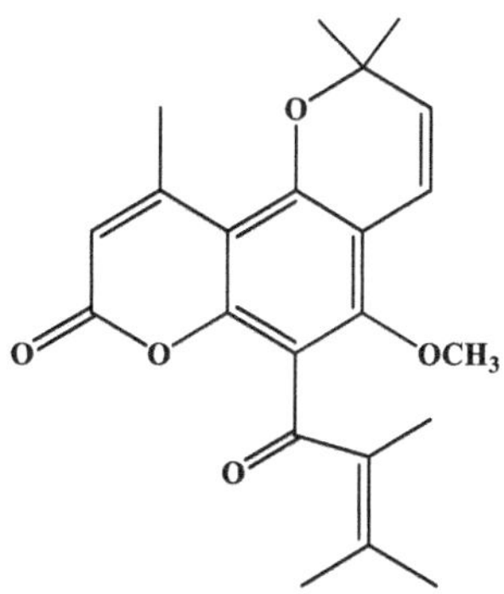

69

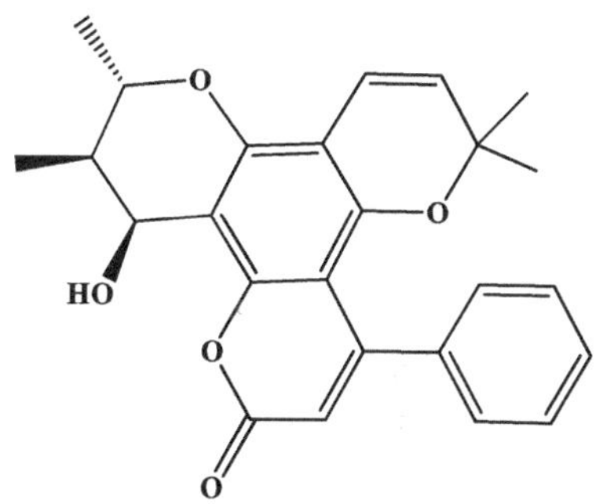

70

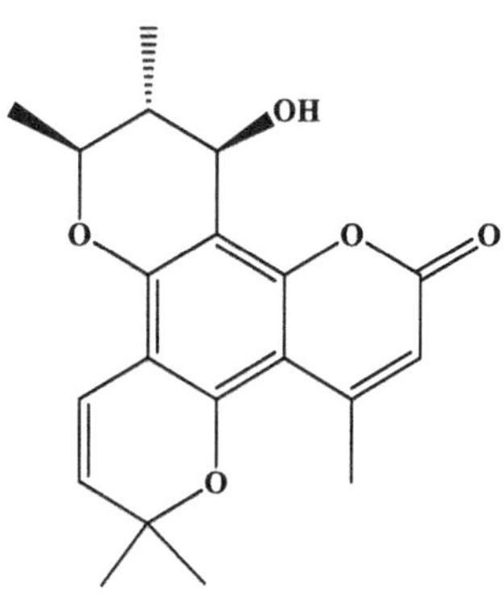

71

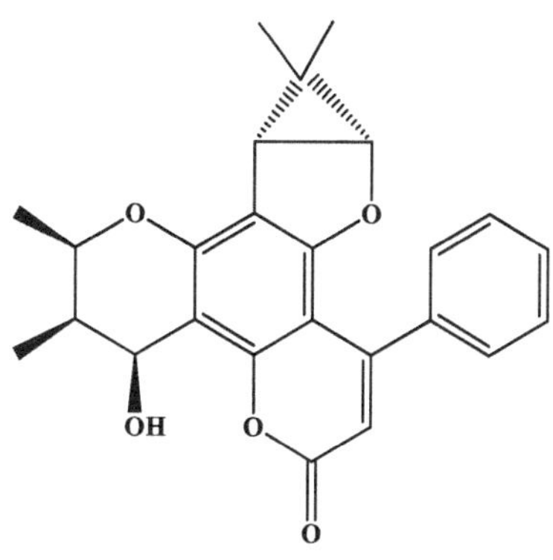

72

FIG. 2 Examples of pyranocoumarin derivatives. Oblongulide (**69**), soulattrolide (**70**), cordatolide A (**71**), inophyllum G-1 (**72**).

Ashok et al., reported the synthesis of hybrid compounds, substituted 4-chloro-8-methyl-2-phenyl-1,5-dioxa-2H-phenanthren-6-ones (**75**), from substituted (E)-1-(7-hydroxy-4-methyl-8-coumarinyl)-3-phenyl-2-propen-1-ones, in a good yield using the microwave-assisted Vilsmeier-Haack reaction [53].

The synthetic route is shown in Scheme 20. The condensation of 8-acetyl-7-hydroxy-4-methyl coumarin with aromatic or hetero aromatic aldehydes in the presence of piperidine under the microwave irradiation produced the substituted (*E*)-1-(7-hydroxy-4-methyl-8-coumarinyl)-3-phenyl-2-propen-1-ones in excellent yields. Afterward, these chalcones on reaction with Vilsmeier-Haack reagent (DMF/$POCl_3$) yielded the desired compounds, the substituted 4-chloro-8-methyl-2-phenyl-1,5-dioxa-2*H*-phenanthren-6-ones. The optimum results were obtained by the microwave irradiation at 160 W for 4–5 min. All the synthesized compounds were tested in vitro for their antimicrobial activity, and some compounds were found to be potent against the tested fungal and bacterial strains.

Using the microwave-assisted palladium-catalyzed Heck coupling reaction, a series of flavonoid derivatives were synthesized from 6- and 8-iodo-3,7,3′4′-tetramethoxy-quercetin with terminal alkenes [54]. Alkenes such as *n*-butyl acrylate, 2-methyl-3-buten-2-ol, acrylonitrile, and styrene were considered for the reaction, and the substituted flavonoid compounds were obtained with moderate to good yields. The reaction is shown in Scheme 21.

Synthesis of 2-acyl-5,6-dihydro-1,4-dioxines (**81**) through the ring expansion of oxiranes and diazo-dicarbonyl compounds is shown in Scheme 22. For the synthesis, 1.5 equivalents of diazo-dicarbonyl compounds reacted with one equivalent of oxiranes under the catalysis of 10 mol% $Cu(hfacac)_2{\cdot}xH_2O$ at 150°C in DCE with the assistance of microwave irradiation for 20 min. Four selected synthesized compounds are shown in Table 5.

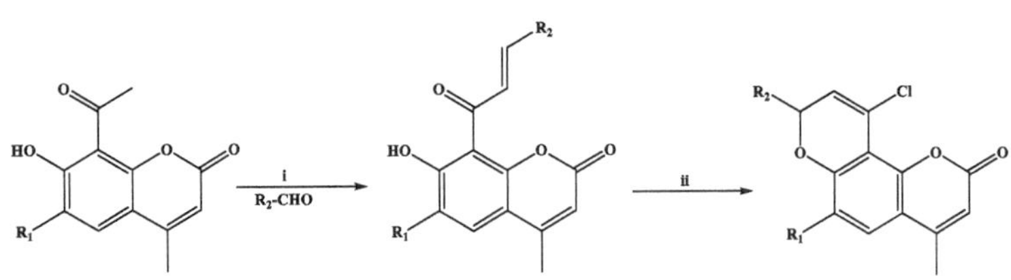

73 **74** **75**

R_1 = H,Cl; R_2 = HPhenyl; Phenyl; 4-Methoxyphenyl; 4-Methoxyphenyl; 3,4-Dimethoxyphenyl; 3,4-Dimethoxyphenyl; 4-Fluorophenyl; 4-Fluorophenyl; 4-Methylphenyl; 4-Methylphenyl; 2-Chlorophenyl; 2-Chlorophenyl; 1-Naphthyl; 1-Naphthyl; 2-Thienyl

SCHEME 20 Microwave-assisted synthesis of substituted 4-chloro-8-methyl-2-phenyl-1,5-dioxa-2Hphenanthren-6-ones. Reagents and conditions: (i) Piperidine, MW (ii) DMF, $POCl_3$, MW, 160 W, 4–5min.

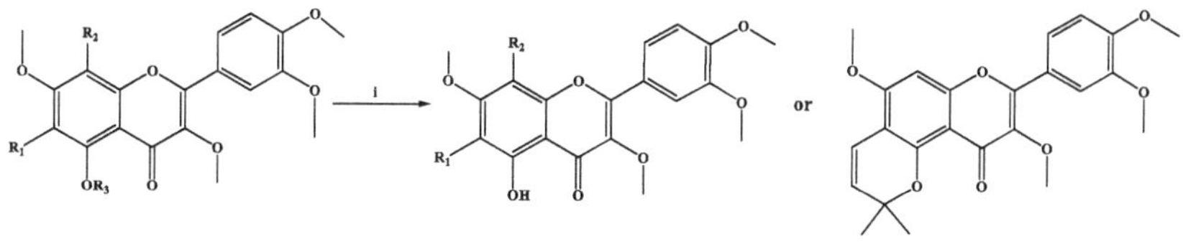

76 **77** **78**

R_1=I, H, $CHCO_2nBu$, CH_2, CHCN, CHPh; R_2=I, H, $CHCO_2nBu$, CH_2, CHCN, CHPh;

R_3=Ac

SCHEME 21 Microwave-assisted synthesis of 3,7,3′,4′-tetramethoxy-quercetin derivatives. Reagents and conditions: (i) alkene, base, $Pd(OAc)_2$, $P(o\text{-}tol)_3$, DMF, MW, 150 W.

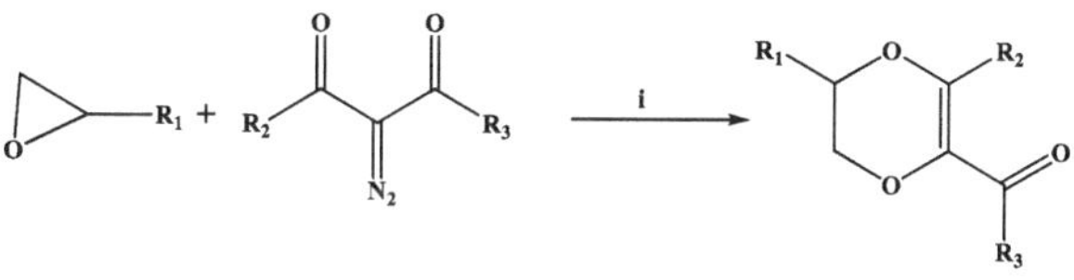

79 **80** **81**

SCHEME 22 Synthesis of 2-acyl-5,6-dihydro-1,4-dioxines from oxiranes and diazo-dicarbonyl compounds. Reagents and conditions: (i) $Cu(hfacac)_2{\cdot}xH_2O$, DCE, MW, 20 min, 150°C.

TABLE 5 Four selected 2-acyl-5,6-dihydro-1,4-dioxines.

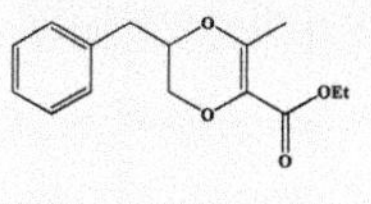

81a

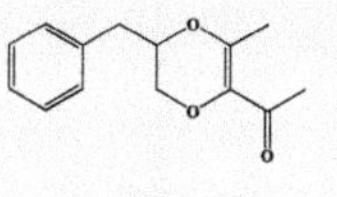

81b

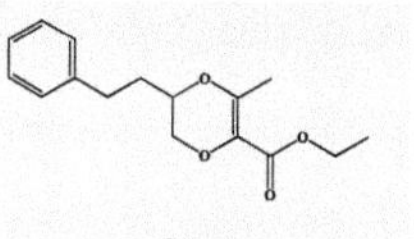

81c

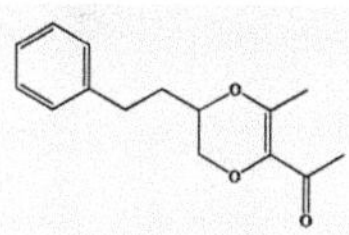

81d

The synthesized derivatives were tested for the in vitro urease inhibition profile and observed to exhibit an excellent urease inhibition. All the compounds were also evaluated for the cytotoxic activity, and one compound showed an activity against lung carcinoma (H157) with 66.32% inhibition.

4.3 Organosulfur compounds

Sulfur-containing molecules are biologically important compounds [55–57]. A wide variety of sulfur-containing scaffolds are present in natural products as well as in drugs. For example, important drugs such as penicillin (**82**), discorhabdin A (**83**), and disulfiram (**84**) contain sulfur scaffolds (Fig. 3).

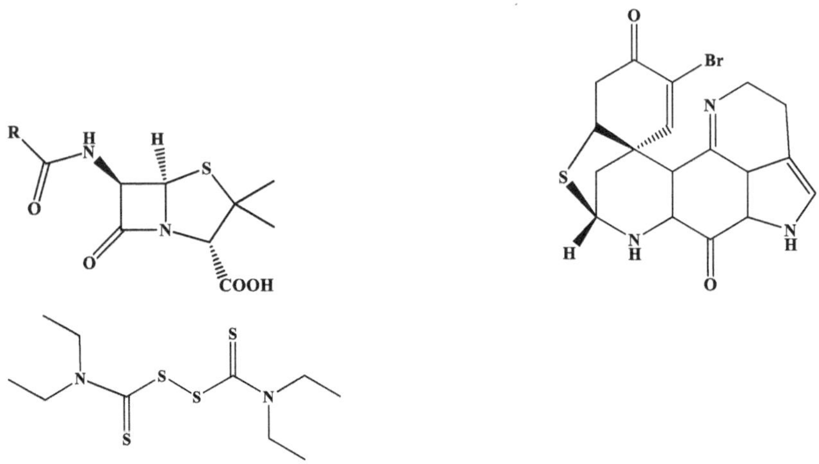

82 **83** **84**

FIG. 3 Penicillin, discorhabdin A, and disulfiram.

Microwave-assisted efficient and facile solution-phase parallel synthesis for a 38-member library of *N*-aroyl-*N*′-aryl thioureas (**87**) was realized successfully [58]. It was noticed during the synthesis that the reaction time is drastically reduced from 8 to 12 h for conventional methods to only 10–15 min in the microwave method.

The synthesis of *N*-aroyl-*N*′-aryl thioureas is shown in Scheme 23. Stoichiometric amounts of 2-cholorobenzoyl isothiocyanate (synthesized by the conventional manner and purified by steam distillation) and the respective

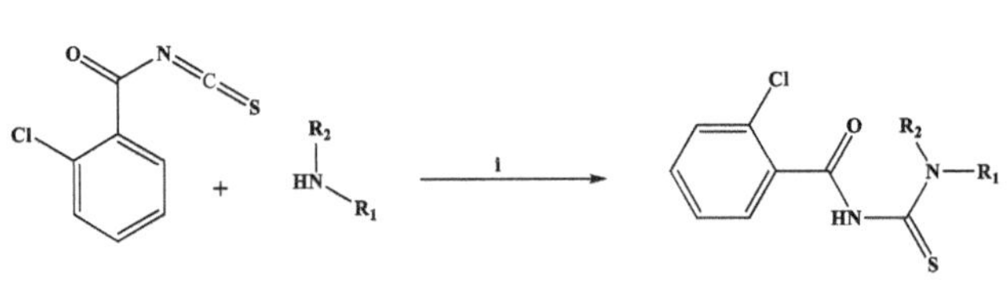

85 **86** **87**

5 selected $\mathbf{R_1}$ substituents:

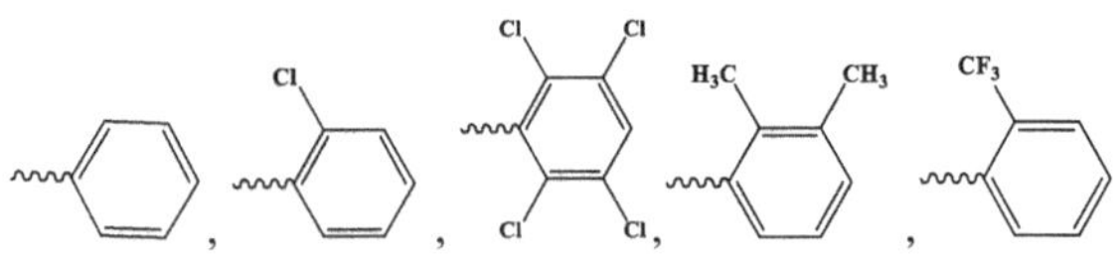

$\mathbf{R_2}$= H, CH_3,

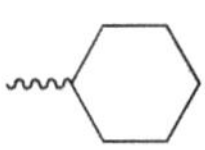

SCHEME 23 Synthesis of *N*-aroyl-*N*′-aryl thioureas. Reagents and conditions: (i) THF, MW, 10 min, 60–65°C, 1 atm.

primary or secondary amines were treated in a glass vials along with THF as the solvent and then irradiated the vials under optimal conditions (10 min, 60–65°C and 1 atm pressure). The yields of the final products obtained through the microwave-assisted reactions were strongly influenced by the nature of the amine substrate.

Microwave-assisted copper-catalyzed ring expansions of the three-membered heterocycles with a-diazo-b-dicarbonyl compounds were reported [59]. This study provided a direct and simple strategy in an efficient method for the synthesis of medicinally and agriculturally important compounds such as 3-acyl-5,6-dihydro-1,4-oxathiines and 2-acyl-5,6-dihydro-1,4-dioxines from readily available thiiranes and oxiranes, respectively.

Synthesis of 3-acyl-5,6-dihydro-1,4-oxathiines (**90**) through the ring expansion of thiiranes (**88**) and diazo-dicarbonyl compounds (**89**) is shown in Scheme 24. Thiiranes (0.6 mmol) and diazo compounds (0.3 mmol) were added to DCE (1.0 mL) in a 10-mL microwave tube. After that, $CuSO_4{\cdot}5H_2O$ (0.6 mmol) was added, and the reaction mixture was stirred at 100°C for 20 mins under the microwave irradiation in a sealed vessel and provided the desired products in good yields. Four selected synthesized compounds are shown in Table 6.

Recently, a series of thiophene hydrazones, namely, acetyl thiophene benzohydrazide (ATBH), acetyl thiophene amino benzohydrazide (ATABH), propanoyl thiophene benzohydrazide (PTBH), and propanoyl thiophene amino benzohydrazide (PTABH), were synthesized by the microwave-assisted

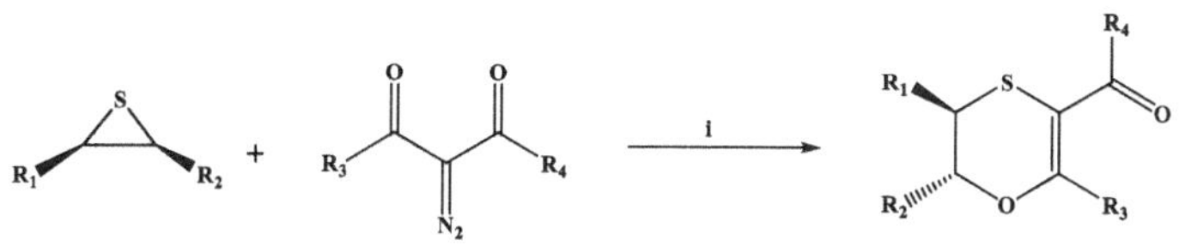

88 **89** **90**

SCHEME 24 Synthesis of 3-acyl-5,6-dihydro-1,4-oxathiines through the ring expansion of thiiranes with a-diazo-b-dicarbonyl compounds. Reagents and conditions: (i) 2 equiv. $CuSO_4{\cdot}5H_2O$, DCE, MW, 20 min, 100°C.

TABLE 6 Four selected 3-acyl-5,6-dihydro-1,4-oxathiines.

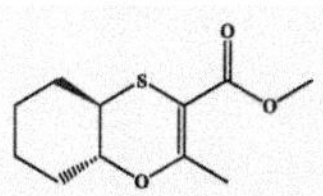

90a

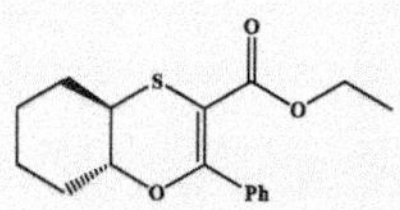

90b

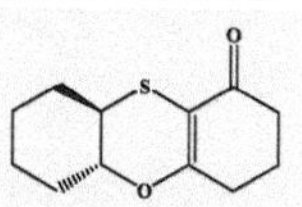

90c

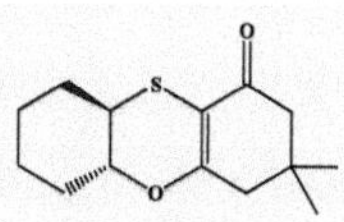

90d

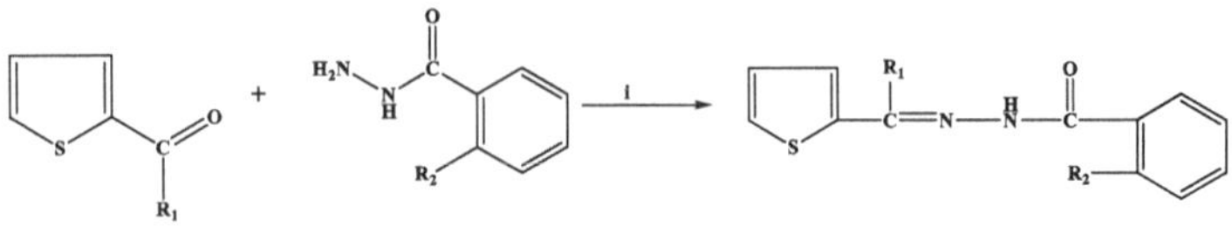

91 92 93

For ATBH: R_1=CH_3, R_2=H; For ATABH: R_1=CH_2CH_3, R_2=H; For PTBH: R_1=CH_3, R_2=NH_2; For PTABH: R_1=CH_2CH_3, R_2=NH_2.

SCHEME 25 Microwave-assisted synthesis of thiophene hydrazones. Reagents and conditions: (i) Acid clay, MW, 3–5 min.

methods [60]. Schematic representation of the synthesis of thiophene hydrazones (**93**) is given in Scheme 25.

The synthesis of the thiophene hydrazones was achieved by exposure of microwave irradiation to a mixture of reactants in the presence of acid clay. After 3–5 min exposure to microwave irradiation, the product was separated in reasonably good yield (90%). The corrosion inhibition potential of the synthesized compounds was investigated by different methods. The compounds were found to have an excellent potential to mitigate the corrosion of mild steel in different acid media.

Because of the potential bioactivity of sulfur compounds and coumarin derivatives, several chemists made attempts to synthesize sulfanyl-substituted chromenones. Anjaiah et al. demonstrated the microwave-assisted synthesis of 3-(arylsulfanyl)-4-hydroxy-2*H*-chromen-2-ones (**96**) by condensation of arenesulfonohydrazides (**94**) with 4-hydroxy-2H-chromen-2-one (**95**) in the presence of iodine [61]. The synthetic route is shown in Scheme 26.

For the synthesis, dimethyl sulfoxide (5 mL) was added to a mixture of arenesulfonohydrazide (10 mmol) and 4-hydroxy-2*H*-chromen-2-one. After that, a catalytic amount of iodine was added to the mixture. The reaction mixture was then irradiated with microwaves at 160 W for 10 min.

The synthesized compounds were screened for in vitro antimicrobial activity against bacterial (*E. coli* and *S. aureus*) and fungal (*Aspergillus niger* and *Candida metapsilosis*) organisms.

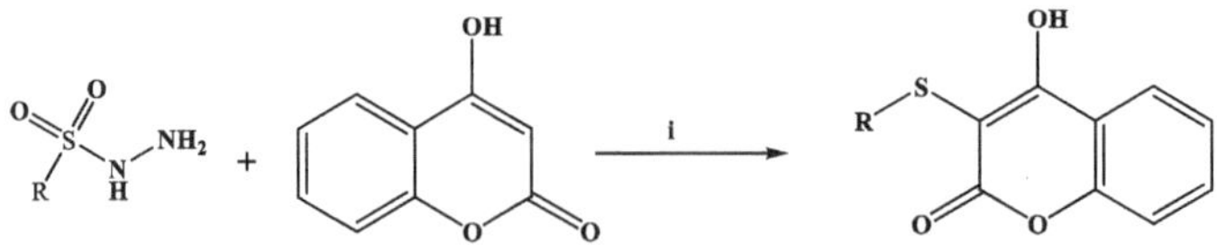

94 95 96

R=Ph, 4-MeC_6H_4, 4-MeOC_6H_4, 4-BrC_6H_4, 4-ClC_6H_4, 4-H_2NC_6H_4, 4-HOC_6H_4, Naphthalen-2-yl

SCHEME 26 Microwave-assisted synthesis of 3-(arylsulfanyl)-4-hydroxy-2*H*-chromen-2-ones. Reagents and conditions: (i) DMSO, iodine, MW, 160 W, 10 min.

The introduction of a sulfur atom(s) in biologically active compounds could enhance their properties. A series of 1-(arylthio)naphthalen-2-ols (**99**) were synthesized by condensation of naphthalen-2-ol (**97**) with 1,2-diaryldisulfanes (**98**) in the presence of iodine under the microwave irradiation method [62].

For the synthesis, a mixture of naphthalen-2-ol with 1,2-diaryldisulfides was added to DMSO followed by iodine, and the reaction mixture was subjected to a microwave irradiation at 160 W for 5–6 min (Scheme 27). All synthesized compounds were tested for in vitro antibacterial activities against *E. coli* and *S. aureus* and antifungal activities against *Aspergillus niger* and *Candida metapsilosis*.

An efficient and simple microwave-assisted protocol for the preparation of hetaryl and ferrocenyl functionalized thioketones (**101**) via the treatment of the corresponding ketones (**100**) with Lawesson's reagent was demonstrated [63].

Thioketones were prepared via oxygen/sulfur exchange starting with the corresponding ketones by treatment in the toluene solution with Lawesson's reagent (L.R.) as shown in Scheme 28. The reaction mixtures were irradiated using the microwave for 2 min, at a power of 150 W. Table 7 shows the six selected thioketones.

Microwave-assisted one-pot synthesis of 2-nitroalkylidene-1,3-oxathiolane derivatives was reported by Kermani [64].

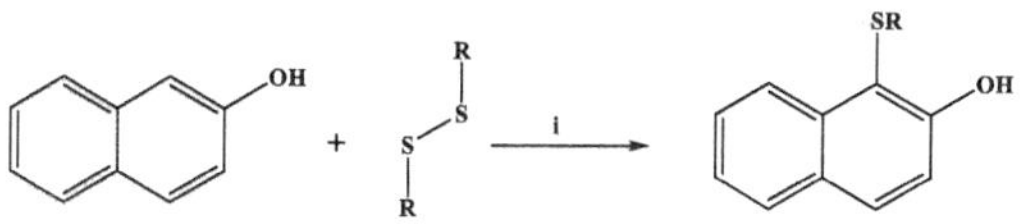

97 **98** **99**

R=Phenyl, 4-Methyl phenyl, 4-Methoxy phenyl, 4-Bromo phenyl, 4-Chloro phenyl, 2-Methyl phenyl, Naphthyl

SCHEME 27 Synthesis of 1-(arylthio)naphthalen-2-ols. Reagents and conditions: (i) I_2, DMSO, MW, 160 W, 5–6 min.

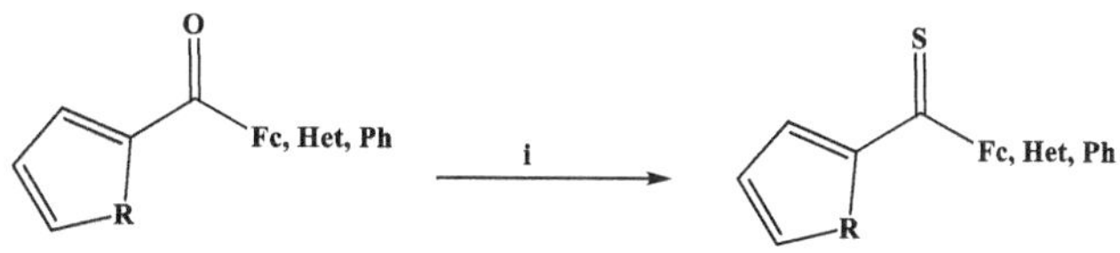

100 **101**

R=O, S, Se, CH=CH

SCHEME 28 Microwave-assisted synthesis of aryl, hetaryl, and ferrocenyl thioketones. Reagents and conditions: (i) L.R., MW.

TABLE 7 Six selected synthesized thioketones.

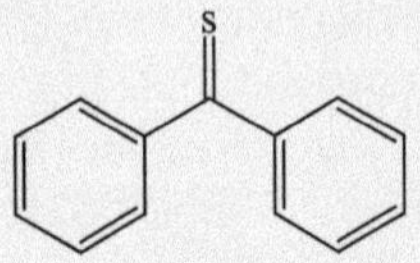 101a	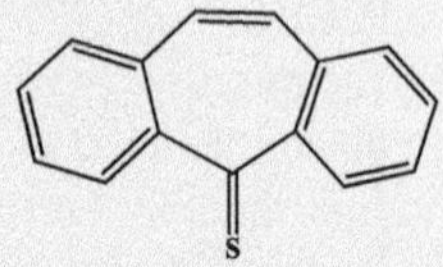 101b	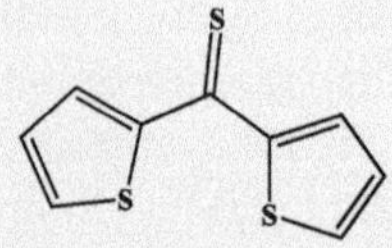 101c
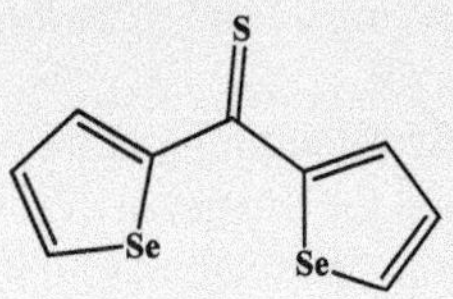 101d	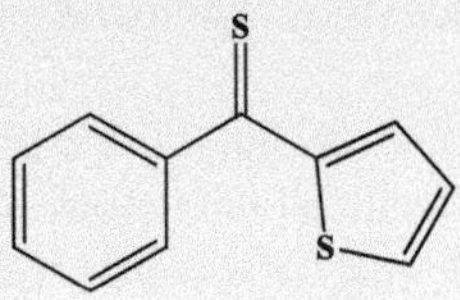 101e	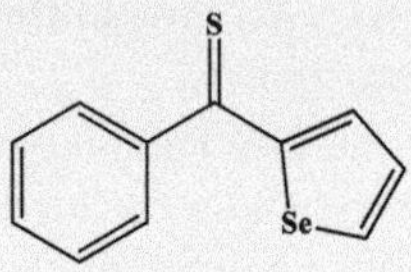 101f

For the synthesis of 2-(nitromethylidene)-1,3-oxathiolane (**105**), a mixture of nitromethane (1.0 mmol, 0.061 g), CS_2 (2.0 mmol, 0.162 g), oxirane (1.0 mmol), and Et_3N (20 mol%, 0.02 g) were treated under the solvent-free condition in a microwave with a power of 800 W at 100°C for 40 min (Scheme 29).

Under the microwave irradiation, differently substituted hetaryl thioketones reacted with less-reactive diazoketones in the toluene solution [65]. The reactions were complete after 2 min, and depending on the type of the used diazoketone, the products α, β-unsaturated ketones, acyl substituted thiiranes, or 1,3-oxathioles were obtained. When azibenzil and di(thiophen-2-yl) thioketone were considered, a new type of 1,5-dipolar electrocyclization of the intermediate thiocarbonyl ylide involving a thiophene ring led to a fused sulfur heterocycle. The reaction of phenyl

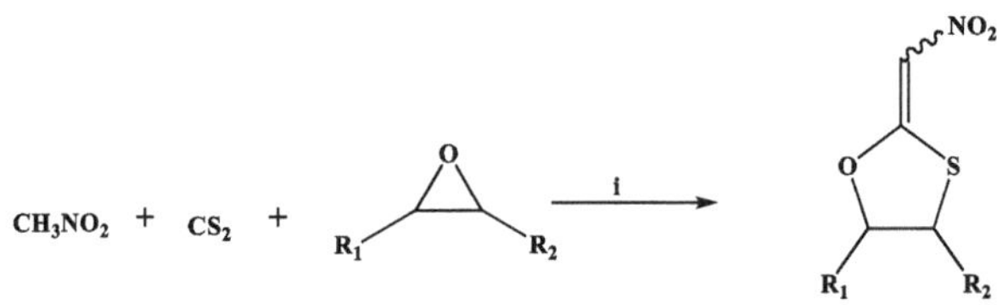

102 103 104 105

R1 =Me, ClCH2, Et, Ph, PhOCH2, CH2=CHCH2OCH2, Me2CHOCH2, CH2=C(Me)CO2CH2, –(CH2)4–, Ph, –(CH2)3–, –(CH2)6–; R2 =H, –(CH2)4– , Ph (cis), Ph (trans), –(CH2)3– , –(CH2)6–, 4-Nitrophenyl.

SCHEME 29 Synthesis of 1,3-oxathiane derivatives under MW irradiation. Reagents and conditions: (i) Et_3N, MW, 100°C, 40 min–1 h.

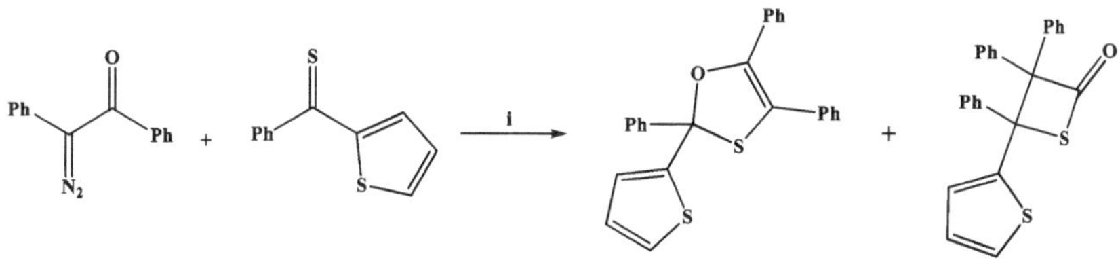

106 107 108 109

SCHEME 30 Reaction of hetaryl thioketone with azibenzil under MW irradiation. Reagents and conditions: (i) Toluene, MW, 150°C, 2 min.

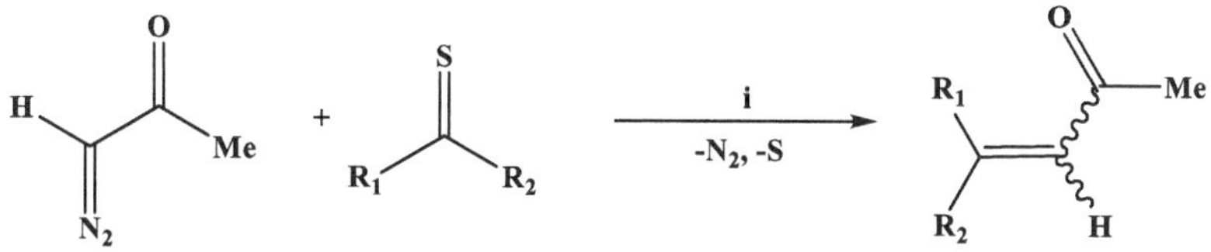

110 111 112

R_1=4-ClC$_6$H$_4$, Ph, Thi, Sel; R_2=Sel, Fc

SCHEME 31 Microwave-assisted synthesis of α,β-unsaturated ketones from the reactions of diazopropanone with hetaryl and ferrocenyl thioketones (Fc = ferrocenyl, Sel = selenophen-2-yl, Thi = thiophen-2-yl). Reagents and conditions: (i) Toluene, MW or THF, $LiClO_4$, reflux.

(thiophen-2-yl) thioketone with 2-diazo-1,2-diphenylethanone (azibenzil), in the presence of toluene under the microwave irradiation (200 W, 150°C), produced two products, 2,4,5-triphenyl-2-(thiophen-2-yl)-1,3-oxathiole and 3,3,4-triphenyl-4-(thiophen-2-yl)thietan-2-one (Scheme 30).

Studies also showed that ferrocenyl analogues decompose under the microwave irradiation. They reacted with diazopropanone and 2-diazo-1-phenylethanone in boiling THF in the presence of $LiClO_4$ to give α,β-unsaturated ketones (**112**) as sole products. The reaction of diazopropanone (**110**) with thioketones (**111**) is shown in Scheme 31.

4.4 Other heterocyclic compounds

4.4.1 Benzoselenophenes

Even though the benzoselenophene system has not been found in natural compounds so far, it is considered to be a bioisoster of benzofuran, naphthalene, benzothiophene, and indole [66]. Benzoselenophenes have attracted increasing attention due to their wide range of applications in both materials science and medicinal chemistry [67–70].

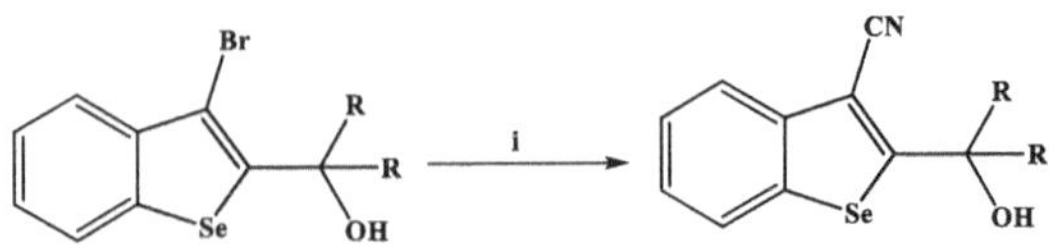

113 **114**

R=Me, H; for third compound R+R=$(CH_2)_5$

SCHEME 32 Microwave-assisted cyanation of 3-bromo-2-(1-hydroxyalkyl)benzo[b]selenophene. Reagents and conditions: (i) $Zn(CN)_2$ (1.2 equiv.), $Pd(PPh_3)_4$ (10 mol%), DMF, MW, 140°C, Ar, 3 min.

Arsenyan et al. reported the microwave-assisted palladium-catalyzed cyanation of 3-bromo-2-(1-hydroxyalkyl)benzo[b]selenophene derivatives [71]. Their study was focused on the introduction of the cyano group in the 3-position of 2-hydroxymethyl-3-bromobenzo[b]selenophenes. The substrate was activated by employing the microwave irradiation and Pd(PPh3)4 was used as a catalyst. The reaction was completed at 140°C for 3 min (Scheme 32) and provided the desired 3-cyano derivative (**114**). Under the elevated temperature, the product 3-cyano derivative underwent a rearrangement to afford amide 3 through an intramolecular transfer of a water molecule.

4.4.2 Phenanthrene

Yadav et al., reported a facile method for the dinitration of 9, 10-dihydrophenanthrene using the domestic microwave and bismuth nitrate-impregnated clay [72].

Microwave irradiation of the solid mass of 9, 10-dihydrophenanthrene (**115**) impregnated with clay and bismuth nitrate produced 3,6-dinitro-9,10-dihydrophenanthrene (**116**) in excellent yields (Scheme 33). This nitration method has numerous advantages. Aromatic nitration using other methods requires highly corrosive reagents such as concentrated nitric acid/sulfuric acid mixture or nitronium tetrafluoroborate. In this method, a domestic microwave-induced nitration was carried out with bismuth nitrate impregnated with clay.

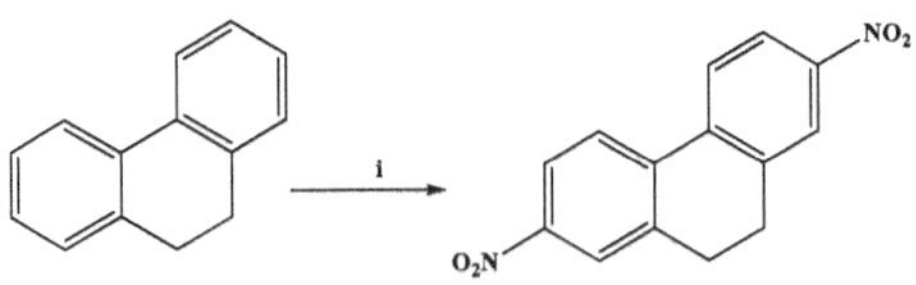

115 **116**

SCHEME 33 Microwave-induced nitration of 9, 10-dihydrophenantheren. Reagents and conditions: (i) $Bi(NO_3)_3{\cdot}5H_2O$, KSF Clay (500 mg), DCM, MW, 1 min.

4.4.3 Arylphosphonates

Attempts were made to prepare phosphinic and phosphonic esters, as well as phosphine oxides by P–C coupling reactions of aryl- and vinyl halides with dialkyl phosphites, *H*-phosphinates, and secondary phosphine oxides in the presence of catalysts using the microwave technique [73]. These conditions allowed the realization of the Hirao reaction more economically, efficiently, and environmentally friendly way.

Microwave-assisted techniques can be used for the synthesis of arylphosphonates using the cyclodiphosphazane-containing Pd complex [74]. The P-arylation of diphenylphosphine oxide under microwave conditions is shown in Scheme 34, and the desired product triarylphosphine oxide (**119**) was obtained in moderate to good yields (43%–95%).

Keglevich et al. demonstrated that under MW conditions, Hirao reaction is possible with the excess P-reagents and there is no need for an expensive or sensitive P-ligand [75]. The reaction of bromobenzene and dialkyl phosphites and diphenylphosphine oxide is shown in Scheme 35.

Jablonkai and Keglevich demonstrated a catalyst and ligand-free C–P coupling under microwave conditions [76]. This reaction of iodo-, and

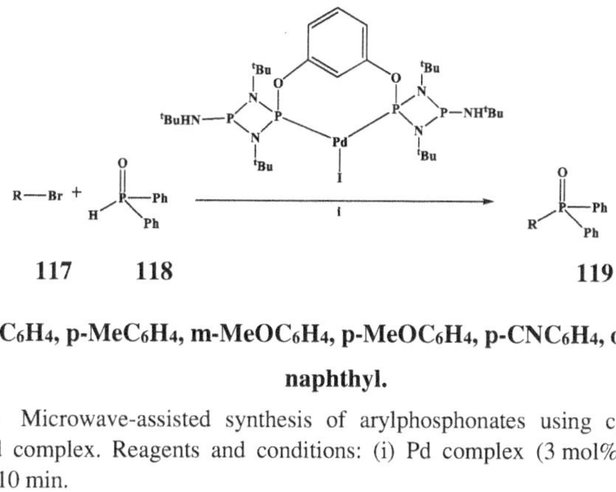

117 **118** **119**

R=Ph, o-MeC_6H_4, p-MeC_6H_4, m-$MeOC_6H_4$, p-$MeOC_6H_4$, p-CNC_6H_4, o-$HOCH_2C_6H_4$, 2-naphthyl.

SCHEME 34 Microwave-assisted synthesis of arylphosphonates using cyclodiphosphazane-containing Pd complex. Reagents and conditions: (i) Pd complex (3 mol%), Cs_2CO_3, MeCN, MW, 100°C, 10 min.

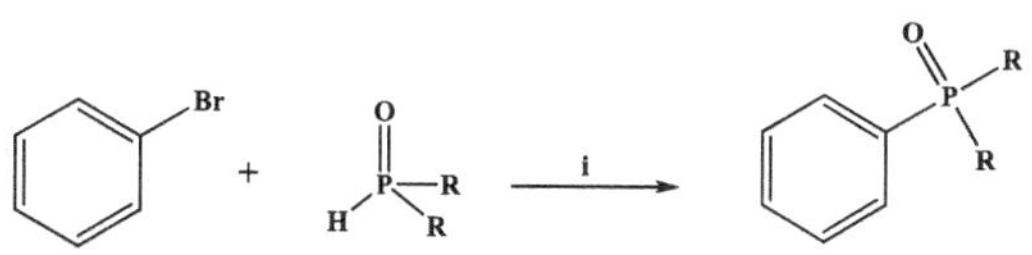

120 **121** **122**

R=OEt, OBu, OBn, Ph

SCHEME 35 Microwave-assisted P–C coupling of bromobenzene with dialkyl phosphites and diphenylphosphine oxide using the P-reagents in excess. Reagents and conditions: (i) Pd $(OAc)_2$(2.5-10%), 1.1equiv. Et3N, solvent (MeCN, EtOH, BuOH, BnOH) MW, 120°C, 30–60 min.

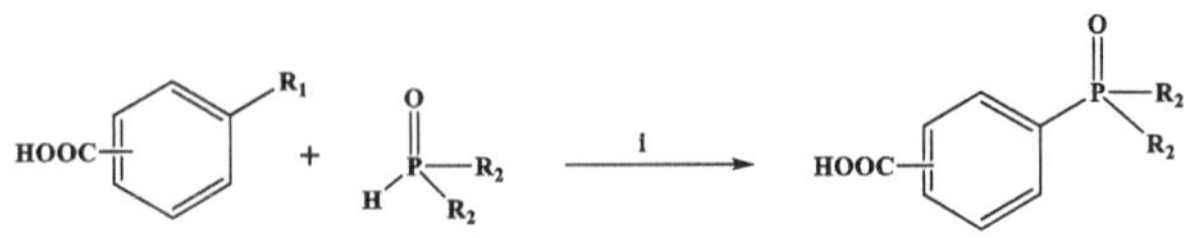

123 124 125

R_1=I, Br; R_2=Ph, p-MeC_6H_4

SCHEME 36 Catalyst and ligand-free C-P coupling under MW conditions. Reagents and conditions: (i) K_2CO_3, H_2O, MW, 180°C, 1–6 h.

bromobenzoic acids, and diaryl phosphine oxides was attained in the absence of any catalyst in water as the solvent under microwave conditions (Scheme 36), and this method is applicable for the substitution of only halobenzoic acids.

Microwave-assisted synthesis of dialkyl ω-azidoalkylphosphonates (**127**) via the nucleophilic substitution of the appropriate dialkyl ω-bromoalkylphosphonates (**126**) by sodium azide in water or dimethylformamide (DMF) as solvent was developed [77] as shown in Scheme 37.

Most of these reactions were accomplished in water using 1.5 equivalents of sodium azide as a source of azide anion. But for more hydrophobic and/or hindered diisobutyl ω-bromoalkylphosphonates, it was essential to perform reactions in DMF because in water or aqueous EtOH, only a partial conversion of starting bromophosphonates into azides was found. This microwave-assisted method was compatible with a range of structurally diverse dialkyl ω-bromoalkylphosphonates, with different steric hindrance around the phosphonato groups and with unbranched side carbon chains of various lengths and the desired products obtained in high yields. Isomerization of a series of 7-substituted cycloheptatrienes into the corresponding 2-isomers had been attained, under the controlled microwave heating in the presence of easily available DABCO and t-BuOK [78]. For instance, the starting compound, allyltriphenylphosphonium bromide (**128**), was obtained by the reaction of triphenylphosphine with allyl bromide in dichloromethane. To carry out the isomerization of allyltriphenylphosphonium bromide, a solution of allyltriphenylphosphonium bromide (1 g) in dry acetonitrile (5 mL) was added to a

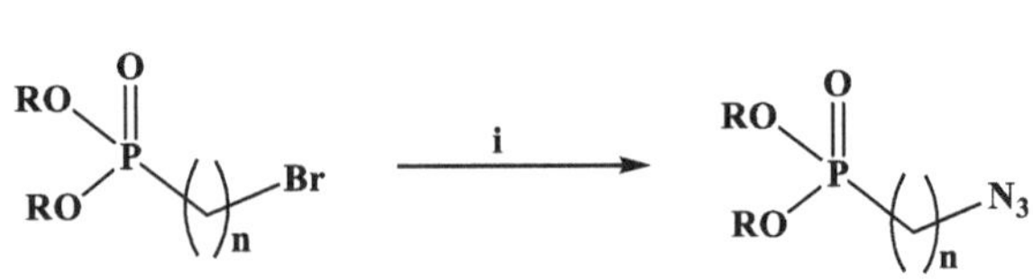

126 127

R=Me, Et, i-Pr, i-Bu; n=2, 3, 4 ,5, 6

SCHEME 37 Microwave-assisted synthesis of dialkyl ω-azidoalkylphosphonates. Reagents and conditions: (i) NaN_3 (3.0 mmol), solvent (3 mL), MW.

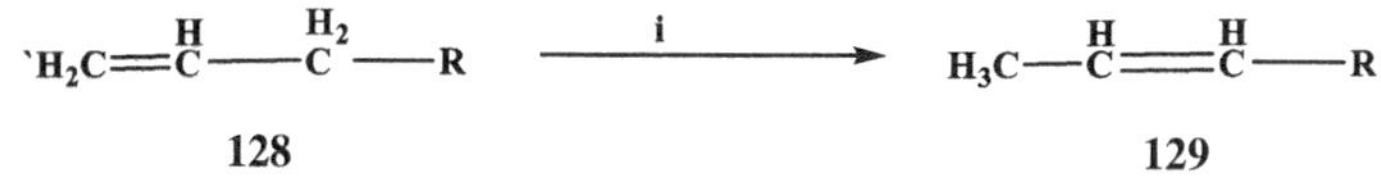

R=PPh3Br

SCHEME 38 Isomerization of allyltriphenylphosphonium bromide. Reagents and conditions: (i) CH_3CN, Et_3N in MeOH, MW, 500 W, 80°C, 20 min.

methanolic solution of triethylamine and irradiated with microwave (500 W), and the reaction temperature was set up at 80°C (Scheme 38). The isomerization was completed after 20 min.

The other starting compounds (**130**) (R $=^{+}PPh3BF4^{-}$, $SO_2C_6H_4CH_3$, $OCOC_6H_4CH_3$, and CN) were obtained by the reaction of triphenylphosphine, methyl diphenylphosphinite, p-toluenesulphinic acid sodium salt, sodium toluate, and sodium cyanide, respectively, with tropylium fluoroborate in dichloromethane and methanol (Scheme 39). Isomerization of these compounds was carried out in CH_3CN and t-BuOH with DABCO and t-BuOK under the microwave irradiation (500 W). The isomerization was completed after 15–20 min, whereas the same reactions required 18–24 h by conventional procedures.

4.4.4 Alcohols and acetones

The AuNP-supported Gd complex was found to be a highly effective catalyst for the acetylation of various alcohols and phenol in the presence of acetic anhydride under microwave [79]. Using 0.4 mol% of RS-Au-L-Gd, the almost complete transformation was achieved in 60 s under microwave irradiation conditions (Scheme 40).

With a catalyst loading of 0.4 mol%, the microwave-assisted acetylation of primary alcohols and phenol was carried out heterogeneously in dimethylformamide (DMF) and provided 91%–99% yields of acetate products in 60 s (Table 8). Even for bulky alcohols and less reactive alcohols, the catalyst (0.4 mol%) still produced 49%–80% yields of products in 60 s.

The microwave-assisted Cannizzaro reaction was described in order to develop fully reproducible synthetic protocols for the transformation of aldehydes to carboxylic acid and alcohols [80]. The reactions were optimized in terms of power, temperature, and time. This microwave-assisted method was applicable to aromatic, heteroaromatic, and aliphatic aldehydes. The results showed that furfural, thiophene-2-carbaldehyde, pyridinecarboxaldehyde, and aromatic aldehydes reacted even under mild conditions, though 1-methyl-pyrrole-2-carboxaldehyde derivatives and aliphatic aldehydes necessitated more drastic reaction conditions and longer reaction time.

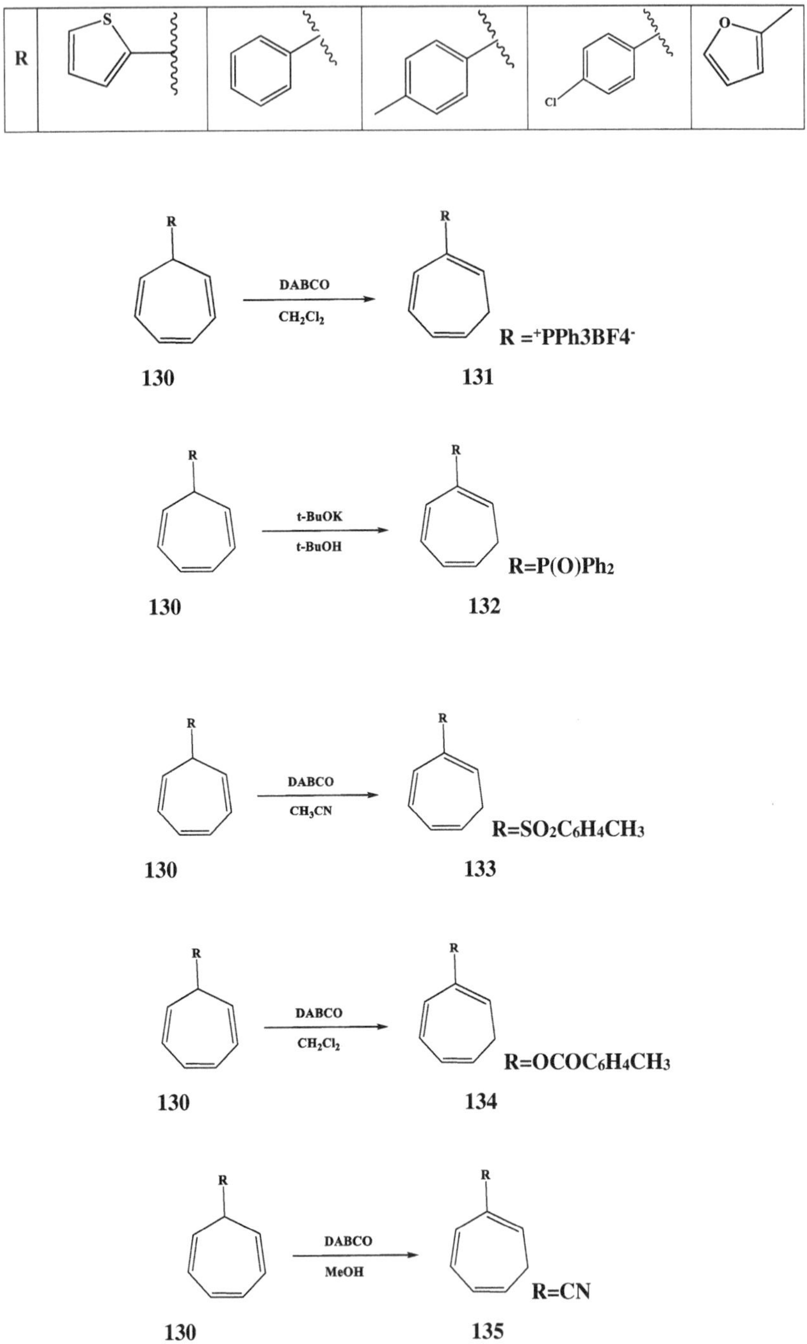

SCHEME 39 Microwave-assisted reaction conditions for the isomerization of compounds (**130**).

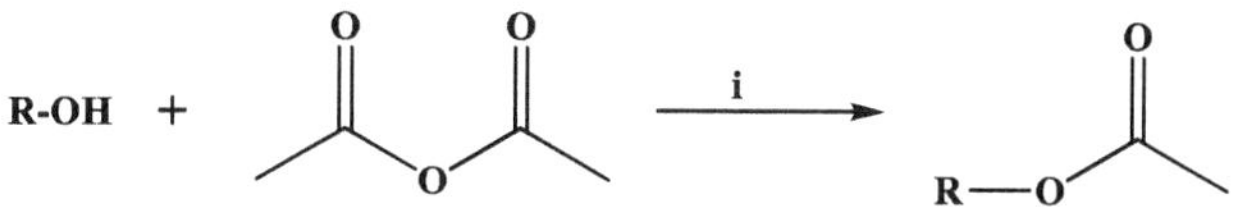

136 137 138

SCHEME 40 Microwave-assisted acetylation of alcohols and phenol catalyzed by RS-Au-L-Gd. Reagents and conditions: (i) Alcohol 0.5 mmol, acetic anhydride 5 mmol, solvent DMF (80 μL), RS-Au-L-Gd, MW, 300 W, 60 s.

TABLE 8 Eight selected products.

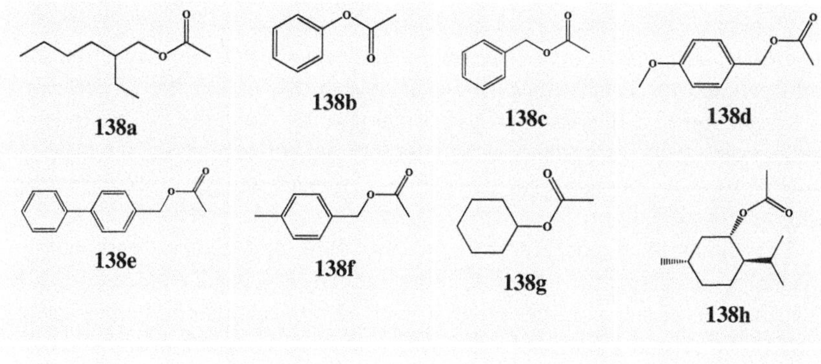

The schematics of the synthesis and optimal conditions found for furfural aldehyde, thiophene-2-carbaldehyde, and aromatic aldehydes are shown in Scheme 41. The same reaction conditions were applicable for the heteroaromatic analogues of pyridinecarboxaldehyde and pyrrole-2-carboxaldehyde.

A simple and direct microwave-assisted method for the Claisen-Schmidt reaction to prepare the functionalized α,β-unsaturated ketones was reported [81]. Microwave irradiation of aldehydes with acetone produces benzalacetones in very short reaction times and good yields, and the synthesis was selective without self-condensation product (Scheme 42).

To synthesize benzalacetone, reactions were performed on the same aldehyde with 1.5 equiv of NaOH under microwave-assisted solvent-free

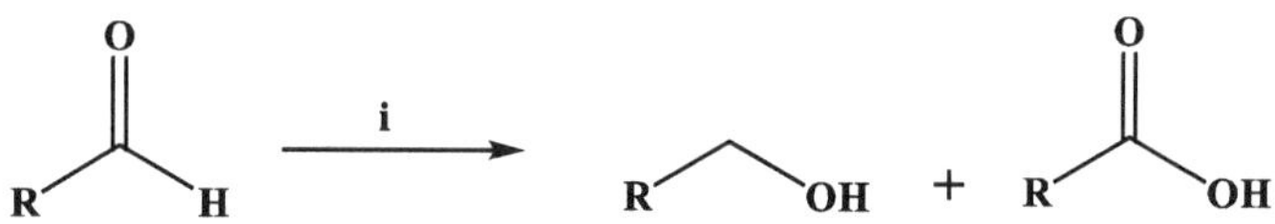

139 140 141

SCHEME 41 Synthesis of alcohols and carboxylic acids. Reagents and conditions: (i) NaOH (0.5 g, 12.5 mmol), Al_2O_3 (2 g, 19.6 mmol), H_2O (400 μL), MW (standard mode), 100°C, 2 min.

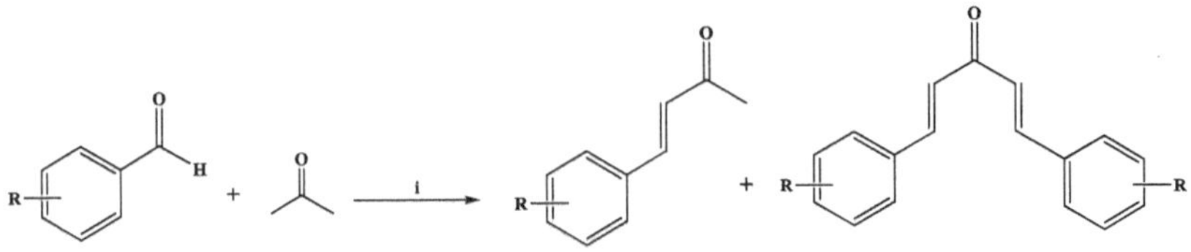

142 143 144 145

R=H, 4-Me, 4-*t*-Bu, 4-F, 4-Br, 4-OMe, 3-Cl, 3,4-diOMe, 2,5-NO_2, Cl, Phenyl

SCHEME 42 Synthesis of benzalacetones using microwave conditions. Reagents and conditions: (i) NaOH, MW, 5 W, 50°C, 10–15 min.

conditions. The compounds were mixed in a sealed microwave reaction tube and irradiated for 10–30 min (5 W) with stirring at 50°C for the synthesis.

To prepare 2,3-dihydroxypropyl decanoate, the solvent-free microwave-assisted synthesis was carried out by esterification of decanoic acid in the presence of two distinct glycerol derivatives, glycidol and glycerol carbonate, respectively [82].

To synthesize monoglycerides, decanoic acid, and glycerol carbonate or glycidol were mixed at a 1:1 molar ratio and TBAI was added to the mixture (Scheme 43). In this case, the amount of TBAI was calculated as a percentage of the molar content of decanoic acid. After microwave irradiation, diethyl ether was added to the reaction mixture, and to eliminate the quaternary ammonium salt, the resulting solution was washed with water. The reaction time was 1 min, which was 100 times lower than that of the classical heating method.

The α-aryl malonates are the key intermediates for the synthesis of a variety of heterocyclic compounds, such as benzodiazepines, isoquinolines, and pyrrolopyridine scaffolds. Synthesis of α-aryl malonates has been executed using the microwave-assisted method [83]. The coupling of aryl halides with diethyl malonate carried on effectively in a short reaction time in the presence of a catalytic amount of $Cu(OTf)_2$, Cs_2CO_3, and 2-picolinic acid in toluene using the microwave irradiation.

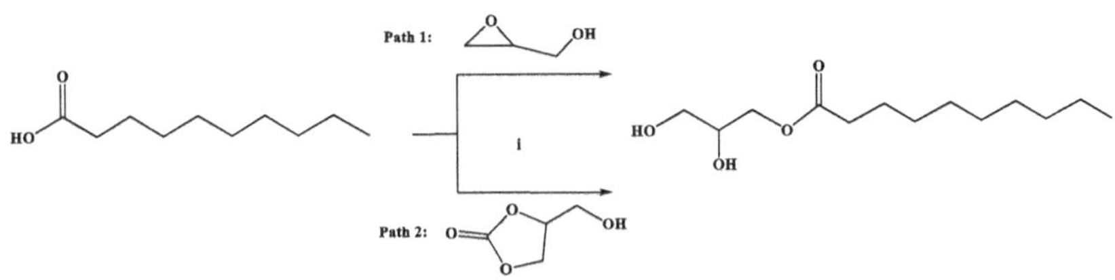

146 147

SCHEME 43 Solvent-free microwave-assisted synthesis of 2,3-dihydroxypropyl decanoate from glycidol and glycerol carbonate (path1 and 2, respectively), in solvent-free conditions, using TBAI as organocatalyst. Reagents and conditions: (i) TBAI, solvent-free, MW.

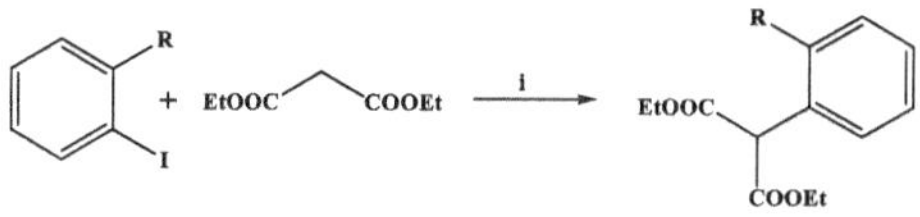

148 **149** **150**

R=COOCH3, benzoyl, 4-methoxy benzoyl, nitrile

SCHEME 44 Arylation of diethyl malonate with substituted iodobenzene. Reagents and conditions: (i) $Cu(OTf)_2$, Cs_2CO_3,2-picolinic acid, toluene, MW, 30 min, 90°C.

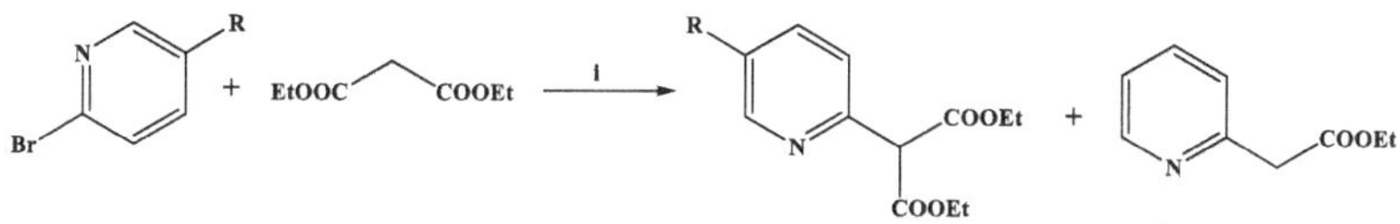

151 **152** **153** **154**

R=H, Methyl

SCHEME 45 Arylation of diethyl malonate with substituted bromopyridine. Reagents and conditions: (i) $Cu(OTf)_2$, Cs_2CO_3, 2-picolinic acid, toluene/$MgSO_4$, MW, 30 min, 100°C.

In this method, the desired aryl halides (1 mmol), copper triflate (0.1 mmol), cesium carbonate (3 mmol), and picolinic acid (0.2 mmol) were mixed and flushed with argon. Then, anhydrous toluene was added followed by diethyl malonate (2 mmol), and the mixture was then kept inside the microwave and irradiated for 30 min at 90°C. α-Arylation of diethyl malonate with substituted iodo benzene (methyl 2-iodobenzoate, 2-iodobenzophenone, 2-iodo 4-methoxybenzophenone, and 2-iodobenzonitrile) is shown in Scheme 44.

Arylation of diethyl malonate with *N*-heterocyclic compounds (2-bromopyridine and 2-bromo-5-methylpyridine) is shown in Scheme 45. The reaction proceeded well in the presence of cesium carbonate, $Cu(OTf)_2$, and picolinic acid in toluene/$MgSO_4$ under microwave irradiations at 100°C for 30 min and provided the desired product in 91% yield.

4.5 Conclusion

In this chapter, we discussed the importance of microwave-assisted synthesis in organic chemistry. Mainly, we considered the examples of microwave-assisted synthesis of pharmacologically significant compounds which contain oxygen and sulfur atoms. The features such as the possibility of solvent-free reactions, reduction of reaction time for a chemical/pharmaceutical reaction, the possibility of parallel chemical reaction, and instantaneous and uniform heating caused the microwave technology to stay dominant over conventional heating methods. The researchers involved in drug discovery and development processes like high-speed combinatorial and medicinal chemistry mostly prefer and adopt microwave technology. The microwave technology has encouraged the scientists to initiate new unexplored areas of complex systems both in the organic and in the inorganic chemistry field.

Acknowledgments

AD and BKB are grateful to Prince Mohammad Bin Fahd University for support. BKB is also grateful to US NIH, US NCI, and Kleberg Foundation of Texas for financial support.

References

[1] (a) S. Shi, Y. Zhang, K. Huang, S. Liu, Y. Zhao, Application of preparative high-speed countercurrent chromatography for separation and purification of lignans from Taraxacum mongolicum, Food Chem. 108 (2008) 402–406. (b) F. Zhang, L. Li, S. Niu, Y. Si, L. Guo, X. Jiang, Y. Che, A thiopyranchromenone and other chromone derivatives from an Endolichenic fungus, Preussia Africana, J. Nat. Prod. 75 (2012) 230–237.

[2] Z.M. Yang, J. Huang, J.K. Qin, Z.K. Dai, W.L. Lan, G.F. Su, H. Tang, F. Yang, Design, synthesis and biological evaluation of novel 1-hydroxyl-3-aminoalkoxy xanthone derivatives as potent anticancer agents, Eur. J. Med. Chem. 85 (2014) 487–497.

[3] D.U. Ganihigama, S. Sureram, S. Sangher, P. Hongmanee, T. Aree, C. Mahidol, S. Ruchirawat, P. Kittakoop, Antimycobacterial activity of natural products and synthetic agents: pyrrolodiquinolines and vermelhotin as anti-tubercular leads against clinical multidrug resistant isolates of Mycobacterium tuberculosis, Eur. J. Med. Chem. 89 (2015) 1–12.

[4] W.C. Song, W. Sangwook, H.P. Jung, K. Youngjoo, N. Younghwa, H.J. Lee, Xanthone analogues as potent modulators of intestinal P-glycoprotein, Eur. J. Med. Chem. 93 (2015) 237–245.

[5] (a) C.G. Knight, T. Stephens, Xanthene-dye-labelled phosphatidylethanolamines as probes of interfacial pH. Studies in phospholipid vesicles, Biochem. J. 258 (1989) 683. (b) O. Sirkecioglu, N. Talinli, A. Akar, Chemical aspects of santalin as a histological stain, J. Chem. Res. (1995) 502.

[6] (a) R.W. Lambert, J.A. Martin, J.H. Merrett, K.E.B. Parkes, G.J. Thomas, PCT Int. Appl. WO 9706178, Chem. Abstr. 126 (1997) 212377. (b) T. Hideo, Jpn. Tokkyo Koho JP 56005480, Chem. Abstr. 95 (1981) 80922.

[7] S.M. Menchen, S.C. Benson, J.Y.L. Lam, W. Zhen, D. Sun, B.B. Rosenblum, S.H. Khan, M. Taing, US Patent, 6,583,168, Chem. Abstr. 139 (2003) 54287.

[8] D. Ashok, M.G. Devulapally, V.K. Aamate, S. Gundu, S. Adam, S.D.S. Murthy, S. Balasubramanian, B. Naveen, T. Parthasarathy, Novel pyrano [3,2-b]xanthen-7(2*H*)-ones: Synthesis, antimicrobial, antioxidant and molecular docking studies, J. Mol. Struct. 1177 (2019) 215–228.

[9] A.N. Dadhania, V.K. Patel, D.K. Raval, Ionic liquid promoted facile and green synthesis of 1,8-dioxo-octahydroxanthene derivatives under microwave irradiation, J. Saudi Chem. Soc. 21 (2017) 163–169.

[10] A. Chavez, J. Cruz, A. Munoz, R.N. Yadav, D. Bandyopadhyay, B.K. Banik, Microwave-induced bismuth iodide-catalyzed synthesis of octahydroxanthenes, Heterocycl. Lett. 7 (2017) 507–511.

[11] D. Ashok, R.S. Kumar, D.M. Gandhi, M. Sarasija, A. Jayashree, S. Adam, Solvent-free microwave-assisted synthesis and biological evaluation of 2, 2-dimethylchroman-4-one based benzofurans, Heterocycl. Commun. 22 (2016) 363–368.

[12] S.S. Zambre, S.N. Darandale, J.N. Sangshetti, D.B. Shinde, Microwave-assisted solvent-free one pot synthesis of isobenzofuran-1(3H)-ones using sulphamic acid catalyst, Arab. J. Chem. 9 (2016) S1416–S1419.
[13] B.A. Trofimov, A.V. Stepanov, A.G. Mal'kina, O.G. Volostnykh, O.A. Shemyakina, I.A. Ushakov, An expedient access into functionalized furan/3(2H)-furanone ensembles via microwave-assisted domino reactions, Synth. Commun. 1 (45) (2015) 2718–2729.
[14] A. Detsi, M. Majdalani, C.A. Kontogiorgis, D. Hadjipavlou-Litina, P. Kefalas, Natural and synthetic 2'-hydroxy-chalcones and aurones: synthesis, characterization and evaluation of the antioxidant and soybean lipoxygenase inhibitory activity, Bioorg. Med. Chem. 17 (2009) 8073–8085.
[15] M.J. Jung, H.Y. Chung, S.S. Kang, J.H. Choi, K.S. Bae, J.S. Choi, Antioxidant activity from the stem bark of Albizzia julibrissin, Arch. Pharm. Res. 6 (2003) 458–462.
[16] H.M. Sim, K.Y. Loh, W.K. Yeo, C.Y. Lee, M.L. Go, Aurones as modulators of ABCG2 and ABCB1: synthesis and structure–activity relationships, Chem. Med. Chem. 6 (2011) 713–724.
[17] H. Cheng, L. Zhang, Y. Liu, S. Chen, H. Cheng, X. Lu, Z. Zheng, G.C. Zhou, Design, synthesis and discovery of 5-hydroxyaurone derivatives as growth inhibitors against HUVEC and some cancer cell lines, Eur. J. Med. Chem. 45 (2010) 5950–5957.
[18] M. Roussaki, S.C. Lima, A.M. Kypreou, P. Kefalas, A.S.D. Cordeiro, A. Detsi, Aurones: a promising heterocyclic scaffold for the development of potent antileishmanial agents, Int. J. Med. Chem. 2012 (2012) 196921.
[19] D. Kerboeuf, M. Riou, F. Guégnard, Flavonoids and related compounds in parasitic disease control, Mini-Rev. Med. Chem. 8 (2008) 116–128.
[20] S. Shrestha, S. Natarajan, J.H. Park, D.Y. Lee, J.G. Cho, G.S. Kim, Y.J. Jeon, S.W. Yeon, D.C. Yang, N.I. Baek, Potential neuroprotective flavonoid-based inhibitors of CDK5/p25 from Rhus Parviflora, Bioorg. Med. Chem. Lett. 23 (2013) 5150–5154.
[21] J. Schoepfer, H. Fretz, B. Chaudhuri, L. Muller, E. Seeber, L. Meijer, O. Lozach, E. Vangrevelinghe, P. Furet, Structure-based design and synthesis of 2-benzylidene-benzofuran-3-ones as Flavopiridol Mimics, J. Med. Chem. 45 (2002) 1741–1747.
[22] I.T. Kim, Y.M. Park, K.M. Shin, J.H. Ha, J.W. Choi, H.J. Jung, H.J. Park, K.T. Lee, Anti-inflammatory and anti-nociceptive effects of the extract from Kalopanax pictus, Pueraria thunbergiana and Rhus verniciflua, J. Ethnopharmacol. 94 (2004) 165–173.
[23] K.Y. Park, G.O. Jung, K.T. Lee, J.W. Choi, M.Y. Choi, G.T. Kim, H.J. Jung, H.J. Park, Antimutagenic activity of flavonoids from the heartwood of Rhus verniciflua, J. Ethnopharmacol. 90 (2004) 73–79.
[24] A. Boumendjel, Aurones: a subclass of flavones with promising biological potential, Curr. Med. Chem. 10 (2003) 2621–2630.
[25] E.H. Lee, D.G. Song, J.Y. Lee, C.H. Pan, B.H. Um, S.H. Jung, Inhibitory effect of the compounds isolated from Rhus verniciflua on aldose reductase and advanced glycation end products, Biol. Pharm. Bull. 31 (2008) 1626–1630.
[26] M.Y. Song, G.S. Jeong, K.B. Kwon, S.O. Ka, H.Y. Jang, J.W. Park, Y.C. Kim, B.H. Park, Sulfuretin protects against cytokine-induced beta-cell damage and prevents streptozotocin-induced diabetes, Exp. Mol. Med. 42 (2010) 628–638.
[27] D. Ashok, K. Rangu, S. Gundu, A. Lakkadi, P. Tigulla, Microwave-assisted synthesis, molecular docking, and biological evaluation of 2-arylidene-2H-furo[2,3-f]chromen-3 (7H)-ones as antioxidant and antimicrobial agents, Med. Chem. Res. 26 (2017) 1735–1746.
[28] X. Hu, H. Lai, F. Zhao, S. Hu, Q. Sun, L. Fang, Microwave-assisted synthesis of benzofuran-3 (2H)-ones, Acta Chim. Slov. 66 (2019) 745–750.
[29] A.B. Gawade, G.D. Yadav, Microwave assisted synthesis of 5-ethoxymethylfurfural in one pot from D-fructose by using deep eutectic solvent as catalyst under mild condition, Biomass Bioenergy 117 (2018) 38–43.

[30] A. Thakur, R. Singla, V. Jaitak, Coumarins as anticancer agents: a review on synthetic strategies, mechanism of action and SAR studies, Eur. J. Med. Chem. 101 (2015) 476–495.

[31] A. Vogel, De l'existence de l'acide bezoïque dans la feve de tonka et dans les fleurs de melilot (On the existence of benzoic acid in the tonka bean and in the flowers of melilot), J. de Pharm. 6 (1820) 305–309.

[32] W.H. Perkin, On the artificial production of coumarin and formation of its homologues, J. Chem. Soc. 21 (1868) 53–63.

[33] M.E. Riveiro, N.D. Kimpe, A. Moglioni, R. Vazquez, F. Monczor, C. Shayo, C. Davio, Coumarins: old compounds with novel promising therapeutic perspectives, Curr. Med. Chem. 17 (2010) 1325–1338.

[34] K.N. Venugopala, V. Rashmi, B. Odhav, Review on natural coumarin lead compounds for their pharmacological activity, Biomed. Res. Int. 2013 (2013) 1–14.

[35] S. Emami, S. Dadashpour, Current developments of coumarin-based anticancer agents in medicinal chemistry, Eur. J. Med. Chem. 102 (2015) 611–630.

[36] R.S. Keri, B.S. Sasidhar, B.M. Nagaraja, M.A. Santos, Recent progress in the drug development of coumarin derivatives as potent antituberculosis agents, Eur. J. Med. Chem. 100 (2015) 257–269.

[37] M.M. Liu, X.Y. Chen, Y.Q. Huang, P. Feng, Y.L. Guo, G. Yang, Y. Chen, Hybrids of phenylsulfonylfuroxan and coumarin as potent antitumor agents, J. Med. Chem. 57 (2014) 9343–9356.

[38] A. Stefanachi, N. Hanke, L. Pisani, F. Leonetti, O. Nicolotti, M. Catto, S. Cellamare, R.W. Hartmann, A. Carotti, Discovery of new 7-substituted-4-imidazolylmethyl coumarins and 4'-substituted-2-imidazolyl acetophenones open analogues as potent and selective inhibitors of steroid-11beta-hydroxylase, Eur. J. Med. Chem. 89 (2015) 106–114.

[39] D. Konrádová, H. Kozubíková, K. Doležal, J. Pospíšil, Microwave-assisted synthesis of phenylpropanoids and coumarins: total synthesis of osthol, Eur. J. Org. Chem. 2017 (2017) 5204–5213.

[40] A.R. Kaneria, R.R. Giri, V.G. Bhila, H.J. Prajapati, D.I. Brahmbhatt, Microwave assisted synthesis and biological activity of 3-aryl-furo[3,2-c]coumarins, Arab. J. Chem. 10 (2017) S1100–S1104.

[41] Y. Shi, B. Zhang, X.J. Chen, D.Q. Xu, Y.X. Wang, H.Y. Dong, S.R. Ma, R.H. Sun, Y.P. Hui, Z.C. Li, Osthole protects lipopolysaccharide-induced acute lung injury in mice by preventing down-regulation of angiotensin-converting enzyme 2, Eur. J. Pharm. Sci. 48 (2013) 819–824.

[42] H.J. Liang, F.M. Suk, C.K. Wang, L.F. Hung, D.Z. Liu, N.Q. Chen, Y.C. Chen, C.C. Chang, Y.C. Liang, Osthole, a potential antidiabetic agent, alleviates hyperglycemia in db/db mice, Chem. Biol. Interact. 181 (2009) 309–315.

[43] L. Wu, X. Huang, J. Li, R. Zhang, H. Hu, Recent advances in the multifunction of a natural occurring coumarin: osthole, J. Intercult. Ethnopharmacol. 2 (2013) 57–66.

[44] Y.W. Liu, Y.T. Chiu, S.L. Fu, Y.T. Huang, Osthole ameliorates hepatic fibrosis and inhibits hepatic stellate cell activation, J. Biomed. Sci. 22 (2015) 63.

[45] S. Vogl, M. Zehl, P. Picker, E. Urban, C. Wawrosch, G. Reznicek, J. Saukel, B. Kopp, Identification and quantification of coumarins in Peucedanum ostruthium (L.) Koch by HPLC-DAD and HPLC-DAD-MS, J. Agric. Food Chem. 59 (2011) 4371–4377.

[46] Y. Wei, T. Zhang, Y. Ito, Preparative isolation of osthol and xanthotoxol from Common Cnidium Fruit (Chinese traditional herb) using stepwise elution by high-speed counter-current chromatography, J. Chromatogr. A 1033 (2004) 373–377.

[47] X. Cao, Y. Xu, Y. Cao, R. Wang, R. Zhou, W. Chu, Y. Yang, Design, synthesis, and structure-activity relationship studies of novel thienopyrrolidone derivatives with strong antifungal activity against Aspergillus fumigates, Eur. J. Med. Chem. 102 (2015) 471–476.

[48] A.K. Bhattacharya, H.R. Chand, J. John, M.V. Deshpande, Clerodane type diterpene as a novel antifungal agent from Polyalthia longifolia var. pendula, Eur. J. Med. Chem. 94 (2015) 1–7.

[49] A.R. Villalva, D. Gonzalez-Calderon, C. Gonzalez-Romero, M. Morales-Rodríguez, B. Jauregui-Rodríguez, E. Cuevas-Yanez, A facile synthesis of novel miconazole analogues and the evaluation of their antifungal activity, Eur. J. Med. Chem. 97 (2015) 275–279.

[50] M.Z. Zhang, R.R. Zhang, J.Q. Wang, X. Yu, Y.L. Zhang, Q.Q. Wang, W.H. Zhang, Microwave-assisted synthesis and antifungal activity of novel fused Osthole derivatives, Eur. J. Med. Chem. 124 (2016) 10–16.

[51] R.R. Zhang, J. Liu, Y. Zhang, M.Q. Hou, M.Z. Zhang, F. Zhou, W.H. Zhang, Microwave-assisted synthesis and antifungal activity of novel coumarin derivatives: pyrano[3,2-c]chromene-2,5-diones, Eur. J. Med. Chem. 116 (2016) 76–83.

[52] K. Ostrowska, E. Hejchman, D. Maciejewska, A. Włodarczyk, K. Wojnicki, D. Matosiuk, A. Czajkowska, I. Młynarczuk-Biały, Ł. Dobrzycki, Microwave-assisted preparation, structural characterization, lipophilicity, and anti-cancer assay of some hydroxycoumarin derivatives, Monatsh. Chem. 146 (2015) 89–98.

[53] D. Ashok, B.V. Lakshmi, S. Ravi, A. Ganesh, Microwave-assisted synthesis of substituted 4-chloro-8-methyl-2-phenyl-1,5-dioxa-2H-phenanthren-6-ones and their antimicrobial activity, Med. Chem. Res. 24 (2015) 1487–1495.

[54] I.T.L. Ramos, T.M.S. Silva, C.A. Camara, Synthesis of new 6- and 8-Alkenyl-3,7,3',4'-tetramethoxyquercetin derivatives by microwave-assisted heck coupling, Synth. Commun. 49 (2019) 2583–2589.

[55] M.D. McReynolds, J.M. Dougherty, P.R. Hanson, Synthesis of phosphorus and sulfur heterocycles via ring-closing olefin metathesis, Chem. Rev. 104 (2004) 2239–2258.

[56] T. Nash, W.G. Rice, Efficacies of zinc-finger-active drugs against Giardia lamblia, Antimicrob. Agents Chemother. 42 (1998) 1488–1492.

[57] N.B. Perry, J.W. Blunt, M.H.G. Munro, Cytotoxic pigments from new zealand sponges of the genus latrunculia: discorhabdins a, b and c, Tetrahedron 44 (1988) 1727–1734.

[58] M.R. Khan, S. Zaib, M.K. Rauf, M. Ebihara, A. Badshah, M. Zahid, M.A. Nadeem, J. Iqbal, Solution-phase microwave assisted parallel synthesis, biological evaluation and in silico docking studies of 2-chlorobenzoyl thioureas derivatives, J. Mol. Struct. 1164 (2018) 354–362.

[59] S. Li, X. Chen, J. Xu, Microwave-assisted copper-catalyzed stereoselective ring expansion of three-membered heterocycles with a-diazo-b-dicarbonyl compounds, Tetrahedron 74 (2018) 1613–1620.

[60] A.K. Singh, S. Thakur, B. Pani, B. Chugh, H. Lgaz, I.-M. Chung, P. Chaubey, A.K. Pandey, J. Singh, Solvent-free microwave assisted synthesis and corrosion inhibition study of a series of hydrazones derived from thiophene derivatives: experimental, surface and theoretical study, J. Mol. Liq. 283 (2019) 788–803.

[61] C. Anjaiah, M. Nagamani, C. Abraham Lincoln, D. Ashoka, Microwave-assisted synthesis and antimicrobial activity of 3-(arylsulfanyl)-4-hydroxy-2H-chromen-2-ones1, Russ. J. Gen. Chem. 88 (2018) 2149–2153.

[62] C. Anjaiah, M. Nagamani, C. Abraham Lincoln, D. Ashok, Microwave assisted synthesis of 1-(arylthio)naphthalen-2-ols and their antimicrobial activity1, Russ. J. Gen. Chem. 87 (2017) 2930–2932.

[63] R.H. Fałdyga, P. Grzelak, P. Pipiak, G. Mloston, Microwave assisted synthesis of ferrocenyl and hetaryl functionalized thioketones, Phosphorus Sulfur Silicon 192 (2017) 197–198.

[64] A.S. Kermani, Microwave-assisted one-pot synthesis of 2-nitroalkylidene-1,3-oxathiolane derivatives, J. Sulfur Chem. 37 (2016) 105–113.

[65] G. Mloston, R.H. Fałdyga, M. Jeske, M. Godziszewska, K. Urbaniak, H. Heimgartner, Microwave-assisted reactions of α-diazoketones with hetaryl and ferrocenyl thioketones, J. Sulfur Chem. 39 (2018) 47–63.

[66] R. Lisiak, J. Mochowski, Facile method for conversion of 2-(chloroseleno)benzoyl chloride into 2-substituted 3-hydroxybenzo[b]selenophenes, Synth. Commun. 39 (2009) 4271–4281.
[67] M.K. Staples, R.L. Grange, J.A. Angus, J. Ziogas, N.P.H. Tan, M.K. Taylor, C.H. Schiesser, Tandem free-radical addition/substitution chemistry and its application to the preparation of novel AT1 receptor antagonists, Org. Biomol. Chem. 9 (2011) 473–479.
[68] K. Takimiya, Y. Kunugi, Y. Konda, H. Ebata, Y. Toyoshima, 2,7-Diphenyl[1]benzoselenopheno[3,2-b][1]benzoselenophene as a stable organic semiconductor for a high-performance field-effect transistor, T. Otsubo, J. Am. Chem. Soc. 128 (2006) 3044–3050.
[69] H. Ebata, E. Miyazaki, T. Yamamoto, K. Takimiya, Synthesis, properties, and structures of benzo[1,2-b:4,5-b‘]bis[b]benzothiophene and benzo[1,2-b,4,5-b‘]bis[b]benzoselenophene, Org. Lett. 9 (2007) 4499–4502.
[70] T. Yamamoto, K. Takimiya, Facile synthesis of highly π-extended heteroarenes, dinaphtho [2,3-b,2‘,3‘-f]chalcogenopheno[3,2-b]chalcogenophenes, and their application to field-effect transistors, J. Am. Chem. Soc. 29 (2007) 2224–2225.
[71] P. Arsenyan, E. Paegle, S. Belyakov, Microwave-assisted cyanation of 3-bromo-3-(1-hydroxyalkyl)benzo[b]selenophene derivatives, Mendeleev Commun. 25 (2015) 119–120.
[72] R.N. Yadav, L. Salazar, A.K. Singh, B.K. Banik, Microwave-induced bismuth nitrate-impregnated claymediated novel dinitration OF 9,10- dihydrophenanthrene: a precursor for new heterocycles, Heterocyclic Lett. 8 (2018) 273–276.
[73] R. Henyecz, Microwave-assisted synthesis of phosphonic and phosphinic esters and phosphine oxides by the Hirao reaction, Phosphorus Sulfur Silicon Relat. Elem. 194 (2019) 372–376.
[74] G.S. Ananthnag, J.T. Mague, M.S. Balakrishna, A cyclodiphosphazane based pincer ligand, [2,6-{1-$(tBuN)_2P$ (tBuHN)PO}$2C_6H_3I$]: Ni^{II}, Pd^{II}, Pt^{II} and Cu^{I} complexes and catalytic studies, Dalton Trans. 44 (2015) 3785–3793.
[75] G. Keglevich, R. Henyecz, Z. Mucsi, N.Z. Kiss, The palladium acetate-catalyzed microwave-assisted Hirao reaction without an added phosphorus ligand as a "green" protocol: a quantum chemical study on the mechanism, Adv. Synth. Catal. 359 (2017) 4322–4331.
[76] E. Jablonkai, G. Keglevich, Catalyst-free P–C coupling reactions of halobenzoic acids and secondary phosphine oxides under microwave irradiation in water, Tetrahedron Lett. 56 (2015) 1638–1640.
[77] Ł. Janczewski, A. Gajda, J. Braszczyńska, T. Gajda, Microwave-assisted synthesis of dialkyl ω-azidoalkylphosphonates, Synth. Commun. 46 (2016) 1625–1633.
[78] A. Abdulwadood, A.S.H. Alsamarrai, Microwave assisted base-catalyzed 1,3-isomerization of 7-substituted cycloheptatriene bearing electron withdrawing groups, J. King Saud Univ. Sci. 32 (2020) 332–336.
[79] T.-C. Chang, S.J. Yu, Microwave-assisted catalytic acetylation of alcohols by gold-nanoparticle–supported gadolinium complex, Synth. Commun. 45 (2015) 651–662.
[80] Ł. Janczewski, M. Walczak, J. Frączyk, Z.J. Kamiński, B. Kolesińska, Microwave-assisted Cannizzaro reaction—optimisation of reaction conditions, Synth. Commun. 49 (2019) 3290–3300.
[81] A. Rayar, M.S.-I. Veitía, C. Ferroud, An efficient and selective microwave-assisted Claisen-Schmidt reaction for the synthesis of functionalized benzalacetones, Springerplus 4 (2015) 221.
[82] A. Mhanna, L. Chupin, C.H. Brachais, D. Chaumont, G. Boni, L. Brachais, J.P. Couvercelle, L. Lecamp, L. Plasseraud, Efficient microwave-assisted synthesis of glycerol monodecanoate, Eur. J. Lipid Sci. Technol. 120 (2018) 1700133.
[83] M.A. Ibrahim, Microwave-assisted synthesis of a-aryl malonates: key intermediates for the preparation of azaheterocycles, Arab. J. Chem. 9 (2016) S1973–S1983.

Chapter 5

Microwave-assisted synthesis of *N*-heterocycles

Aparna Das* and Bimal Krishna Banik*
Department of Mathematics and Natural Sciences, College of Sciences and Human Studies, Prince Mohammad Bin Fahd University, Al Khobar, Kingdom of Saudi Arabia
**Corresponding authors: E-mails: aparnadasam@gmail.com (Aparna Das); bimalbanik10@gmail.com, bbanik@pmu.edu.sa (Bimal Krishna Banik)*

5.1 Introduction

Heterocyclic molecules are an important family of compounds because of their broad range of pharmaceutical and material science applications. Among them, nitrogen-containing heterocyclic (*N*-heterocyclic) compounds have attained the peculiar interest of researchers, owing to its various applications in different areas. Due to widespread interest in *N*-heterocycles, the synthesis of these compounds has always been among the most important research areas in synthetic chemistry. In recent years, numerous studies were exposed for the synthesis of *N*-heterocycles under various reaction conditions such as solvent-free conditions, catalytic reaction conditions, reactants immobilized on a solid support, one-pot synthetic and microwave irradiation conditions. Chemists have successfully conducted a wide range of organic reactions under microwave irradiation, such as the Ene reaction [1], Diels-Alder reaction [2], Heck reaction [3], Mannich reaction [4], Suzuki reaction [5], hydrogenation of lactam [6], esterification [7], dehydration [8], hydrolysis [9], cycloaddition [10], reduction [11], epoxidation [12], cyclization [13], and condensation [14] reactions.

This book chapter is focused on the microwave-assisted catalytic and noncatalytic synthesis of *N*-heterocycles, mainly five-membered, six-membered, fused, and polycyclic *N*-heterocycles. Examples from several families are considered. Mostly, we explore the microwave-assisted synthesis of pharmacologically significant compounds, such as pyrimidines, steroidal derivatives, thiazoles, imines, tetrazoles, triazoles, quinolines, indolizine, β-lactams, pyrroles, and quinoxalines. Because of the limitations, we included only the selected examples available in the literature from 2015 to early 2020.

Microwaves in Chemistry Applications. https://doi.org/10.1016/B978-0-12-822895-1.00006-0

5.2 Pyrimidines

Pyrimidine is a heterocyclic aromatic organic compound containing two nitrogen atoms at positions 1 and 3 of the 6-membered ring. The pyrimidine core is present in numerous natural products and structurally diverse synthetic derivatives [15–18]. Because of their biological importance, pyrimidines have attracted much attention.

Acosta et al. reported the microwave-assisted synthesis of novel pyrazolo [4′,3′:5,6]pyrido[2,3-d]pyrimidines [19]. This synthesis was carried out under solvent-free conditions via cyclocondensation reaction between heterocyclic *o*-aminonitriles and cyanopyridines in the presence of tBuOK as a catalyst.

The reaction between the *o*-aminonitrile (R = Cl) and 4-cyanopyridine **2** is shown in Scheme 1. Several conditions were tested to identify the optimum conditions for the synthesis of pyrimidine derivatives (**3**) by changing the solvents, temperatures, and power of the microwave source. When ethanol or DMF was used as the solvent and the mixture was subjected to reflux, the product was obtained in low yield (30%–31%) after 8–9 h. Improvement in yields was obtained when the reaction was performed under MWI using DMF as the solvent (11 min, yield = 36%) or under solvent-free MWI (10 min, yield = 41%). The use of tBuOK as a catalyst was found to improve the reaction. When the MW potency was raised for shorter times, in the presence of tBuOK, the yields also increased. A higher yield of 61% was obtained, when the reaction was performed for 5 min under solvent-free MWI (100°C, 250 W) in the presence of tBuOK.

All the synthesized pyrimidine derivatives were tested for antifungal properties against two clinically important fungi: *Candida albicans* and *Cryptococcus neoformans*. Most of the compounds showed moderate activity against both

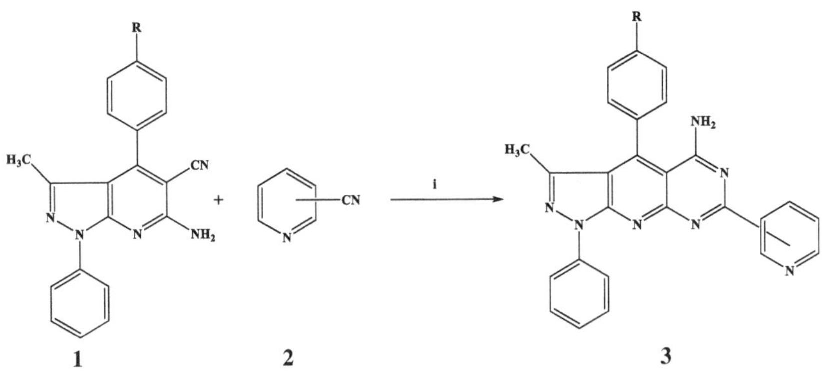

R=CH3, OCH3, 3,4-OCH2O, Cl

SCHEME 1 Microwave-assisted synthesis of pyrazolo[4′,3′:5,6]pyrido[2,3-d]pyrimidine derivatives. Reagents and conditions: (i) tBuOK, MW.

fungi. The antifungal analysis of the compounds showed that the position of the N in the pyrimidyl moiety per se does not play a role in the activity, but the type of 4-R substituent appears to influence the activity.

Synthesis of novel ring D- and A-fused pyrimidines in the androstane series was reported by Baji et al [20]. The synthesis occurred efficiently within 10–15 min in polar protic solvents under microwave irradiation via two kinds of multicomponent heterocyclization reactions followed by spontaneous or promoted oxidation. The schematics of the reactions are shown in Schemes 2 and 3.

The compounds 16-hydroxymethylene-dehydroepiandrosterone (1), benzaldehyde (2a), or *p*-chlorobenzaldehyde (2b) and ammonium acetate were thoroughly mixed in 1:2:2 molar ratios with silica gel as solid support and irradiated in a closed vessel at 120°C for 6 min. The product was obtained in low yield. The same reaction was repeated in the polar protic solvent (EtOH) medium by applying an irradiation time of 15 min, and a 10% improvement in the yield was observed.

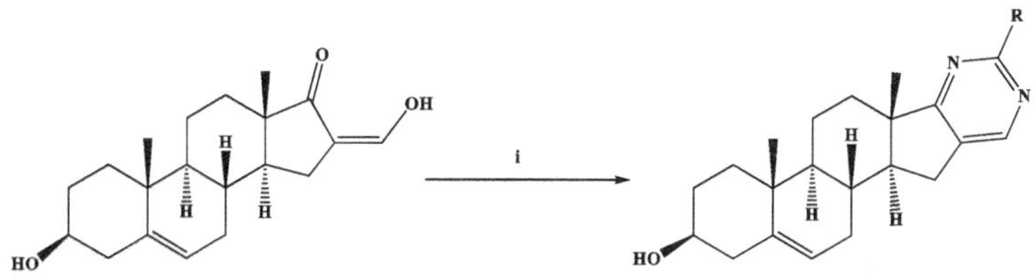

4 **5**

R=Ph, *p*-Cl-C_6H_4

SCHEME 2 MW-assisted synthesis of ring D-fused arylpyrimidines in the androstane series. Reagents and conditions: (i) R-CHO, NH_4OAc, silica gel, MW, 120°C, 6 min (solvent-free), or R-CHO, NH_4OAc, EtOH, MW, 120°C, 15 min.

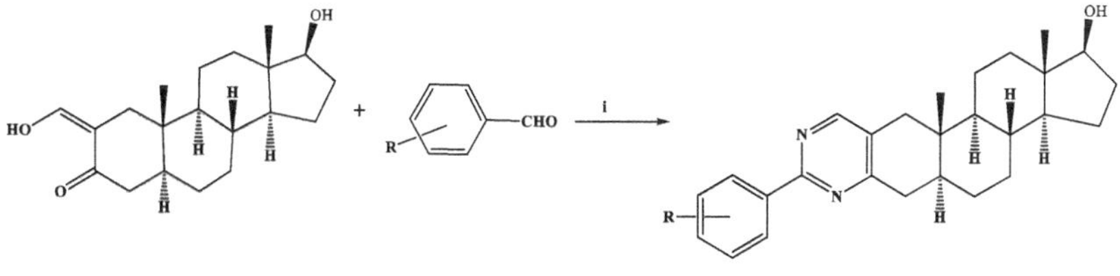

6 **7** **8**

R=H, *p*-Cl, *p*-Br, *p*-F, *p*-NO_2, *p*-OMe, *p*-CH_3, *m*-CH_3, *o*-CH_3

SCHEME 3 Microwave-assisted synthesis of ring A-fused arylpyrimidines in the androstane series. Reagents and conditions: (i) NH_4OAc, EtOH, MW, 120°C, 15 min.

To investigate the biological effects, all the compounds were tested in vitro on human cancer cell lines as well as on noncancerous fibroblast cells by the MTT assay. The results showed that acetylated ring A-fused 4′-arylpyrimidin-2′-ones possessed prominent cytotoxicity. Moreover, in three 2′-arylpyrimidine derivatives, a strong prostate cancer cell-specific activity had been identified.

Bhat et al. described microwave-assisted one-pot synthesis of methyl 7-amino-4-oxo-5-phenyl-2-thioxo-2, 3, 4,5-tetrahydro-1*H*-pyrano[2,3-d]pyrimidine-6-carboxylates (**12**) [21]. The synthesis was carried out using various benzaldehyde derivatives, methylcyanoacetate, and thio-barbituric acid in water as a green solvent. The one-pot three-component domino Knoevenagel-Michael addition reaction under microwave irradiation is shown in Scheme 4.

During the process, a mixture of benzaldehyde derivatives (1 mmol), methylcyanoacetate, (1.2 mmol), thio-barbituric acid, (1 mmol), and water (3.0 mL) was placed into a Teflon vessel and subjected to microwave irradiation under catalyst-free conditions at the power of 250 W and 120°C. To identify the biological activities, all the synthesized compounds were tested against different strains. The compounds showed comparatively good in vitro antimicrobial and antifungal activities. Some compounds showed maximum antimicrobial activity against *Staphylococcus aureus* and *Bacillus cereus* (gram-positive bacteria), and *Escherichia coli*, *Klebshiella pneumonia*, and *Pseudomonas aeruginosa* (gram-negative bacteria).

Chavan et al. described an efficient, high-yielding, and rapid synthesis of *N*-tert-butyl-4-(4-substitutedphenyl)-2-((substituted-2-oxo-2*H*-chromen-4-yl) methylthio)-6-oxo-1,6-dihydropyrimidine-5-carboxamides under microwave irradiation [22]. The synthetic method is outlined in Scheme 5.

They also carried out the synthesis of tert-butyl amides under the conventional method. The results showed that the microwave approach proved to be extremely fast. The reactions were completed within 6–10 min, which is 12–23 times faster than the conventional method. In addition, the microwave approach provided good to excellent yields (80%–92%) of the products as compared to the conventional method (55%–71%).

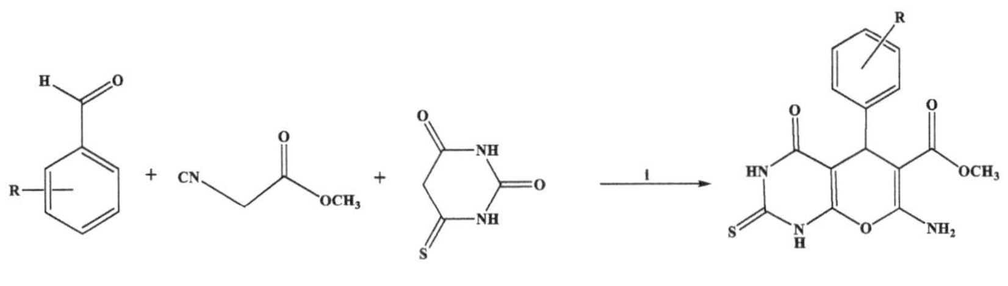

9 10 11 12

SCHEME 4 Microwave-assisted synthesis of methyl 7-amino-4-oxo-5-phenyl-2-thioxo-2,3,4,5-tetrahydro-1*H*-pyrano[2,3-d]pyrimidine-6-carboxylate derivatives. Reagents and conditions: (i) MW, 120°C.

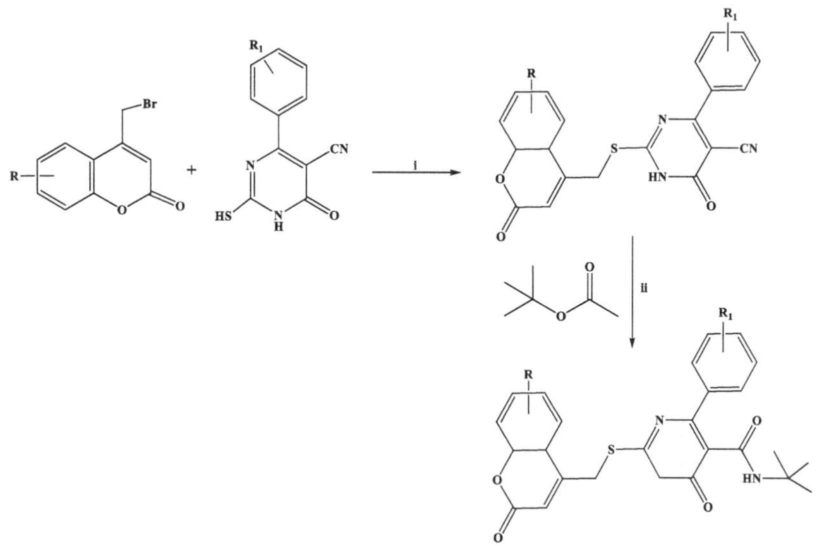

13

R=6-CH$_3$, 7-CH$_3$, 5,7-diCH$_3$, 6-ter-butyl, 6-OCH$_3$, 5,6-Benzo, 7,7-Benzo; R$_1$=H

R=6-CH$_3$, 7-CH$_3$, 5,7-diCH$_3$, R$_2$=4-OCH3

SCHEME 5 Synthesis of coumarin-pyrimidine amide derivatives by using Ritter reaction under microwave irradiation. Reagents and conditions: (i) Acetone, K_2CO_3, RT; (ii) MW, 65–70°C, 200 W, 6–10 min.

The synthesized compounds were evaluated for their antibacterial activity. They used the agar-well diffusion method and antiinflammatory activity by egg albumin denaturation method for activity test. One compound exhibited an antibacterial effect with MIC-2.5 gm/mL against gram-positive *Staphylococcus aureus* bacterial strain compared to standard ciprofloxacin drug (MIC-10 lg/mL). Another compound showed inhibition of heat-induced protein denaturation 75.42% at a concentration of 31.25 lg/mL and is about ten times more active compared to standard aceclofenac drug (5.5%).

5.3 Steroidal derivatives

Steroids are natural compounds occurring in living organisms and possess a broad array of biological activities [23]. To enhance their valuable biological activities, chemists were directed toward the modification in the structure of steroids. That led to the introduction of more specific and potent therapeutic agents [24].

A series of steroidal thiazole derivatives by one-step reaction methodology were synthesized by Asif et al. A reaction of steroidal ketones with thiosemicarbazide and phenacyl bromide in ethanol under microwave irradiation was performed [25]. The reactions are shown in Scheme 6.

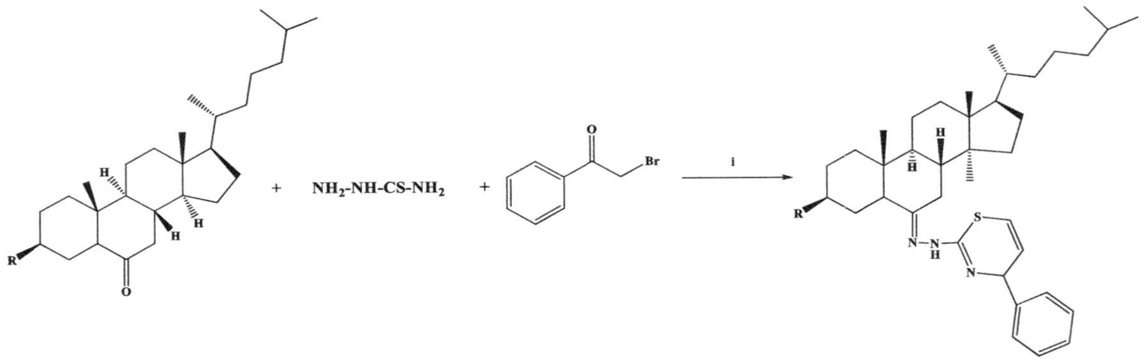

14 **15** **16** **17**

R=OAc, Cl, H

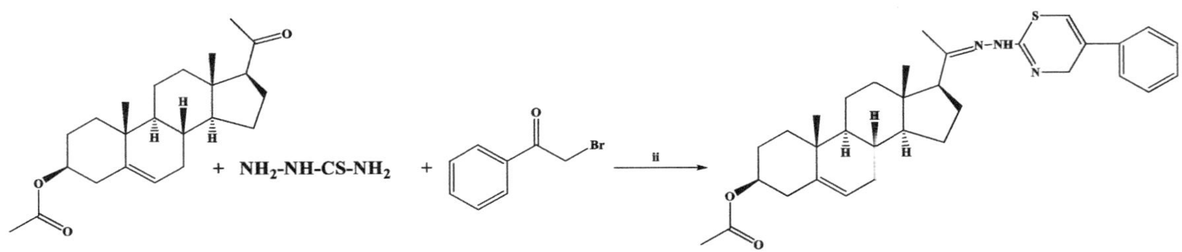

18 **19** **20** **21**

SCHEME 6 Synthesis of steroidal thiazoles. Reagents and conditions: (i) EtOH, 35–40 min, MW; (ii) EtOH, 45 min, MW.

The synthesized compounds were analyzed for in vitro antioxidant activity molecular property prediction, pBR322DNA cleavage activity, reactive oxygen species (ROS) production, and RBC hemolysis.

5.4 Thiazoles

Thiazoles are very important functional groups for application in medicinal chemistry. They act as ligands toward a great variety of biological substrates. Thiazoles are used in a wide range of therapeutic applications such as antimicrobial, antiviral, antifungal, antiallergic, antihypertension, pain treatment, and also for controlling schizophrenia symptoms [26–29].

Cagide et al. reported the synthesis of 4-oxo-*N*-(substituted-thiazol-2-yl)-4*H*-chromene-2-carboxamides (**24**) from chromone-2-carboxylic acid by two different amidation methods (with conventional heating and/or microwave irradiation) [30]. These new series of chromone-thiazole hybrids were designed as potential ligands for human adenosine receptors. The synthesis of chromone-thiazole hybrids is shown in Scheme 7. In this method, the chromone carboxamide derivatives were obtained by a condensation reaction that requires the previous activation of the chromone-2-carboxylic with phosphorus oxychloride and following the addition of the aminothiazole derivative to the formed acyl chloride in a process assisted by microwave irradiation.

The same research group also reported another method for the intended chromone without microwave irradiation. The chromone-2-carboxylic acid was activated by reaction with benzotriazol-1-yloxytripyrrolidinophosphonium hexafluorophosphate (PyBOP), and the in situ formed intermediate reacted with the aminothiazole or its derivatives giving rise to the chromone carboxamides. Even though both the methods provided moderate yields, the PyBOP reactions were more time-consuming, and also, the coupling reagent was quite expensive.

Chinnaraja et al. described a microwave-assisted synthesis of hydrazinyl thiazoles under the solvent- and catalyst-free conditions [31]. A series of aryl

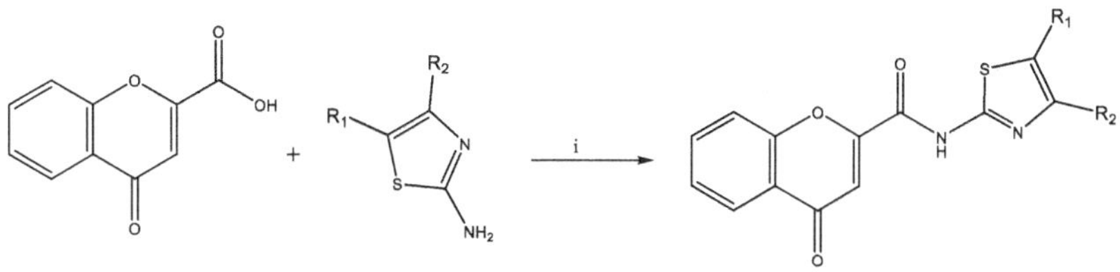

22 **23** **24**

R_1=CH_3, COOEt; R_2=CH_3

SCHEME 7 Synthesis of 4-oxo-*N*-(substituted-thiazol-2-yl)-4*H*-chromene-2-carboxamides. Reagents and conditions: (i) PCl_3O, DMF, MW.

ketones/4-benzoyl pyridine thiosemicarbazone, thiosemicarbazide, and α-haloketones were used for the synthesis.

To optimize the conditions, they examined the reaction of acetophenones with thiosemicarbazide and α-haloketones under different conditions using MW activation. The best result was obtained when the reaction mixture was irradiated at a power level of 300 W for 30–175 s and 400 W for 50–120 s. Schemes 8 and 9 show the synthesis of a series of hydrazinyl thiazoles (**28**, **32**, and **35**) using substituted acetophenones/benzophenones under optimized conditions. This study showed that the reaction of acetophenone having the electron-donating methoxy group was faster compared to that one has electron-withdrawing groups. Along with that, a high yield of the product was obtained when there is the electron-donating group in the para position.

The reaction of hydrazones of 4-benzoyl pyridine with α-haloketones is shown in Scheme 10.

Bhoi et al. described a one-pot, three-component microwave-assisted synthesis of novel ethyl 2-methyl-4-(pyridin-2-yl)-4*H*-benzo[4,5]thiazolo[3,2-a] pyrimidine-3-carboxylate derivatives by the reaction of 2-aminobenzothiazole derivatives with pyridine 2-aldehyde and ethyl acetoacetate in the presence of $PdCl_2$ as an expeditious catalyst under solvent-free condition (Scheme 11) [32].

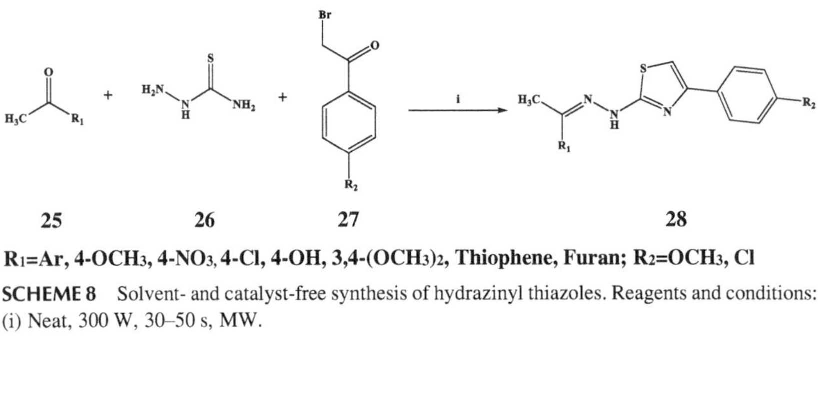

25 **26** **27** **28**

R_1=Ar, 4-OCH_3, 4-NO_3, 4-Cl, 4-OH, 3,4-$(OCH_3)_2$, Thiophene, Furan; R_2=OCH_3, Cl

SCHEME 8 Solvent- and catalyst-free synthesis of hydrazinyl thiazoles. Reagents and conditions: (i) Neat, 300 W, 30–50 s, MW.

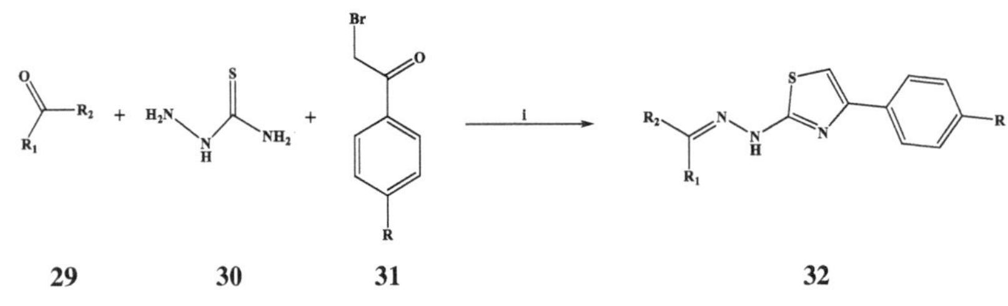

29 **30** **31** **32**

R_1= 4-F, 4-NH_2, 4-NH_2-5-NO_2, 4-Cl, 4-NH_2-5-Cl, Ar; R_2=4,4'$(CH_3)_2$; R=OCH_3, Cl, NO_2.

SCHEME 9 Solvent- and catalyst-free synthesis of hydrazinyl thiazoles. Reagents and conditions: (i) Neat, 300 W, 30–175 s, MW.

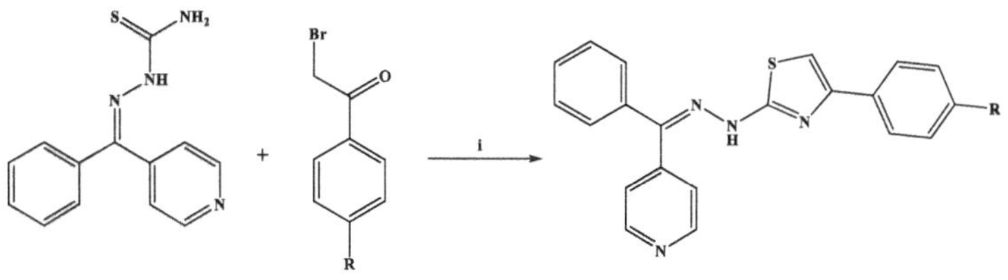

33 34 35

R=OCH_3, Cl, NO_2, H.

SCHEME 10 Solvent- and catalyst-free synthesis of hydrazinyl thiazoles. Reagents and conditions: (i) Neat, 400 W, 50–120 s, MW.

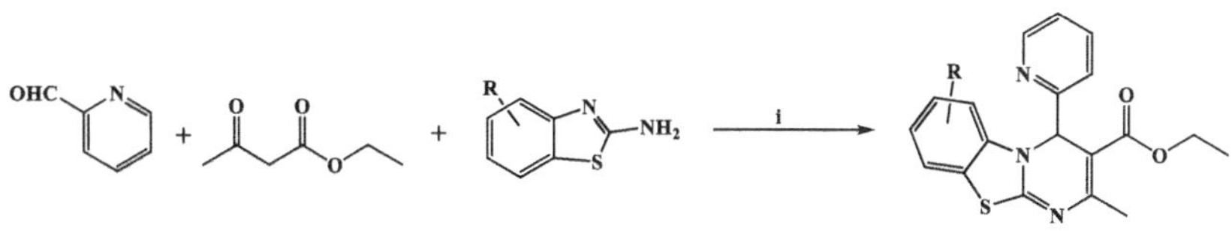

36 37 38 39

R=H, Br, Me, NO_2, Cl, F, OCH_3, $OCCH_3$, OCF_3, OH

SCHEME 11 Synthesis of tri-heterocyclic benzothiazole derivatives. Reagents and conditions: (i) $PdCl_2$(10%), Neat, 90°C, 3–5 min, MW.

All the synthesized compounds were screened for antibacterial and antioxidant activities. These compounds were proved to be active against the two gram-negative and two gram-positive references of bacterial strains. The compounds also showed moderate to good antioxidant activity, when the compounds were evaluated as antioxidants according to a DPPH radical scavenging activity, superoxide anion scavenging assay, and $ABTS^+$ radical scavenging activity. The compounds were also screened for antitubercular activity against *Mycobacterium tuberculosis* $H_{37}RV$ strain. Two compounds showed very good activity against *M. tuberculosis* H37RV, and other compounds showed moderate to good antitubercular activity.

5.5 Imines

A series of imine-derived triazines in good to excellent yields (47%–92%) using green conditions, microwave irradiation in the absence of an acid catalyst, were prepared by Corrochano et al [33].

Scheme 12 shows the reaction of 6-(4-aminophenyl)-1,3,5-triazine-2,4-diamine with the appropriate aldehyde using microwave irradiation in the presence of a minimum amount of solvent (1 mL/mmol). The optimized conditions are shown in Table 1. **43a** was obtained at 80°C with irradiation at 90 W for 30 min and with a maximum pressure of 10 bar using ethanol as

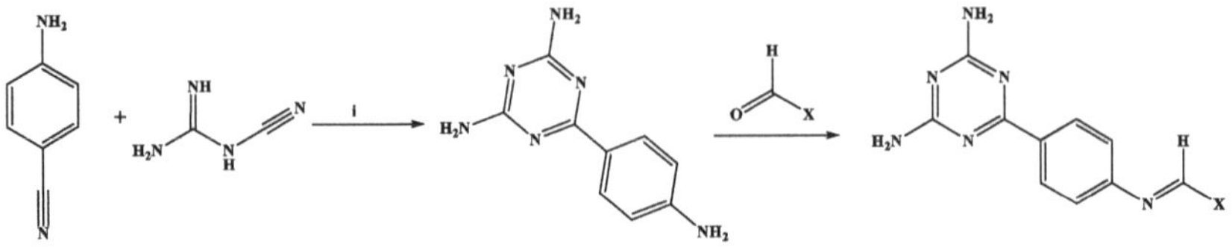

40 **41** **42** **43**

SCHEME 12 Preparation of imine-derived triazine amines. Reagents and conditions: (i) KOH, DMSO.

TABLE 1 Microwave-assisted conditions for the preparation of imines.

Compounds	X	Solvent	MW conditions (W)	*T* (°C)	Time (h)	Yield (%)
43a	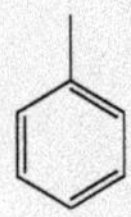	Ethanol	90	80	0.5	86
43b	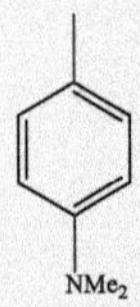	Pentanol	80	130	2	47
43c	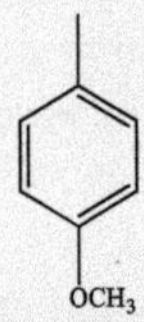	Pentanol	80	130	0.5	92
43d	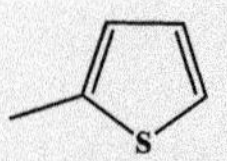	Pentanol	80	130	2	89
43e	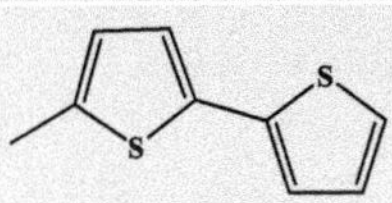	Pentanol	80	130	2	64

the solvent. These conditions were not suitable for the complete conversion of the starting materials with less reactive aldehydes, even on using acid catalysis (*p*-toluenesulfonic acid). Therefore, a high-boiling solvent, pentanol, and neutral conditions were used to produce a complete conversion of the starting materials. The compounds were observed to be violet-blue emitters and show large Stokes shifts which are a good characteristic for the preparation of the most appreciated blue LEDs.

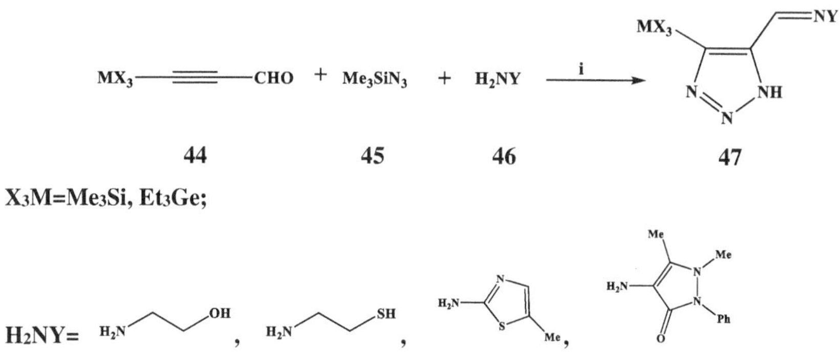

SCHEME 13 Synthesis of 1*H*-1,2,3-triazoloimines. Reagents and conditions: (i) Solvent-free, 100°C, 10 min, MW, propynals (2 mmol), trimethylsilyl azide (4 mmol), amine (2 mmol).

Medvedeva et al. reported a microwave-assisted solvent- and catalyst-free three-component synthesis of *NH*-1,2,3-triazoloimines [34]. They described mostly the synthesis of 5-*R*-imino-[4-trialkylsilyl(germyl)]-1*H*-1,2,3-triazoles (**47**) from Si/Ge-substituted propynals, trimethylsilyl azide, and functionalized primary amines (Scheme 13).

5.6 Oxadiazoles

Oxadiazoles are an important class of heterocyclic aromatic compounds of the azole family. Four isomers of oxadiazole are identified. The oxadiazole derivatives may act as ester and amide bioisosteres, and hence, they are important in pharmaceutical and agrochemical fields [35]. The broad range of activities associated with oxadiazoles includes antimicrobial [36], antiviral [37], antineoplastic [38], inhibition of tyrosinase [39], fungicidal [40], and cathepsin K [41].

Farshori et al. reported a rapid and efficient solvent-free synthesis of 2,5-disubstituted-1,3,4-oxadiazoles (**50**) from fatty acid hydrazides under microwave irradiation [42]. The reaction is shown in Scheme 14. During this method, the reactions were carried out by irradiating a mixture of fatty acid hydrazide withthe appropriate carboxylic acid, supported on neutral alumina, using phosphorus oxychloride as the cyclizing agent. To optimize the reaction conditions for the synthesis of oxadiazole bearing an alkanyl/alkenyl/hydroxyalkenyl

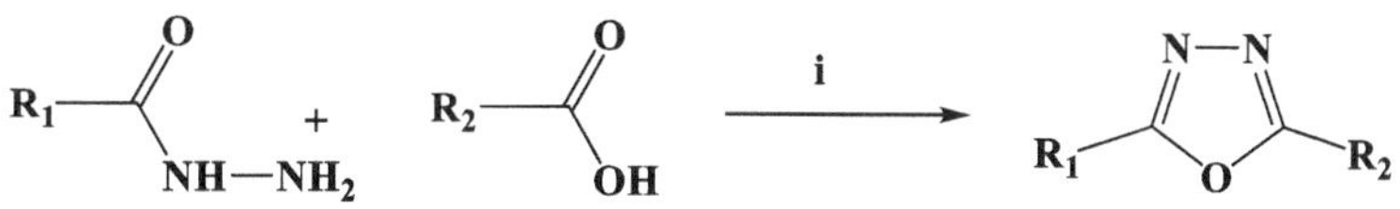

48 49 50

SCHEME 14 Synthesis of 2,5-disubstituted-1,3,4-oxadiazoles under microwave irradiation. Reagents and conditions: (i) $POCl_3$, Al_2O_3, MW.

chain substituent, variations in molar ratios of reagents, the microwave irradiation time, and the power level of microwave setup were investigated during this study.

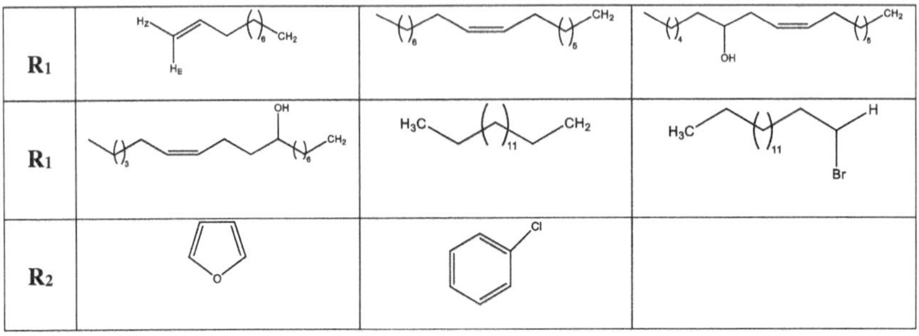

All the synthesized compounds were screened for antibacterial and antifungal activities. Three compounds were found to be the most potent antimicrobial agents.

5.7 Tetrazoles and triazoles

Tetrazole derivatives are important therapeutic scaffolds that are present in many natural products and pharmaceutical compounds [43–46]. Besides, they are considered as a bioisosteric substituent of carboxylic acid during the synthesis of biologically active compounds [47]. It also possesses excellent metabolic stability, hydrogen bonding properties, and lipophilicity. Because of these significant properties, these scaffolds facilitate interactions between ligands and receptors to accelerate transport across cell membranes [48, 49]. Even though different methods are available for the synthesis of tetrazoles, the most common method is the reaction of NaN_3 with organic nitriles.

Padmaja et al. described an environmentally friendly, microwave-assisted synthesis of 5-substituted 1*H*-tetrazoles via (3 + 2) dipolar cycloaddition of nitriles and NaN_3 using recyclable CuO nanoparticles [50].

Scheme 15 shows the nano-CuO-catalyzed synthesis of 5-substituted 1*H*-tetrazoles (**53**) that involve the (3 + 2) cycloaddition reaction using DMF as the solvent with microwave irradiation (75°C, 15–18 min). After the reaction, an acidic treatment produced excellent yields (89%–99%) of the substituted 1*H*-tetrazoles. They also reported that the CuO nanoparticles were recycled 5 times without any marked loss of catalytic activity.

Cycloaddition of 1,2,3-triazole moiety with an already bioactive organophosphorus compound through the click synthesis is important. This process can further enhance the biological activity of the resulting compounds, which have good polar hosting sites for efficient chelation with the metal or metal complexes in the biofluids in the living systems.

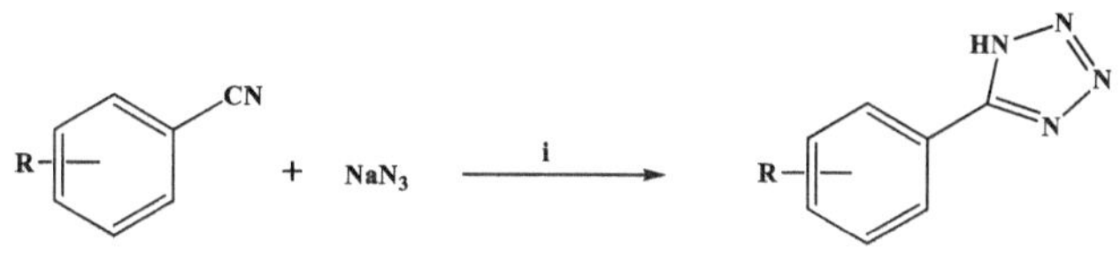

51 **52** **53**

R=H, CH$_3$, OCH$_3$, NO$_2$, Cl, Br, NH$_2$, CHO, COOCH$_3$

SCHEME 15 Microwave-assisted nano-CuO-catalyzed synthesis of 1*H*-tetrazoles. Reagents and conditions: (i) Nano-CuO, DMF, MW, 75°C, 15–18 min.

An efficient and simple one-pot microwave-assisted click formation of 1-(substituted)-1*H*-1,2,3-triazol-4-yl)methyl)diphenylphosphineoxide derivatives (**56**) was described by Chinthaparthi et al [51]. The synthesis occurred via Huisgen regioselective [3 + 2]-cycloaddition of an in situ generated organic azides and diphenyl(prop-2-yn-1-yl)phosphine oxide in a highly polar DMSO-H_2O medium.

R=

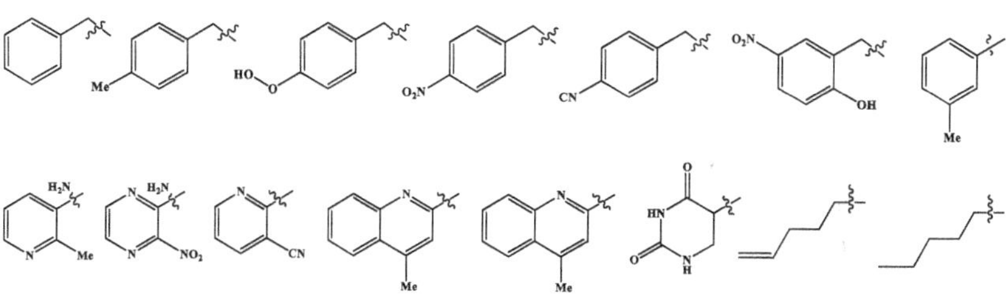

The synthesis is depicted in Scheme 16. The 1,3-cycloaddition reaction was performed between diphenyl(prop-2-yn-1-yl)phosphine oxide (4) precursor and benzyl azide in the presence of CuBr(PPh$_3$)$_3$ as a catalyst in DMSO. To optimize the reaction conditions, different reaction conditions were considered

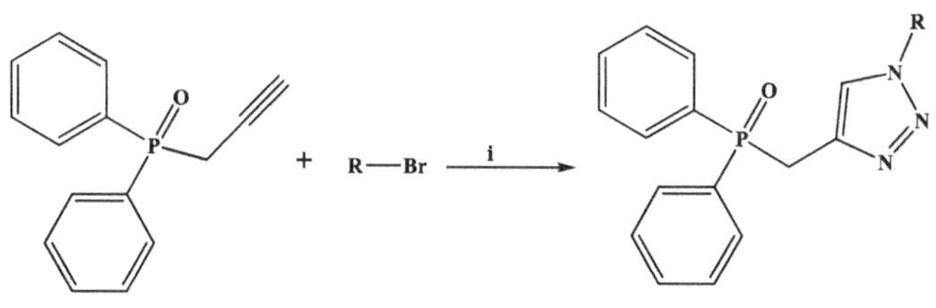

54 **55** **56**

SCHEME 16 Microwave-assisted CuAAC synthesis of 1-(substituted)-1*H*-1,2,3-triazol-4-yl) methyl)diphenylphosphine oxides. Reagents and conditions: (i) NaN$_3$/H$_2$O, CuBr(PPh$_3$)$_3$, DMSO, MW, 120–170°C, 12–15 min.

and it is observed that an increase of MW power, reaction time and temperature, has a significant effect on the rate of the reaction and yield of the product. And the optimum experimental conditions were observed with a high polar solvent medium of DMSO/H_2O (1:1) and 280 W of microwave irradiation at 100°C for 10 min using 0.5 mol% of $CuBr(PPh_3)_3$ as a catalyst. This synthetic procedure allowed for a one-pot three-component commercial preparation of a wide variety of bio-active 1,2,3-triazole diphenylphosphine oxides in a very good yield.

5.8 Quinolines

Quinoline is a heterocyclic aromatic compound. Quinoline and many of its derivatives are useful in diverse applications. The important compounds such as quinine, chloroquine, amodiaquine, primaquine, cryptolepine, neocryptolepine, and isocryptolepine belong to the quinoline family. Also, these compounds have antimalarial activity and that has made them attractive synthetic targets.

Synthesis of the tetracyclic ring system of isocryptolepine and regioisomers was reported by Håheim et al [52]. Suzuki-Miyaura cross-coupling reaction between bromoquinolines (2-bromoquinoline-8-bromoquinoline and 5-bromo-3-methoxyquinoline) and 2-aminophenylboronic acid hydrochloride produced the desired biaryl systems in good yields. The synthesized biaryls were then subjected to microwave-assisted palladium-catalyzed C—H activation/C—N bond formation reaction utilizing $PdCl_2$(dppf) as shown in Scheme 17.

Selected tetracyclic ring systems formed during this process are shown in Table 2, which is required for the formation of isocryptolepine. The results showed that the quinoline scaffold is the most active, and it appears that the electron densities at the various carbons determined the regioselectivity of the cyclization process.

Parvatkar et al. reported the microwave-induced $Bi(NO_3)_3$-catalyzed one-pot synthesis of a series of linear indoloquinolines under mild reaction conditions [53]. The advantages of this methodology include short reaction time; use

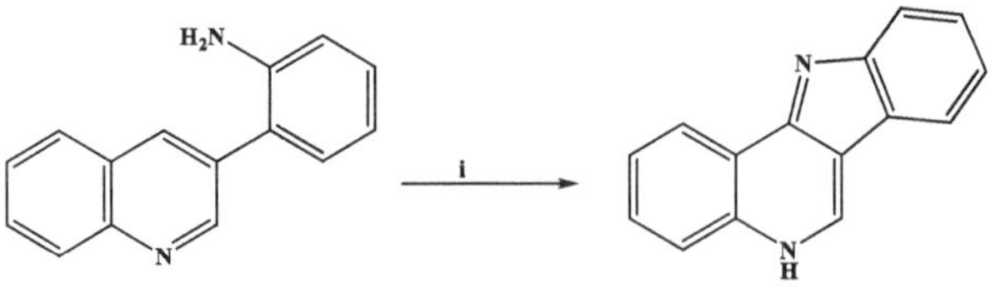

57 58

SCHEME 17 Intramolecular cyclization via tandem C—H activation and C—N bond formation. Reagents and conditions: (i) $PdCl_2$(dppf) (20 mol%), 1,3-bis(2,4,6-trimethylphenyl)-imidazolim (IMes) (5 mol%), H_2O_2 (35 wt%, 29 mol%), AcOH, MW sealed reactor tube 118°C.

TABLE 2 Selected tetracyclic ring systems of isocryptolepine.

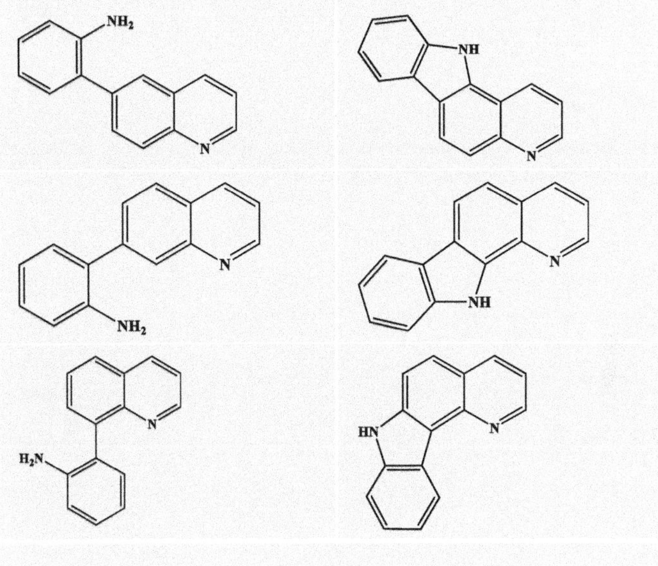

of inexpensive, nontoxic, and commercially available catalyst; ease of product isolation; good yields; and mild reaction conditions.

During this process, a mixture of indole-3-carboxyaldehyde and aniline in presence of 10 mol% of Bi(III) salt as a catalyst in a tightly sealed vessel under solvent-free condition were irradiated in the microwave at 60°C. To optimize the reaction conditions, different analyses were performed and the results revealed that, in the absence of catalyst and with Bi_2O_3 catalyst, no product is formed, whereas in the presence of other Bi(III) salts ($Bi(NO_3)_3 \cdot 5H_2O$, $BiCl_3$, BiI_3), the desired product was obtained within a short period. The reaction in the presence of $Bi(NO_3)_3 \cdot 5H_2O$ gave the best results in terms of both yield and time. These experiments revealed that 10 mol% loading of the $Bi(NO_3)_3 \cdot 5H_2O$ catalyst provided the highest yield of desired products in just 3 min.

The schematics of the protocol are shown in Scheme 18. A series of compounds were prepared using various aromatic/hetero-aromatic amines and indole-3-carboxyaldehyde, and six selected synthesized compounds are shown in Table 3.

5.9 Indolizines

Indolizine is a heterocyclic compound and is an isomer of indole. Indolizine derivatives have found a wide range of applications in natural products and synthetic pharmaceuticals. The broad spectrum of pharmacological properties

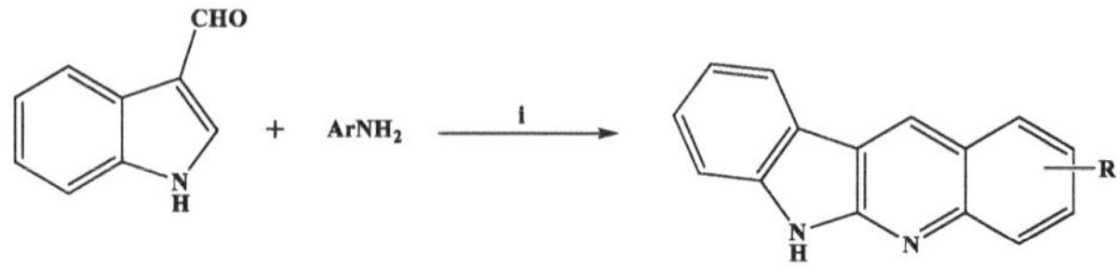

59 60 61

SCHEME 18 Microwave-mediated synthesis of indoloquinolines. Reagents and conditions: (i) Bi $(NO_3)_3{\cdot}5H_2O$ (10 mol%, 0.1 equiv.), MW, 300 W, 60°C.

TABLE 3 Six selected synthesized indoloquinolines.

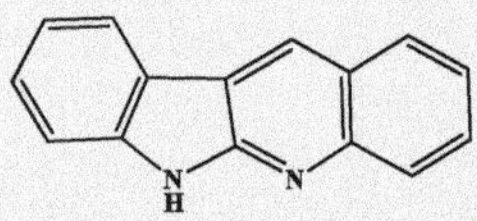 61a	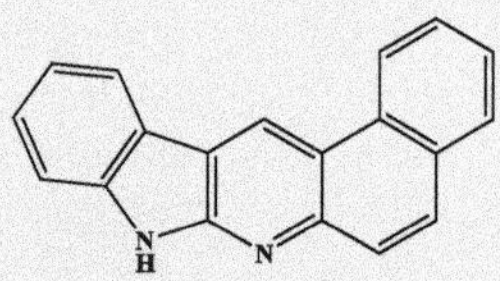 61b	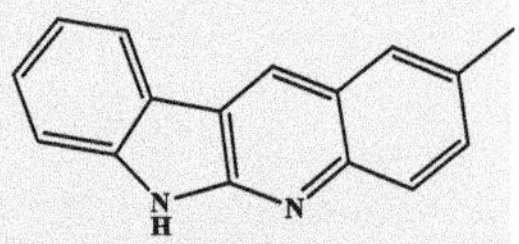 61c
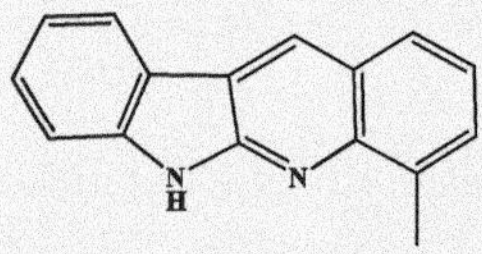 61d	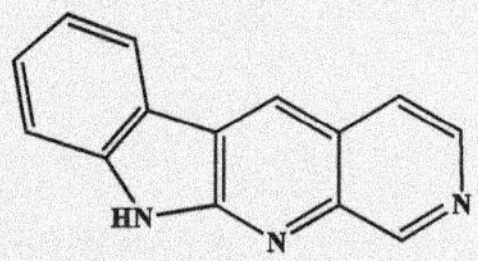 61e	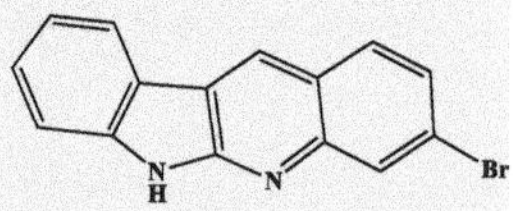 61f

of indolizines includes anticancer [54, 55], analgesic [56], antidiabetic [57], antiinflammatory [58], antihistaminic [59], antileishmanic [60], antimutagenic [61], antimicrobial [62], antioxidant [63], antiviral [64], antitubercular [65], larvicidal [66], in vitro COX-2 inhibition [67], and herbicidal activities [68]. There are different methods available for rapid construction and functionalization of indolizines, especially toward differently substituted indolizines. The methods include Tschitschibabin reaction, Schotz reaction, dipolar cycloadditions of pyridinium and related heteroaromatic ylides with electron-deficient alkynes or alkenes, C—H functionalization reaction, cyclization of pyridines with alkenyldiazoacetates, transmetal-catalyzed cycloisomerizations of alkynylpyridine derivatives, trans-annulations of pyridotriazoles with alkynes, and 1,3-dipolar cyclocondensation using potassium dichromate as oxidant.

Chandrashekharappa et al. investigated the synthesis of a series of novel ethyl 3-substituted-7-methylindolizine-1-carboxylates by the microwave-assisted one-pot method [69].

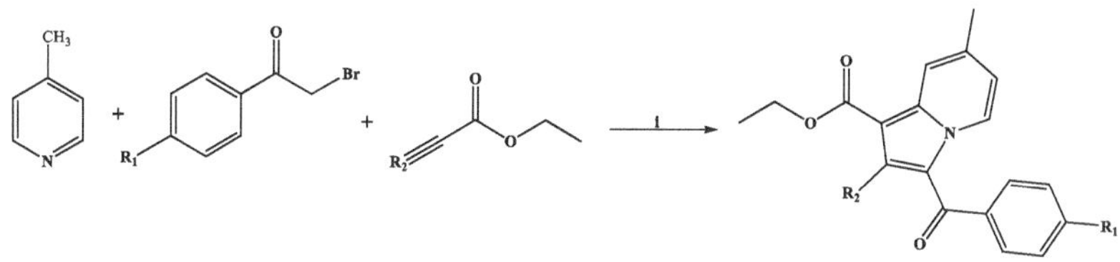

62 **63** **64** **65**

R_1= Cl, F, CN, H, Br; R_2=H, CH_3, C_2H_5

SCHEME 19 Microwave-assisted one-pot synthesis of ethyl 3-substituted-7-methylindolizine-1-carboxylates. Reagents and conditions: (i) Acetonitrile, 100°C, 5 min, MW.

The compounds were obtained by running the reaction at a 1.07 mmol scale between 4-methyl pyridine, 4-chlorophenacyl bromide, ethylbutynoate, and triethylamine in acetonitrile in 8 mL microwave tube under nitrogen atmosphere for 5 min as shown in Scheme 19. To optimize the reaction conditions, different bases and solvents were considered. Trimethylamine along with acetonitrile emerged as a good combination for the construction of ethyl 3-substituted-7-methylindolizine-1-carboxylate scaffolds in less reaction time with good yields. Larvicidal properties of the synthesized compounds were evaluated against *Anopheles arabiensis,* and it was found that indolizine pharmacophore influences larvicidal activity. The structure-activity relationship of ethyl 3-substituted-7-methylindolizine-1-carboxylates showed that different functional groups on the para position of the phenyl ring and second carbon on indolizine pharmacophore are very essential for potential larvicidal activity.

Panda et al. developed an efficient method for the synthesis of indole derivatives via microwave-assisted technology [70]. In this method, the Schiff bases were produced by the reaction of 1*H*-indole-2,3-dione with various substituted anilines in presence of acetic acid under microwave irradiation. After that, the Mannich bases were synthesized by the condensation reaction of Schiff bases with different secondary amines in the presence of formaldehyde. All the synthesized compounds were evaluated for anthelmintic activity against *Pheretima posthuma*. The compounds were screened for anthelmintic activity at the concentration of 10, 20, and 50 mg/mL along with the standard drug albendazole at 10 mg/mL. The results showed that some of the synthesized compounds display significant anthelmintic activity in a dose-dependent manner.

5.10 β-Lactams

Synthesis of enantiopure C-3/C-4 disubstituted β-lactams through Staudinger cycloaddition reaction of acid chloride and imine under microwave irradiation was reported by Indrani et al. [71].

In the conventional method, the reaction of the imine with acetoxyacetyl chloride in the presence of trimethylamine in dry dichloromethane produced a single optically active β-lactam in good yield. Using the microwave-induced method, the same reaction for β-lactam occurred within a few minutes. Thus, the reaction of acetoxyacetyl chloride with imine obtained through a 5-membered carbohydrate derivative in dichloroethane in the presence of *N*-methylmorpholine under microwave irradiation produced an identical β-lactam in good yield (Scheme 20). Higher boiling *N*-methylmorpholine was used instead of triethylamine in the microwave-induced method, and also, dichloroethane was used instead of dichloromethane in the microwave method. After the reaction, the protective group in lactam was removed with methanolic iodine solution and the intermediate was then oxidized with ruthenium-mediated sodium metaperiodate-mediated oxidation to the carboxylic acid. Using diazomethane, the carboxylic acid was then esterified to methyl ester without any isolation. These series of reactions produced a compound with 3S, 4R absolute stereochemistry at the ring system.

The same group had conducted a similar reaction with the nitrogen-containing group at the C-3 center. In that reaction, chiral imine and phthalimidoacetyl chloride were used to produce a single optically active β-Lactam. The reaction was performed in the presence of dichloroethane and *N*-metyhlmorpholine under microwave irradiation. The phthalimido group in lactam was then removed by ethylenediamine to produce an amino compound with 3R, 4S absolute stereochemistry at the ring system as shown in Scheme 21.

It is also possible to invert the stereochemistry at the C-3 center of the lactam compound (10). The method is shown in Scheme 22. The compound was first treated with a strong base DBN under microwave irradiation method and the

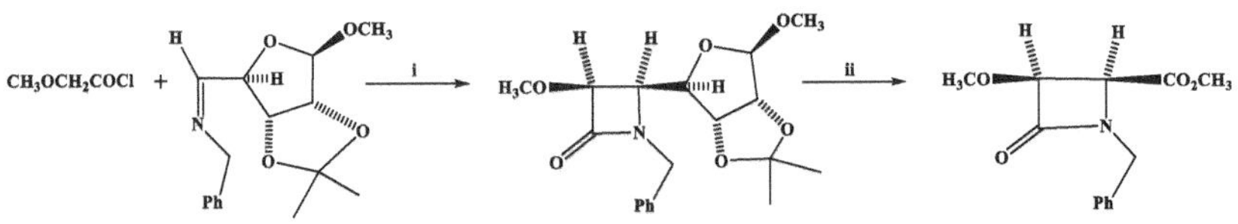

66 67 68 69

SCHEME 20 Stereospecific synthesis of optically active β-lactam. Reagents and conditions: (i) NMM, CH_2ClCH_2Cl, MW; (ii) I_2/CH_3OH, $RuCl_3/NaIO_4$, CH_2N_2.

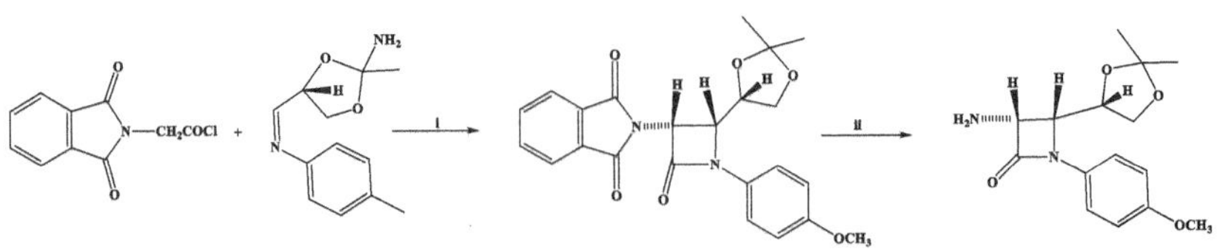

70 71 72 73

SCHEME 21 Stereospecific synthesis of optically active β-lactam. Reagents and conditions: (i) NMM, CH_2ClCH_2Cl, MW; (ii) $NH_2(CH_2)_2NH_2$, MW.

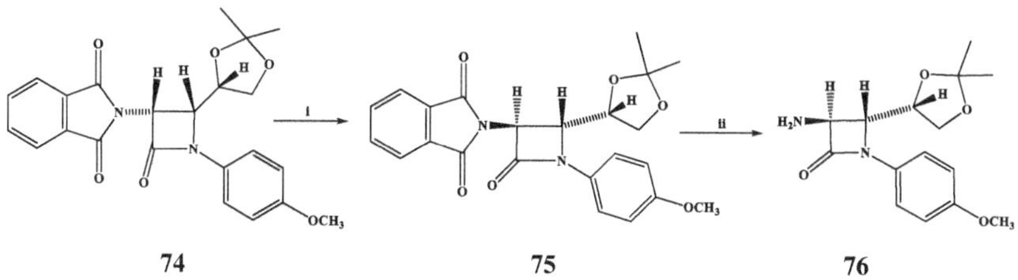

SCHEME 22 Enantiospecific synthesis of optically active β-lactam. Reagents and conditions: (i) DBN, MW; (ii) $NH_2(CH_2)_2NH_2$, MW.

intermediate compound, and then, treatment of ethylenediamine under microwave irradiation produced the lactam compound with 3S, 4S absolute stereochemistry at the ring junction.

Sulfonamide-β-lactam molecular conjugates were synthesized through the microwave-assisted method [72]. β-Lactams reacted with sulfonamides under microwave irradiation in the presence of KI and mediated by K_2CO_3 to produce the corresponding conjugates.

Microwave-assisted synthesis of β-lactam-sulfonamide conjugates (**79**) is shown in Scheme 23. During this reaction, nucleophilic coupling via the primary aromatic amine function of representative sulfonamides with *N*-haloalkyl β-lactams has occurred. Six selected synthesized β-lactam-sulfonamide conjugates are shown in Table 4.

A robust microwave-assisted synthesis for the diastereoselective generation of 3-saccharinyl trans-β-lactams was reported by Janssen et al [73]. The method was optimized for combinatorial library production in which decoration of the scaffold was altered on both the β-lactam and the saccharine moiety.

To optimize the Staudinger reaction, saccharinyl acetic acid and imine were considered as benchmark substrates. In the initial experiment, saccharinyl acetic acid was treated first with Mukaiyama's salt in CH_2Cl_2 for 8 h at reflux,

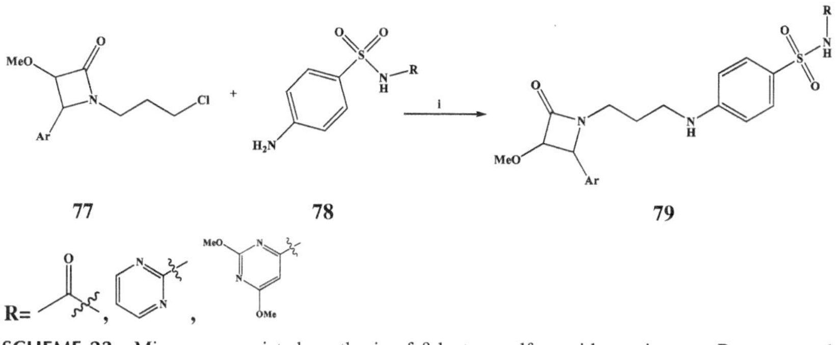

SCHEME 23 Microwave-assisted synthesis of β-lactam-sulfonamide conjugates. Reagents and conditions: (i) KI, K_2CO_3, MeCN, MW, 1 h.

TABLE 4 Six selected synthesized β-lactam-sulfonamide conjugates.

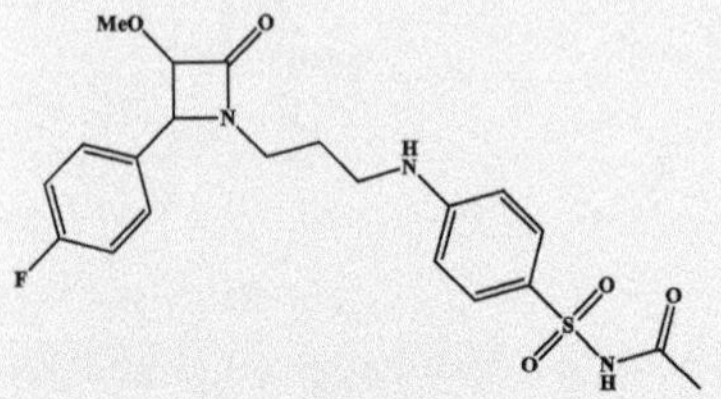 **79a**	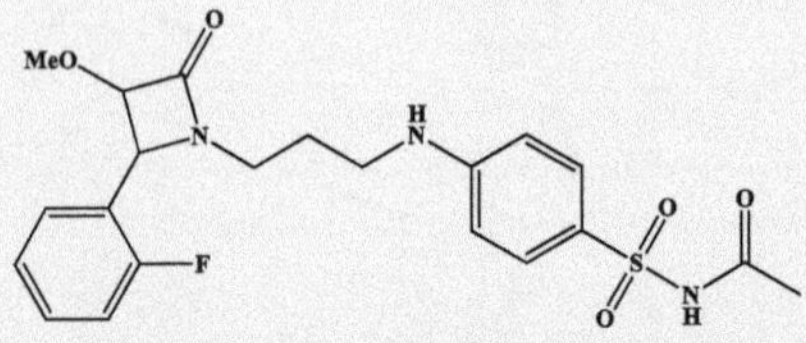 **79b**
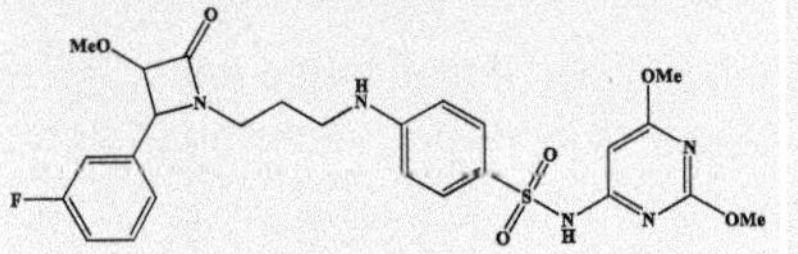 **79c**	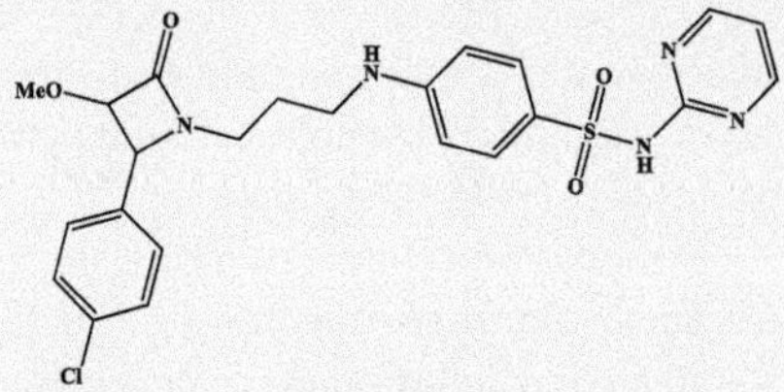 **79d**
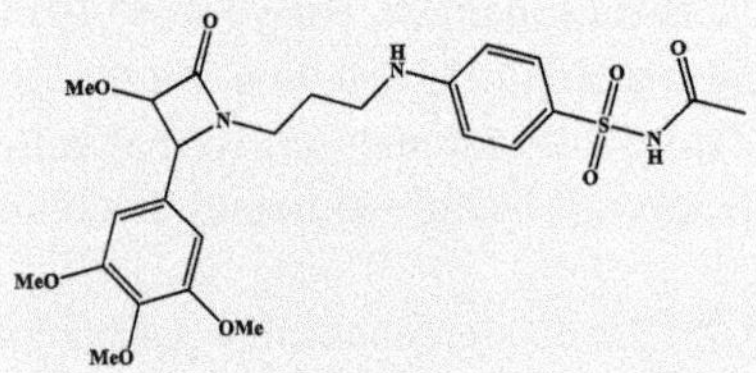 **79e**	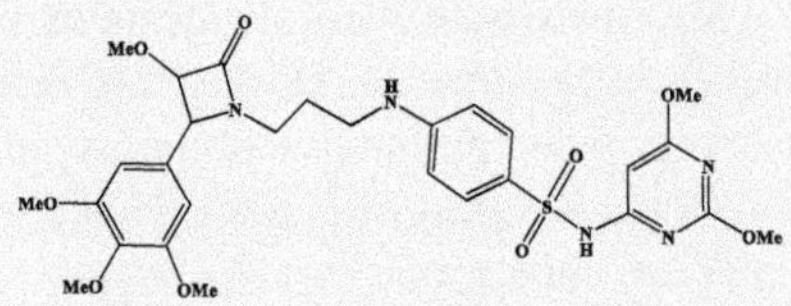 **79f**

after that imine and Et_3N were added, and was refluxed for 16 h. It produced the desired β-lactam in 85% yield with excellent diastereoselectivity. Refluxing of saccharinyl acetic acid, imine, and Mukaiyama's salt in CH_2Cl_2 with Et_3N as the base for 16 h also provided a comparable yield of 82%. A dramatic change in the reaction time was observed when the same synthesis was carried out in the microwave. The reaction time is reduced from 16 h to 10 min. The reaction of saccharinyl acetic acid and imine in a microwave at 100°C for 30 or 10 min afforded the corresponding β-lactam in a good yield of 87% and 86%, respectively, and selectively the *trans* compounds. The reaction of the synthesis is depicted in Scheme 24.

To explore the scope of the reaction, acetic acid was treated with a variety of imines. The results exhibited that electron-donating as well as electron-withdrawing aromatic rings on the R_2- and R_3-position yielded the products in *trans* configuration. At the same time, the reactions of aliphatic imines (derived from neohexanal and *n*-hexanal) resulted in mixtures of unidentified

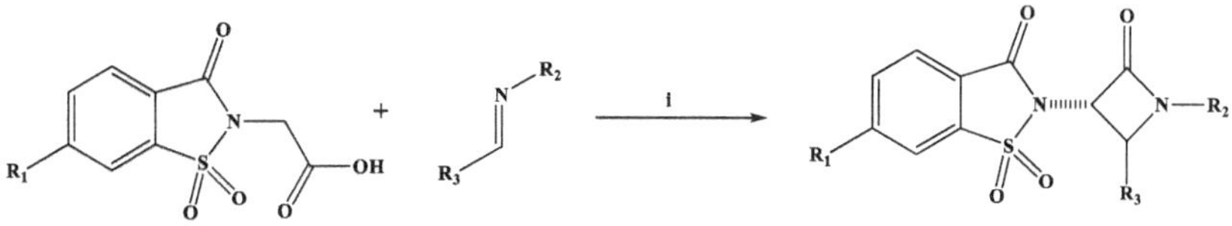

80 **81** **82**

R_1=H, NO_2.

R_2=4-BrC_6H_4, 4-BrC_6H_4, 4-BrC_6H_4, 3-$OMeC_6H_4$, 3-$OMeC_6H_4$, 3-$OMeC_6H_4$; R_3= 4-$OMeC_6H_4$, 4-MeC_6H_4, 3-$NO_2C_6H_4$, 4-$OMeC_6H_4$, 4-MeC_6H_4, 3-$NO_2C_6H_4$ (seven selected R_2 and R_3 substituents).

SCHEME 24 Synthesis of β-lactams. Reagents and conditions: (i) Mukaiyama's salt, CH_2Cl_2, Et_3N, MW, 10 min, 100°C.

products. The reaction with imines derived from isopropylamine synthesized the corresponding products in reasonable yields with full diastereoselectivity. However, the reactions with 4-aminomethylpyridine and β-alanine derived imines or *N*-aliphatic imines were more challenging, because the reaction did not afford any product or only in a minimal yield. On the other hand, the reaction of the 6-nitrosaccharinyl acetic acid with imines produced desired products in acceptable yields as single diastereomers. This methodology was useful to produce a 263 compound library.

In the absence of a tertiary base, synthesis of β-lactams in *N*, *N*-dimethylformamide was carried out by cycloaddition of acid chlorides with imines in a domestic microwave oven [74]. Synthesis of enantiopure 3-hydroxy β-lactams was reported by Indrani et al. [75]. Microwave-mediated bismuth nitrate-induced cleavages of *O*-glycosidic bond in β-lactams were attained to produce their desired hydroxyl derivatives in excellent yields. Yadav et al. described the microwave-induced stereoselective synthesis of novel bis-β-lactams substituted at the 2,7 position of the phenanthrene via ketene-imine [2 + 2] cycloaddition reaction [76].

5.11 Pyrroles

Pyrrole is a heterocyclic aromatic organic compound and has a wide range of applications. It is present in many natural and nonnatural products, and they have various medicinal properties too. Numerous methods are available for the synthesis of pyrroles [77–82].

An efficient microwave-induced synthesis of pyrroles in excellent yield using commercially available 2,5-dimethoxytetrahydrofuran and primary amines in the presence of ammonium chloride (Paal-Knorr reaction) was reported by Bandyopadhyaya et al [82]. The reaction between the 2,5-dimethoxytetrahydrofuran (1.1 mmol) and the aromatic amine (1 mmol) in the presence of ammonium chloride (200 mg, 3.73 mmol) under microwave irradiation (50°C and 300 W) in the absence of any solvent produced N-substituted pyrrole as the only product in excellent yield within 5–8 min. A series of N-substituted pyrrole derivatives such as *N*-phenylpyrrole (93% yield), *N*-4-methoxyphenylpyrrole (90% yield), *N*-4-methylphenylpyrrole (89% yield), *N*-allylpyrrole (90% yield), and *N*-benzylpyrrole (90% yield) were synthesized employing this method.

5.12 Aza cyclo-octanes

The octane ring has served as an important precursor for natural product synthesis. For example, it is found in the Iboga family of alkaloids [83, 84]. It also has other synthetic applications.

Microwave-assisted molecular iodine-catalyzed one-pot three-component aza-Diels-Alder for efficient synthesis of 2-azabicyclo-[2, 2, 2]-cyclooctanones reactions was investigated by Yadav et al [85]. To optimize the conditions for the aza-Diels-Alder reaction, aromatic aldehydes, amines, and cyclohexanone were treated in various solvents under conventional heating. For example, the reaction of the equimolar proportion of aniline (1 mmol), benzaldehyde (1 mmol), and cyclohexenone (1 mmol) was investigated in the presence of 10 mol% of molecular iodine in different solvents such as THF, DCM, ethanol, toluene, benzene, and DME with the various range of temperatures. The reaction was unsuccessful under these conditions. At the same time, the microwave-induced reaction of the same reaction mixtures in DMF and the presence of 10 mol% of molecular iodine provided desired products with high stereochemical control of diastereoselectivity in reasonably good yield.

The reaction is shown in Scheme 25. This method was applied to various aromatic aldehydes, amines, and cyclohexenone to synthesize 2,3-diaryl-substituted nitrogen-containing bicyclic compounds.

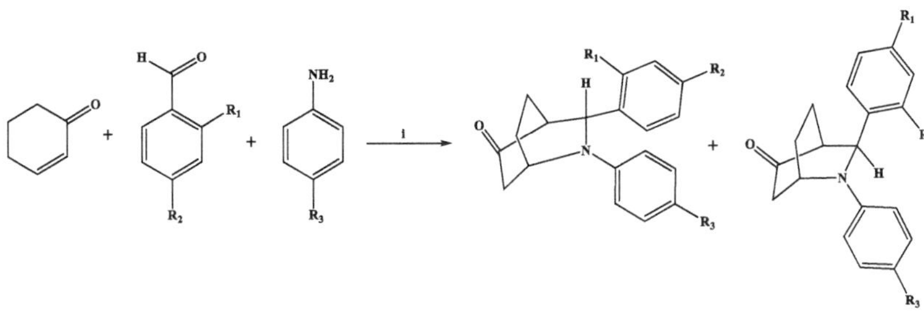

83 **84** **85** **86** **87**

R_1=H, F, CH_3, OMe; R_2=H, OMe, CH_3; R_3=H, OMe, CH_3, 2 OMe.

SCHEME 25 Microwave-induced molecular iodine-catalyzed aza-Diels-Alder Reaction. Reagents and conditions: (i) I_2 (10 mol%), THF, MW, 25–50 min, 30–60°C.

5.13 Schiff bases

Schiff bases are the condensation products of primary amines with carbonyl compounds. Schiff bases own imine or azomethine (—C=N—) functional group in their structures and are considered as a synthon to design and synthesize various important molecules. These compounds possess various biological activities such as analgesic, antimicrobial, antiinflammatory, anticancer, antioxidant, antitubercular, anticonvulsant, and anthelmintic.

A series of pyrazole, triazole based benzohydrazones were synthesized via microwave methods in the presence of acetic acid catalyst [86]. The reaction is shown in Scheme 26. (*E*)-*N*′-((1,3-diaryl-1*H*-pyrazol-4-yl)methylene)-3-(4-phenyl-1*H*-1,2,3-triazol-1-yl) benzohydrazides (**104**) were obtained by the condensation of pyrazole aldehydes with 3-(4-Phenyl-1*H*-1,2,3-triazol-1-yl) benzohydrazide.

During the synthesis, a suspension of pyrazole aldehyde (50 mg, 0.2 mmol) and 3-(4-Phenyl-1*H*-1,2,3-triazol-1-yl) benzohydrazide (56 mg, 0.2 mmol) in 2 mL of ethanol and two drops of glacial acetic acid were irradiated using the microwave at 300 W for 4 min with an interval of 30 s.

Synthesis of the same benzohydrazones was carried out under the conventional method. From the comparison of the results from the conventional and microwave methods, it is clear that the reaction time is reduced from 1–3 h to 2–5 min and the yield increased from 65%–80% to 88%–98% when the microwave method was employed in the synthesis. A series of benzohydrazones were synthesized via microwave methods, and Table 5 shows the four selected compounds.

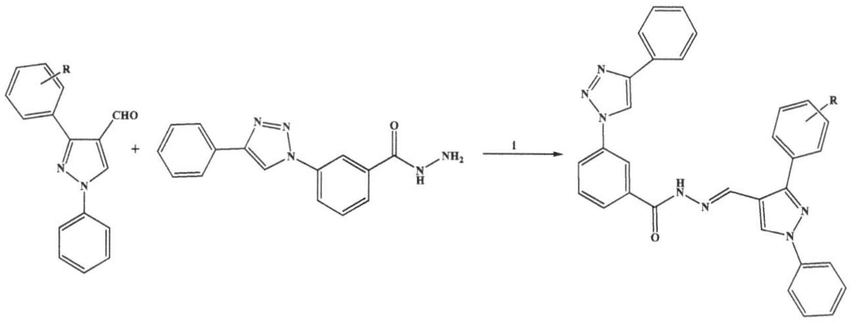

88 **89** **90**

R= H; 4-CH$_3$, 4-OCH$_3$, 4-F, 4-Cl, 4-Br, 4-NO$_2$, 2-Thiophene, 3-OH, 2,4-diOH, diOCH$_3$, 1-Naphthyl.

SCHEME 26 Synthesis of benzohydrazones under microwave irradiation. Reagents and conditions: (i) EtOH, cat. AcOH, MW, 4–6 min.

TABLE 5 Four selected benzohydrazones.

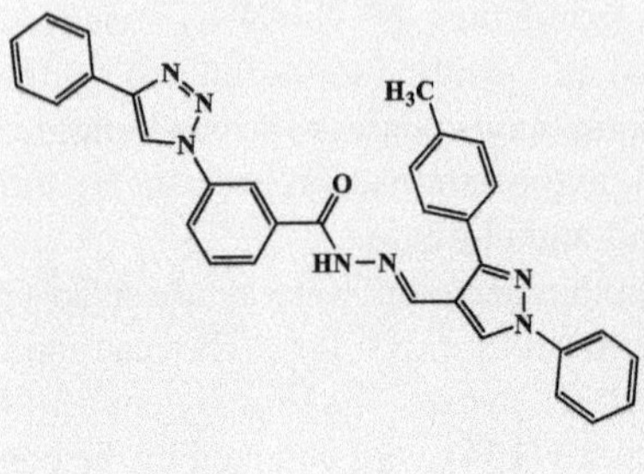

90a

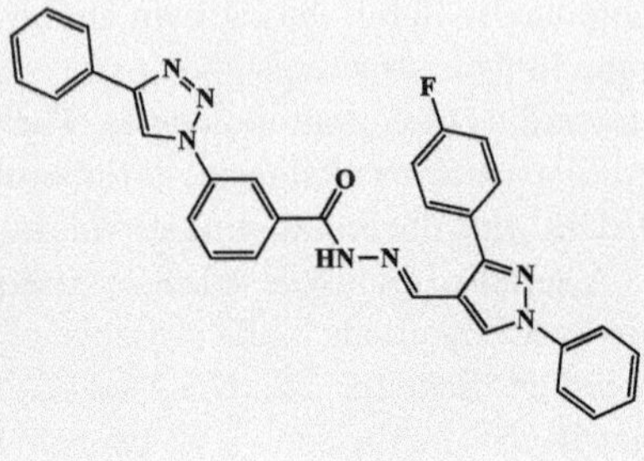

90b

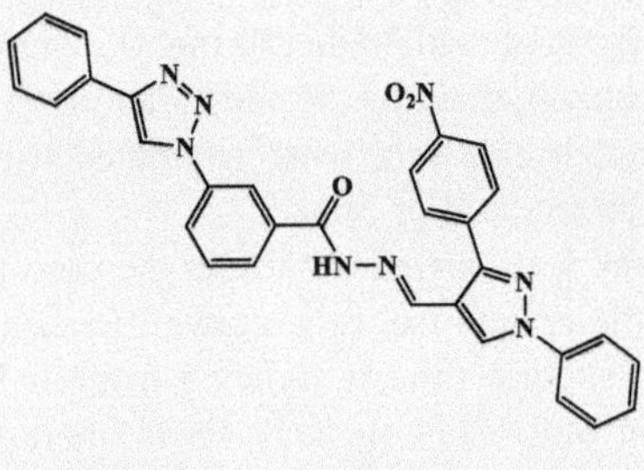

90c

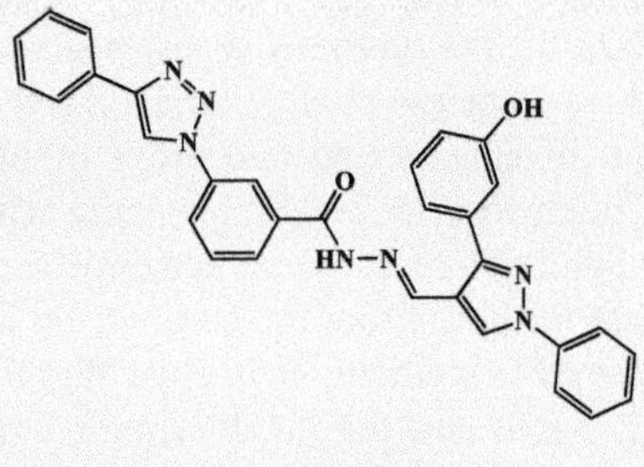

90d

All the synthesized compounds were screened for their in vitro antibacterial activity against gram-positive bacterial strains (*Bacillus cereus* and *Staphylococcus aureus*) and gram-negative bacterial strains (*E. coli* and *Pseudomonas aeruginosa*) at a concentration of 10 and 20 μg/mL. The in vitro antibacterial studies exhibited that 4-chloro and 4-nitro derivatives showed a significant broad spectrum of activities against both gram-negative and gram-positive bacterial strains.

Sahoo et al. reported the microwave-assisted synthesis of Schiff's bases of 1,2,4-triazole derivatives [87]. This synthesis of Schiff's bases of 1,2,4-triazole-3-thiol includes multistep reactions. Benzohydrazide was used as the starting material. The reaction of benzohydrazide with carbon disulfide and potassium hydroxide produced potassium dithiocarbazinate, potassium dithiocarbazinate then undergoes a cyclization reaction in the presence of hydrazine hydrate to yield 4-amino-5-phenyl-4*H*-1,2,4-triazole-3-thiol. The reaction of 4-amino-5-phenyl-4*H*-1,2,4-triazole-3-thiol with substituted benzaldehydes under microwave irradiations for 5–10 min provided the desired Schiffs bases. All the synthesized compounds were tested for their anthelmintic activity.

The microwave-assisted synthesis of (*E*)-3-(4-or3-Aminophenylimino) quinoxaline-2(3*H*)-one oxime Schiff base derivatives was reported by Gopi et al [88]. The synthesis was carried out by the condensation reactions of (*E*)-3-(4-or3-Aminophenyl imino)quinoxaline-2(3*H*)-one and aromatic aldehydes.

The synthesis route is depicted in Scheme 27. Irradiating of (*E*)-3-(Phenylimino) quinoxaline-2(3*H*)-one and benzene 1,2 diamine in 10 mL of ethanol and a few drops of glacial acetic acid produced the key intermediate (*E*)-3-(4-or3-Aminophenyl imino)quinoxaline-2 (3*H*)-one oxime. The condensation reactions of (*E*)-3-(4-or3 Aminophenyl imino) quinoxaline-2(3*H*)-one oxime and aromatic aldehydes provided the (*E*)-3-(4-or3-Aminophenylimino)quinoxaline-2(3*H*)-one oxime Schiff base derivatives.

The synthesized compounds were screened for sporicidal activity against *Leptospira icterohaemorrhagiae* using both in vitro and in vivo methods.

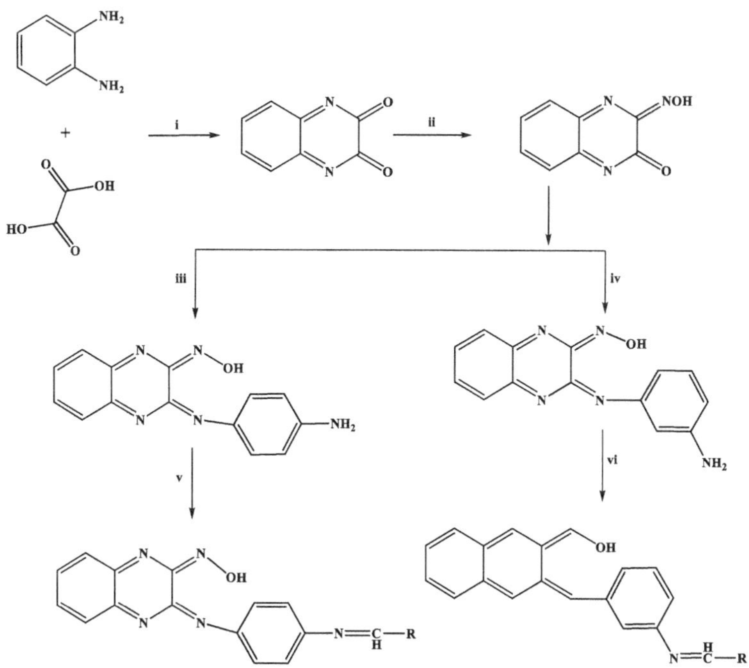

91 **92**

R = C_6H_5, 4Cl-C_6H_4, $3NO_2C_6H_4$, $C_6H_7O_2$, $C_7H_7O_3$, $4OHC_6H_4$

SCHEME 27 Synthesis of a new series of Schiff base derivatives. Reagent and condition:(i) microwave irradiation at 400 W for 3 min; (ii) $NH_2OH{\cdot}HCl$, microwave irradiation at 400 W for 5 min; (iii) Benzene 1,4-diamine, microwave irradiation at 400 W for 2 min; (iv) Benzene 1,3-diamine, microwave irradiation at 400 W for 2 min; and (v and vi) Ar-CHO, microwave irradiation at 400 W for 45 s.

The result displayed that most of the compounds were exhibiting considerable activity against *L. icterohaemorrhagiae*. Interestingly, as compared to the standard drugs, one compound demonstrated remarkable activity at low concentration against the *L. icterohaemorrhagiae*.

5.14 Chalcone

Sahoo et al. reported the microwave-mediated green chemistry approach to prepare chalcone derivatives [89]. Using this method, a series of chalcone derivatives were synthesized by Claisen-Schmidt condensation of ethyl 4-acetamidobenzoate with various aromatic aldehydes in the presence of ethanolic sodium hydroxide solution followed by dehydration. Due to the use of microwave irradiation, the rate of reaction was increased with less by-product formation and better yield of the product. All the synthesized compounds were screened for their anthelmintic activity, and significant activity was found as compared to the standard drug. The results showed that the compounds with electron-withdrawing groups were found to be highly potent among the series.

Microwave-assisted condensation of 3-acetyl-2,5-dimethylthiophene and 9-ethyl-9*H*-carbazole-3- carbaldehyde produced 1-(2,5-Dimethylthiophen-3-yl)-3-(9-ethyl-9*H*-carbazol-3-yl)prop-2-en-1-one (chalcone) [90]. The synthetic route of chalcone is outlined in Scheme 28.

The microwave-assisted cyclization reaction of chalcone with phenylhydrazine/thiosemicarbazide/guanidine hydrochloride/thiourea produced the compounds pyrazolines and pyrimidines (Scheme 29). All the synthesized compounds were screened for antibacterial activity against two gram-positive and two gram-negative bacterial strains and were tested for any possible toxic effect on the mammalian cells. The results displayed that pyrazoline ring

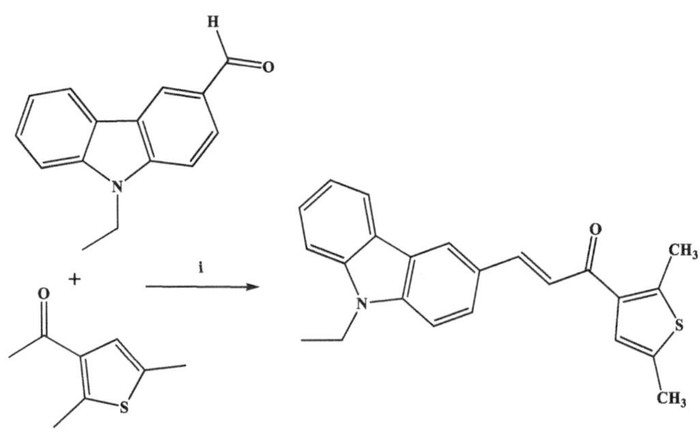

93

SCHEME 28 Synthesis of chalcone. Reagent and condition: (i) EtOH, MW.

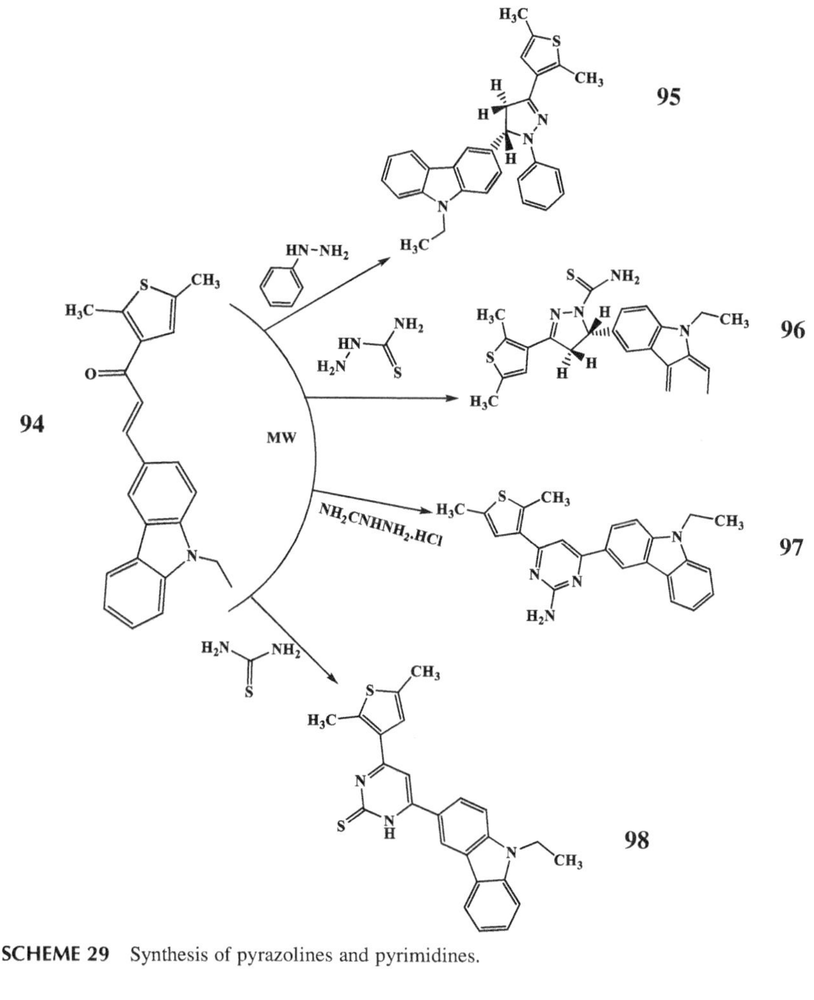

SCHEME 29 Synthesis of pyrazolines and pyrimidines.

bearing compounds display better antibacterial values compared to their pyrimidine homologs.

Microwave-assisted synthesis of 2-[(*E*)-2-(1*H*-indol-3-yl)vinyl]hetarenes was also reported [91]. The condensation of (indol-3-yl)carbaldehydes with 2-methylazoles and 2-methylazines was executed under activation with microwave.

To optimize the reaction conditions, the mixture of aldehyde and quinaldine, without solvent, in the presence of catalytic amounts of piperidine was heated in a microwave reactor at varying temperatures. Good yield of the condensation product was obtained from isothermic heating of the neat mixtures of starting materials in the presence of catalytic amounts of piperidine (20 mol%) at 160°C for 4 h (Scheme 30). This method was important

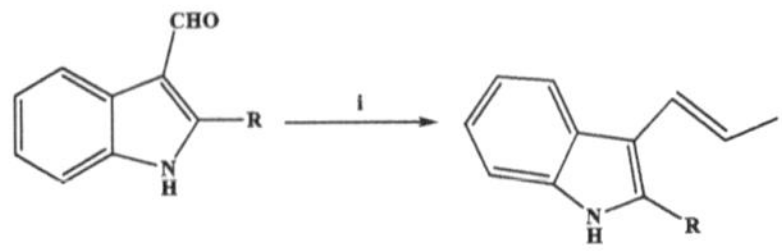

99

100

R=H, Me, Ph, 2-naphthyl;

SCHEME 30 Synthesis of 2-[(*E*)-2-(1*H*-indol-3-yl)vinyl]hetarenes. Reagents and conditions: (i) Piperidine (20 mol%), 160°C, 4 h, MW.

for the efficient preparative synthesis of compounds that contain nonsymmetrical bishetarylethylene structural motif, which is of great interest for bioorganic and medicinal chemistry.

5.15 Quinoxalines

A quinoxaline (benzopyrazine) is a heterocyclic compound containing a ring complex made up of a benzene ring and a pyrazine ring. Compounds possessing quinoxaline derivatives were bestowed with a variety of significant biological properties such as antiviral [92], antimalarial [93], anticancer [94], DNA intercalation [95], DNA duplex stabilization [96], and many others [97].

Efficient and convenient microwave synthesis of a new series of 2-{(6*H*-indolo[2,3-b]quinoxalin-6-yl)methyl}-5-aryl-1,3,4-oxadiazoles from readily available 1,2-diaminobenzene and isatins was revealed [98].

The reaction procedure is shown in Scheme 31. In the first step, isatin (1a) was treated with *o*-phenylenediamine (2) in glacial acetic acid, to obtain corresponding 6*H*-indolo[2,3-b]quinoxaline (3a) under microwave irradiation at 360 W for 3 min at a temperature of 75°C. Then, the 6*H*-indolo[2,3-b] quinoxalines reacted with ethyl chloroacetate (4) to provide the corresponding ethyl 2-(6*H*-indolo [2,3-b]quinoxalin-6-yl)acetates (5a–c) in the presence of anhydrous potassium carbonate in dry acetone using the microwave for 10–20 min. Microwave irradiation of the ethyl ester (5a–c) and hydrazine hydrate (6) in ethanol for 2–3 min at 75°C had resulted in respective 2-(6*H*-indolo[2,3-b]quinoxalin-6-yl)acetohydrazides (7a–c). In the final step, under microwave irradiation 2-(9-H/Br/F-6Hindolo[2,3-b]quinoxalin-6-yl) acetohydrazides (7a–c) were treated with a series of carboxylic acids (8) in the presence of $POCl_3$ for 5–18 min at 75°C to obtain 2-{(9-H/Br/F-6*H*-indolo[2,3-b]quinoxalin-6-yl)methyl}-5-substituted-1,3,4-oxadiazoles (9a–p).

The synthesized compounds were screened for their in vitro antibacterial activity against *B. subtilis*, *Staphylococcus aureus*, *Staphylococcus epidermidis*, *E. coli*, *Pseudomonas aeruginosa*, and *K. pneumoniae* using the broth dilution method. The standard drug penicillin was used for comparison.

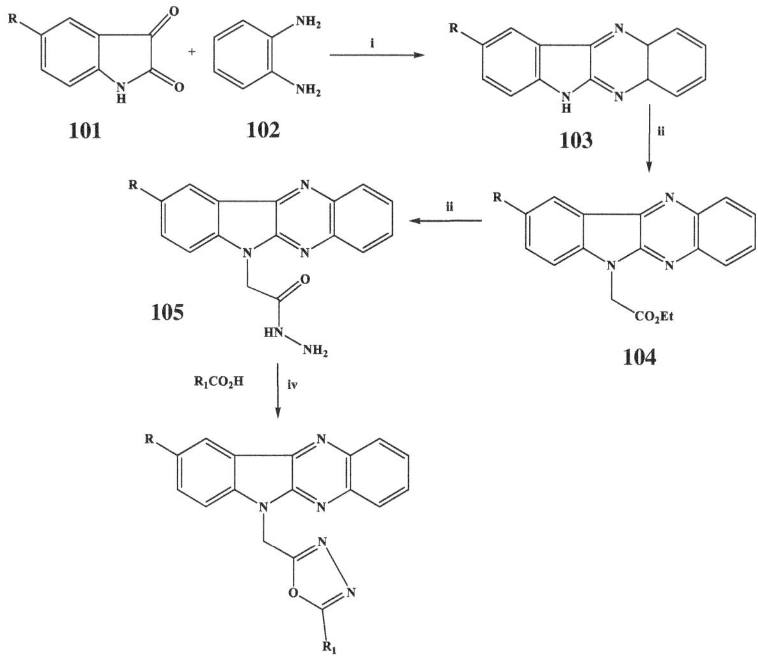

SCHEME 31 Microwave-assisted synthesis of 2-{(6*H*-indolo[2,3-b]quinoxalin-6-yl)methyl}-1,3,4-oxadiazoles (9a–p). Reagents and conditions: (i) Glacial acetic acid, MW, 360 W, 75°C, 3–4 min; (ii) Ethyl chloroacetate, dry acetone, MW, 360 W, 65°C, 10–20 min; (iii) Hydrazine hydrate, EtOH, MW, 320 W, 80°C; and (iv) Carboxylic acid, $POCl_3$, MW, 360 W, 75°C, 5–10 min.

Compounds, with their minimum inhibitory concentration values of over 150 μg/mL, did not display any promising antibacterial activity.

5.16 Carbazoles

Carbazole is a polycyclic aromatic hydrocarbon, having a tricyclic structure. The structure consisting of two six-membered benzene rings fused on either side of a five-membered nitrogen-containing ring. There are several carbazoles isolated from natural sources. It possesses a wide range of biological activities such as antiproliferative [99], antifungal [100, 101], antituberculosis [102], antibacterial [103, 104], antiinflammatory [105], antiviral [106], antitumor [107], antioxidant [108], and antihistaminic activities [109].

Molecular hybridization of different bioactive scaffolds is proved to be effective in the synthesis of new drugs with improved activity [110]. For example, a series of carbazole aminothiazoles were shown potent and promising antimicrobial activity [111].

New derivatives of carbazole incorporated with thiazole moiety were synthesized via the reaction of carbazole thiosemicarbazone with hydrazonoyl chloride under microwave irradiation [112].

The condensation reaction of carbazole-carbaldehyde derivative 1 with thiosemicarbazide 2 in ethanol in the presence of drops of HCl under microwave irradiation produced the thiosemicarbazone, namely, 2-((9-ethyl-9*H*-carbazol-3-yl)methylene)-hydrazine-1-carbothioamide 3 (Scheme 32).

The same reaction was also carried out under conventional heating for comparison. The reaction time under microwave heating was 15 min, and the yield was 92%. For conventional heating, the reaction time and yield were 5 h and 80%, respectively. The observations of the yield and the reaction time from the two heating conditions revealed that microwave irradiation is more promising than thermal heating.

Synthesis of a series of heterocyclic compounds having carbazole incorporated with thiazole moiety is shown in Scheme 33. Carbazole derivative 3 was treated with *N*-aryl-2-oxopropanehydrazonoyl chlorides 4a–f in dioxane and triethylamine as a base catalyst under microwave irradiation for the synthesis of the corresponding 2(((9-ethyl-9*H*-carbazol-3-yl)methylene)hydrazinyl)-4-methyl-5-(aryldiazenyl) thiazoles 6a–f. The reactions of the thiosemicarbazone 3 with *N*-aryl Cethoxycarbonyl methanehydrazonoyl chlorides 7a–e are depicted in Scheme 34.

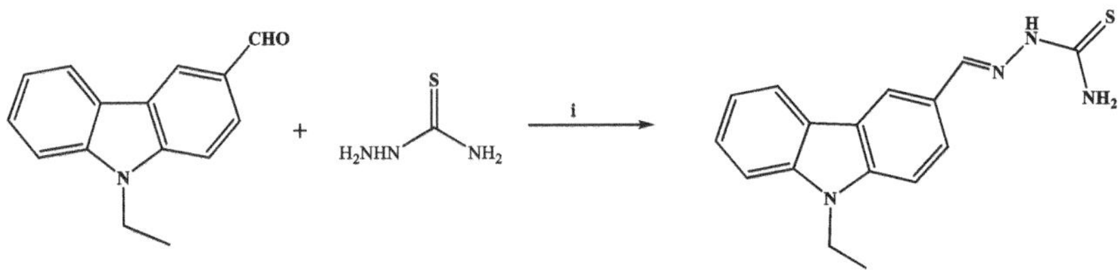

107 **108** **109**

SCHEME 32 Synthesis of thiosemicarbazone derivative 3. Reagents and conditions: (i) EtOH, HCl, MW.

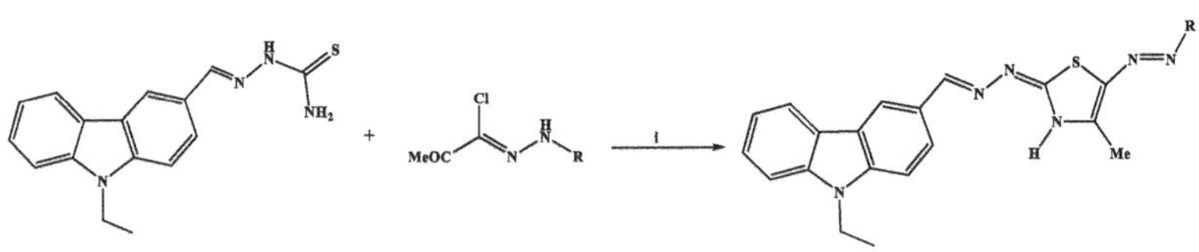

110 **111** **112**

R=XC$_6$H$_4$

X=4-Me; 4-Cl; 4-NO$_2$; 3-NO$_2$; 3-Cl; 4-OMe.

SCHEME 33 Synthesis of thiazole derivatives 6. Reagents and conditions: (i) Et_3N, Dioxane, MW.

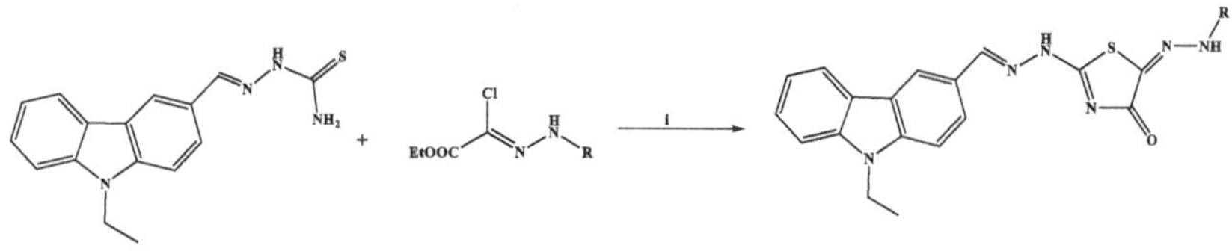

113 114 115

R=XC_6H_4

X=4-Me; H; 4-Cl; 3-Cl; 4-NO_2

SCHEME 34 Synthesis of thiazole derivatives 9. Reagents and conditions: (i) Et_3N, Dioxane, MW.

The thiosemicarbazone and the thiazole derivatives were screened in vitro for their antimicrobial activity against two fungi species (*Aspergillus niger* and *Geotrichum candidum*), gram-positive bacteria (*Staphylococcus aureus*, *Staphylococcus epidermidis*, *B. subtilis*, and *Streptococcus pyogenes*), and gram-negative bacteria (*Pseudomonas aeruginosa*, *E. coli*, *K. pneumoniae*, and *Salmonella typhimurium*). The antimicrobial activity results revealed that the hybridization of carbazole with the thiazole ring increases significantly their antimicrobial activity. It has been also found that the substituent on the thiazole ring plays an important and critical role in the activity as the most active derivatives contained a phenyl ring or substituted phenyl group with electron-donating substituents (methyl or methoxy groups) as in derivatives.

Under microwave irradiation in the presence of a catalytic amount of copper powder and a base, β-bromo-α,β-unsaturated amides were coupled and cyclized with amidine hydrochlorides to produce the corresponding pyrimidinones [113].

Coupling and cyclization reaction of β-bromo-α,β-unsaturated amides with amidine hydrochlorides is shown in Scheme 35. To optimize the reaction,

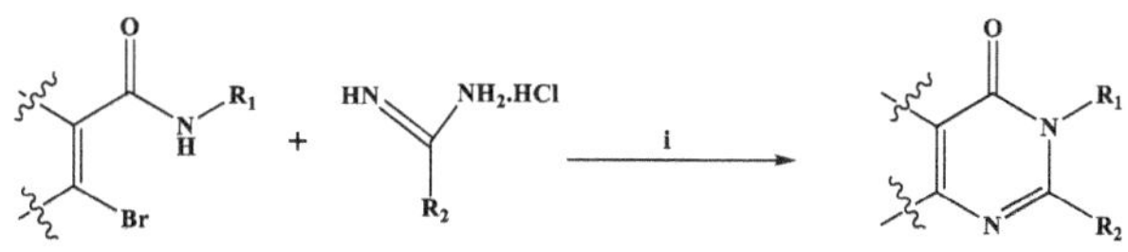

116 117 118

R1= various rings (eg: Ph, benzyl)

R2=Me, Ph

SCHEME 35 Microwave-assisted copper powder-catalyzed synthesis of pyrimidinones from β-bromo-α,β-unsaturated amides, and amidine hydrochlorides. Reagents and conditions: (i) 1a (0.5 mmol), 2 (0.75 mmol), copper powder (0.05 mmol), NaOtBu (1.5 mmol), DMF (3 mL), 150°C, for 2 h, MW (200 W of initial power).

various reaction conditions such as the molar ratio of reactants, several kinds of base and solvent, and reaction time were considered. Treatment in *N*,*N*-dimethylformamide (DMF) at 150°C for 2 h in the presence of 10 mol% of copper powder (0.05 mmol) along with sodium tert-butoxide under microwave irradiation (200 W of initial power) afforded the pyrimidinones in good yield. The reaction provided a new method for synthesizing the pyrimidinone scaffold from readily available ketones using the copper powder-microwave irradiation system.

5.17 Ipsapirones

One of the known compounds that can be used in the treatment of CNS diseases is ipsapirone, belonging to the class of azapirones. Ipsapirone, 2-(4-(4-(pyrimidin-2-yl)piperazin-1-yl) butyl)benzo[d]isothiazol-3(2*H*)-one1,1-dioxide, is a known ligand with anxiolytic, antidepressive, and antiaggressive effects, and a partial antagonist of the 5-HT1A receptor. It can also be used to treat alcohol dependence. From a chemical point of view, the compound has a long-chain arylpiperazine, composed of a pyrimidylpiperazine linked via a butyl moiety to saccharin.

Microwave-assisted solvent-free synthesis of ipsapirone was also possible [114]. Ipsapirone was obtained in two different methods: microwave-assisted *N*-alkylation of bromobutyl saccharin with 1-(2-pyrimidyl)piperazine dihydrochloride and microwave-assisted one-pot method. In both cases, around 10 min of microwave radiation produced the final product ipsapirone with a yield of 85% and 67% in the first and the second method, respectively.

N-Alkylation of bromobutyl saccharin with 1-(2-pyrimidyl)piperazine dihydrochloride is shown in Scheme 36.

Ipsapirone can also be synthesized in a one-pot approach by treating saccharin, 1,4-dibromobutane, and 1-(2-pyrimidyl) piperazine dihydrochloride under the microwave (Scheme 37).

Recently, microwave-promoted synthesis of highly luminescent *s*-tetrazine-1,3,4-oxadiazole and *s*-tetrazine-1,3,4-thiadiazole hybrids was reported [115]. From *s*-tetrazine-3,6-dicarbohydrazide, triethyl orthoesters, and aroyl chlorides, two series of new s-tetrazine derivatives containing 1,3,4-oxadiazole

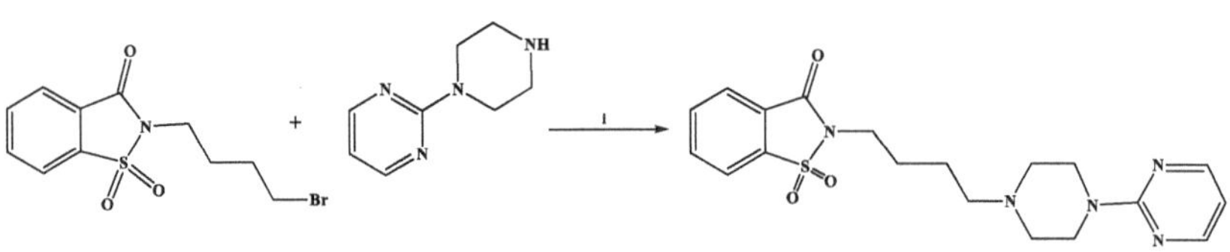

119 120 121

SCHEME 36 Synthesis of ipsapirone. Reagents and conditions: (i) HCl, Base, TBAB, MW, 50 W.

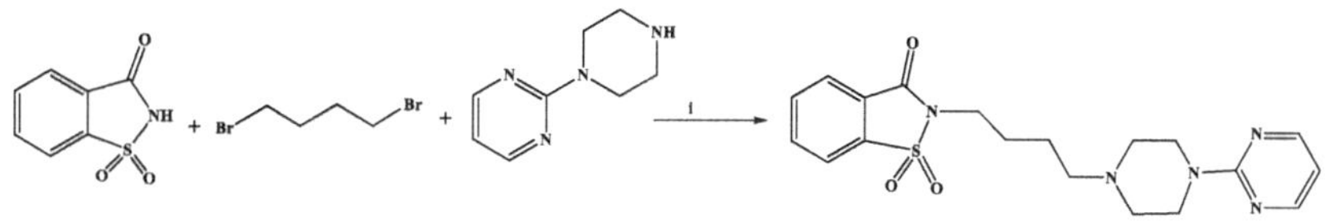

122 **123** **124** **125**

SCHEME 37 Synthesis of ipsapirone according to one-pot method. Reagents and conditions: (i) HCl, Base, TBAB, MW, 50 W.

and 1,3,4-thiadiazole rings were synthesized in good yield. Under microwave irradiation, cyclocondensation reactions for both groups of conjugated arrangements proceeded speedily and expeditiously. The synthesized compounds displayed efficient optical properties and a large quantum yield of emitted photons.

The synthesis of *s*-tetrazine derivatives containing 1,3,4-oxadiazole ring was carried out using triethyl orthobenzoate via microwave heating in glacial acetic acid, as shown in Scheme 38. After heating in the microwave for 7.5 min, the product 3,6-di(1,3,4-oxadiazol-2-yl)-1,2,4,5-tetrazine was obtained in good yield.

Similar microwave-assisted approaches were applied for the synthesis of 1,3,4-thiadiazoles (Scheme 39). The s-tetrazine-3,6-dicarbohydrazide, triethyl orthoesters, and Lawesson's reagent (LR) were used as the starting materials for the synthesis. Microwave-assisted heating of the substrates in glacial acetic acid resulted in the formation of the desired s-tetrazine-1,3,4-thiadiazoles with satisfactory yields.

A series of hybrid isatin-aminorhodanine molecules were prepared by microwave activation under mild conditions [116].

Microwave-assisted Knoevenagel condensation of *N*-aminorhodanine with isatin derivatives is shown in Scheme 40. Irradiation of an equimolar mixture of reactants in ethanol at 100°C (50 W) for 5 min allowed their full conversion into the product, isatin-3-aminorhodanine derivatives. An assembly of 12

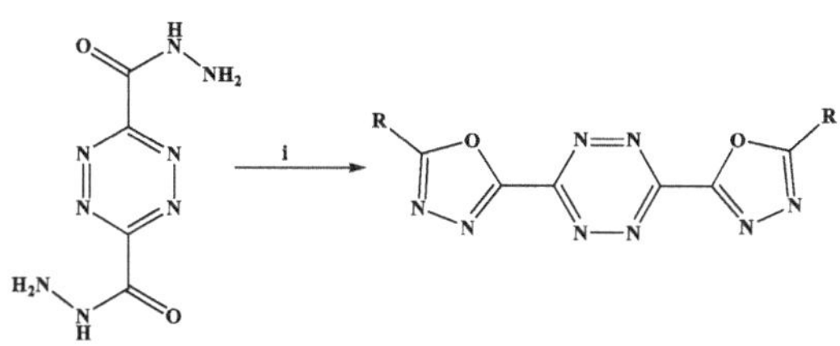

126 **127**

R=H, CH_3, CH_2CH_3, C_6H_5

SCHEME 38 Synthesis of 3,6-di(1,3,4-oxadiazol-2-yl)-1,2,4,5-tetrazine. Reagents and conditions: (i) Glacial acetic acid, MW, 7.5 min, 150 W, 50–100°C.

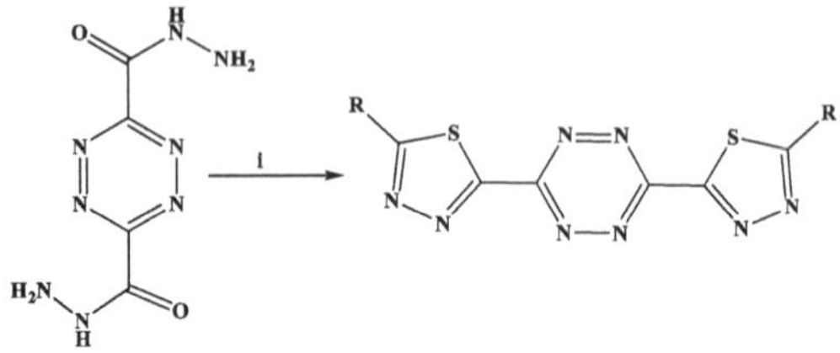

128 129

R=H, CH_3, CH_2CH_3, C_6H_5

SCHEME 39 Synthesis of 3,6-di (1,3,4-thiadiazol-2-yl)-1,2,4,5-tetrazine. Reagents and conditions: (i) Glacial acetic acid, Lawesson's reagent, MW, 7.5 min, 150 W, 50–100°C.

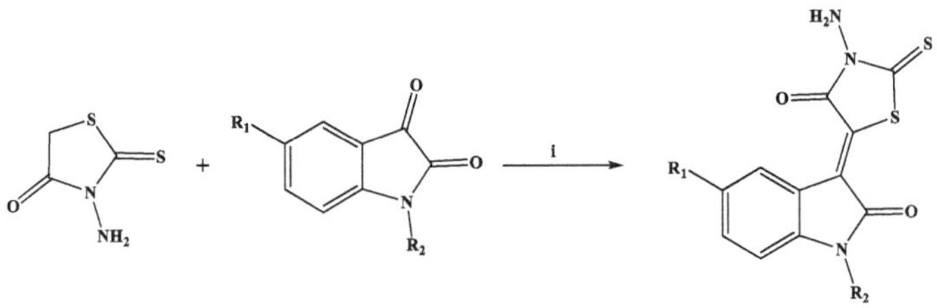

130 131 132

R_1= H, F, OCH_3, NO_2, Cl, Br; R_2= H, CH_3

SCHEME 40 Synthesis of isatin-3-aminorhodanine derivatives. Reagents and conditions: (i) AcOH, EtOH, MW, 5 min, 50 W, 100°C.

compounds was synthesized in good to excellent yields using these optimized conditions and using different isatin derivatives.

The synthesized compounds were tested for their ability to inhibit eight kinases: MmCLK1, HsHASPIN, HsCDK2/CyclinA, HsCDK5/p25, HsCDK9/CyclinT, SscGSK3α/β, RnDYRK1A, and HsPIM1. The evaluation data indicated that the brominated hybrids are the most active compounds against the PIM1 and CLK1 kinases.

5.18 Thiazoles

Aromatic diamides, derivative of 2,6-pyridinedicarboxylic acid and isophthalic acid, bearing 1,3,4-thiadiazole residue were prepared with satisfactory yields using microwave-assisted methods [117].

The synthesis of diamides using the respective acid dichlorides and 2-amino-1,3,4-thiadiazole is shown in Scheme 41. The reactions were carried out with the use of imidazole under solvent-free conditions. Microwave-assisted solvent-free

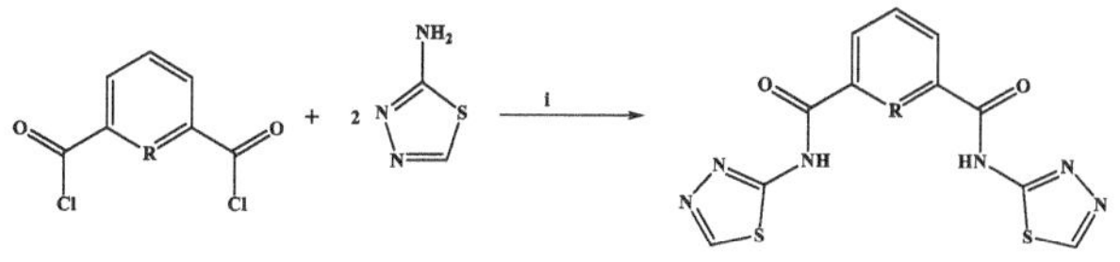

133 134 135

R=CH, N.

SCHEME 41 Microwave-assisted synthesis of 2-amino-1,3,4-thiadiazole diamides from acid dichlorides. Reagents and conditions: (i) Imidazole, solvent-free, MW, 400 W, 15 min.

conditions and the use of imidazole allowed for more than a twofold increase of reaction yield and shortening the reaction time from 12 h to 15 min compared to the conventional method.

A hyperfluorescence 2-pyridone derivative (D-TPCA) was synthesized from the aqueous mixture of citric acid and 2-aminoethanethiol hydrochloride within 5 min by the microwave-assisted method [118]. Because of the fluorescence of the synthesized compound, it can be effectively quenched by picric acid and it can be used as a "turn-off" fluorescence sensor for picric acid in an aqueous solution. The synthetic route of D-TPCA is shown in Scheme 42.

A series of 6-(5-aryl-4,5-dihydro-1*H*-pyrazol-3-yl)-2,2-dimethylchroman-7-ol and 6-(5-aryl-4,5-dihydro-1*H*-pyrazol-3-yl)-2,2-dimethylchroman-5-ol derivatives were synthesized under microwave irradiation [119]. Corresponding chalcones and hydrazine hydrate in the presence of sodium acetate were used for the synthesis, and Michael addition followed by cycloaddition was involved during the reaction. The synthesis of pyrazoline compounds containing chroman moiety by microwave irradiation methods is shown in Scheme 43.

The reactants 1-(7-hydroxy-2,2-dimethylchroman-6-yl)ethanone and 1-(5-hydroxy-2,2-dimethylchroman-6-yl)ethanone were synthesized from acetophenone and isoprene using Amberlyst-15 as a catalyst in THF and heptane [120]. The Claisen-Schmidt condensation of these compounds with the appropriate aryl aldehydes in the presence of powdered KOH under microwave irradiation produced desired chalcones: (*E*)-3-aryl-1-(7-hydroxy-2,2-dimethylchroman-6-yl)prop-2-en-1-ones and (*E*)-1-(5-hydroxy-2,2-dimethylchroman-6-yl)-3-phenylprop-2-en-1-ones, respectively [121]. These chalcones and

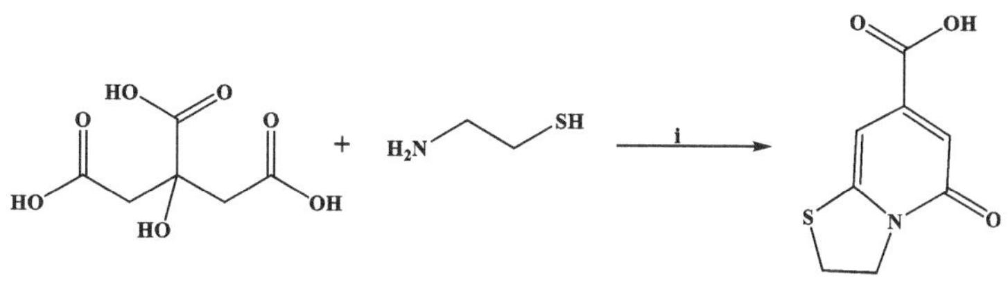

136 137 138

SCHEME 42 Synthetic route of hyperfluorescence 2-pyridone derivative. Reagents and conditions: (i) MW, 5 min, $(-4H_2O)$.

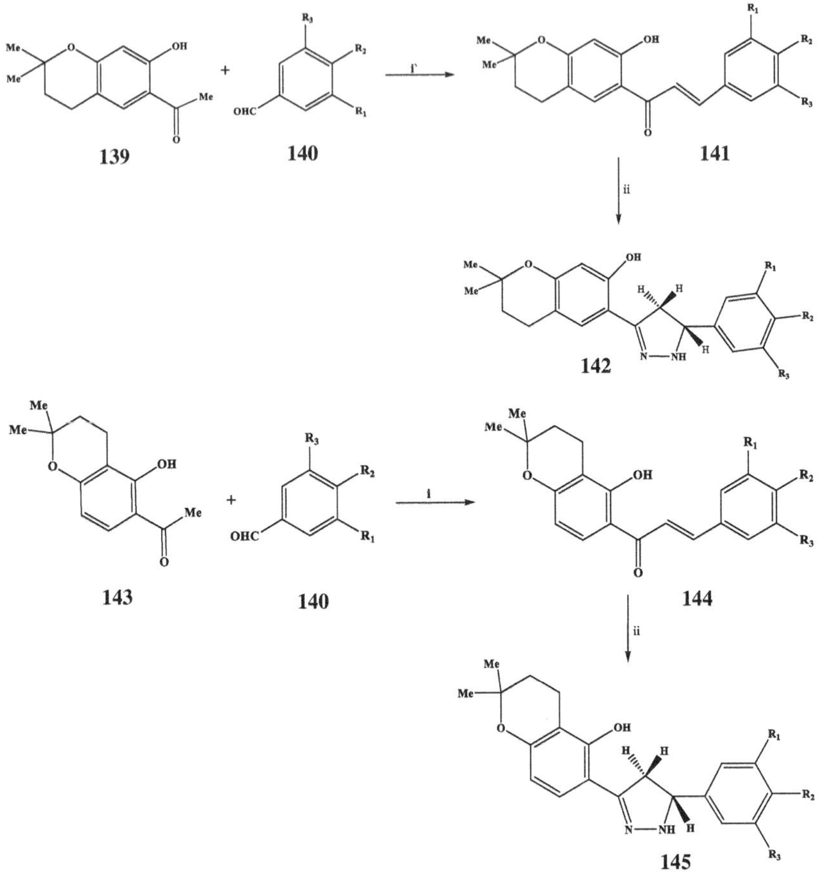

R1=H, OMe; R2=H, Me, OMe, Cl, *i*-Pr, NMe2; R3=H, OMe, Cl, NO2

SCHEME 43 Microwave-assisted synthesis of pyrazolines. Reagents and conditions: (i) KOH, EtOH, MW, 11–15 min; (ii) $NH_2NH_2{\cdot}H_2O$, NaOAc (1.0 mmol), DMF (5 mL), MW, 180 W, 2–5 min.

hydrazine hydrate in DMF in the presence of sodium acetate were irradiated for 2–5 min at 180 W power to afford pyrazole derivatives in high yields.

All the newly synthesized 18 compounds were screened for their in vitro antimicrobial activity by the cup-plate agar diffusion method [122]. Antibacterial activity was tested against two gram-positive bacteria (*B. subtilis* and *Staphylococcus aureus*) and two gram-negative bacteria (*Pseudomonas aeruginosa* and *E. coli*). The activity data were compared with standard drug streptomycin sulfate. Antifungal activity was carried out against three fungi (*Aspergillus niger*, *Candida albicans*, and *Aspergillus foetidus*), and Nystatin was used as standard. Most of the compounds were displayed very good to excellent antimicrobial activity in comparison with standard drugs.

The synthesis of N-substituted 4-aryloxy/4-thiophenoxy/thioisopropyloxy-phthalimides via microwave-assisted aromatic nucleophilic substitution reaction of 4-fluoro-phthalimides with sodium arylate/thiophenoxide/thioisopropoxide was reported [123]. This method allowed for the introduction of substituents through an atom-efficient one-step synthesis and provided good to excellent yields and tolerated a wide range of substituents.

For the synthesis of phthalimides, a mixture of N-substituted-4-fluorophthalimide [11] and sodium phenoxide was heated at 175°C for one hour under microwave conditions (Scheme 44).

A series of compounds such as fused quinolinyl and quinolonyl pyrans were synthesized via a one-pot microwave-assisted reaction of quinolinyl and quinolonyl carbaldehydes, malononitrile, and a 1,3-diketone [124]. These microwave-assisted reactions were catalyzed by a new humic acid supported 1-butyl-3-methyl imidazolium thiocyanate ionic liquid.

The three-component reaction of the mixtures of 2-methoxy quinoline-3-carbaldehyde, malononitrile, and 1,3-diketone catalyzed by HASIL in ethanol as a medium is shown in Scheme 45. Antimicrobial, antioxidant, and toxicity studies of the synthesized compounds displayed various biological activities depending on the structure of the pyrans.

The catalyst- and solvent-free synthesis of benzo[d]thiazole-[1,3]-thiazine hybrids from benzo[d]-thiazol-2-amines, isothiocyanates, and dialkyl acetylenedicarboxylates with no more than 30 min of microwave irradiation was reported [125].

The synthesis of benzo[d]thiazole-[1,3]-thiazine hybrids by successive cyclization of 2-aminobenzo[d]thiazoles, isothiocyanates, and dialkyl acetylenedicarboxylates under microwave irradiation is shown in Scheme 46. During the synthesis, 2-aminobenzo[d]thiazoles (0.50 mmol) and isothiocyanates (0.50 mmol) were irradiated with microwaves at 120°C for 10 min, and then, dialkyl acetylenedicarboxylates (0.50 mmol) were added, followed by microwave irradiation for 10 min. Six selected (*Z*)-alkyl 3-(benzo[d]thiazol-2-yl)

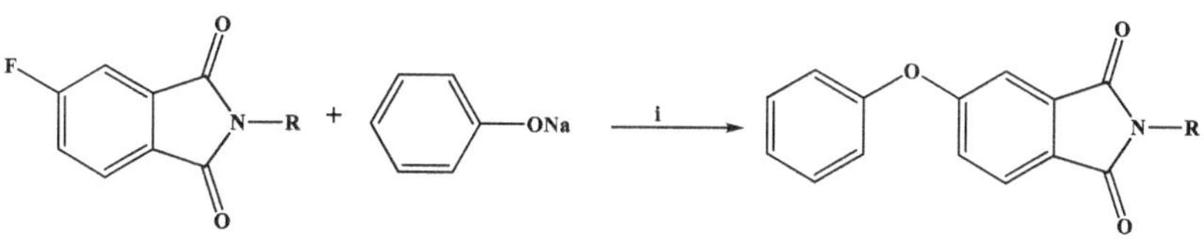

146 **147** **148**

R= 4-C_2H_5Ph, 4-OCH_3Ph, 3-OCH_3Ph, 2,6-di-CH_3Ph, 4-CF_3Ph (5 selected substituents)

SCHEME 44 Microwave-assisted synthesis of phthalimides. Reagents and conditions: (i) DMSO, 175°C, microwave, 1 h.

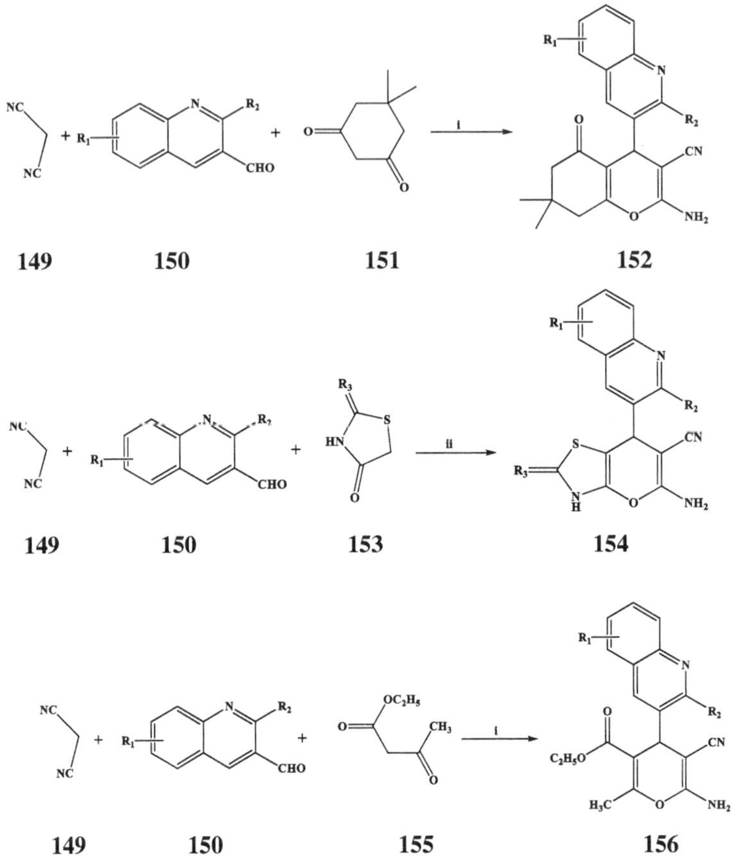

R_1=H, 7-F, 6-CH_3, 6-NO_2; R_2=OCH_3, 2-$CH_3C_6H_4O$, OCH_3, Cl, OCH_3; R_3=O, S.

SCHEME 45 Synthesis of quinolinyl-4*H*-pyrans in the presence of HASIL under microwave irradiation. Reagents and conditions: (i) HASIL, MW, 120 W, 80°C, ethanol, 10–12 min; (ii) HASIL, MW, 120 W, 80°C, ethanol, 10–15 min; and (iii) HASIL, MW, 120 W, 80°C, ethanol, 10–15 min.

-4-oxo-2-(cyclohexyl/arylimino)-3,4-dihydro-2*H*-1,3-thiazine-6-carboxylates are shown in Table 6.

An efficient microwave-assisted method for the preparation of an arylindoline unit containing heterocyclic derivatives was developed [126]. This method was based on ligand-free, palladium-catalyzed cross-coupling of 7-bromo- and 5,7-dibromoimidazo[1,2-a]indolone and 8-bromopyrimido[1,2-a] indolone derivatives with (hetero)aryl boronic acids. A list of (hetero)aryl boronic acids used for Suzuki-Miyaura cross-coupling is shown in Table 7.

The microwave-assisted cross-coupling reactions of brominated indoline substrates with (hetero)aryl boronic acids was carried on smoothly in aqueous media

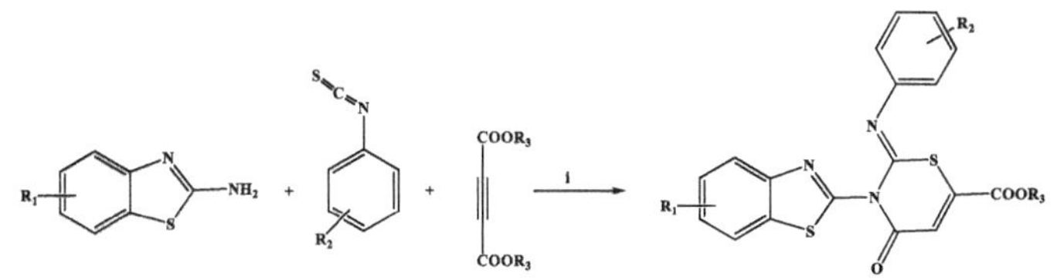

157 **158** **159** **160**

R1=H, 6-Cl, 4-Me, 6-F, 6-OEt; R2=Ph, 2-Cl-Ph, 3-Cl-Ph, cyclohexyl, 4-NO2-Ph, 4-F-Ph; R3=Me, Et.

SCHEME 46 Microwave-assisted one-pot protocol for the synthesis of (*Z*)-alkyl 3-(benzo[d]thiazol-2-yl)-4-oxo-2-(cyclohexyl/arylimino)-3,4-dihydro-2*H*-1,3-thiazine-6-carboxylates. Reagents and conditions: (i) Catalyst-free, solvent-free, MW, 120°C, 20 min.

TABLE 6 Six selected (*Z*)-alkyl 3-(benzo[d]thiazol-2-yl)-4-oxo-2-(cyclohexyl/arylimino)-3,4-dihydro-2*H*-1,3-thiazine-6-carboxylates.

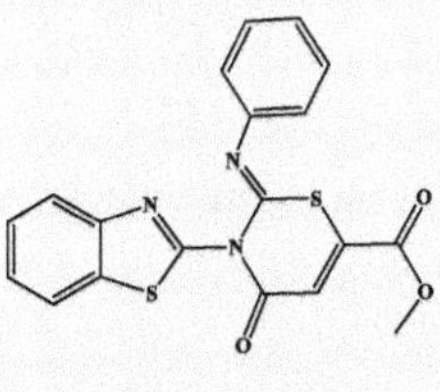

160a

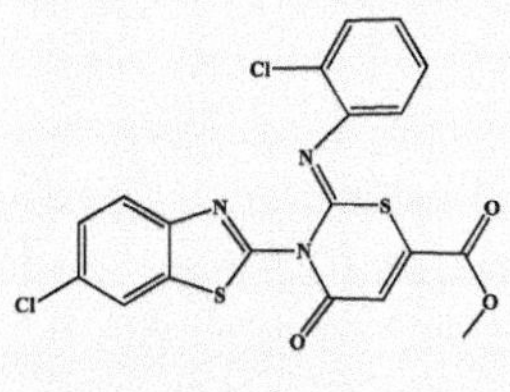

160b

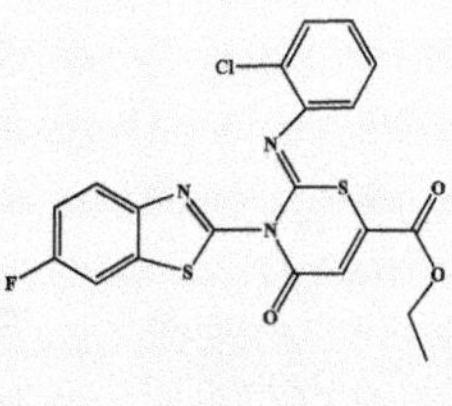

160c

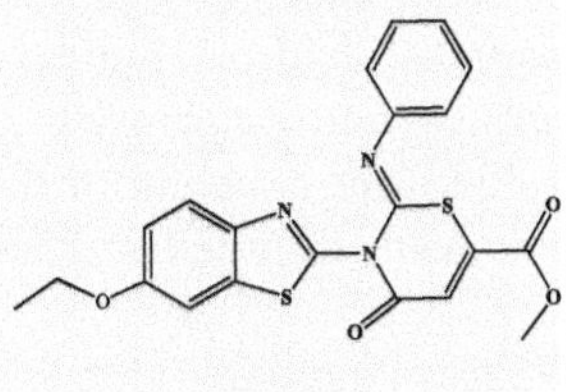

160d

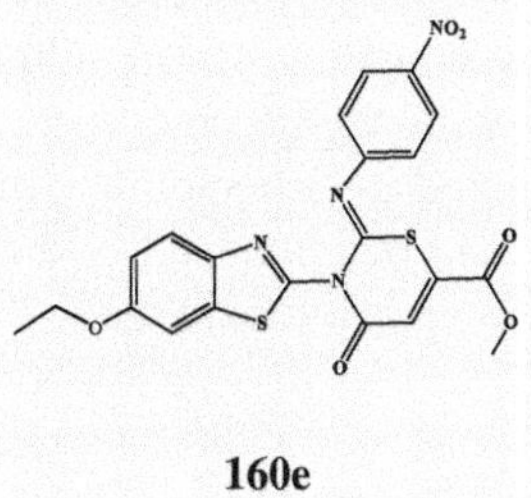

160e

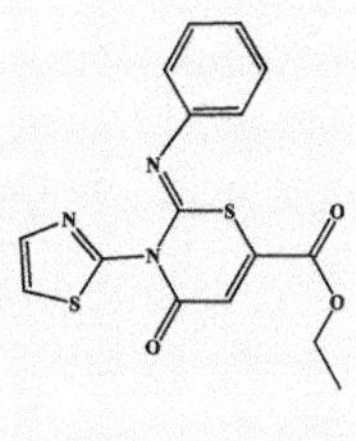

160f

in the presence of the $Pd(OAc)_2$ catalyst and the Cs_2CO_3 base. The Suzuki-Miyaura coupling of 7-bromo-9,9a-dihydro-1*H*-imidazo[1,2-a]indol-2(3*H*)-ones with arylboronic acids is shown in Scheme 47. In detail, Suzuki-Miyaura cross-coupling of compounds with phenylboronic, (4-methoxyphenyl)boronic, naphthalen-2-ylboronic, pyren-1-ylboronic and dibenzo[b,d]thiophen-4-ylboronic acids was investigated, and the subsequent coupling afforded arylated products in 70%, 59%, 63%, 88%, and 93% yields, respectively.

TABLE 7 List of (hetero)aryl boronic acids used for Suzuki-Miyaura cross-coupling.

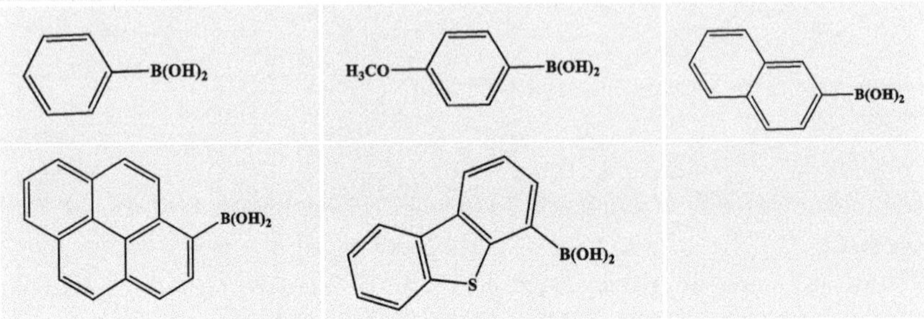

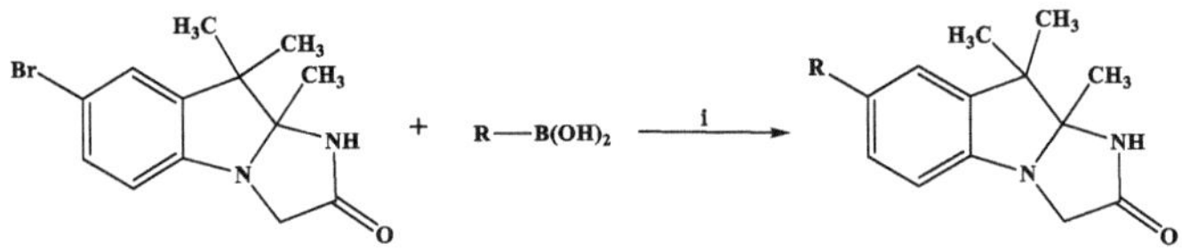

161 162 163

R=Phenyl, 4-methoxyphenyl, naphthalen-2-yl, pyren-1-yl, dibenzo[b.d]thiophen-4-yl

SCHEME 47 Suzuki-Miyaura coupling of 7-bromo-9,9a-dihydro-1Himidazo[1,2-a]indol-2(3*H*)-ones with arylboronic acids. Reagents and conditions: (i) Cs_2CO_3, $Pd(OAc)_2$, $EtOH/H_2O$, MW, 50 W, 100°C.

The Suzuki-Miyaura coupling of 5,7-dibromo-9,9a-dihydro-1Himidazo [1,2-a]indol-2(3*H*)-ones with arylboronic acids (phenyl- and (naphthalen-2-yl)boronic acids) is depicted in Scheme 48. The subsequent coupling of the compound with phenyl- and (naphthalen-2-yl)boronic acids afforded arylated products in 80% and 70% yields, respectively.

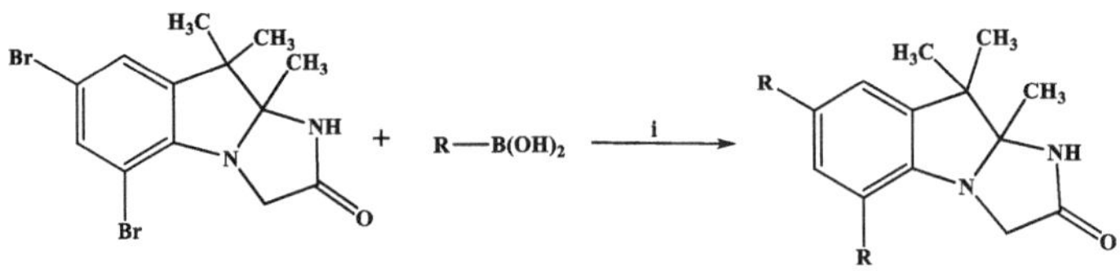

164 165 166

R=Phenyl, naphthalen-2-yl

SCHEME 48 Suzuki-Miyaura coupling of 5,7-dibromo-9,9a-dihydro-1Himidazo[1,2-a]indol-2 (3*H*)-ones with arylboronic acids. Reagents and conditions: (i) Cs_2CO_3, $Pd(OAc)_2$, $EtOH/H_2O$, MW, 50 W, 100°C.

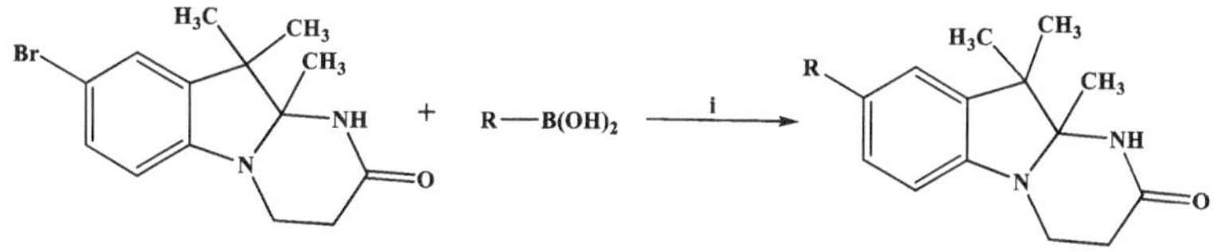

167 **168** **169**

R=Phenyl, naphthalen-2-yl

SCHEME 49 Suzuki-Miyaura cross-coupling of 8-bromo-10,10,10a-trimethyl-3,4,10,10a-tetrahydropyrimido [1,2-a]indol-2(1*H*)-one. Reagents and conditions: (i) Cs_2CO_3, $Pd(OAc)_2$, $EtOH/H_2O$, MW, 50 W, 100°C.

The Suzuki-Miyaura coupling of 8-bromo-10,10,10a-trimethyl-3,4,10,10a-tetrahydropyrimido[1,2-a]indol-2(1*H*)-one (10) with arylboronic acids is shown in Scheme 49. The subsequent coupling of the compound with phenyl- and (naphthalen-2-yl)boronic acids using the optimal reaction conditions afforded arylated products in 78% and 76% yields, respectively.

The synthesized indoline derivatives were incorporated by a π-conjugated biaryl-type fluorophore with an attached amino group-like moiety at the terminal position of the unit structure. The compounds were also characterized by high quantum yields and demonstrated intense fluorescence with significant Stokes shifts making them promising building blocks for diverse biomedical and technical applications.

The microwave-assisted synthesis of pyrazole-3-one containing Schiff base as a push-pull chromophore was reported recently [127]. The synthesis is shown in Scheme 50.

The compound 4-((9-ethyl-9*H*-carbazol-3-yl) methylene) amino-1,5-dimethyl-2-phenyl-1,2-dihydro-3*H*-pyrazol-3-one (EDPO) was synthesized, in 94.8% yield, from the reaction of 9-ethyl-3-carbazole carboxaldehyde with 4-amino phenazone. First, ethyl-3-carbazole carboxaldehyde (0.011 mol,

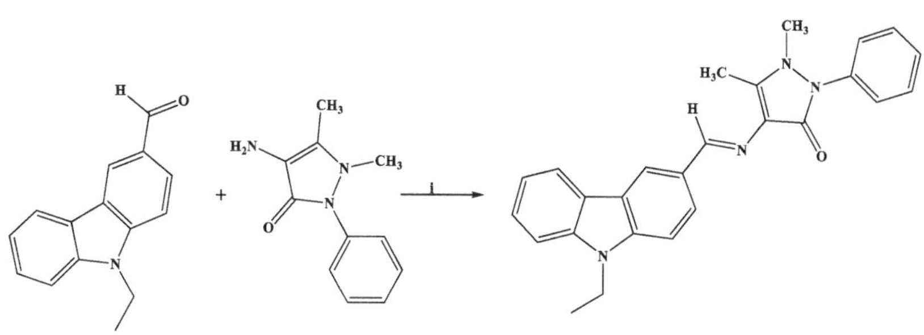

170 **171** **172**

SCHEME 50 Synthesis of 4-((9-ethyl-9*H*-carbazol-3-yl) methylene) amino-1,5-dimethyl-2-phenyl-1,2-dihydro-3*H*-pyrazol-3-one. Reagents and conditions: (i) EtOH, HCl (one drop), MW, 4 min.

2.50 g) and 4-amino phenazone (0.011 mol, 2.27 g) in the minimum amount of (10 mL) EtOH were taken in a 50-mL conical flask and one drop of HCl was added to the reaction mixture. After that, the conical flask was taken in a domestic microwave oven and the mixture was irradiated under 210 W microwave power for 4 min.

The synthesized chromophore EDPO showed good photophysical properties. It showed a bathochromic shift when the polarity of the solvent increased *n*-hexane to DMSO. Along with this, the emission intensity, Stokes shift, transition dipole moments, oscillator strength, and fluorescence quantum yields of the chromophore compound EDPO increased when the polarity of solvents increased.

5.19 Acridine-acetazolamide conjugates

Acridines represent one of the most important nitrogen-containing heterocyclic systems and use as the building block for several medicinally important compounds. These groups of compounds have wide range of applications such as antioxidant [128], cytotoxic [129], antitumor [130], antimicrobial [131–133], antimultidrugresistant [134], fungicides [135], b-channel opener in cardiovascular disease [136] and as carbonic anhydrase inhibitor for glaucoma treatment [137–139], anticancer agents [140, 141], and also used as photosensitizer in organic dye laser applications [142].

Sulfonamides and isoesters (sulfamides, sulfamates) are an important class of drugs. They are possessing various types of biological activities such as anticancer [143], antibacterial [144], carbonic anhydrase inhibitor for glaucoma treatment [145–147], acetylcholinesterase inhibitor agents for Alzheimer's disease [148], high-ceiling diuretic [149], and antiobesity [150].

Conjugates of these two groups of compounds, the acridine-acetazolamide conjugates, may possess significant potential applications.

A series of acridine-acetazolamide conjugates were synthesized from the condensation reaction of 4-Amino-*N*-(5-sulfamoyl-1,3,4-thiadiazol-2-yl)benzamide with cyclic-1,3-diketones (dimedone and cyclohexane-1,3-dione) and aromatic aldehydes under microwave irradiation [151].

The synthesis of acridine-acetazolamide compounds was performed in ethanol with a microwave-assisted method (Scheme 51). During the synthesis, the amino-derivative compound was condensed with substituted 1,3-diketones and substituted benzaldehydes, in the molar range of 1:2:1, under microwave irradiation with a power of 300 W and at 120°C, and this led to the formation of acridine-acetazolamide compounds. Several series of compounds were synthesized by changing the substituents on the diketone and/or aldehyde components.

The synthesized compounds were investigated for inhibition of three cytosolic human isoforms, hCA I, II, and VII, and one membrane-bound hCA IV. All inquired isoforms were inhibited in low micromolar and nanomolar range by these compounds.

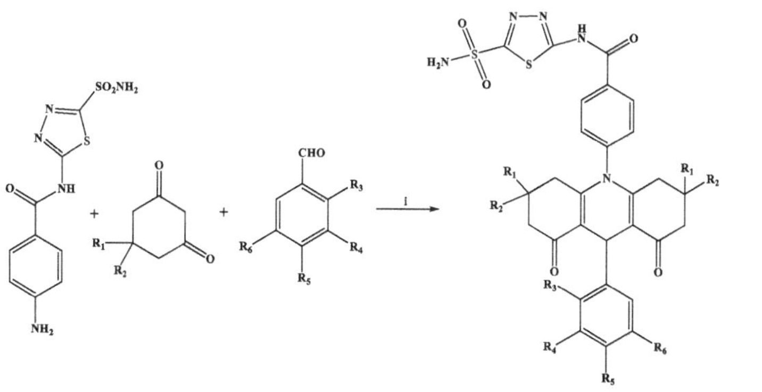

173 174 175 176

R_1=CH_3, H; R_2=CH_3, H; R_3=H, OH, Cl; R_4=H, CN, R_5=H, CN, F, Cl, Br, OCH_3, SCH_3, CH_3, CH_2CH_3, Ph; R_6=H, Br

SCHEME 51 Microwave-assisted synthesis of acetazolamide-based hybrid acridine derivative compounds. Reagents and conditions: (i) Ethanol, MW, 300 W, 120°C.

A series of (*E*)-2-(1-(4-methyl-2-(substituted-phenylamino)thiazol-5-yl) ethylidene) hydrazinecarbothio-amide and (*E*)-*N*-cyclohexyl-2-(1-(4-methyl-2-(substitutedphenylamino) thiazol-5-yl)ethylidene)hydrazinecarbothioamide compounds were synthesized using microwave irradiation methods [152].

The thiazole fused thiosemicarbazones, (*E*)-2-(1-(4-methyl-2-(substituted-phenylamino)thiazol-5-yl)ethylidene)hydrazinecarbothio-amide and (*E*)-*N*-cyclohexyl-2-(1-(4-methyl-2-(phenylamino)thiazol-5-yl)ethylidene) hydrazinecarbothioamide, were prepared by condensation reaction of substituted 1-(4-methyl-2-(phenylamino)thiazol-5-yl)ethanone with hydrazinecarbothioamide and *N*-cyclohexylhydrazinecarbothioamide, respectively (Scheme 52). The condensation reaction was carried out under microwave irradiation in the presence of a catalytic amount of glacial acetic acid. Six selected synthesized thiazole fused thiosemicarbazones are shown in Table 8.

The synthesized compounds were evaluated for their in vitro antimicrobial, antimalarial, and antituberculosis activity as well as in silico study. The results displayed that thirteen out of twenty compounds presented well in vitro antibacterial activity against *Staphylococcus aureus*.

The microwave-assisted synthesis of rhodamine-derived imines was reported by Abebe and co-workers [153]. The reaction involved the condensation of rhodamine hydrazide with various aromatic aldehydes in ethanol under microwave irradiation.

Using this method (Scheme 53), a total of 12 rhodamine-derived imine derivatives were synthesized under microwave using ethanol as a solvent. The reactant, rhodamine B hydrazide, was synthesized from the parent

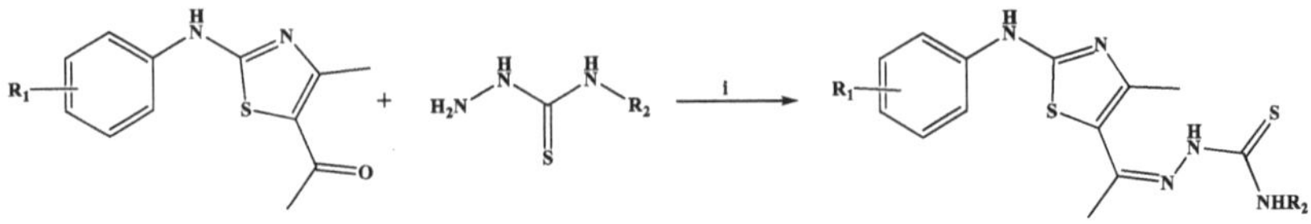

177 178 179

For (E)-2-(1-(4-methyl-2-(substituted-phenylamino)thiazol-5-yl)ethylidene) hydrazine carbothio-amide, **R_2=H.**

For (E)-N-cyclohexyl-2-(1-(4-methyl-2-(phenylamino)thiazol-5-yl)ethylidene) hydrazine carbothioamide, **R_2=Cyclohexyl.**

SCHEME 52 Microwave-assisted synthesis of thiazole fused thiosemicarbazones. Reagents and conditions: (i) EtOH (5 mL), Gla. CH_3COOH (1.5 mL), MW, 300 W, 190°C, 4–5 min.

TABLE 8 Three selected (*E*)-2-(1-(4-methyl-2-(substituted-phenylamino) thiazol-5-yl)ethylidene)hydrazinecarbothio-amides and three selected (*E*)-*N*-cyclohexyl-2-(1-(4-methyl-2-(phenylamino)thiazol-5-yl)ethylidene) hydrazinecarbothioamides.

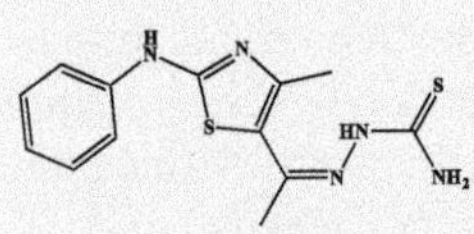 179a	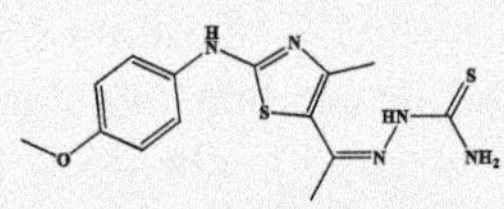 179b	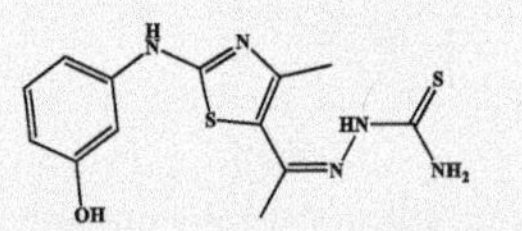 179c
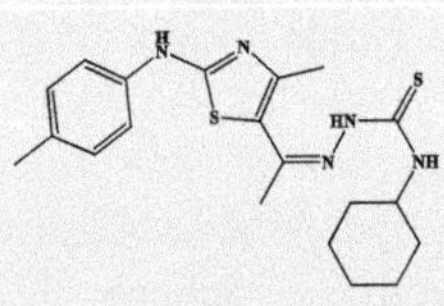 179d	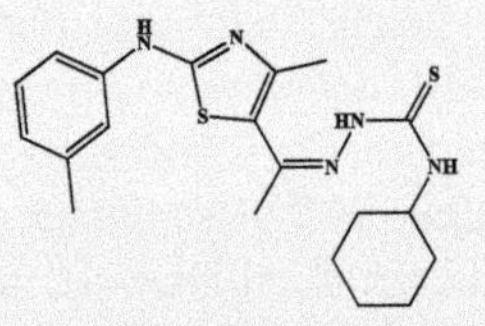 179e	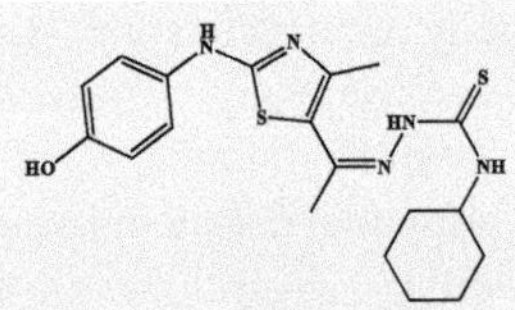 179f

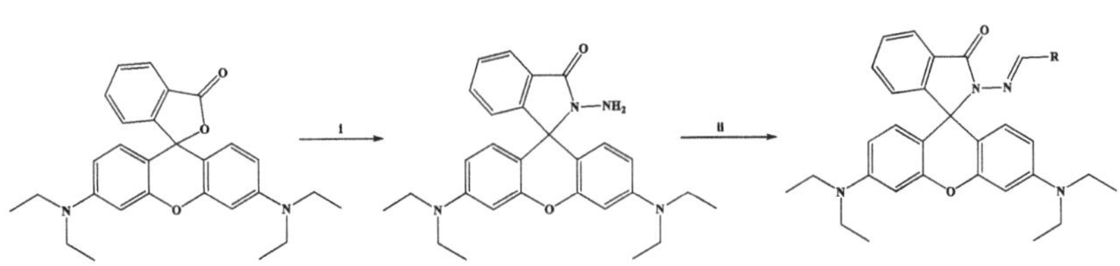

180 181 182

SCHEME 53 Microwave-assisted synthesis of rhodamine-derived imines. Reagents and conditions: (i) NH_2NH_2, Ethanol, MW. (ii) RCHO, Ethanol, MW.

rhodamine B and hydrazine in the one-step process using microwave irradiation and conventional oil-bath heating. The same reaction conditions were applicable for various substituted aromatic aldehydes and rhodamine hydrazide to get imines in high yields (Table 9). The analysis showed that no catalyst or water trapping was necessary for the reaction along with that purification was accomplished by simple filtration and washing the isolated product.

Microwave-assisted copper-catalyzed cascade reactions between 2-halobenzoic acids and amidines to synthesize quinazolinone derivatives in water were demonstrated [154]. The desired products were obtained in good to excellent yields.

A variety of 2-halobenzoic acids and amidines were used for the synthesis (Scheme 54). For instance, the catalytic reactions were executed smoothly with a variety of substrates bearing different functional groups including methoxyl, methyl, nitro, and bromo groups (Table 10) and observed that aryl iodides were more reactive than aryl bromides and aryl chlorides.

TABLE 9 Six selected rhodamine-derived imines.

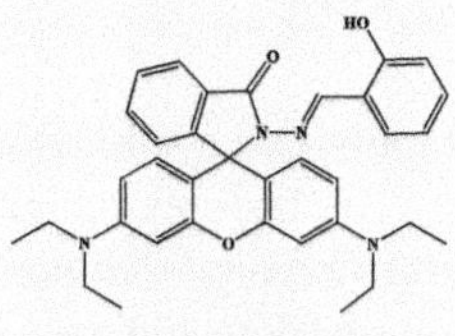

182a

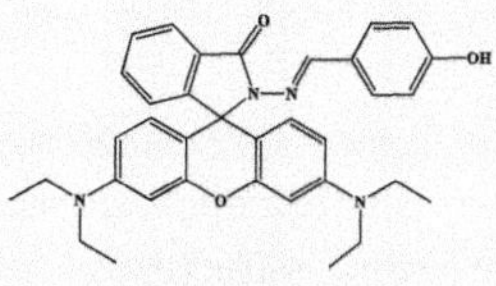

182b

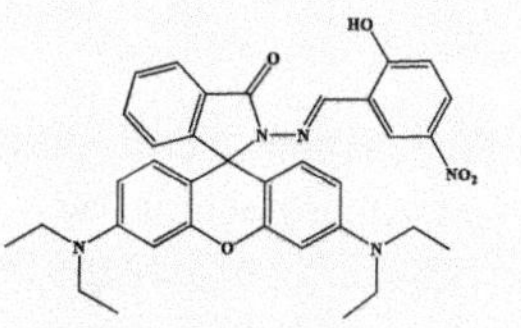

182c

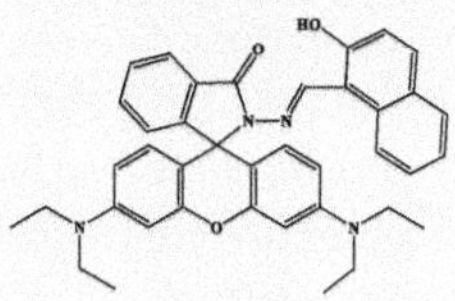

182d

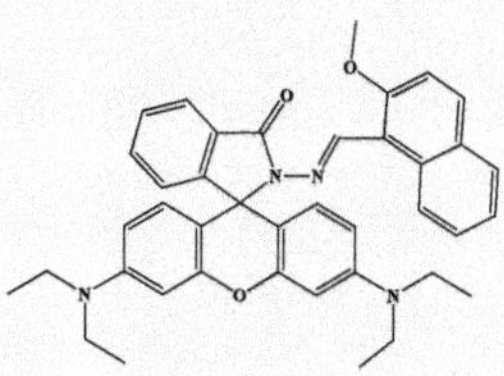

182e

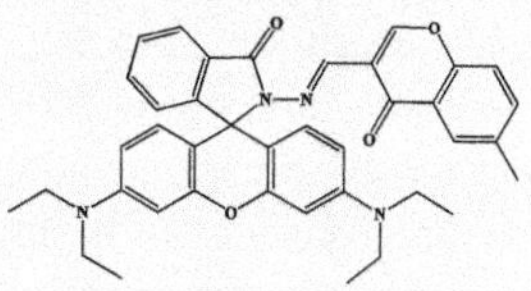

182f

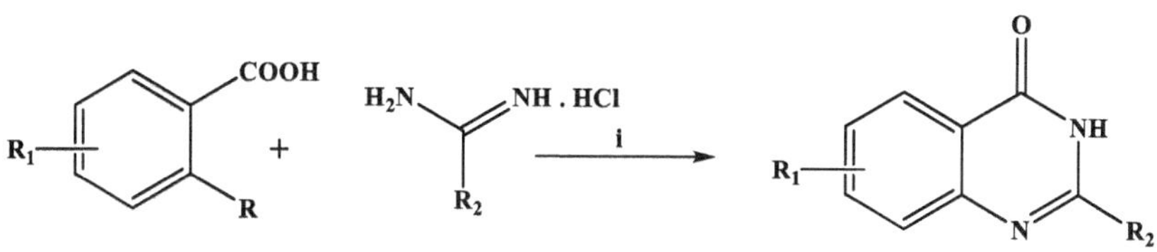

183 **184** **185**

SCHEME 54 Microwave-assisted copper-catalyzed synthesis of quinazolinone derivatives. Reagents and conditions: (i) $CuCl_2$, L2, NaOH, H_2O, MW, 120 W, 20 min, r.t.

TABLE 10 Six selected synthesized quinazolinone derivatives.

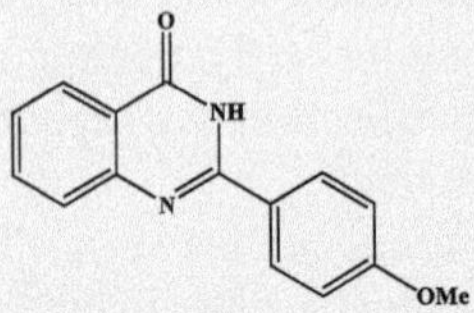 185a	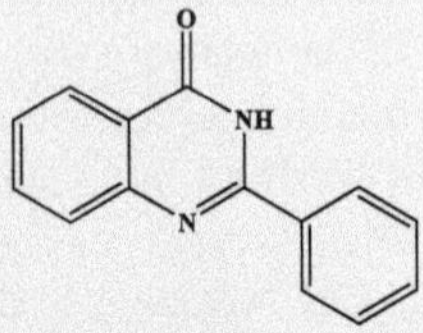 185b	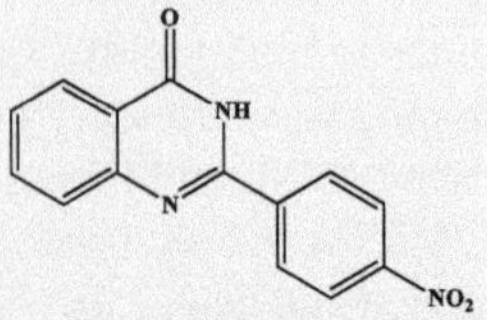 185c
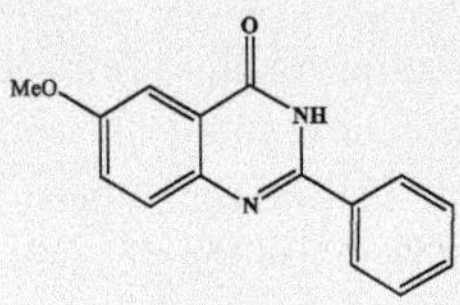 185d	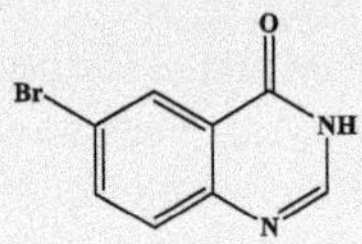 185e	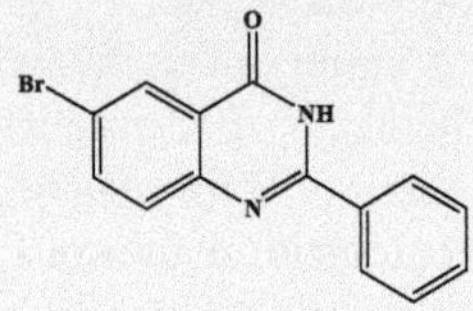 185f

5.20 Conclusion

In this chapter, we discussed the importance of microwave-assisted synthesis of several pharmacologically significant organic compounds. The features such as the possibility of solvent-free reactions, reduction of reaction time for a chemical/pharmaceutical reaction, the possibility of parallel chemical reaction, and instantaneous and uniform heating caused the microwave technology to stay dominant over conventional heating methods. The researchers involved in drug discovery and development processes like high-speed combinatorial and medicinal chemistry mostly prefer and adopt microwave technology. The microwave method has encouraged scientists to initiate new unexplored areas of complex systems both in the organic and inorganic chemistry field and in the drug discovery.

Acknowledgments

AD and BKB are grateful to Prince Mohammad Bin Fahd University for support. BKB is also grateful to US NIH, US NCI, and Kleberg Foundation of Texas for finance support.

References

[1] S. Montel, D. Bouyssi, G. Balme, An efficient and general microwave-assisted copper-catalyzed conia-ene reaction of terminal and internal alkynes tethered to a wide variety of carbonucleophiles, Adv. Synth. Catal. 352 (2010) 2315–2320.

[2] M.G. Victoria, A.I. Aranda, A. Moreno, F.P. Cossio, A.D. Cozar, A. Diaz-Ortiz, A.D.L. Hoz, P. Prieto, Microwave-assisted reactions of nitroheterocycles with dienes. Diels–Alder and tandem hetero Diels–Alder/[3, 3] sigmatropic shift, Tetrahedron 65 (2009) 5328–5336.

[3] V. Declerck, J. Martinez, F. Lamaty, Microwave-assisted copper-catalyzed Heck reaction in PEG solvent, Synlett 18 (2006) 3029–3032.

[4] X. Dong, T. Liu, J. Chen, H. Ying, Y. Hu, Microwave-assisted Mannich reaction of 2-hydroxy-chalcones, Synth. Commun. 39 (2009) 733.

[5] C.N. Raut, S.M. Bharambe, Y.A. Pawar, P.P. Mahulikar, Microwave-mediated synthesis and antibacterial activity of some novel 2-(substituted biphenyl) benzimidazoles via Suzuki-Miyaura cross coupling reaction and their N-substituted derivatives, J. Heterocyclic Chem. 48 (2011) 419–425.

[6] K. Ajay, B.K. Banik, K.J. Barakat, M.S. Manhas, Simplified rapid hydrogenation under microwave irradiation: selective transformations of B lactams, Synlett 8 (1993) 575–576.

[7] H. Shi, W. Zhu, H. Li, H. Liu, M. Zhang, Y. Yan, Z. Wang, Microwave-accelerated esterification of salicylic acid using Brönsted acidic ionic liquids as catalysts, Catal. Commun. 11 (2010) 588–591.

[8] K. Tanji, J. Koshio, O. Sugimoto, Microwave-assisted dehydration and chlorination using phosphonium salt, Synth. Commun. 35 (2005) 1983–1987.

[9] G.V. Baelen, U.W.M. Bert, Study of the microwave-assisted hydrolysis of nitriles and esters and the implementation of this system in rapid microwave-assisted Pd-catalyzed amination, Tetrahedron 64 (2008) 5604–5619.

[10] P. Appukkuttan, V.P. Mehta, E.V.V.D. Eycken, Microwave-assisted cycloaddition reactions, Chem. Soc. Rev. 39 (2010) 1467–1477.

[11] C. Schmoeger, A. Stolle, W. Bonrath, B. Ondruschka, Microwave-assisted organic reduction reactions, Curr. Org. Chem. 15 (2011) 151–167.

[12] U.R. Pillai, E. Sahle-Demessie, R.S. Varma, Microwave-expedited olefin epoxidation over hydrotalcites using hydrogen peroxide and acetonitrile, Tetrahedron Lett. 43 (2002) 2909–2911.

[13] H.M. Ibrahim, H. Behbehani, S. Makhseed, M.H. Elnagdi, Acylation of heteroaromatic amines: facile and efficient synthesis of a new class of 1,2,3-triazolo[4,5-b]pyridine and pyrazolo[4,3-b]pyridine derivatives, Molecules 16 (2011) 3723–3739.

[14] Y. Zhang, B. Zou, K. Wang, Microwave-assisted synthesis of 4-methyl-coumarin derivatives, HuaxueShiji 32 (2010) 787–789.

[15] A. Lawen, Apoptosis-an introduction, BioEssays 25 (2003) 888–896.

[16] L. Cordeu, E. Cubedo, E. Bandres, A. Rebollo, X. Saenz, H.M. Chuzes, V. Dominiquez, M. Echeverria, B. Mendevil, C. Sanmartin, J.A. Palop, M. Font, Biological profile of new apoptotic agents based on 2,4-pyrido[2,3-d]pyrimidine derivatives, Bioorg. Med. Chem. 15 (2007) 1659–1669.

[17] J.M. Quintela, C. Peinador, L. Botana, M. Estévez, R. Riquera, Synthesis and antihistaminic activity of 2-guanadino-3-cyanopyridines and pyrido[2,3-d]-pyrimidines, Bioorg. Med. Chem. 5 (1997) 1543–1553.

[18] A. Choudhury, H. Chen, C.N. Nilsen, K.L. Sorgi, A chemoselective aniline–chloropyrimidine coupling in a competing electrophilic environment, Tetrahedron Lett. 49 (2008) 102–105.

[19] P. Acosta, B. Insuasty, A. Ortiz, R. Abonia, M. Sortino, S.A. Zacchino, J. Quiroga, Solvent-free microwave-assisted synthesis of novel pyrazolo[4',3',5,6]pyrido[2,3-d]pyrimidines with potential antifungal activity, Arab. J. Chem. 9 (2016) 481–492.

[20] Á. Baji, T. Kiss, J. Wölfling, D. Kovács, N. Igaz, M.K. Gopisetty, M. Kiricsi, É. Frank, Multicomponent access to androstano-arylpyrimidines under microwave conditions and evaluation of their anti-cancer activity in vitro, J. Steroid Biochem. Mol. Biol. 172 (2017) 79–88.

[21] A.R. Bhat, A.H. Shalla, R.S. Dongre, Microwave-assisted one-pot catalyst-free green synthesis of new methyl-7-amino-4-oxo-5-phenyl-2-thioxo-2,3,4,5-tetrahydro-1H-pyrano[2,3-d] pyrimidine-6-carboxylates as potent in vitro antibacterial and antifungal activity, J. Adv. Res. 6 (2015) 941–948.
[22] R.R. Chavan, K.M. Hosamani, B.D. Kulkarni, S.D. Joshi, Molecular docking studies and facile synthesis of most potent biologically active N-tert-butyl-4-(4-substituted phenyl)-2-((substituted-2-oxo-2H-chromen-4-yl)methylthio)-6-oxo-1,6-dihydropyrimidine-5-carboxamide hybrids: An approach for microwave-assisted syntheses and biological evaluation, Bioorg. Chem. 78 (2018) 185–194.
[23] D. Kakati, R.K. Sarma, R. Saikia, N.C. Barua, J.C. Sarma, Rapid microwave assisted synthesis and antimicrobial bioevaluation of novel steroidal chalcones, Steroids 78 (2013) 321–326.
[24] (a) B. Green, B.L. Jenseen, P.L. Lalan, The synthesis of steroidal [16α,17α-d]-2′-pyrazolines and [16,17-d]-pyrazoles, Tetrahedron 34 (1978) 1633–1639. (b) R.O. Clinton, A.J. Manson, F.W. Stonner, H.C. Neumann, R.G. Christiansen, R.L. Clarke, J.H. Ackerman, D.F. Page, J. W. Dean, W.B. Dickinson, C. Carabatea, Steroidal [3,2-c] pyrazoles. III. Androstanes, 19-norandrostanes and their unsaturated analogs, J. Am. Chem. Soc. 83 (1961) 1478–1491.
[25] M. Asif, A. Ali, A. Zafar, M. Farhan, H. Khanam, S.M. Hadi, Shamsuzzaman, Microwave-assisted one pot synthesis, characterization, biological evaluation and molecular docking studies of steroidal thiazoles, J. Photochem. Photobiol. B 166 (2017) 104–115.
[26] K.D. Hargrave, F.K. Hess, J.T. Oliver, N-(4-substituted-thiazolyl)oxamic acid derivatives, new series of potent, orally active antiallergy agents, J. Med. Chem. 26 (1983) 1158–1163.
[27] N. Ergenc, G. Capan, N.S. Gunay, S. Ozkirimli, M. Gungor, S. Ozbey, E. Kendi, Synthesis and hypnotic activity of new 4-thiazolidinone and 2-thioxo-4,5-imidazolidinedione derivatives, Arch. Pharm. Med. Chem. 332 (1999) 343–347.
[28] W.C. Patt, H.W. Hamilton, M.D. Taylor, M.J. Ryan, D.G. Taylor, Structure-activity relationships of a series of 2-amino-4-thiazole-containing renin inhibitors, J. Med. Chem. 35 (1992) 2562–2572.
[29] J.C. Jean, L.D. Wise, B.W. Caprathe, H. Tecle, S. Bergmeier, 4-(1,2,5,6-tetrahydro-1-alkyl-3-pyridinyl)-2-thiazolamines: a novel class of compounds with central dopamine agonist properties, J. Med. Chem. 33 (1990) 311–317.
[30] F. Cagide, F. Borges, L.R. Gomes, J.N. Low, Synthesis and characterisation of new 4-oxo-N-(substituted-thiazol-2-yl)-4H-chromene-2-carboxamides as potential adenosine receptor ligands, J. Mol. Struct. 1089 (2015) 206–215.
[31] D. Chinnaraja, R. Rajalakshmi, A facile, solvent and catalyst free, microwave assisted one pot synthesis of hydrazinylthiazole, derivatives, J. Saudi Chem. Soc. 19 (2015) 200–206.
[32] M.N. Bhoi, M.A. Borad, E.A. Pithawala, H.D. Patel, Novel benzothiazole containing 4H-pyrimido[2,1-b]benzothiazoles derivatives: one pot, solvent-free microwave-assisted synthesis and their biological evaluation, Arab. J. Chem. 12 (2019) 3799–3813.
[33] D.R. Corrochano, A.D.L. Hoz, A.M.S. Migallon, R. Caballero, J.R. Ramírez, Synthesis of imine-derived triazines with donor-acceptor properties, J. Clean. Prod. 118 (2016) 223–228.
[34] A.S. Medvedeva, M.M. Demina, T.V. Kon'kova, T.L.H. Nguyen, A.V. Afonin, I.A. Ushakov, Microwave assisted solvent- and catalyst-free three-component synthesis of NH-1,2,3-triazoloimines, Tetrahedron 73 (2017) 3979–3985.
[35] E.V. Zarudnitskii, I.I. Pervak, A.S. Merkulov, A.A. Yurchenko, A.A. Tolmachev, Trimethylsilyl-1,3,4-oxadiazolesd new useful synthons for the synthesis of various 2,5-disubstituted-1,3,4-oxadiazoles, Tetrahedron 64 (2008) 10431–10442.
[36] A.N. Mayekar, Synthesis and antimicrobial studies on new substituted 1,3,4-oxadiazole derivatives bearing 6-Bromonaphthalene moiety, Int. J. Chem. 2 (2010) 38–54.

[37] T.M. Tan, Y. Chen, K.H. Kong, J. Bai, Y. Li, S.G. Lim, T.H. Ang, Y. Lam, Synthesis and the biological evaluation of 2-benzenesulfonylalkyl-5-substituted-sulfanyl-[1,3,4]-oxadiazoles as potential anti-hepatitis B virus agents, Antivir. Res. 71 (2006) 7–14.
[38] A.S. Aboraia, H.M.A. Rahman, N.M. Mahfouz, M.A. El-Gendy, Novel 5-(2-hydroxyphenyl)-3-substituted-2,3-dihydro-1,3,4-oxadiazole-2-thione derivatives: promising anticancer agents, Bioorg. Med. Chem. Lett. 14 (2006) 1236–1246.
[39] M.T.H. Khan, M.I. Choudhary, K.M. Khan, M. Rani, A. Rahman, Structure–activity relationship of tyrosinase inhibitory combinational library of 2,5-disubstituted-1,3,4-oxidiazole analogues, Bioorg. Med. Chem. 13 (2005) 3385–3395.
[40] Y. Li, J. Liu, H. Zhang, X. Yang, Z. Liu, Bioorg. Stereoselective synthesis and fungicidal activities of (E)-alpha-(methoxyimino)-benzeneacetate derivatives containing 1,3,4-oxadiazole ring, Med. Chem. Lett. 16 (2006) 2278–2282.
[41] J.T. Palmer, B.L. Hirschbein, H. Cheung, J. McCarter, J.W. Janc, W.Z. Yu, G. Wesolowski, Keto-1,3,4-oxadiazoles as cathepsin K inhibitors, Bioorg. Med. Chem. Lett. 16 (2006) 2909–2914.
[42] N.N. Farshori, A. Rauf, M.A. Siddiqui, E.S. Al-Sheddi, M.M. Al-Oqail, A facile one-pot synthesis of novel 2,5-disubstituted-1,3,4-oxadiazoles under conventional and microwave conditions and evaluation of their in vitro antimicrobial activities, Arab. J. Chem. 10 (2017) S2853–S2861.
[43] M.B. Hossain, D.V.D. Helm, R. Sanduja, M. Alam, Structure of 6-azidotetrazolo[5,1-a] phthalazine, $C_8H_4N_8$, isolated from the toxic dinoflaggelate Gymnodinium breve, Acta Crystallogr. C 41 (1985) 1199–1202.
[44] J.H. Toney, P.M. Fitzgerald, N.G. Sharma, S.H. Olson, W.J. May, J.G. Sundelof, D.E. Vanderwall, K.A. Cleary, S.K. Grant, J.K. Wu, J.W. Kozarich, D.L. Pompliano, G.G. Hammond, Antibiotic sensitization using biphenyl tetrazoles as potent inhibitors of Bacteroides fragilis metallo-β-lactamase, Chem. Biol. 5 (1998) 185–196.
[45] S. Berghmans, J. Hunt, A. Roach, P. Goldsmith, Zebrafish offer the potential for a primary screen to identify a wide variety of potential anticonvulsants, Epilepsy Res. 75 (2007) 18–28.
[46] Y. Tamura, F. Watanabe, T. Nakatani, K. Yasui, M. Fuji, T. Komurasaki, H. Tsuzuki, R. Maekawa, T. Yoshioka, K. Kawada, K. Sugita, M. Ohtani, Highly selective and orally active inhibitors of type IV collagenase (MMP-9 and MMP-2): N-sulfonylamino acid derivatives, J. Med. Chem. 41 (1998) 640–649.
[47] R.J. Herr, 5-Substituted-1H-tetrazoles as carboxylic acid isosteres: medicinal chemistry and synthetic methods, Bioorg. Med. Chem. 10 (2002) 3379–3393.
[48] F.H. Allen, C.R. Groom, J.W. Liebeschuetz, D.A. Bardwell, T.S.G. Olsson, P.A. Wood, The hydrogen bond environments of 1H-tetrazole and tetrazolate rings: the structural basis for tetrazole–carboxylic acid bioisosterism, J. Chem. Inf. Model. 52 (2012) 857–866.
[49] (a) R.G. Franz, Comparisons of pKa and log P values of some carboxylic and phosphonic acids: synthesis and measurement, AAPS PharmSci. 3 (2001) 1–13. (b) P.W. Kenny, Hydrogen bonding, electrostatic potential, and molecular design, J. Chem. Inf. Model. 49 (2009) 1234–1244.
[50] R.D. Padmaja, S. Rej, K. Chanda, Environmentally friendly, microwave-assisted synthesis of 5-substituted 1H-tetrazoles by recyclable CuO nanoparticles via (3+2) cycloaddition of nitriles and NaN_3, Chin. J. Catal. 38 (2017) 1918–1924.
[51] R.R. Chinthaparthi, C.S.R. Gangireddy, M.K.R. Kandula, S.K. Balam, S.R. Cirandur, An elegant synthesis of a new class of 1-(substituted)-1H-1,2,3-triazol-4-yl(methyl)diphenylphosphineoxides by microwave irradiation, J. Heterocyclic Chem. 52 (2015) 1876–1882.

[52] K.S. Håheim, I.T. UrdalHelgeland, E. Lindbäck, M.O. Sydnes, Mapping the reactivity of the quinoline ring-system-synthesis of the tetracyclic ring-system of isocryptolepine and regioisomers, Tetrahedron 75 (2019) 2949–2957.
[53] P.T. Parvatkar, P.S. Parameswaran, D. Bandyopadhyay, S. Mukherjee, B.K. Banik, Microwave-induced bismuth(III)-catalyzed synthesis of linear indoloquinolines, Tetrahedron Lett. 58 (2017) 2948–2951.
[54] M.S. Butler, Natural products to drugs: natural product-derived compounds in clinical trials, Nat. Prod. Rep. 25 (2008) 475–516.
[55] C. Sandeep, B. Padmashali, K. Venugopala, R. Kulkarni, R. Venugopala, B. Odhav, Synthesis and characterization of ethyl 7-acetyl-2-substituted 3- (substituted benzoyl)indolizine-1-carboxylates for in vitro anticancer activity, Asian J. Chem. 28 (2016) 1043–1048.
[56] J.L. Vaught, J.R. Carson, R.J. Carmosin, P.S. Blum, F.J. Persico, W.E. Hageman, R.P. Shank, R.B. Raffa, Antinociceptive action of McN-5195 in rodents: a structurally novel (indolizine) analgesic with a nonopioid mechanism of action, J. Pharmacol. Exp. Ther. 255 (1990) 1–10.
[57] W. Mederski, N. Beier, L.T. Burgdorf, R. Gericke, M. Klein, C. Tsaklakidis, Indolizine Derivatives and the Use Thereof as Antidiabetics, US Patent, 8, 2012.
[58] S. Hagishita, M. Yamada, K. Shirahase, T. Okada, Y. Murakami, Y. Ito, T. Matsuura, M. Wada, T. Kato, M. Ueno, Y. Chikazawa, K. Yamada, T. Ono, I. Teshirogi, M. Ohtani, Potent inhibitors of secretory phospholipase A2: synthesis and inhibitory activities of indolizine and indene derivatives, J. Med. Chem. 39 (1996) 3636–3658.
[59] G.M. Cingolani, F. Claudi, M. Massi, F. Venturi, Indolizine derivatives with biological activity VI 1-(2-aminoethyl)-3-benzyl-7-methoxy-2-methylindolizine, benanserin structural analogue, Eur. J. Med. Chem. 25 (1990) 709–712.
[60] P. Jaisankar, B. Pal, K.N. Manna, P.K. Pradhan, S. Medda, M.K. Basu, V.S. Giri, Synthesis of antileishmanial (5R)-(-)-5-carbomethoxy-3-formyl-5,6-dihydroindolo-[2,3-a]-indolizine, ARKIVOC 2003 (2003) 150–157.
[61] P. Olejnikova, L. Birosova, L. Svorc, Antimicrobial and antimutagenic properties of newly synthesized derivatives of indolizine, Sci. Pharm. 77 (2009) 216.
[62] A. Hazra, S. Mondal, A. Maity, S. Naskar, P. Saha, R. Paira, K.B. Sahu, P. Paira, S. Ghosh, C. Sinha, A. Samanta, S. Banerjee, N.B. Mondal, Amberlite-IRA-402(OH) ion exchange resin mediated synthesis of indolizines, pyrrolo [1,2-a]quinolines and isoquinolines: antibacterial and antifungal evaluation of the products, Eur. J. Med. Chem. 46 (2011) 2132–2140.
[63] A.I. Nasir, L.L. Gundersen, F. Rise, O. Antonsen, T. Kristensen, B. Langhelle, A. Bast, I. Custers, G.R. Haenen, H. Wikstrom, Inhibition of lipid peroxidation mediated by indolizines, Bioorg. Med. Chem. Lett. 8 (1998) 1829–1832.
[64] B.B. Mishra, V.K. Tiwari, Natural products in drug discovery: clinical evaluations and investigations, opportunity, Chall. Scope Nat. Prod. Med. Chem. 46 (2011) 4769–4807.
[65] G. Dannhardt, W. Meindl, S. Gussmann, S. Ajili, T. Kappe, Anti-mycobacterial7-hydroxy-2,3-dihydro-1H-indolizin-5-ones, Eur. J. Med. Chem. 22 (1987) 505–510.
[66] C. Sandeep, K.N. Venugopala, R.M. Gleiser, A. Chetram, B. Padmashali, R.S. Kulkarni, R. Venugopala, B. Odhav, Greener synthesis of indolizine analogues using water as a base and solvent: study for larvicidal activity against Anopheles arabiensis, Chem. Biol. Drug Des. 88 (2016) 899–904.
[67] C. Sandeep, K.N. Venugopala, M.A. Khedr, B. Padmashali, R.S. Kulkarni, R. Venugopala, B. Odhav, Design and synthesis of novel indolizine analogues as COX-2 inhibitors: computational perspective and in vitro screening, Ind. J. Pharm. Edu. Res. 51 (2017) 452–460.
[68] S.C. Smith, E.D. Clarke, S.M. Ridley, D. Bartlett, D.T. Greenhow, H. Glithro, A.Y. Klong, G. Mitchell, G.W. Mullier, Herbicidal indolizine-5,8-diones:photosystem I redox mediators, Pest Manag. Sci. 61 (2005) 16–24.

[69] S. Chandrashekharappa, K.N. Venugopala, S.K. Nayak, R.M. Gleiser, D.A. García, H.M. Kumalo, R.S. Kulkarni, F.M. Mahomoodally, R. Venugopala, M.K. Mohan, B. Odhav, One-pot microwave assisted synthesis and structural elucidation of novel ethyl 3-substituted-7-methylindolizine-1-carboxylates with larvicidal activity against Anopheles arabiensis, J. Mol. Struct. 1156 (2018) 377–384.

[70] J. Panda, A. Kumar, B.M. Sahoo, B.K. Banik, Microwave-assisted synthesis and evaluation of indole derivatives as potential anthelmintic agents, J. Indian Chem. Soc. 95 (2018) S1283–S1288.

[71] B. Indrani, F.B. Fredrick, K.B. Bimal, Microwave-induced synthesis of enantiopure β-lactams, microwave-induced synthesis of enantiopure β-lactams, Mod. Chem. appl. 5 (2017) 1–3.

[72] D.P. Brown, P. Saklani, J. Luo, Microwave-assisted synthesis and characterization of novel sulfonamide-β-lactam conjugates, J. Heterocyclic Chem. 55 (2018) 1815.

[73] G.V. Janssen, J.A.C. van den Heuvel, R.P. Megens, J.C.J. Benningshof, H. Ovaa, Microwave-assisted diastereoselective two-step three-component synthesis for rapid access to drug-like libraries of substituted 3-amino-b-lactams, Bioorg. Medi. Chem. 26 (2018) 41–49.

[74] R.N. Yadav, A. Chavez, B.K. Banik, Microwave-induced cycloaddition of imines with acid chlorides in absence of a tertiary amine: unprecedented synthesis of β-lactams in dimethylformamide, J. Indian Chem. Soc. 95 (2018) 1365–1367.

[75] I. Banik, R.N. Yadav, F.F. Becker, B.K. Banik, Bismuth nitrate-induced microwave-mediated deglycosylation of O-glycosides: synthesis of enantiopure 3-hydroxy β-lactams, J. Indian Chem. Soc. 95 (2018) 1373–1376.

[76] R.N. Yadav, I. Banik, B.K. Banik, Microwave-assisted novel stereoselective synthesis of bis-β-lactams with 2,7-phenanthrenyl imines, J. Indian Chem. Soc. 95 (2018) 1377–1380.

[77] R.A. Baidoo, R. Danso, S. Mukherjee, D. Bandyopadhyay, B.K. Banik, Microwave-induced N-bromosuccinimide-mediated novel synthesis of pyrroles via Paal-Knorr reaction, Heterocycl. Lett. 1 (2011) 107–109.

[78] N. Iwasawa, K. Maeyama, M. Saitou, Reactions of propargyl metallic species generated by the addition of alkynyllithiums to Fischer-type carbene complexes, J. Am. Chem. Soc. 119 (1997) 1486–1487.

[79] A. Katrizky, J. Jiang, P.J. Steel, 1-Aza-1,3-bis(triphenylphosphoranylidene)propane: a novel: $CHCH_2N$: synthon, J. Org. Chem. 59 (1994) 4551–4555.

[80] M. Periasamy, G. Srinivas, P.J. Bharati, Conversion of aryl methyl ketimines to 2,5-diarylpyrroles using $TiCl_4/Et_3N$, J. Org. Chem. 64 (1999) 4204–4205.

[81] P. Ruault, J.F. Pilard, B. Touaux, F.T. Boullet, J. Hamelin, Rapid generation of amines by microwave irradiation of ureas dispersed on clay, Synlett 6 (1994) 935–936.

[82] D. Bandyopadhyaya, B.K. Banik, Microwave-induced Paal-Knorr reaction with ammonium chloride: synthesis of pyrroles, Heterocycl. Lett. 7 (2017) 473–474.

[83] R.J. Sundberg, S.Q. Smith, The IBOGA alkaloids and their role as precursors of anti-neoplastic bisindolecatharanthus alkaloids, Alkaloids Chem. Biol. 59 (2002) 281–386.

[84] G.K. Jana, S. Paul, S. Sinha, Progress in the synthesis of ibogaalkaloids and their congeners, Org. Prep. Proced. Int. 43 (2011) 541–573.

[85] R.N. Yadav, A. Bobbala, B.K. Banik, Iodine-catalyzed microwave-induced multicomponent Aza-Diels Alder [4+2] cycloaddition reaction: a versatile approach towards BICYCLO-[2,2,2]-octanones, Mod. Chem. Appl. 6 (2018) 1–4.

[86] K. Sravanthi, P. Snehalatha, N.J.P. Subhashini, Microwave assisted green synthesis of pyrazole, 1, 2, 3- triazole based novel benzohydrazones and their antibacterial activities, J. Heterocyclic Chem. 55 (2018) 508.

[87] B.M. Sahoo, K. Nagamounika, B.K. Banik, Microwave-assisted synthesis of Schiff's bases of 1,2,4-triazole derivatives and their anthelmintic activity, J. Indian Chem. Soc. 95 (2018) 1289–1294.

[88] C. Gopi, V.G. Sastry, M.D. Dhanaraju, Microwave-assisted synthesis, structural activity relationship and biological activity of some new quinoxaline Schiff base derivatives as highly potent spirochete bactericidal agents, Beni-Suef Univ. J. Basic Appl. Sci. 6 (2017) 39–47.

[89] B.M. Sahoo, D.S. Rao, B.K. Banik, K.C. Sahoo, Microwave-mediated green chemistry approach for the synthesis of some chalcone derivatives and evaluation of their anthelmintic activity, J. Indian Chem. Soc. 95 (2018) 1295–1299.

[90] S.A. Khan, A.M. Asiri, N.S.M. Al-Ghamdi, M. Asad, M.E.M. Zayed, S.A.K. Elroby, F.M. Aqlan, M.Y. Wani, K. Sharma, Microwave assisted synthesis of chalcone and its polycyclic heterocyclic analogues as promising antibacterial agents: in vitro, in silico and DFT studies, J. Mol. Struct. 1190 (2019) 77–85.

[91] A.V. Aksenov, O.N. Nadein, N.A. Aksenov, A.A. Skomorokhov, I.V. Aksenova, M.A. Rubin, Microwave synthesis of 2-[(E)-2-(1H-indol-3-yl)vinyl]hetarenes, Chem. Heterocycl. Compd. 51 (2015) 865–868.

[92] M.O. Shibinskaya, S.A. Lyakhov, A.V. Mazepa, S.A. Andronati, A.V. Turov, N.M. Zholobak, N.Y. Spivak, Synthesis, cytotoxicity, antiviral activity and interferon inducing ability of 6-(2-aminoethyl)-6H-indolo[2,3-b]quinoxalines, Eur. J. Med. Chem. 45 (2010) 1237–1243.

[93] B. Zarranz, A. Jaso, I. Aldana, A. Monge, S. Maurel, E. Deharo, V. Jullian, M. Sauvain, Synthesis and antimalarial activity of new 3-arylquinoxaline-2-carbonitrile derivatives, Arzneimittelforschung 55 (2005) 754–761.

[94] S.S. Karki, R. Hazare, S. Kumar, V.S. Bhadauria, J. Balzarini, E.D. Clercq, Synthesis, anticancer and cytostatic activity of some 6H-indolo[2,3-b]quinoxalines, Acta Pharm. 59 (2009) 431.

[95] M.C. Wamberg, A.A. Hassan, A.D. Bond, E.B. Pederson, Intercalating nucleic acids (INAs) containing insertions of 6H-indolo[2,3-b]quinoxaline, Tetrahedron 62 (2006) 11187–11199.

[96] (a) A. Carta, P. Cornona, M. Lorgia, A versatile scaffold endowed with manifold activities, Curr. Med. Chem. 12 (2005) 2259–2272. (b) L.W. Deady, A.J. Kaye, G.J. Finlay, B.C. Baguley, W.A. Denny, Synthesis and antitumor properties of N-[2-(dimethylamino)ethyl]carboxamide derivatives of fused tetracyclic quinolines and quinoxalines: a new class of putative topoisomerase inhibitors, J. Med. Chem. 40 (1997) 2040–2046. (c) R.S. Varma, R.K. Pandey, Potential biologically active agents, XXVIII synthesis of substituted indophenazines, Arch. Pharm. 314 (1981) 307–310. (d) A.H. Abadi, 5-Substituted 2-bromoindolo[3,2-b]quinoxalines. A class of potential antitumor agents with cdc25 phosphatase inhibitory properties, Arch Pharm. 331 (1998) 352–358. (e) A. Gazit, H. App, G. McMahon, J. Chen, A. Levitzki, F.D. Bohmer, Tyrphostins. 5. potent inhibitors of platelet-derived growth factor receptor tyrosine kinase: structure-activity relationships in quinoxalines, quinolines, and indole tyrphostins, J. Med. Chem. 39 (1996) 2170–2177.

[97] (a) J. Harmenberg, B. Wahren, J. Bergman, S. Akerfeldt, L. Lundblad, Antiherpesvirus activity and mechanism of action of indolo-(2,3-b)quinoxaline and analogs, Anti Microb. Agents Chemother. 32 (1988) 1720. (b) L.M. Wilhelmsson, N. Kingi, J. Bergman, Interactions of antiviral indolo[2,3-b]quinoxaline derivatives with DNA, J. Med. Chem. 51 (2008) 7744–7750.

[98] S. Avula, J.R. Komsani, S. Koppireddi, R. Yadla, Microwave-assisted synthesis of 6-{(5-Aryl-1,3,4-oxadiazol-2-yl)methyl}-6H-indolo[2,3-b]quinoxalines, J. Heterocyclic Chem. 52 (2015) 1737.

[99] L. Gupta, A. Talwar, P.M. Chauhan, Bis and tris indole alkaloids from marine organisms: new leads for drug discovery, Curr. Med. Chem. 14 (2007) 1789.

[100] M.M. Rahman, A.I. Gray, A benzoisofuranone derivative and carbazole alkaloids from Murraya koenigii and their antimicrobial activity, Phytochemistry 66 (2005) 1601–1606.

[101] A.C. Karaburun, Z.A. Kaplancıklı, N. Gundogdu-Karaburun, F. Demirci, Synthesis, antibacterial and antifungal activities of some carbazole dithiocarbamate derivatives, Lett. Drug Des. Discov. 8 (2011) 811.

[102] T.A. Choi, R. Czerwonka, W. Fröhner, M.P. Krahl, K.R. Reddy, S.G. Franzblau, H.J. Knolker, Synthesis and activity of carbazole derivatives against mycobacterium tuberculosis, Chem. Med. Chem. 1 (2006) 812–815.

[103] P. Rajakumar, K. Sekar, V. Shanmugaiah, N. Mathivanan, Synthesis of novel carbazole based macrocyclic amides as potential antimicrobial agents, Eur. J. Med. Chem. 44 (2009) 3040.

[104] P. Rajakumar, K. Sekar, V. Shanmugaiah, N. Mathivanan, Synthesis of some novel imidazole-based dicationic carbazolophanes as potential antibacterials, Bioorg. Med. Chem. Lett. 18 (2008) 4416–4419.

[105] J.L. Arbiser, B. Govindarajan, T.E. Battle, R. Lynch, D.A. Frank, M. Ushio-Fukai, B.N. Perry, D.F. Stern, G.T. Bowden, A. Liu, E. Klein, P.J. Kolodziejski, N.T. Eissa, C.F. Hossain, D.G. Nagle, Carbazole is a naturally occurring inhibitor of angiogenesis and inflammation isolated from antipsoriatic coal tar, J. Invest. Dermatol. 126 (2006) 1396–1402.

[106] G. Periyasami, R. Raghunathan, G. Surendiran, N. Mathivanan, Synthesis of novel spiropyrrolizidines as potent antimicrobial agents for human and plant pathogens, Bioorg. Med. Chem. Lett. 18 (2008) 2342–2345.

[107] T. Lemster, U. Pindur, G. Lenglet, S. Depauw, C. Dassi, M.H.E. David-Cordonnier, Photochemical electrocyclisation of 3-vinylindoles to pyrido[2,3-a]-, pyrido[4,3-a]- and thieno [2,3-a]-carbazoles: design, synthesis, DNA binding and antitumor cell cytotoxicity, J. Med. Chem. 44 (2009) 3235–3325.

[108] Y. Tachibana, H. Kikuzaki, N.H. Lajis, N.J. Nakatani, Antioxidative activity of carbazoles from Murraya koenigii leaves, Agric. Food Chem. 49 (2001) 5589–5594.

[109] J.B. Wright, Histamine antagonists. IV. indole and carbazole derivatives, J. Am. Chem. Soc. 71 (1949) 1028–1030.

[110] (a) C. Viegas-Junior, A. Danuello, B.V. da Silva, E.J. Barreiro, C.A. Fraga, Molecular hybridization: a useful tool in the design of new drug prototypes, Curr. Med. Chem. 14 (2007) 1829–1852. (b) S.B.B.B. Bahia, W.J. Reis, G.A.M. Jardim, F.T. Souto, C.A. de Simone, C.C. Gatto, R.F.S. Menna-Barreto, S.L. de Castro, B.C. Cavalcanti, C. Pessoa, M.H. Araujo, E.N.D.S. Ju'nior, Molecular hybridization as a powerful tool towards multitarget quinoidal systems: synthesis, trypanocidal and antitumor activities of naphthoquinone-based 5-iodo-1,4-disubstituted-, 1,4- and 1,5-disubstituted-1,2,3-triazoles, Med. Chem. Commun. 7 (2016) 1555–1563. (c) V.U. Jeankumar, R.S. Reshma, R. Janupally, S. Saxena, J.P. Sridevi, B. Medapi, P. Kulkarni, P. Yogeeswari, D. Sriram, Enabling the (3 + 2) cycloaddition reaction in assembling newer anti-tubercular lead acting through the inhibition of the gyrase ATPase domain: lead optimization and structure activity profiling, Org. Biomol. Chem. 13 (2015) 2423–2431.

[111] D. Addla, S. Wen, W. Gao, S.K. Maddili, L. Zhang, C. Zhou, Design, synthesis, and biological evaluation of novel carbazole aminothiazoles as potential DNA-targeting antimicrobial agents, Med. Chem. Commun. 7 (2016) 1988–1994.

[112] R.S. Jasass, F. Alshehrei, T.A. Farghaly, Microwave-assisted synthesis of antimicrobial agents containing carbazole and thiazole moieties, J. Heterocyclic Chem. 55 (2018) 2099–2106.

[113] Y. Jiao, C.S. Cho, Microwave-assisted copper powder-catalyzed coupling and cyclization of β-bromo-α, β-unsaturated amides with amidine hydrochlorides leading to pyrimidinones, Appl. Organomet. Chem. 29 (2015) 372–375.

[114] D. Kułaga, J. Jaśkowsk, R. Jasiński, Microwave-assisted solvent-free synthesis of ipsapirone, J. Heterocyclic Chem. 56 (2019) 1498–1504.

[115] A. Kędzia, A. Kudelko, M. Świątkowski, R. Kruszyński, Microwave-promoted synthesis of highly luminescent s-tetrazine-1,3,4-oxadiazole and s-tetrazine-1,3,4-thiadiazole hybrids, Dyes Pigments 172 (2020) 107865.

[116] K. Khaldoun, A. Safer, N. Boukabcha, N. Dege, S. Ruchaud, M. Souab, S. Bach, A. Chouaih, S. Saidi-Besbes, Synthesis and evaluation of new isatin-aminorhodanine hybrids as PIM1 and CLK1 kinase inhibitors, J. Mol. Struct. 1192 (2019) 82–90.

[117] N. Łukasik, E. Luboch, J. Chojnacki, E. Wagner-Wysieck, 1,3,4-Thiadiazole-based diamides: synthesis and complexation properties, J. Mol. Struct. 1146 (2017) 713–722.

[118] T. Luo, Y. Li, Y. Xu, S. Zhang, Y. Wanga, X. Kou, D. Xiao, Rapid synthesis of a hyperfluorescence 2-pyridone derivative as a fluorescent molecular sensor for picric acid, Sensors Actuators B 253 (2017) 231–238.

[119] A. Dongamanti, M.G. Devulapally, V.K. Aamate, S. Gundu, Microwave-assisted synthesis and antimicrobial evaluation of novel pyrazolines, Chem. Heterocycl. Compd. 51 (2015) 872–882.

[120] G.P. Kalena, A. Jain, A. Banerji, Amberlyst 15 catalyzed prenylation of phenols: one-step synthesis of benzopyrans, Molecules 2 (1997) 100–105.

[121] Y.L.N. Murthy, R.S. Prabha, S. Seetharam, Synthesis, characterisation of new chromanoisoxazoles and investigation of optical power limiting properties, Indian J. Chem. 43 (2004) 361.

[122] A.M. Revol-Junelles, R. Mathis, F. Krier, A. Delfour, G. Lefebvre, Leuconostoc mesenteroides subsp. mesenteroides FR52 synthesizes two distinct bacteriocins, Lett. Appl. Microbiol. 23 (1996) 120–124.

[123] V. Kumar, K. Chand, E. Chorell, Synthesis of various N-substituted 4-aryloxy/thiophenoxy/thioisopropyloxy-phthalimides, ChemistrySelect 2 (2017) 3293–3296.

[124] M. Thangaraj, B. Ranjan, R. Muthusamy, A. Murugesan, R.M. Gengana, Microwave synthesis of fused pyrans by humic acid supported ionic liquid catalyst and their antimicrobial, antioxidant, toxicity assessment, and molecular docking studies, J. Heterocyclic Chem. 56 (2019) 867–885.

[125] P. Wadhwa, T. Kaur, A. Sharma, Solvent-free pot-, atom- and step-economic synthesis of novel benzo[d]thiazole-[1,3]-thiazine hybrids in a one-pot reaction, Asian J. Org. Chem. 5 (2016) 763–769.

[126] Z. Zukauskaite, V. Buinauskaite, J. Solovjova, L. Malinauskaite, A. Kveselyte, A. Bieliauskas, G. Ragaite, A. Sackus, Microwave-assisted synthesis of new fluorescent indoline-based building blocks by ligand free Suzuki-Miyaura cross-coupling reaction in aqueous media, Tetrahedron 72 (2016) 2955–2963.

[127] M.E.M. Zayed, P. Kumar, S.A. Khan, Microwave assisted synthesis, spectroscopic and photophysical properties of novel pyrazol-3-one containing push -pull chromophore, J. Mol. Struct. 1202 (2020) 127103.

[128] R. Ulus, I. Yesildağ, M. Elmastas, M. Kaya, Rapid synthesis of novel 1,8-dioxoacridine carboxylic acid derivatives by microwave irradiation and their free radical scavenging activity, Med. Chem. Res. 24 (2015) 3752–3759.

[129] I. Antonini, P. Polucci, R.L. Kelland, E. Menta, N. Pescalli, S. Martelli, 2,3-Dihydro-1H,7H-pyrimido[5,6,1-de]acridine-1,3,7-trione derivatives, a class of cytotoxic agents active on

multidrug-resistant cell lines: synthesis, biological evaluation, and structure-activity relationships, J. Med. Chem. 42 (1999) 2535–2541.

[130] Y. Mikata, M. Yokoyama, K. Mogami, M. Kato, I. Okura, M. Chikira, S. Yano, Intercalator-linked cisplatin: synthesis and antitumor activity of cis-dichloroplatinum(II) complexes connected to acridine and phenylquinolines by one methylene chain, Inorg. Chim. Acta 279 (1998) 51–57.

[131] L. Ngadi, A.M. Galy, J.P. Galy, J. Barbe, A. Cremieux, D. Sharples, Some new 1-nitro acridine derivatives as antimicrobial agents Quelques nouveaux dérivés acridiniques nitrés en position 1, agents antimicrobiens, Eur. J. Med. Chem. 25 (1990) 67–70.

[132] M. Wainwright, Acridine—a neglected antibacterial chromophore, J. Antimicrob. Chemother. 47 (2001) 1.

[133] M. Kaya, Y. Yıldırır, G.Y. Çelik, Synthesis and antimicrobial activities of novel bisacridine-1,8-dione derivatives, Med. Chem. Res. 20 (2011) 293–299.

[134] S. Gallo, S. Atifi, A. Mahamoud, C. Santelli-Rouvier, K. Wolfa'rt, J. Molnar, J. Barbe, Synthesis of aza mono, bi and tricyclic compounds. Evaluation of their anti MDR activity, Eur. J. Med. Chem. 38 (2003) 19–26.

[135] M. Kaya, Y. Yıldırır, G.Y. Çelik, Synthesis, characterization, and in vitro antimicrobial and antifungal activity of novel acridines, Pharm. Chem. J. 48 (2015) 722–726.

[136] B. Öcal, B. Saraç, R. Simsek, S. Yıldırım, Y. Sarıoğlu, C. Safak, Vasorelaxing properties of some phenylacridine type potassium channel openers in isolated rabbit thoracic arteries, Eur. J. Med. Chem. 37 (2002) 519–523.

[137] M. Kaya, E. Basar, E. Çakir, E. Tunca, M.J. Bulbul, Synthesis and characterization of novel dioxoacridine sulfonamide derivatives as new carbonic anhydrase inhibitors, J. Enzyme Inhib. Med. Chem. 27 (2012) 509–514.

[138] R. Ulus, I. Yesildağ, M. Tanc, M. Bülbül, M. Kaya, C.T. Supuran, Synthesis of novel acridine and bis acridine sulfonamides with effective inhibitory activity against the cytosolic carbonic anhydrase isoforms II and VII, Bioorg. Med. Chem. 21 (2013) 5799–5805.

[139] I. Yesildağ, R. Ulus, E. Basar, M. Aslan, M. Kaya, M. Bülbül, Facile, highly efficient, and clean one-pot synthesis of acridine sulfonamide derivatives at room temperature and their inhibition of human carbonic anhydrase isoenzymes, Monatsh. Chem. 145 (2014) 1027–1034.

[140] S.A. Gamage, J.A. Spicer, G.J. Atwell, G.J. Finlay, B.C. Baguley, W.A. Denny, Structure-activity relationships for substituted bis(acridine-4-carboxamides): a new class of anticancer agents, J. Med. Chem. 42 (1999) 2383–2393.

[141] M.G.R. Pitta, E.S. Souza, F.W.A. Barros, M.O.M. Filho, O. Pessoa, M.Z. Hernandes, M.D.C. A.D. Lima, S.L. Galdino, I.D.R. Pitta, Synthesis and in vitro anticancer activity of novel thiazacridine derivatives, Med. Chem. Res. 22 (2013) 2421–2429.

[142] M. Kaya, Y. Yıldırır, L. Türker, Synthesis and laser activity of halo-acridinedione derivatives, J. Heterocyclic Chem. 46 (2009) 294–297.

[143] S.M. Monti, C.T. Supuran, G.D. Simone, Anticancer carbonic anhydrase inhibitors: a patent review (2008–2013), Expert Opin. Ther. Patents 23 (2013) 737–749.

[144] M. Kaya, E. Demir, H. Bekci, Synthesis, characterization and antimicrobial activity of novel xanthene sulfonamide and carboxamide derivatives, J. Enzyme Inhib. Med. Chem. 28 (2013) 885–893.

[145] B. Aday, P. Sola, F. Çolak, M. Kaya, Synthesis of novel sulfonamide analogs containing sulfamerazine/sulfaguanidine and their biological activities, J. Enzyme Inhib. Med. Chem. 30 (2015) 1.

[146] E. Basar, E. Tunca, M. Bülbül, M. Kaya, Synthesis of novel sulfonamides under mild conditions with effective inhibitory activity against the carbonic anhydrase isoforms I and II, J. Enzyme Inhib. Med. Chem. 31 (2016) 1356–1361.

[147] I. Esirden, R. Ulus, B. Aday, M. Tanc, C.T. Supuran, M. Kaya, Synthesis of novel acridine bis-sulfonamides with effective inhibitory activity against the carbonic anhydrase isoforms I, II, IX and XII, Bioorg. Med. Chem. 23 (2015) 6573–6580.

[148] S. Bag, R. Tulsan, A. Sood, H. Cho, H. Redjeb, W. Zhou, H. LeVine, B. Török, M. Török, Sulfonamides as multifunctional agents for Alzheimer's disease, Bioorg. Med. Chem. Lett. 25 (2015) 626–630.

[149] F. Carta, C.T. Supuran, Diuretics with carbonic anhydrase inhibitory action: a patent and literature review (2005–2013), Expert Opin. Ther. Patents 21 (2013) 681–691.

[150] R.L. Arechederra, A. Waheed, W.S. Sly, C.T. Supuran, S.D. Minteer, Effect of sulfonamides as carbonic anhydrase VA and VB inhibitors on mitochondrial metabolic energy conversion, Bioorg. Med. Chem. 21 (2013) 1544–1548.

[151] R. Ulus, B. Aday, M. Tanç, C.T. Supuran, M. Kaya, Microwave assisted synthesis of novel acridine–acetazolamide conjugates and investigation of their inhibition effects on human carbonic anhydrase isoforms hCA I, II, IV and VII, Bioorg. Med. Chem. 24 (2016) 3548–3555.

[152] N.P. Prajapati, K.D. Patel, R.H. Vekariya, H.D. Patel, D.P. Rajani, Thiazole fused thiosemicarbazones: microwave-assisted synthesis, biological evaluation and molecular docking study, J. Mol. Struct. 1179 (2019) 401–410.

[153] F. Abebe, T. Sutton, P. Perkins, K. Makins-Dennis, A. Winstead, Microwave-assisted synthesis of rhodamine derivatives, Green Chem. Lett. Rev. 11 (2018) 237–245.

[154] F. Ke, C. Liu, P. Zhang, J. Xu, X. Chen, Efficient and selective microwave-assisted copper-catalyzed synthesis of quinazolinone derivatives in aqueous, Synth. Commun. 48 (2018) 3089–3098.

Chapter 6

Microwave-assisted oxidation and reduction reactions

Aparna Das* and Bimal Krishna Banik*
Department of Mathematics and Natural Sciences, College of Sciences and Human Studies, Prince Mohammad Bin Fahd University, Al Khobar, Kingdom of Saudi Arabia
**Corresponding authors: E-mails: aparnadasam@gmail.com (Aparna Das); bimalbanik10@gmail. com, bbanik@pmu.edu.sa (Bimal Krishna Banik)*

6.1 Introduction

Many scientists have reported the applications of microwave technology. Microwave-assisted oxidation and reduction reaction is a highly efficient and clean method. It provides a safer alternative to conventional thermal reactions. Microwave-induced methods are fast, economical, high yielding, and selective in most of the cases. Several efficient catalysts that are suitable for microwave-induced methods are developed.

This chapter is focused on microwave-assisted oxidation and reduction reactions of important organic compounds.

6.2 Oxidation reactions

6.2.1 Oxidation of polyphenolic molecules

6.2.1.1 Oxidation of a lignin phenolic monomer using Co(salen)/SBA-15 induced by microwave

Lignin is a cross-linked phenolic polymer and is commonly present in plants as an integral part. This polyphenolic compound binds the cellulose together and acts as a gluing material. It also prevents the cell wall from the attack of pathogens [1]. Degradation of lignin, delignification, is an important step in paper manufacturing process [2]. Lignin is needed to be removed during the pulping process to liberate the cellulosic fibrous material. Several hazardous and environmentally friendly delignification methods are available [3–8]. The microwave-assisted oxidative degradation of a lignin model phenolic monomer (1-(4-hydroxy-3-methoxyphenoxy)-ethanol, apocynol) catalyzed by Co(salen)/SBA-15 was reported [9].

Microwaves in Chemistry Applications. https://doi.org/10.1016/B978-0-12-822895-1.00001-1

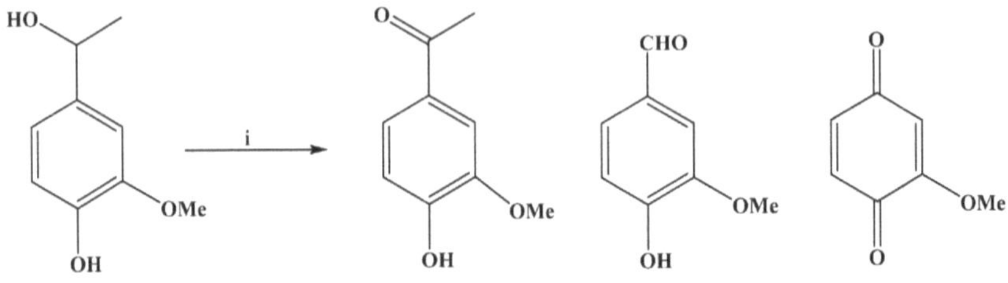

1 2 3 4

SCHEME 1 Product distribution in the oxidation of apocynol under microwave irradiation. Reagents and conditions: (i) H_2O_2, Catalyst (Co(salen)/SBA-15), CH_3CN, MW, 300 W, 5–45 min.

The Scheme 1 shows the product distribution in the oxidation of apocynol (**1**) under microwave irradiation. The oxidation resulted in the formation of acetovanillone (**2**), vanillin (**3**), and 2-methoxybenzoquinone (**4**). The first oxidation product was acetovanillone [10]. Vanillin was formed through side-chain cleavage, and 2-methoxybenzoquinone was the product of oxidation of the phenolic group together with oxidative degradation of the *p*-alkyl side chain [11]. Complete degradation of the apocynol was achieved after 40 min microwave irradiation, and the catalyst was reusable. No conversion was observed in the blank microwave (without catalyst) even after 15 min.

Oxidation of apocynol under microwave irradiation was also possible with mesoporous silica materials [12]. Mesoporous MCM-41, HMS, SBA-15, and amorphous silica were used as catalysts in the study, and different reactivities were observed for the various silica materials.

6.2.1.2 *Oxidation of silibinin assisted by microwave*

Flavonolignans are polyphenolic compounds found copiously in fruits and vegetables and are important for human nutrition. Silibinin is a flavonolignan extracted from milk thistle. Silibinin has shown various pharmacological properties such as antioxidant, hypocholesterolemic, antitumor, cardioprotective, neuroprotective, and antiviral activity [13–18]. Studies showed that the oxidation product of silibinin, 2,3-dehydrosilybin has a more potent biological activity than its parent compound. For instance, this compound appeared to be more effective than silibinin in their antioxidant, antitumor, and antiproliferative activities [19]. In addition, the oxidation product also showed positive activities against some skin diseases [20].

Several synthetic methods were developed to prepare 2,3-dehydrosilybin and its analogues [21–26].

Microwave-assisted base-catalyzed oxidation of silibinin was developed [27]. The reaction produced two different silibinin derivatives, 2,3-dehydrosilybin (**6**) and hemiacetal (**7**) as shown in Scheme 2. For the reaction, the solution of silibinin 1 (1.0 g, 2.1 mmol) and base (KOAc, or TEA 6.22 mmol) in an appropriate solvent (8 mL for THF, THF/H_2O,

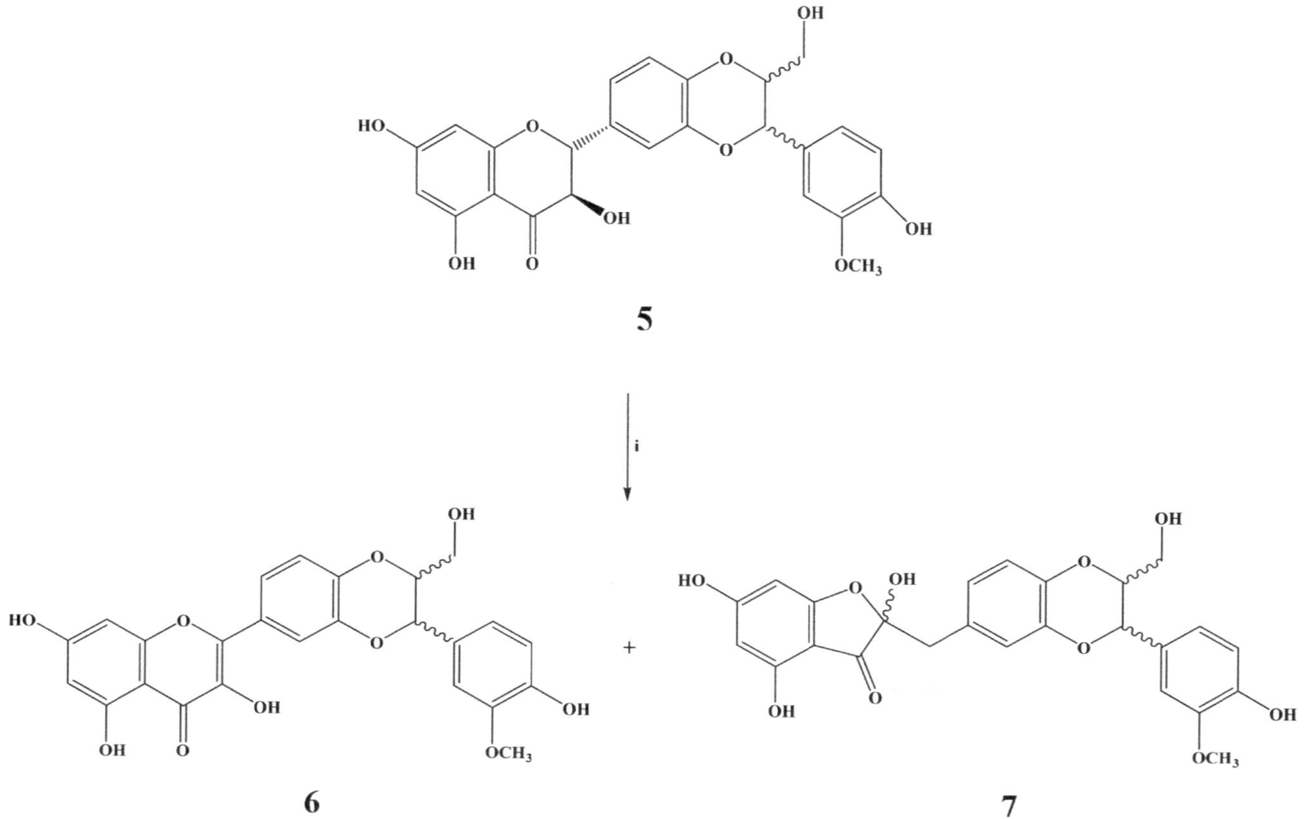

SCHEME 2 Microwave-assisted base-catalyzed oxidation of silibinin. Reagents and conditions: (i) MW, base, solvent.

1,4-dioxane/MeOH and 4 mL for pyridine and DMF) were placed in the microwave cavity and irradiated (0–300 W). The reaction conducted in DMF at 50°C led to the formation of products **6** and **7** in 70% and 15% yield, respectively, within 30 min. Conventional heating methods were compared with results obtained from microwave heating, and the formation of product **7** was not observed under conventional heating.

6.2.2 Oxidation of alcohols

6.2.2.1 *Oxidation of alcohols with hydrogen peroxide assisted by microwave and catalyzed by nitrogen ligands of iron complexes*

The oxidation of alcohols to the corresponding carbonyl compounds is an essential reaction in organic synthesis because these chemicals are crucial precursors and intermediates for the synthesis of a large number of pharmaceuticals, fragrances, and agrochemicals.

Different metal-based catalytic methods are available for the oxidation of alcohol [28–35]. Microwave-assisted alcohol oxidation with iron catalysts was reported [36–42]. Cozzi et al. conducted oxidation of primary and secondary alcohols, including two diols and glycerol, with H_2O_2 catalyzed by the iron (II) complexes $[FeL_3](OTf)_2$ (L=2,2′-bipyridine, 2,2′-bipyrimidine, 1,10-phenanthroline and substituted derivatives) and $[FeL_2](OTf)_2$ (L=bis(2-pyridinylmethyl)amine, 2,6-di(2-pyridyl)pyridine) [36]. The reactions were carried out under microwave at low power in acetonitrile, water, or mixed solvent.

The scheme of the oxidation of glycerol with $[Fe(bipy)_3](OTf)_2$ and H_2O_2 in the microwave reactor is shown in Scheme 3. The main product was formic acid (**9**), and a small amounts of dihydroxyacetone (**10**), hydroxypyruvic acid (**11**), and glycolic acid (**12**) were also obtained.

6.2.2.2 *Solvent-free peroxidative oxidation of alcohols assisted by microwave and catalyzed by Iron(III)-TEMPO*

Solvent-free microwave-assisted peroxidative oxidation of alcohols catalyzed by Iron(III)-TEMPO catalytic systems was demonstrated by Pombeiro et al. [37]. The iron(III) complexes $[H(EtOH)][FeCl_2(L)_2]$, $[H_2bipy]_{1/2}[FeCl_2(L)_2]\cdot DMF$ and $[FeCl_2(L)(2,2'\text{-bipy})]$ (L = 3-amino-2-pyrazinecarboxylate; H_2bipy = doubly protonated 4,4′-bipyridine; 2,2′-bipy = 2,2′-bipyridine, DMF = dimethylformamide) were considered for the study.

The oxidation of 1-phenylethanol using aqueous tertbutylhydroperoxide (*t*BuOOH or TBHP, aq. 70%) as an oxidizing agent under low-power microwave irradiation and in a solvent-free medium was conducted (Scheme 4). The same microwave-assisted oxidation method was also tested toward the oxidation of C_6 secondary alcohols, such as cyclohexanol and the linear 2- and 3-hexanol, and the primary alcohols benzyl alcohol and 1-hexanol. The ketones

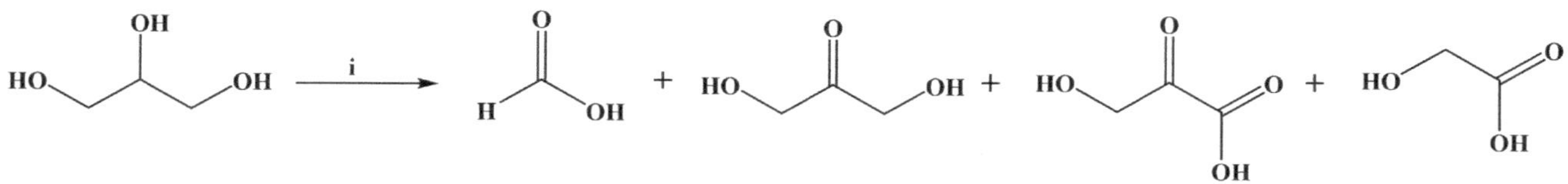

8 9 10 11 12

SCHEME 3 Oxidation of glycerol. Reagents and conditions: (i) H_2O_2, $[Fe(bipy)_3](OTf)_2$, solvent, MW.

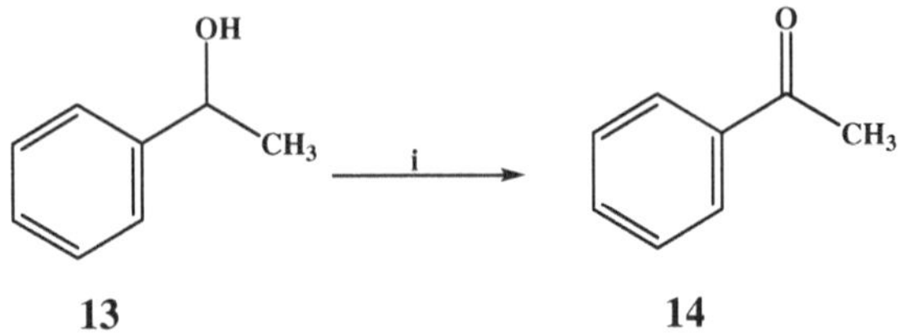

SCHEME 4 Microwave-assisted oxidation of 1-phenylethanol to acetophenone catalyzed by iron (III) complexes. Reagents and conditions: (i) Iron(III) complex, TBHP, MW, 10 W, 0.5–2 h.

were the only products obtained from the oxidation. The low-power microwave irradiation (10 W) provided a more efficient synthetic method than conventional heating and allowed the attainment of similar yields in shorter times. For instance, in the presence of iron(III) complex, a 99% yield of acetophenone was obtained after 1 h reaction under microwave irradiation at 150°C, while the reaction under conventional heating (oil bath) gave only 9% yield. Besides, it took 46 h with conventional heating to achieve the same yield as that obtained with microwave after 1 h.

6.2.2.3 *Peroxidative reaction of 1-phenylethanol and cyclohexane assisted by microwave using N-based additives*

The same researchers reported the microwave-assisted peroxidative oxidation of 1-phenylethanol by *t*-BuOOH in the presence of various N-based additives catalyzed by bis- and tris-pyridyl amino and imino thioether Cu and Fe complexes [38]. The reaction resulted in a maximum yield of acetophenone of 99% after 30 min at 80°C under the very low microwave power of 10 W. The same reaction under conventional oxidation (heating in an oil bath) afforded only 37% of acetophenone.

6.2.2.4 *Solvent-free peroxidative reaction of secondary alcohols induced by microwave and catalyzed by copper(II) 2,4-alkoxy-1,3,5-triazapentadienato complexes*

A solvent-free microwave-assisted peroxidative oxidation of secondary alcohols to the corresponding ketones catalyzed by copper(II) 2,4-alkoxy-1,3,5-triazapentadienato complexes (Fig. 1) was also demonstrated [41]. Tert-butylhydroperoxide (TBHP) was used as the oxidant, and the reaction provided the ketones in high yields (up to 100%). The schematics of the reaction are shown in Scheme 5.

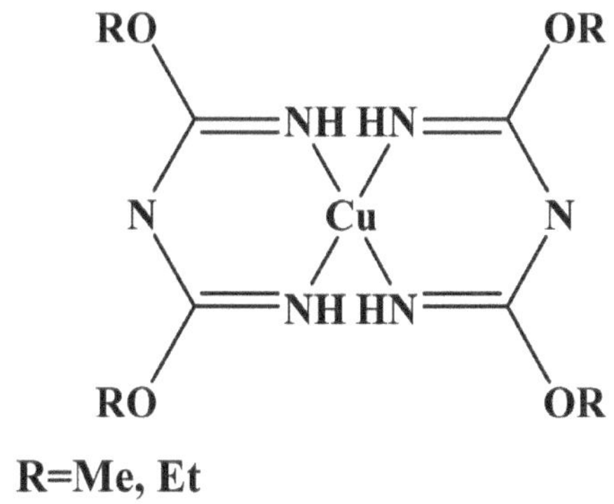

FIG. 1 Molecular structure of copper(II) 2,4-alkoxy-1,3,5-triazapentadienato complexes.

SCHEME 5 Microwave-assisted oxidation of secondary alcohols to ketones. Reagents and conditions: (i) TBHP, copper(II) 2,4-alkoxy-1,3,5-triazapentadienato complexes, 80°C, MW, 30–240 min.

6.2.2.5 Peroxidative reaction of toluene and 1-phenylethanol assisted by microwave with keto and enol aroylhydrazone Cu(II) complexes

Benzaldehyde is an important starting material for the synthesis of dyes, perfumes, and pharmaceuticals [42]. Benzaldehydes are mainly prepared by the oxidation of toluene. Several studies, using different catalytic systems including metalloporphyrins, Au-Pd nanoparticles, Cu-Mn oxides, and graphite carbon nitride nanocomposites (g-C_3N_4), were reported the solvent-free aerobic oxidation of toluene [43–46]. Sutradhar et al. reported the microwave-assisted peroxidative oxidation of toluene and 1-phenylethanol with Cu(II) complexes with tert-butyl hydroperoxide (Scheme 6) [47].

The partial oxidation of toluene gave several products with low selectivities, such as benzaldehyde, benzyl alcohol, benzyl acetate, benzoic acid, or cresols [48, 49]. The microwave-assisted peroxidative oxidation of toluene by *t*-BuOOH produced benzaldehyde as the major product, while benzyl alcohol and benzoic acid were minor products. Benzoic acid was detected only when the concentration of catalyst was greater than 5 mM. Besides, no oxidation of the aromatic ring was observed.

(a)

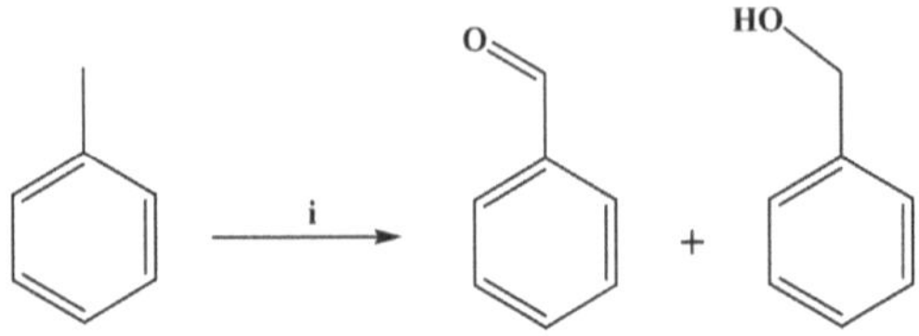

17 18 19

(b)

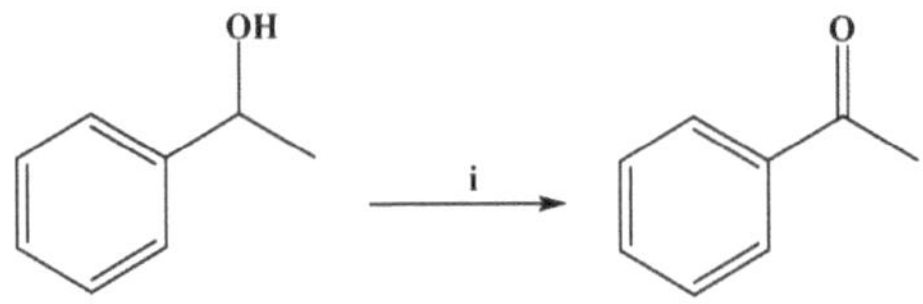

20 21

SCHEME 6 Microwave-assisted peroxidative oxidations of (A) toluene and (B) 1-phenylethanol under mild conditions. Reagents and conditions: (i) Cu(II) complexes, *t*-BuOOH, 50°C, MW, 5 W, 1 h.

6.2.2.6 Cinnamyl alcohol oxidation mediated by microwave using iron and palladium catalysts

Wang et al. explored the microwave-assisted oxidation of cinnamyl alcohol in acetonitrile by using supported iron and palladium catalysts and 1.6–2.8 eq. of aqueous H_2O_2 as a green oxidant [50]. Schematics of the oxidation of cinnamyl alcohol, cinnamaldehyde, and dihydrocinnamyl alcohol are shown in Schemes 7–9.

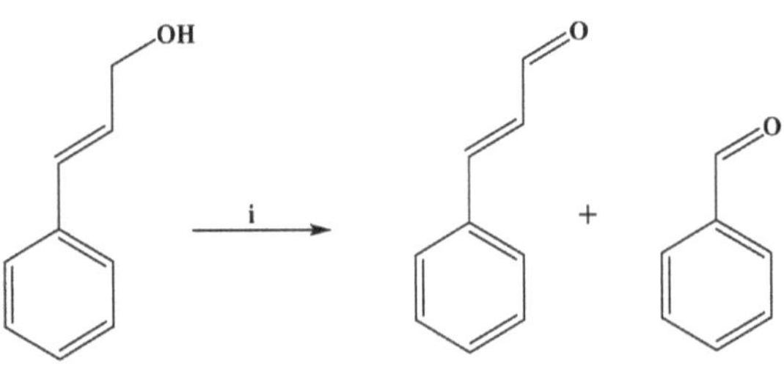

22 23 24

SCHEME 7 Microwave-assisted oxidation of cinnamyl alcohol to cinnamaldehyde and benzaldehyde. Reagents and conditions: (i) Catalyst, H_2O_2, MW, 300 W, 5 min.

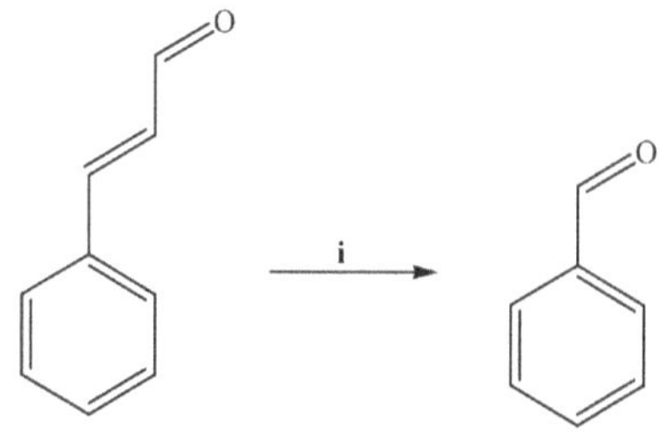

25 **26**

SCHEME 8 Microwave-assisted oxidation of cinnamaldehyde to benzaldehyde. Reagents and conditions: (i) Catalyst, H_2O_2, MW, 300 W, 5 min.

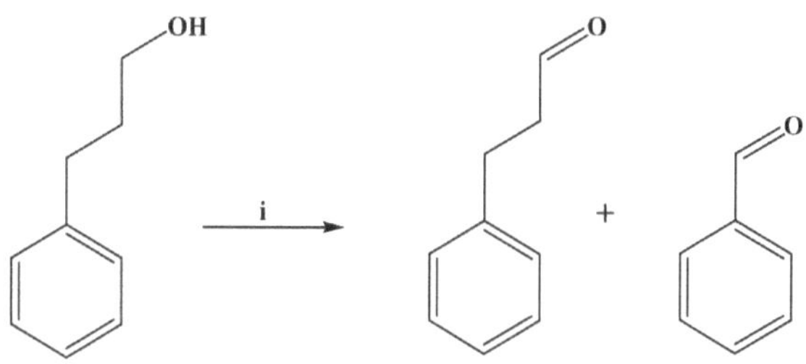

27 **28** **29**

SCHEME 9 Microwave-assisted oxidation of dihydrocinnamyl alcohol to dihydrocinnamaldehyde and benzaldehyde. Reagents and conditions: (i) Catalyst, H_2O_2, MW, 300 W, 5 min.

6.2.2.7 Oxidation of cyclohexane and 1- and 2-phenylethanol induced by microwave and using C-scorpionate iron(II) complex on nanostructured carbon materials

Ribeiro et al. demonstrated C-scorpionate iron(II) complex on nanostructured carbon materials as recyclable catalyst for microwave-assisted oxidation of cyclohexane and 1- and 2-phenylethanol [51].

Microwave-assisted oxidation of 1- and 2-phenylethanol is shown in Scheme 10. Under the optimized conditions, 1 h of microwave irradiation 97% yield of 2-phenylacetaldehyde was obtained. No traces of byproducts were detected in the final reaction mixtures for the optimized conditions. Furthermore, control experiments in the absence of the catalyst led to only 5% or 8% alcohol conversion, for 1-phenylethanol or 2-phenylethanol, respectively. This result confirmed the crucial role of the catalyst to efficiently catalyze the oxidation of the alcohols.

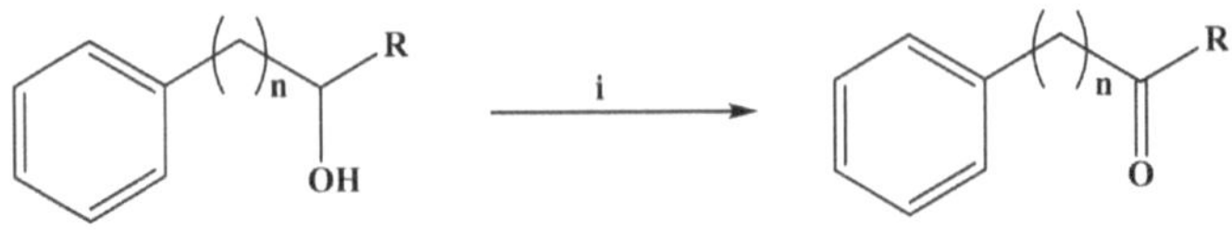

30

n=0,1; R=Me, H

31

SCHEME 10 Microwave-assisted oxidation of 1- and 2-phenylethanol with aqueous tert-butyl hydroperoxide catalyzed by $[FeCl_2(Tpm)]$ supported on functionalized nanodiamonds (NDoxNa). Reagents and conditions: (i) $[FeCl_2(Tpm)]$NDoxNa, aq.TBHP, 80°C, MW, 25 W, 1 h.

6.2.2.8 Oxidation of cyclohexane mediated by microwave and using copper(II) complexes

Jlassi et al. demonstrated microwave-assisted oxidation of cyclohexane using copper(II) complexes with an arylhydrazone of methyl 2-cyanoacetate as effective catalysts [52].

Under mild conditions and low-power microwave irradiation, in the presence of tert-butyl hydroperoxide (TBHP, 70% aqueous solution) as oxidant, copper(II) complexes acted as effective catalysts in the oxidation of cyclohexane to cyclohexanol and cyclohexanone (Scheme 11). At 50°C, after 2 h the activity of the catalyst has reached a turnover number of 1.44×10^3 and a turnover frequency of $1.98 \times 10^3\ h^{-1}$ without a promoter.

6.2.3 Oxidation of carbohydrates

6.2.3.1 Oxidation of sugars induced by microwave in water and zinc-vanadium oxide

Organic acids such as citric acid, acetic acid, lactic acid, galacturonic acid, and gluconic acid are compounds that are essential in many applications, which includes food and beverages, petrochemical, pharmaceutical, textile, adhesive industries, and automotive and construction industries [53–55]. Organic acids are generally produced using carbohydrate substrates [56–66]. Among the organic acids, galacturonic or uronic acids are efficient, biocompatible, and nonionic surfactant molecules and these are widely studied [67–69].

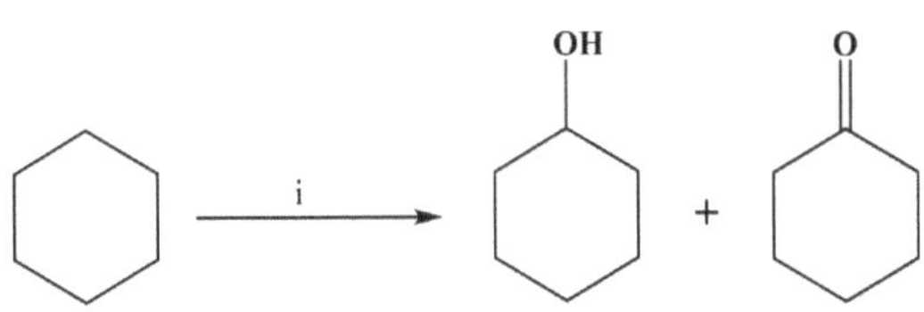

32 33 34

SCHEME 11 Microwave-assisted peroxidative oxidation of cyclohexane catalyzed by copper(II) complexes. Reagents and conditions: (i) TBHP, copper(II) complex, CH_3CN, 50°C, MW.

Recently, Khallouk et al. reported a microwave-assisted selective oxidation of glucose and xylose to galacturonic and glycolic acids, respectively, catalyzed by zinc-vanadium ($Zn_3V_2O_8$) nanostructured mixed oxide (ZVO) [70]. The ZVO was found to be an efficient catalyst under microwave activation for oxidation of sugar. Using this protocol, a 60% selectivity of galacturonic acid and 46% of glycolic acid were obtained from glucose and xylose, respectively. The optimized conditions, to oxidize glucose and xylose, were obtained using $H_2O_2 = 0.207$ mmol, power = 500 W, temperature = 100°C, and time = 30 min and obtained a total (100%) conversion of glucose.

Microwave-assisted selective oxidation of glucose to galacturonic acid is shown in Scheme 12. The catalytic process was exclusively selective into galacturonic acid (60%). This oxidative process produced initially two conformational isomers: galacturonic acid and glucuronic acid. The galacturonic acid is the most stable isomer, and glucuronic acid is the least stable conformational isomer. The rest of the process was a succession of reactions from glucuronic acid (Scheme 12). The other acids formed with low and variable selectivities during the process include glucuronic acid (8%), malic acid (7%), succinic acid (1.5%), malonic acid (12%), lactic acid (1%), acetic acid (1.3%), oxalic acid (3.6%), and formic acid (1.6%).

The route of the xylose oxidation is shown in Scheme 13. Microwave-assisted oxidation of xylose under optimum conditions selectively yielded glycolic acid (46%) and acetic acid (39%). The other acids obtained with very low selectivities include malic acid (0.50%), succinic acid (0.2%), lactic acid (2.4%), oxalic acid (2%), and formic acid (7.8%).

The use of microwave irradiation to catalyze the oxidation of glucose, molasses, and sucrose into levulinic acid was described by Kumar et al. [71]. Microwave-assisted oxidation of glucose either in the presence of HCl alone or both HCl and $ZnBr_2$ as catalysts yielded the formation of levulinic acid and formic acid (Scheme 14). It was observed that the conversion of glucose to levulinic acid and formic acid is much faster (6 min) when both HCl and $ZnBr_2$ are used together. As the microwave irradiation time was increased from 3 to 6 min, a steady increase in the yield of levulinic acid from 34% to 53% was observed.

6.2.3.2 *Base-free oxidation of glucose induced by microwave on gold nanoparticle*

Microwave-assisted base-free oxidation of glucose to gluconic acid using gold nanoparticle catalysts was reported [72]. During the process, high conversion and selectivity were obtained in a short time (10 min) and gluconic acid was obtained in good yields with 0.09 mol% Au/Al_2O_3 and hydrogen peroxide as oxidant (Scheme 15). Conversions obtained by the microwave method were considerably higher than on conventional oil bath. For instance, at 120°C only 62% conversion was obtained on an oil bath in contrast to 83% with the microwave.

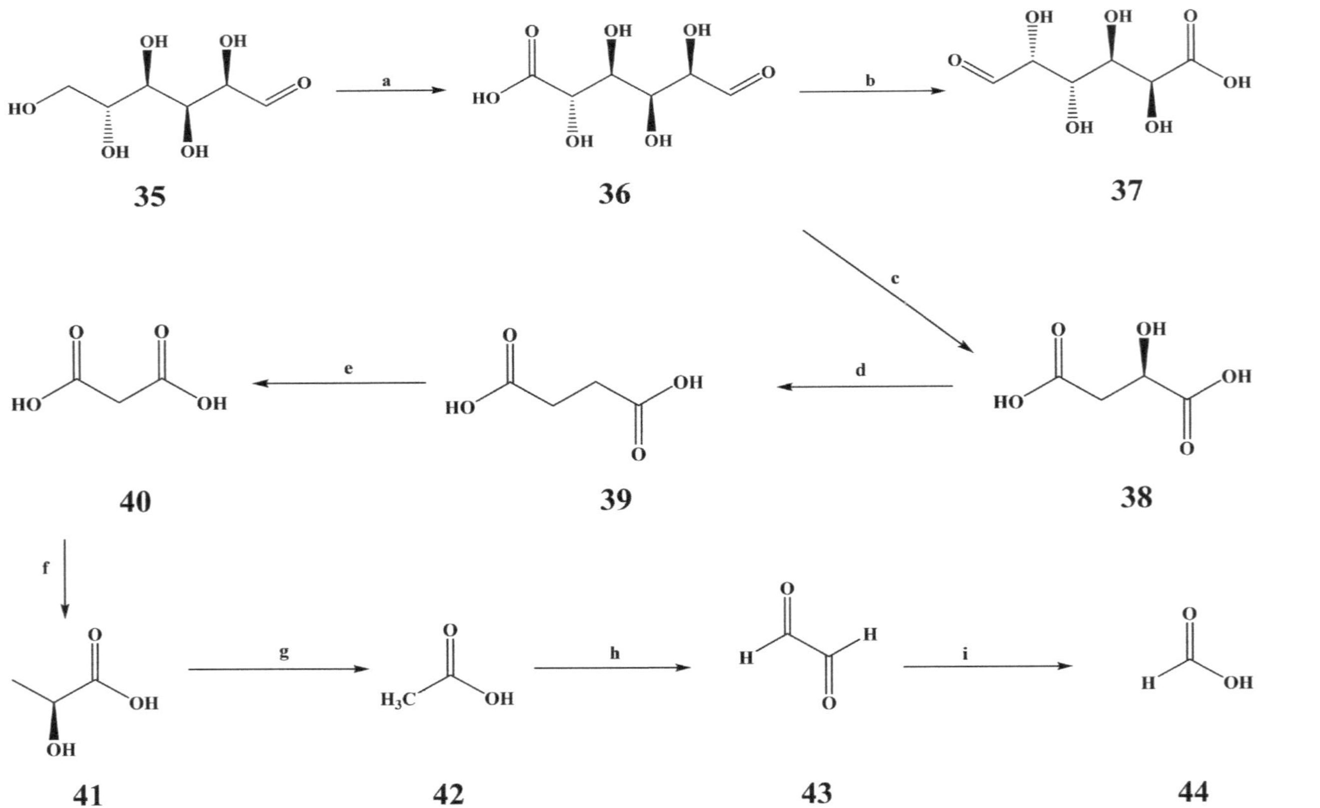

SCHEME 12 Selective oxidation of glucose to galacturonic acid by H_2O_2 over ZVO under microwave irradiation. Reagents and conditions: H_2O_2, ZVO, 100°C, MW, 500 W, 30 min. (a) Oxidation, (b) isomerization, (c) oxidative decarboxylation, (d) dehydration, (e) oxidative decarboxylation, oxidation, (f) reduction, (g) oxidative decarboxylation, oxidation, (h) oxidation, and (i) oxidative decarboxylation.

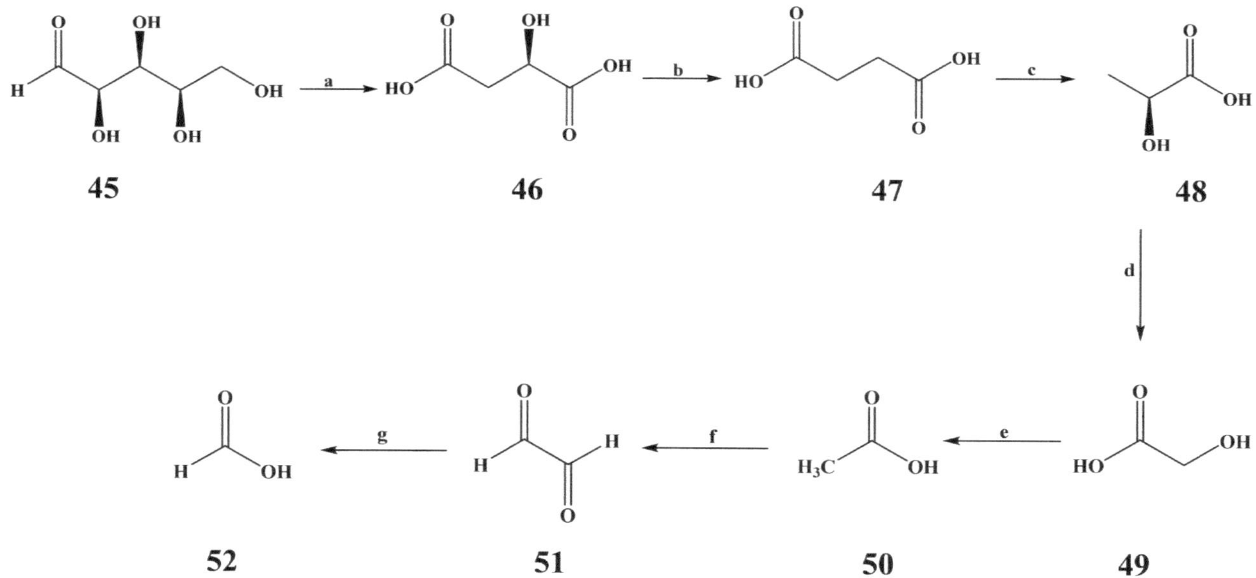

SCHEME 13 Selective oxidation of xylose to glycolic acid by H_2O_2 over ZVO under microwave irradiation. Reagents and conditions: H_2O_2, ZVO, 100°C, MW, 500 W, 30 min. (a) Oxidative decarboxylation, oxidation/dehydration, (b) dehydration, (c) oxidative decarboxylation, hydroxylation, (d) oxidative decarboxylation, (e) dehydration, (f) oxidation, and (g) oxidative decarboxylation.

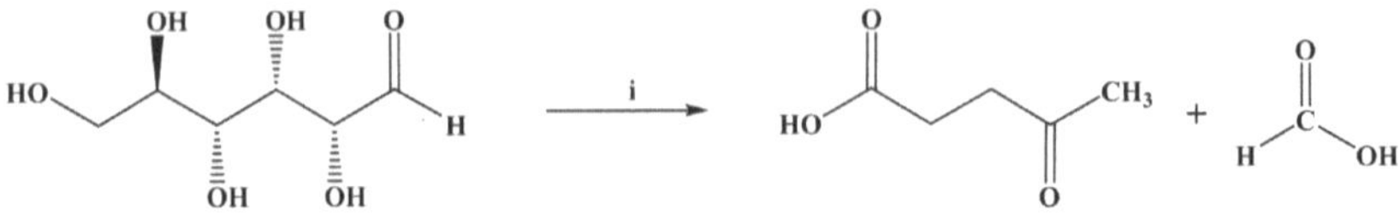

53 54 55

SCHEME 14 Microwave-assisted oxidation of glucose to levulinic acid and formic acid. Reagents and conditions: (i) HCl, $ZnBr_2$, MW, 6 min.

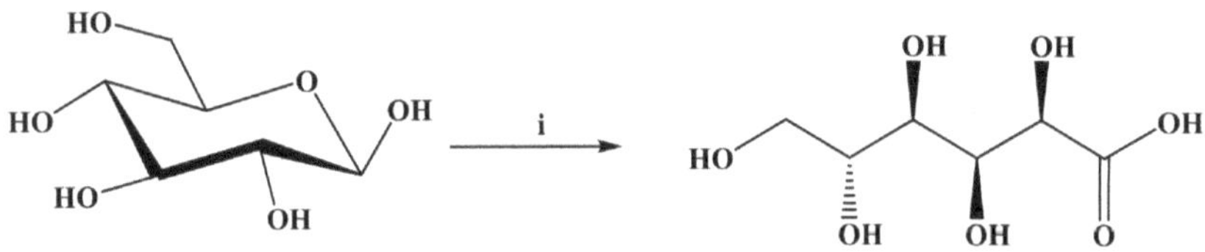

56 57

SCHEME 15 Microwave-assisted base-free oxidation of glucose with Au NP catalysts. Reagents and conditions: (i) Au Nanoparticle, H_2O_2, MW, 10 min.

6.2.3.3 *Oxidation of starch assisted by microwave*

Oxidation of starch can be carried out using different oxidants, and the most common oxidants are sodium hypochlorite, potassium periodate, and hydrogen peroxide. Among them, hydrogen peroxide seems to be very promising as the oxidant of organic molecules in terms of nontoxic side products and high oxidation potential. Several studies were reported on the oxidation of starch by using hydrogen peroxide and catalysts based on the metal compounds: Cu (II), Fe(II), Fe(III), Co(II), Ti(III), W(VI), and V(V) compounds [73]. Lukasiewicz et al. presented a green method of oxidizing starch [74]. Microwave irradiation was used instead of a metal catalyst to activate hydrogen peroxide. The study showed that the application of microwave heating for the oxidation of potato starch leads to high conversion by means of carboxyl content and high polysaccharide depolymerization. The product had generally similar or better properties compared to commercially available ones.

6.2.3.4 *Oxidation of indan-1-ones into ninhydrins mediated by microwave*

Ninhydrin (2,2-dihydroxyindane-1,3-dione) is a well-known chemical, mainly used to detect ammonia or primary and secondary amines [75]. Several ninhydrin derivatives have been studied for their use as latent fingerprint detectors in forensic science [76, 77]. It is also used for the synthesis of numerous heterocyclic compounds [78, 79]. Ninhydrins are usually synthesized by oxidation of indan-1-ones or indane-1,3-diones. Different methods are reported in the literature to oxidize indan-1-ones into ninhydrin [80–85].

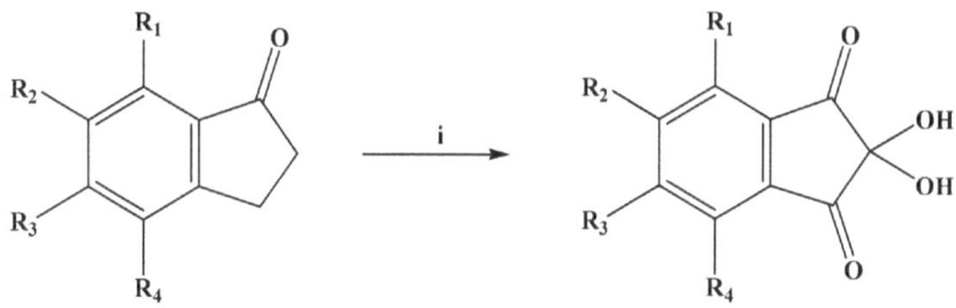

58 59

R_1=H, F, OMe; R_2=H, Me, F, NO_2, OMe;
R_3=H, Br, OH, OMe; R_4=H, Me, Br, CF_3, NO_2, OH, OMe

SCHEME 16 Microwave-assisted oxidation of indan-1-ones. Reagents and conditions: (i) SeO_2, diox/water, MW, 5 min, 180°C.

Marminon et al. explored the oxidation of indan-1-one to ninhydrin under microwave heating by using selenium dioxide [86]. The reaction of the method is shown in Scheme 16. The same yield of the product was obtained even if the microwave exposure is reduced. It was also noticed that the increase in microwave irradiation time improved the yields from 72% to 81%.

6.2.3.5 Oxidation of molecules with cetyltrimethylammonium chlorochromate assisted by microwave

An efficient and mild microwave-assisted methodology for the oxidation of organic compounds such as alcohols, carbohydrate derivatives, and polycyclic arenes was demonstrated using cetyltrimethylammonium chlorochromate (CTMACC) [87]. The oxidation of the carbohydrate derivatives to corresponding ketones is shown in Scheme 17. Table 1 shows the selected list of microwave-assisted oxidation of some organic compounds with CTMABC.

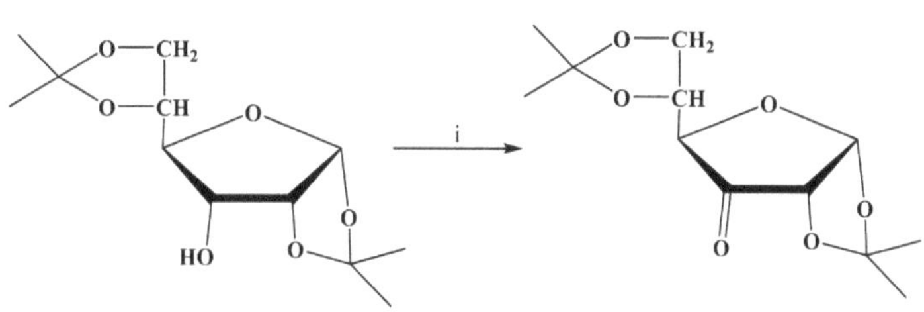

60 61

SCHEME 17 Microwave-assisted oxidation of carbohydrate derivatives with CTMACC. Reagents and conditions: (i) CTMACC, MW, 21 min.

TABLE 1 Microwave-assisted oxidation of some organic compounds with CTMABC.

Substrate	Product	Time (min)	Yield (%)
Primary alcohols			
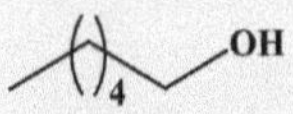	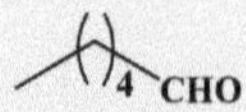	8	92
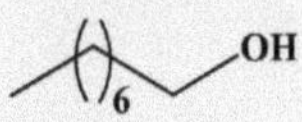	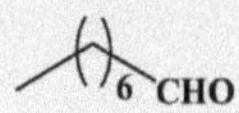	7	92
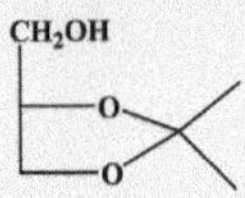	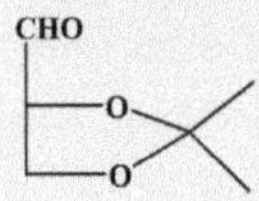	14	87
Secondary alcohols			
Substrate	*Product*	*Time (min)*	*Yield (%)*
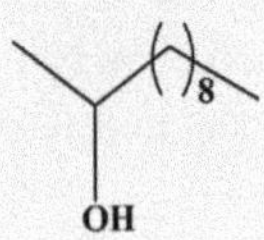	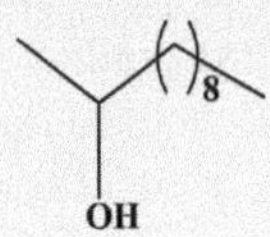	10	90
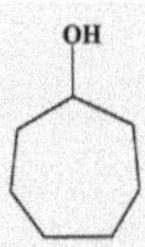	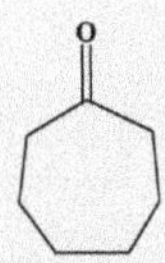	12	95
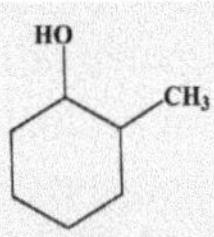	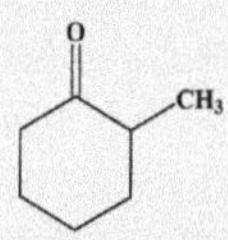	14	90
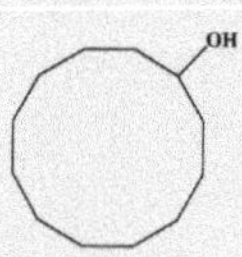	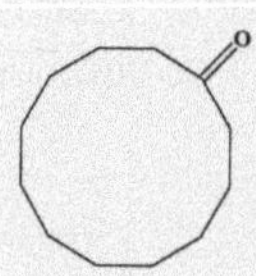	15	89
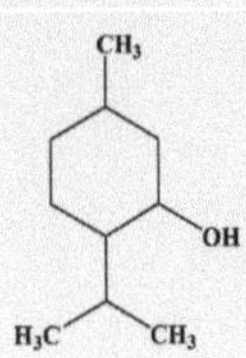	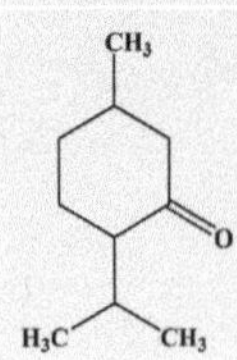	13	90

Polycyclic arenes			
Substrate	*Product*	*Time (min)*	*Yield (%)*
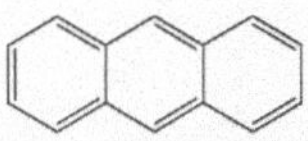	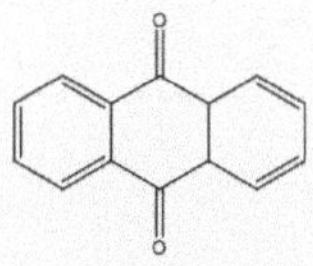	23	87
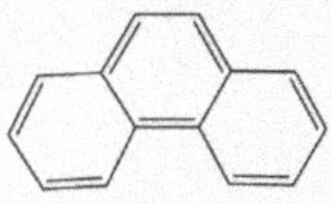	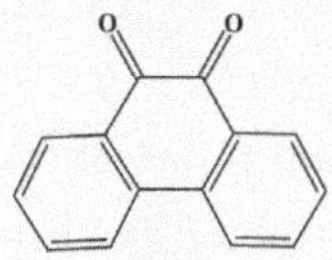	20	85

6.2.3.6 *Manganese(III) acetate-based oxidative cyclizations of olefines with β-ketosulfones mediated by microwave*

Synthesis of dihydrofuran derivatives substituted with a sulfonyl group is important because of their promising utility as building blocks in the asymmetric synthesis of diverse compounds [88–90]. The reaction of keto sulfones with alkenes and Mn(III) acetate provided 2,3-dihydrofurans in good yield [91–93].

Curti et al. performed a microwave-assisted synthesis of 5-(4-nitrophenyl)-2-phenyl-4-(phenylsulfonyl)-2,3-dihydrofuran via manganese(III) acetate-based oxidative cyclization of 1-(4-nitrophenyl)-2-(phenylsulfonyl)ethanone with vinylbenzene [94]. This protocol was applicable to other sulfone derivatives using vinylbenzene and diphenylethene, and the reaction provided a series of 2,3-dihydrofurans in moderate to good yields (26%–55%). Besides, the application of a similar method on allylbenzene led to the formation dehydronaphthalene derivatives in moderate yields.

The Mn(III)-mediated oxidative cyclization of sulfones with two alkenes, vinylbenzene and 1,1-diphenylethene, is shown in Scheme 18. The reactions of sulfones with vinylbenzene provided 2,3-dihydrofurans in a moderate yield of 26%–32%. But the reaction with 1,1-diphenylethylene gave the 2,3-dihydrofuran derivatives in significantly improved yields (48%–55%). The same reaction of sulfones with allylbenzene did not produce dihydrofuran derivatives, instead produced tetraline derivatives with moderate yields (Scheme 19).

The microwave-assisted synthesis decreased the reaction times of Mn(III)-based oxidative free-radical cyclizations.

6.2.3.7 *Fenton's oxidation of amoxicillin assisted by microwave*

Studies depicted that the use of microwave increases the efficiency of traditional processes, such as Fenton's oxidation of pharmaceutical effluents

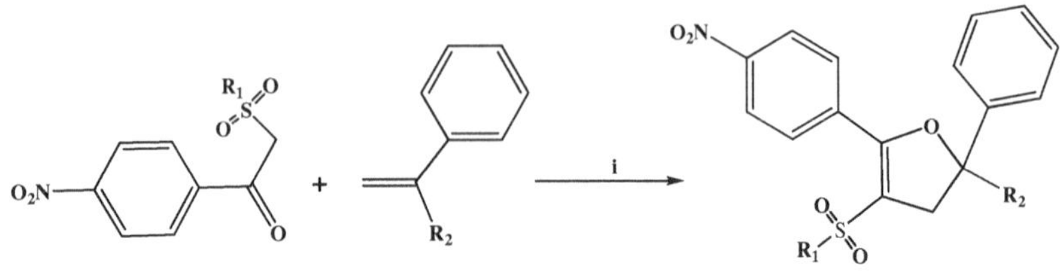

62 **63** **64**

R_1=C_6H_5, *p*-CH_3-C_6H_4, *p*-Cl-C_6H_4, n-C_4H_9; R_2=H, C_6H_5

SCHEME 18 Microwave-assisted Mn(III)-mediated oxidative cyclization of sulfones with alkenes. Reagents and conditions: (i) $Mn(OAc)_3$, $Cu(OAc)_2$, AcOH, 80°C, MW, 200 W, 45 min.

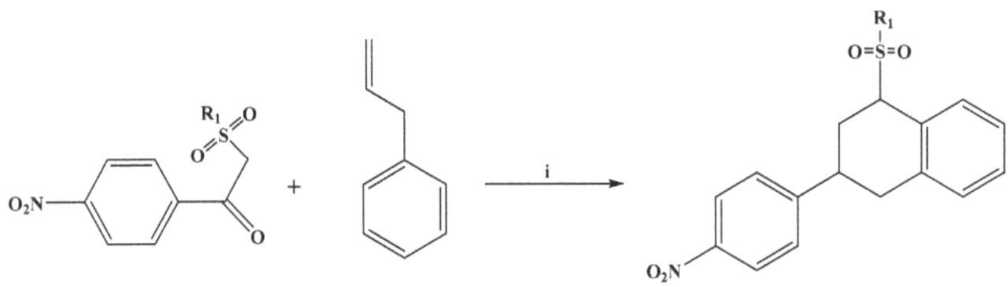

65 **66** **67**

R_1=C_6H_5, *p*-CH_3-C_6H_4, *p*-Cl-C_6H_4, n-C_4H_9;

SCHEME 19 The reaction of allylbenzene with sulfones. Reagents and conditions: (i) $Mn(OAc)_3$, $Cu(OAc)_2$, AcOH, 80°C, MW, 200 W, 45 min.

[95–97]. Homem et al. proposed an innovative solution to degrade amoxicillin by combining microwave with Fenton's reaction [98]. Total degradation of amoxicillin was reached in 5 min, with a reduced amount of iron catalyst.

Different factors, such as irradiation power, initial concentration of H_2O_2, and Fe^{2+}, were evaluated in the degradation of amoxicillin by microwave-assisted Fenton process. The optimum condition was obtained at $[AMOX]_0 = 450\ \mu g\ L^{-1}$, $pH_0 = 3.5$, $P = 162\ W$, $[H_2O_2]_0 = 2.35\ mg\ L^{-1}$, $[Fe^{2+}]_0 = 95\ \mu g\ L^{-1}$. No amoxicillin was remained in solution in less than 5 min under this condition. The study also verified that the use of microwave-assisted Fenton's oxidation led to better results than the application of traditional Fenton's reaction.

6.3 Reduction reactions

6.3.1 Pd/C-catalyzed reduction of levulinic acid induced by microwave

Amarasekara et al. reported the microwave-assisted reduction of levulinic acid to γ-valerolactone in 83%–86% yield with 10 mg of 5% Pd/C per mmol

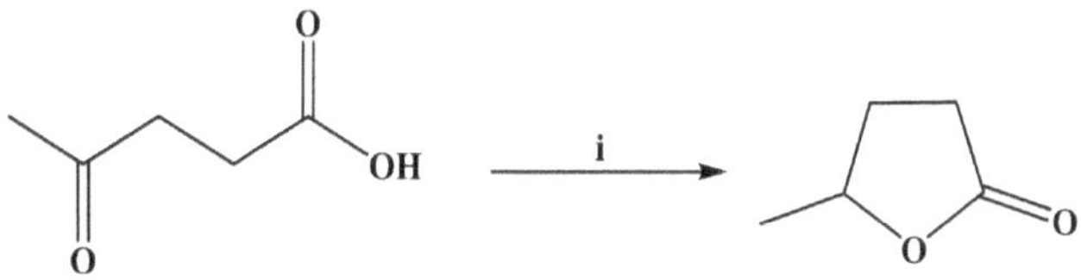

68 69

SCHEME 20 Microwave-assisted Pd/C-catalyzed reduction of levulinic acid to γ-valerolactone using alcohol as a hydrogen donor. Reagents and conditions: (i) Pd/C, KOH, ROH, MW, 50 s.

levulinic acid as the catalyst and 2.0 equivalents of KOH/NaOH in ethanol or iso-propanol [99]. The scheme of the reaction is shown in Scheme 20. In this reaction, alcohol was used as the solvent as well as the hydrogen source. It was possible to reuse the catalyst (Pd/C) for five catalytic cycles without appreciable loss in activity.

For the reaction, a mixture of levulinic acid (232 mg, 2.00 mmol), KOH/NaOH (4.00 mmol), and 20 mg of 5% Pd/C in 4.00 mL of ethanol or iso-propanol was subjected to microwaves for 50 s. The reaction provided γ-valerolactone in good yield, and no α-alkylation was observed. Unexpectedly, the thermal heating reaction failed to produce any products and starting materials were recovered.

The proposed reaction pathway for the conversion of levulinic acid to γ-valerolactone under microwave heating is shown in Scheme 21. The reaction occurred via cyclization to hydroxy lactone and elimination of water to give α-angelica lactone, which then undergoes a hydrogen transfer reduction.

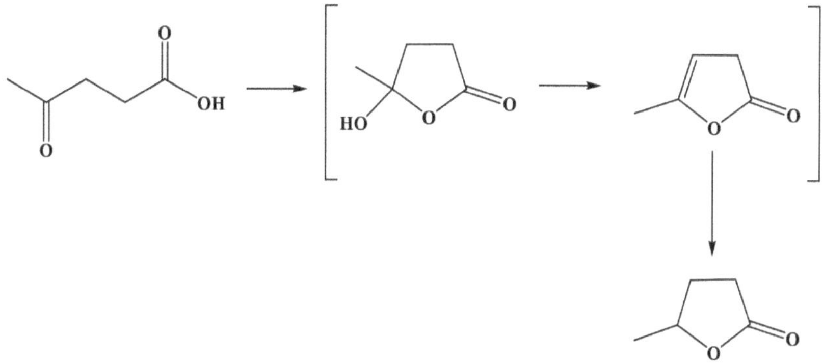

SCHEME 21 Proposed reaction pathway for the conversion of levulinic acid to γ-valerolactone under microwave heating using ethanol as the hydrogen source and Pd/C as the catalyst.

6.3.2 Reduction of nitroarenes assisted by microwave using NiO nanoparticles

As arylamines are important starting materials and intermediates for the synthesis of numerous organic compounds like dyestuffs, pharmaceutical products, agricultural chemicals, surfactants, and polymers [100, 101], the reduction of nitroarenes to arylamines is a synthetically important transformation [102]. Different methods are available for the reduction reaction. The reduction reaction is traditionally performed using metals such as Fe, Zn, and Sn in acidic or basic media, and the process generates a large amount of metal wastes [103–105]. Later, numerous reagents have been developed for the reduction of nitroarenes to their corresponding amines [106–110].

An efficient and selective reduction of aromatic nitro compounds into their corresponding amines by using ethanol as a hydrogen donor (reducing agent) and KOH as a promoter under microwave irradiation was reported Farhadi et al. [111]. The nanosized transition metal oxide (NiO nanoparticles) was used as a heterogeneous catalyst for the reduction of nitroarenes under microwave irradiation (Scheme 22).

This highly regio- and chemoselective microwave-assisted reduction method was fast, simple, inexpensive, high yielding, and clean. This method was compatible also with several sensitive functionalities, such as halogens, —OH, —OCH_3, —CHO, —$COCH_3$, —COOH, —COOEt, —$CONH_2$, —CN, —CH=CH_2, and —$NHCOCH_3$. Also, large-scale preparation of different substituted anilines, as well as other arylamines, was possible using this method. In addition, the catalytic activity of nanosized NiO in the microwave-assisted reduction process was higher than that of the bulk sample.

As an example, microwave-assisted reduction of nitrobenzene with ethanol is shown in Scheme 23. To optimize the reaction conditions, a mixture of nitrobenzene (5 mmol) and KOH (5 mmol) dissolved in ethanol (20 mL) was stirred in the presence of different loadings of the NiO catalyst under microwave irradiation (180 W). It was observed that the yield of aniline increased with the increase in the catalyst amount from 10 to 50 mg/5 mmol of nitrobenzene, and no improvement in the reaction time or the yield was observed by increasing the amount of the catalyst further. The optimized reaction was observed with a catalyst load of 50 mg/5 mmol of nitrobenzene, which resulted in aniline as the only product in good yield (98%) after 13 min irradiation. No reduction reaction occurred when nanosized NiO catalyst or KOH was omitted in the

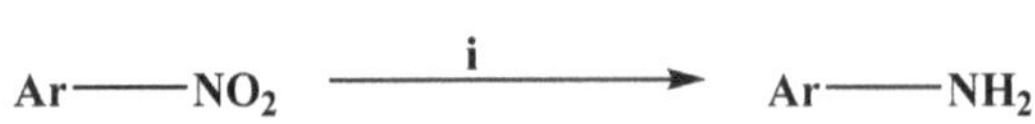

70 71

SCHEME 22 Microwave-assisted reduction of nitroarenes with ethanol. Reagents and conditions: (i) NiO nanoparticles, EtOH, KOH, MW.

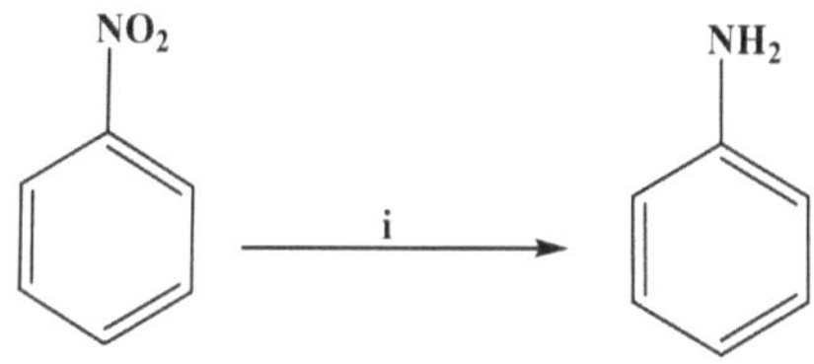

72 **73**

SCHEME 23 Microwave-assisted reduction of nitrobenzene with ethanol. Reagents and conditions: (i) NiO nanoparticles, ethanol, KOH, MW.

system, pointing that NiO and base are essential for this reaction. Besides, no reaction was observed in the absence of microwave irradiation at room temperature.

6.3.3 Wolff-Kishner reduction assisted by microwave

Wolff-Kishner reduction reaction of carbonyl groups is conducted with hydrazine hydrate in the presence of a strong base at elevated temperatures, nearly 200°C, for 3–4 h. Even though a simple Wolff-Kishner reduction reaction of isatin under mild condition was reported [112], the method still required a 3–4 h time and the base, sodium ethoxide. Several possibilities were considered to reduce the reaction time of the Wolff-Kishner reduction. Among those, the microwave-assisted reaction acquired the greatest attraction. The microwave-assisted Wolff-Kishner reduction gave the Wolff-Kishner reduction product in high purity and in a very short period of time without a complicated workup. For instance, microwave-assisted Wolff-Kishner reduction reaction of isatin was demonstrated [113]. The scheme of the two-step processes is shown in Scheme 24.

In the first step, isatin, 55% hydrazine, and ethylene glycol were added to a beaker. The beaker was then irradiated in the microwave oven at medium power for 30 s. In the second step, ethylene glycol and potassium hydroxide were irradiated in the microwave oven. Isatin 3-hydrazone was added to the beaker and irradiated in the microwave oven. This two-step synthesis offered several advantages such as very short reaction time with no need for special microscale glassware, mild experimental conditions, and easy-to-handle reagents.

6.3.4 Reduction of trimethylsilyl carbonyl compounds mediated by microwave in the presence of sodium borohydride

Silylated alcohols are important compounds due to their potential medicinal effect with less toxicity [114–117]. Several methods were reported for the synthesis of silylated alcohols. One of the widely used approaches is the reduction of trimethylsilyl carbonyl compounds by a reducing agent [118]. Numerous

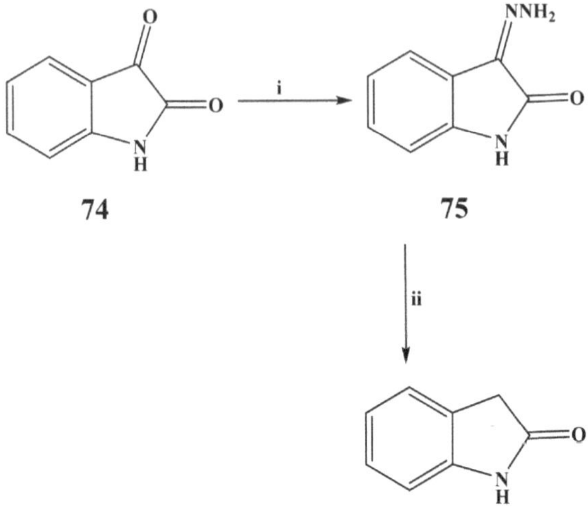

74 75 76

SCHEME 24 Microwave-assisted Wolff-Kishner reduction reaction of isatin. Reagents and conditions: (i) NH_2NH_2, ethylene glycol, MW, 30 s; (ii) KOH, ethylene glycol, MW, 10 s.

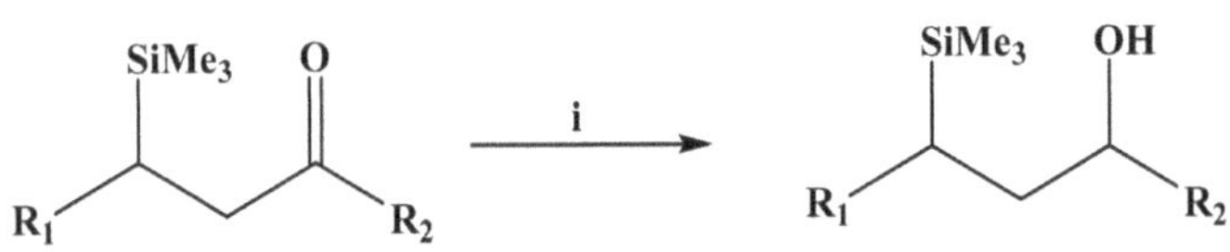

77 78

SCHEME 25 Microwave-assisted reduction of trimethylsilyl carbonyl compounds. Reagents and conditions: (i) $NaBH_4$ $-Al_2O_3$, MW, 3 min.

reagents are available for the reduction of carbonyl compounds to the corresponding alcohols. Among them, relatively inexpensive sodium borohydride ($NaBH_4$) is the extensively used reducing agent [119]. Zadmard et al. reported a mild, convenient, and heterogeneous catalytic methodology for the synthesis of β-silylated alcohols [120]. The scheme of the reaction is depicted in Scheme 25.

During the reaction process, β-trimethylsilyl aldehyde or ketone was mixed with 10% $NaBH_4$/neutral alumina, and then, the mixture was irradiated in a conventional microwave oven for 2–3 min. Table 2 shows different β-trimethylsilyl carbonyl compounds considered during this study.

6.3.5 Reduction of carbonyl compounds induced by microwave in glycerol

Glycerol's high dielectric constant and high boiling point make it an attractive solvent, especially for the microwave-assisted reaction. The

TABLE 2 Considered β-trimethylsilyl carbonyl compounds and corresponding silylated alcohols.

Substrates	Products
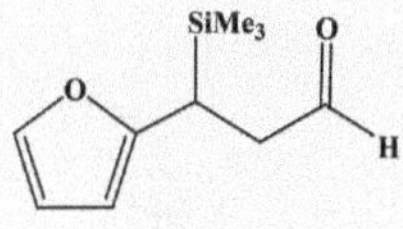 **77a**	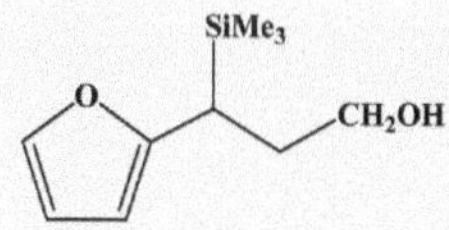 **78a**
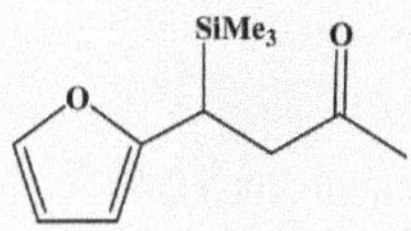 **77b**	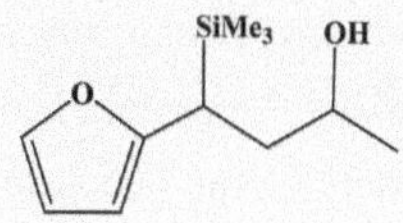 **78b**
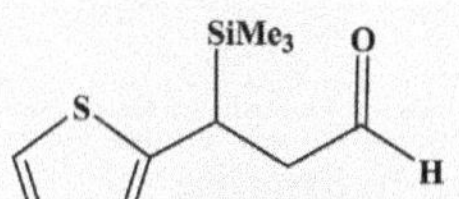 **77c**	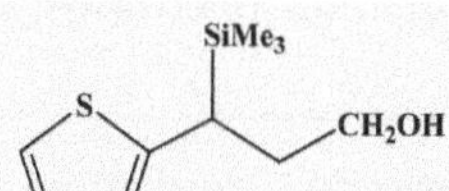 **78c**
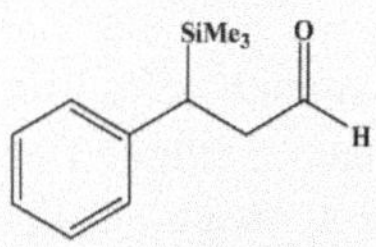 **77d**	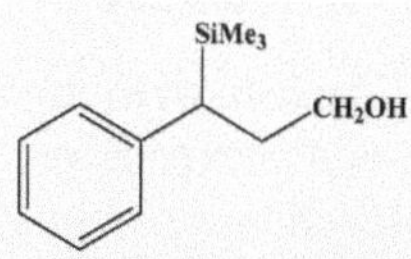 **78d**
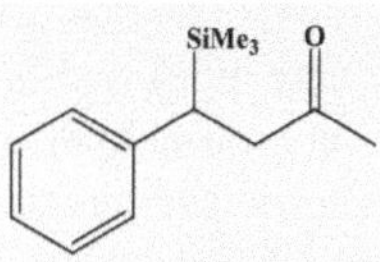 **77e**	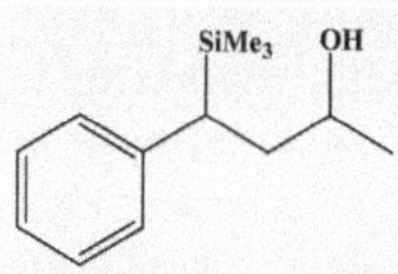 **78e**
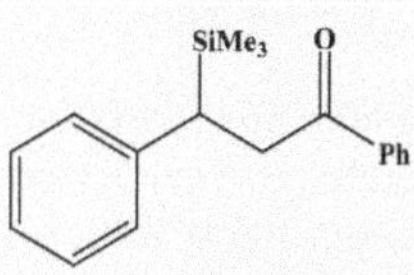 **77f**	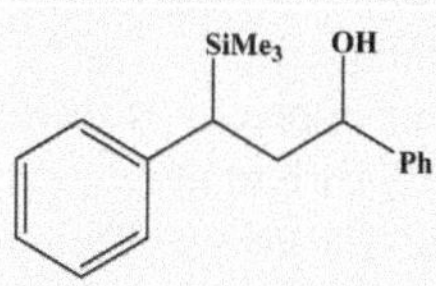 **78f**

microwave-assisted reduction of benzaldehyde to toluene using glycerol was executed in two steps using a conventional microwave oven [121]. In the first step, benzaldehyde and hydrazine were dissolved in glycerol and heated in the microwave oven for 5 min from room temperature to about 110°C. After that, the reaction mixture was cooled to room temperature in an ice bath, and then,

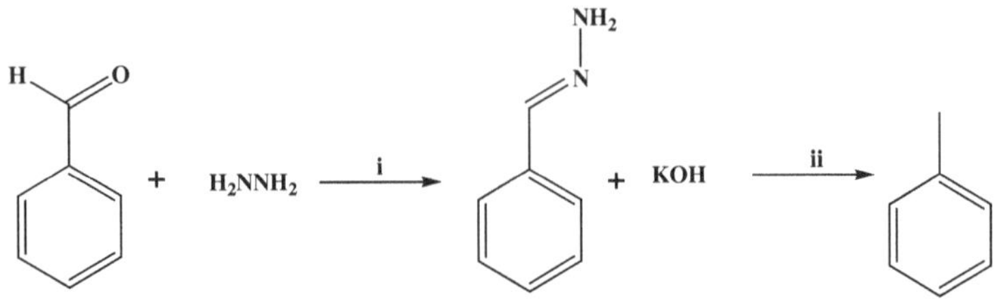

79 80 81

SCHEME 26 Microwave-assisted Wolff-Kishner reduction of benzaldehyde to toluene in glycerol. Reagents and conditions: (i) Glycerol, MW, 5 min; (ii) MW, 5 min.

KOH was added. In the last step, the mixture was again heated for 5 min in the microwave oven. The conversion of benzaldehyde to toluene was completed after 10 min in the microwave (Scheme 26).

6.3.6 Crossed Cannizzaro reaction using barium hydroxide mediated by microwave

The crossed Cannizzaro reaction [122] to produce alcohol in higher yields using a scavenger and inexpensive formaldehyde is an important method. Even though this reaction was conducted on a variety of mineral oxide surfaces [123], the reaction is normally conducted in the solution phase. Cannizzaro reaction is also explored in the solid state using microwaves [124]. The reaction was failed with calcium hydroxide and remained incomplete with sodium hydroxide [123]. Amazingly, the reaction went on promptly on barium hydroxide, $Ba(OH)_2{\cdot}8H_2O$, in a relatively short period of time. Microwave-assisted facile reduction of aldehydes to alcohols underwent smoothly using a formaldehyde equivalent, paraformaldehyde, under neat conditions in the absence of solvent (Scheme 27).

As an example, a mixture of benzaldehyde and paraformaldehyde is admixed with barium hydroxide octahydrate and then irradiated in a microwave (100–110°C) for the specified time (0.5 min) at its full power of 900 W to produce corresponding alcohol. The same reaction at the oil bath condition took nearly 10 min to complete. The different substrate considered during this study is summarized in Table 3. The results showed that benzaldehydes bearing an

$$\underset{\mathbf{82}}{RCHO} + \underset{\mathbf{83}}{(CH_2O)_n} \xrightarrow{\text{i}} \underset{\mathbf{84}}{RCH_2OH} + \underset{\mathbf{85}}{RCOOH}$$

SCHEME 27 Microwave-accelerated crossed Cannizzaro reaction using barium hydroxide under solvent-free conditions. The reaction produced corresponding alcohol (80%–99%) and acid (1%–20%). Reagents and conditions: (i) $Ba(OH)_2{\cdot}8H_2O$, MW.

TABLE 3 List of aldehydes considered during the study.

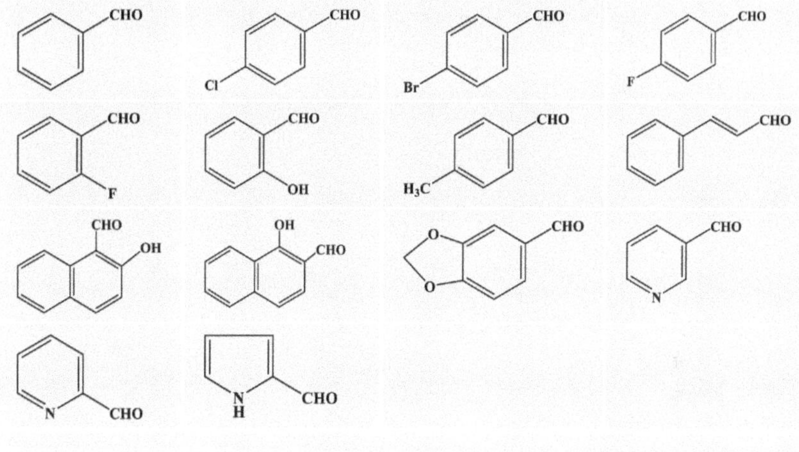

electron-withdrawing substituent underwent reaction at a much faster rate as compared to benzaldehydes with electron-releasing groups. It is also depicted that among the naphthaldehydes, the reaction was slower for 2-naphthaldehyde compared to 1-naphthaldehyde. Likewise, the reaction rate was slower for 3-pyridinecarboxaldehyde in comparison with 2-pyridinecarboxaldehyde. More importantly, the overall reaction was much slower in an oil bath compared to microwave conditions.

6.3.7 Microwave-induced reduction of levulinic acid with alcohols in the presence of Ru/C

Al-Shaal et al. demonstrated the microwave-assisted reduction of levulinic acid with alcohols to produce γ-valerolactone in the presence of a Ru/C catalyst [125]. The reaction rate was increased when microwave heating was employed. A very good yield of γ-valerolactone was observed within 25 min at 160°C under microwave heating based on levulinic acid and isopropanol. The microwave-assisted reduction reaction appeared to proceed via a dehydrogenation-hydrogenation sequence (Scheme 28).

In the first step, dehydrogenation of alcohol generates molecular hydrogen in situ, and in the second step, hydrogenation of a substrate with H_2 into γ-valerolactone occurs via γ-hydroxyvalerates. Metal Ru^0 species remain active for the transformation reaction and in situ activation of the catalyst by reduction of oxidized $Ru^{\delta+}$ centers under reaction conditions are also noticed.

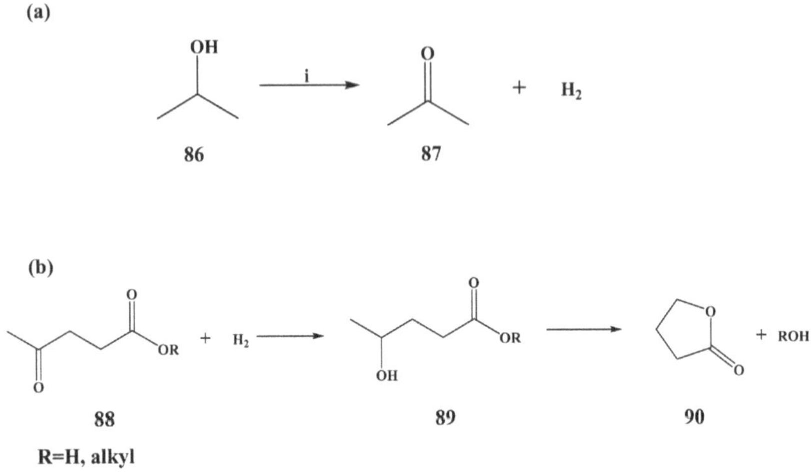

SCHEME 28 Proposed dehydrogenation-hydrogenation mechanism of the synthesis of γ-valerolactone. (A) Dehydrogenation and (B) hydrogenation. (i) Ru^0.

6.3.8 Microwave-assisted reduction of acetophenones using Ni-Al alloy in water

The microwave-assisted reduction of acetophenones using a Ni-Al alloy catalyst in water (H_2O and D_2O) was studied by Akira et al. [126]. The reaction proceeded effectively to give the corresponding (deuterated) alkylbenzenes in good yields (Scheme 29).

6.3.9 Microwave-assisted one-pot selective azido reduction/tandem cyclization in condensed and solid phase with nickel boride

An efficient and inexpensive method using microwave-assisted irradiation with Ni_2B for the syntheses of aromatic amines was demonstrated [127]. When the reduction reactions were carried out under thermal heating using azido

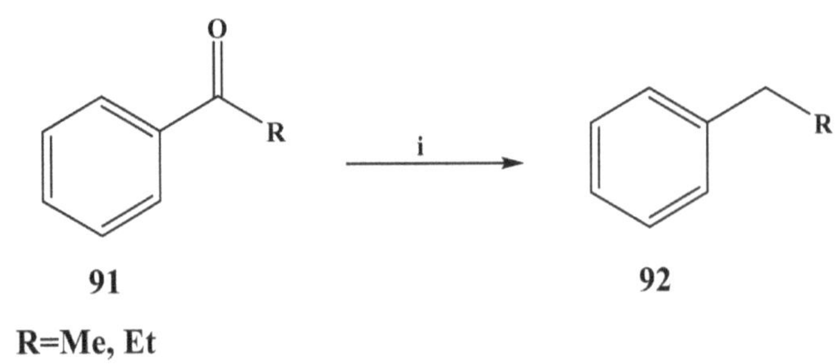

SCHEME 29 Microwave-assisted reduction of acetophenones. Reagents and conditions: (i) Ni-Al alloy, water, MW, 10 min.

derivatives and an excess of Ni_2B (3 equiv) in MeOH-HCl (1 M) for 10–20 min at 60–70°C, the corresponding amines were obtained in good yields (72%–88%). When the same reactions were carried out under microwave, the reaction time was reduced to one minute and the yields were increased to 80%–95% (Table 4). The reaction of the process is shown in Scheme 30.

6.3.10 Reduction of β-carboline imines by chiral auxiliaries induced by microwave

Numerous methods are known for the synthesis of optically active amines. Among those methods, the most popular one is the asymmetric hydrogenation of ketimines or enamides using chiral Rh(I), Ir(I), or Ru(II) complexes [128]. Moraga et al.

TABLE 4 Six selected azido derivatives and the corresponding amines.

Substrate	Product	Microwave irradiation Time (min)	Microwave irradiation Yield (%)	Thermal heating Time (min)	Thermal heating Yield (%)
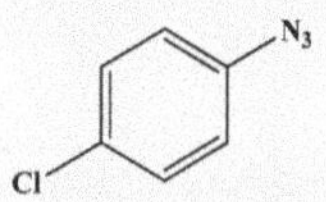	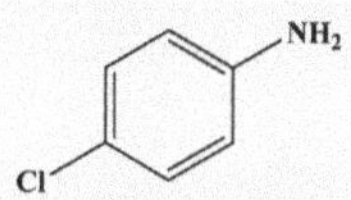	1	93	10	82
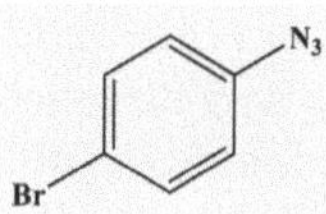	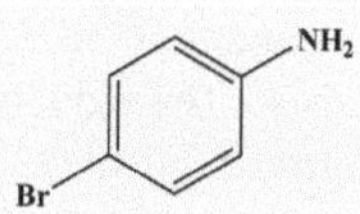	1	84	10	77
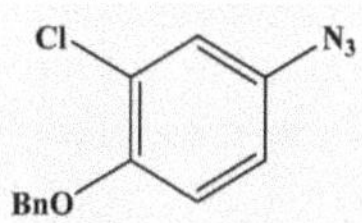	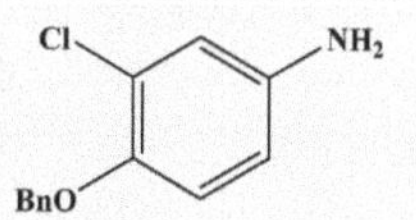	1	90	10	80
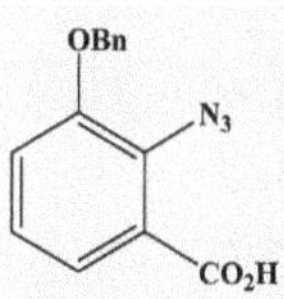	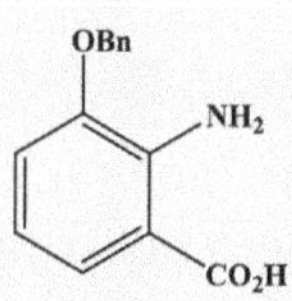	1	82	15	75
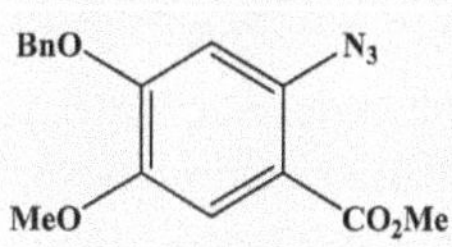	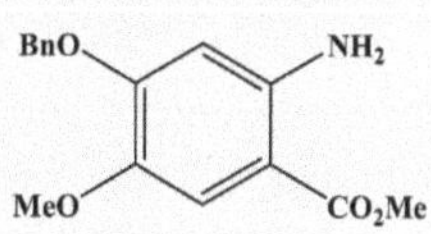	1	90	10	80
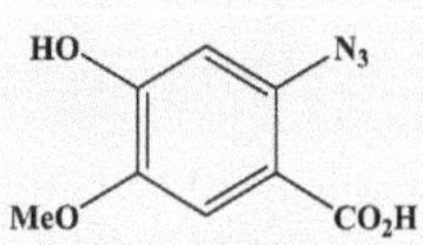	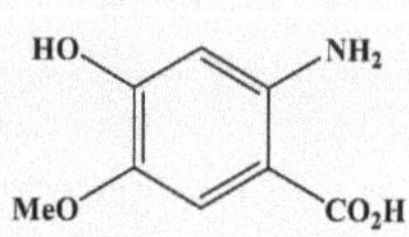	1	80	20	72

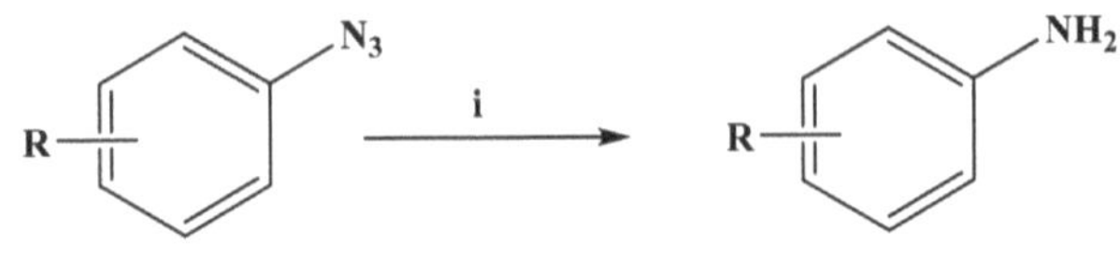

93 **94**

SCHEME 30 Microwave-assisted Ni_2B reduction of azido derivatives to the corresponding amines. Reagents and conditions: (i) Ni_2B, MeOH-HCl, MW, 70 W.

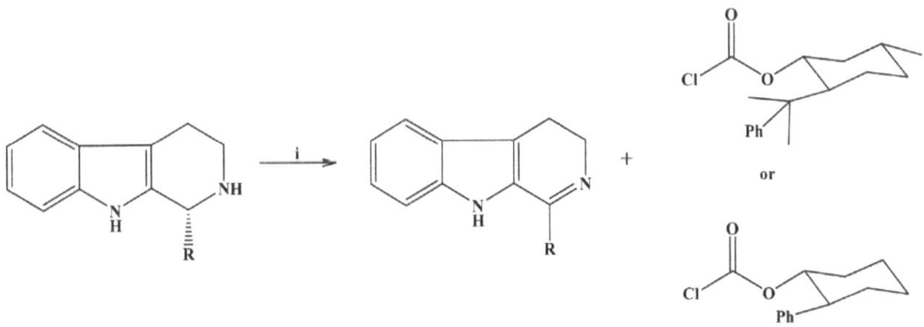

95 **96** **97**

SCHEME 31 Palladium asymmetric reduction of β-carboline imines mediated by chiral auxiliaries assisted by microwave irradiation. Reagents and conditions: (i) $PdCl_2/Et_3SiH$, MW.

investigated the microwave-assisted reduction of imines mediated by chiral auxiliaries in a one-pot manner [129]. The reduction methodology was based on the use of chloroformates as chiral auxiliaries for the reduction of dihydro-β-carbolines [130]. Moreover, the $PdCl_2/Et_3SiH$ protocol [130, 131] was employed as a palladium hydride source. The reaction of the process is depicted in Scheme 31.

This microwave-assisted method decreased reaction times to minutes, reducing energy consumption and increasing efficiency when compared to the conventional method [132].

6.3.11 Microwave-induced reduction of α, β-unsaturated carbonyl compounds using sodium borohydride supported on magnesium sulfate

Ahangar et al. demonstrated a mild, convenient, and simple microwave-assisted method for fast and efficient reduction of carbonyl compounds to their corresponding alcohols with $NaBH_4/MgSO_4{\cdot}7H_2O$ under solvent-free conditions [133].

Solid-state reduction of 4-hydroxy-3,4-diphenylcyclopent-2-enone with $NaBH_4$ under microwave irradiation is shown in Scheme 32. The use of an excessive amount of $NaBH_4$ without any support resulted in only 30% conversion under microwave irradiation. Different supports such as clay, SiO_2, Al_2O_3, and $MgSO_4{\cdot}7H_2O$ were considered during the study. The reduction reaction was more facile and proceeded to give the highest yield (90%) with

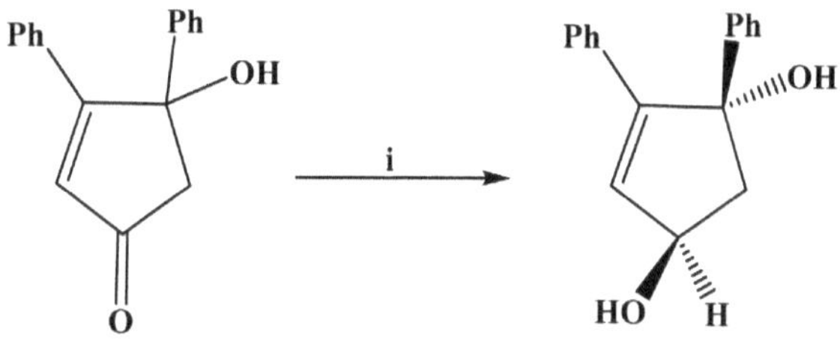

98 99

SCHEME 32 Solid-state reduction of 4-hydroxy-3,4-diphenylcyclopent-2-enone with $NaBH_4$ under microwave irradiation. Reagents and conditions: (i) $NaBH_4/MgSO_4{\cdot}7H_2O$, MW, 8 min, 800 W, solvent-free.

TABLE 5 Five selected carbonyl compounds to their corresponding alcohols.

Substrate	Product	Time (min)	Yield (%)
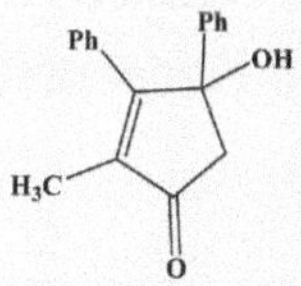	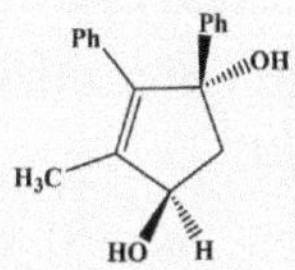	7	95
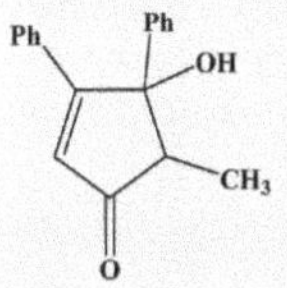	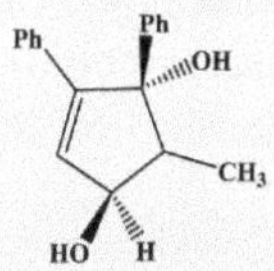	8	90
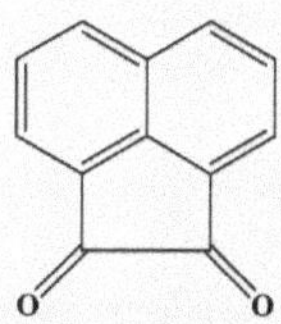	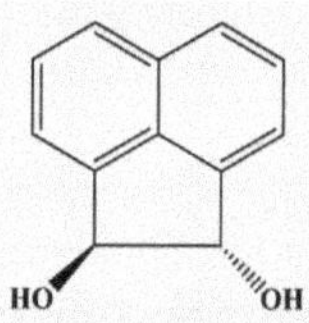	8	82
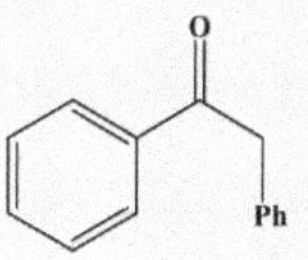	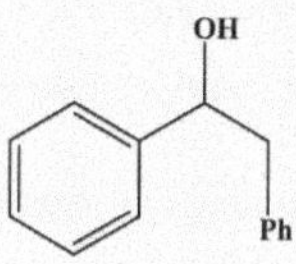	5	95
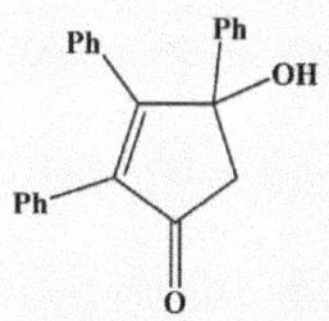	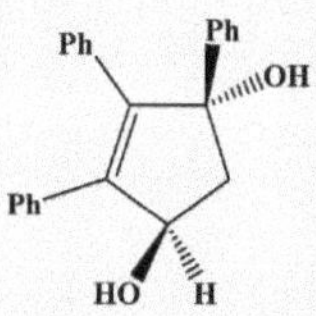	9	80

$NaBH_4/MgSO_4{\cdot}7H_2O$ under solvent-free conditions. Various carbonyl compounds were employed in the reaction to investigate the usefulness of this microwave irradiation protocol (Table 5).

6.3.12 Reduction of carbonyl compounds using sodium borohydride supported on alumina induced by microwave

A microwave-assisted manipulatively simple and rapid method for the reduction of carbonyl compounds conducted under solventless dry conditions using $NaBH_4$-Alumina was also demonstrated [134]. The schematics of the reaction are shown in Scheme 33.

6.3.13 Microwave-assisted bioreduction of aromatic aldehydes using aloe vera

Leyva et al. reported microwave-assisted bioreduction of aromatic aldehydes using aloe vera [135]. This green chemistry methodology was fast and simple for the synthesis of aromatic alcohols and was achieved in good yield using an extract of aloe vera in aqueous suspension (Scheme 34). Details of the aromatic aldehydes examined during this study are shown in Table 6.

6.3.14 Microwave-induced reduction of aromatic nitro compounds with ammonium chloride and zinc dust

An easy and rapid environmentally friendly method for the reduction of aromatic nitro compounds to corresponding amines by using microwave irradiation was demonstrated in the presence of zinc dust ammonium chloride and solvent-free condition [136]. Scheme 35 shows the reaction method.

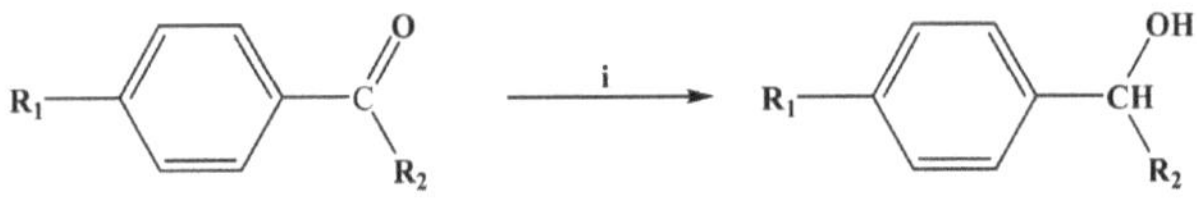

100 101

R_1=CH_3, Cl, NO_2, H, OCH_3; R_2=H, CH_3, C_6H_5, $CH(OH)C_6H_5$, $CH(OH)C_6H_4$-OCH_3-*p*

SCHEME 33 Microwave-assisted solid-state reduction of carbonyl compounds using $NaBH_4$-Al_2O_3. Reagents and conditions: (i) $NaBH_4$-Al_2O_3, MW.

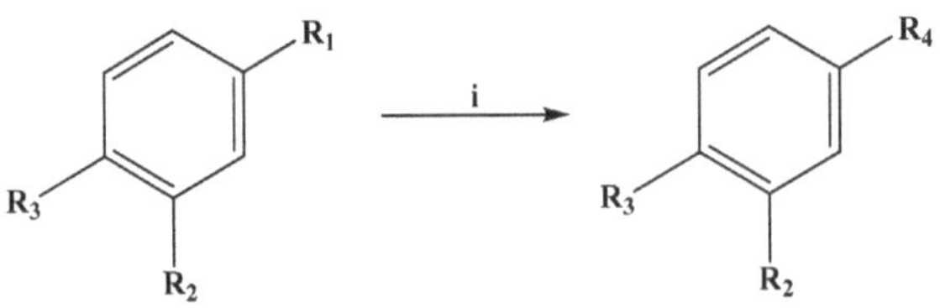

102 103

SCHEME 34 Microwave-assisted bioreduction of aromatic aldehydes. Reagents and conditions: (i) Aloe vera, MW.

TABLE 6 Selected carbonyl compounds.

R_1	R_2	R_3	R_4
CHO	H	H	CH_2OH
CHO	NO_2	H	CH_2OH
CHO	H	NO_2	CH_2OH
CHO	OCH_3	H	CH_2OH
CHO	H	OCH_3	CH_2OH
CHO	CH_3	H	CH_2OH
CHO	H	CH_3	CH_2OH
CHO	Cl	H	CH_2OH
CHO	H	Cl	CH_2OH
CHO	OH	OCH_3	CH_2OH
CH=CH—CHO	H	H	CH=CH—CH_2OH
CH=C(CH_3)—CHO	H	H	CH=C(CH_3)—CH_2OH

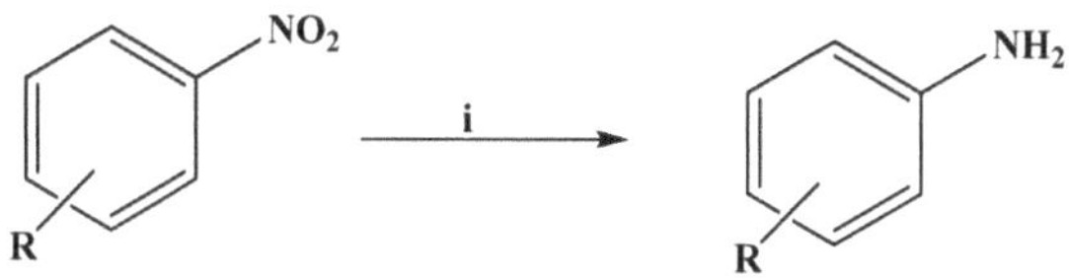

104 105

R=p-OCH_3, m-NH_2, o-CH_3, m-OCH_3, o-NH_2, p-NH_2, p-CH_3, p-Br, p-Cl.

SCHEME 35 Microwave-assisted reduction of aromatic nitro compounds with ammonium chloride and zinc dust. Reagents and conditions: (i) NH_4Cl, zinc dust, MW, 800 W, 8–15 min.

6.3.15 Molybdenum hexacarbonyl and DBU-mediated reduction of nitro compounds induced by microwave

Anilines can be prepared by the reduction of nitro compounds under the microwave. Spencer et al. reported an ethanolic mixture of molybdenum hexacarbonyl ($Mo(CO)_6$) and 1,8-diazabicyclo[5.4.0]undec-7-ene (DBU)-mediated reduction of nitroarenes to the corresponding anilines in excellent yields in 15–30 min under microwave irradiation [137]. The scheme of the reaction is shown in Scheme 36.

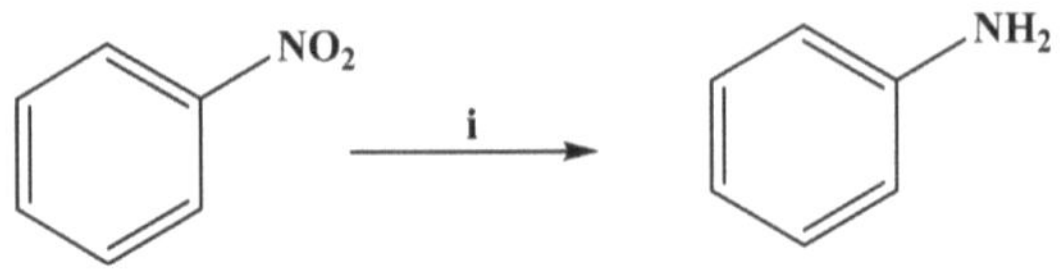

106 **107**

SCHEME 36 Microwave-assisted reduction of nitrobenzene to aniline. Reagents and conditions: (i) $Mo(CO)_6$, DBU, MW, 15–30 min.

6.3.16 Microwave-induced solvent-free reduction of aromatic nitro compounds with alumina-supported hydrazine

Aromatic amines are generally prepared by the reduction of aromatic nitro compounds using reagents in solution-phase reactions [138]. Hydrazine hydrate in the presence of heterogeneous catalysts was extensively used for the reduction of aromatic nitro compounds. The reduction reaction conducted in refluxing alcoholic solvents or dioxane requires several hours of reaction time. Vass et al. reported a solvent-free microwave reduction protocol [139]. The method led to a facile synthesis of aromatic amines from nitro compounds with hydrazine hydrate supported on solid materials such as alumina, silica gel, and clays. The reduction of 4-nitroaniline to 1,4-phenylenediamine is shown in Scheme 37.

During the reduction reaction, aromatic nitro compound was mixed with inorganic solid support or alumina and the mixture was added to hydrazine hydrate and $FeCl_3{\cdot}6H_2O$. After that, the reaction mixture was irradiation for a specified period (5–10 min).

The reaction was examined on various inorganic solid supports to analyze the effect of supports on the reduction of 4-nitroaniline to 1,4-phenylenediamine. The data showed that the efficiency of alumina-supported hydrazine hydrate in the reduction of 4-nitroaniline was good compared to other support surfaces and provided a very good yield (97%). The traditional solid supports with soft acidic surfaces were found to be less effective. In contrast, nontraditional solid

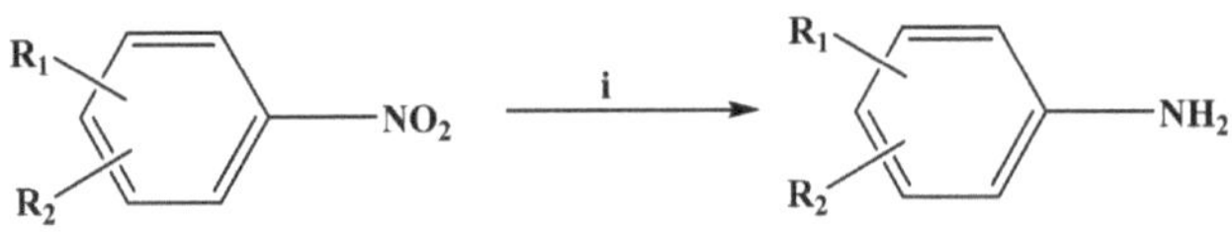

108 **109**

R_1= H, 4-OCH_3, 4-CH_3, 4-Cl, 4-I, 3-OH, 2-OH, 2-NH_2, 4-NH_2;
R_2= H, 5-CH_3, 5-CF_3

SCHEME 37 Microwave-assisted reduction of 4-nitroaniline with hydrazine on various inorganic solid supports in the presence of catalytic $FeCl_3{\cdot}6H_2O$. Reagents and conditions: (i) $H_2NNH_2{\cdot}H_2O$/$FeCl_3{\cdot}6H_2O$, MW, 30–70 W.

supports (NaCl, NaBr, NaI, KCl, KBr, KI, Na_2SO_4, K_2SO_4, $CaCO_3$, and K_2CO_3) with the soft basic or neutral surface are moderately effective. Some of them, such as K_2CO_3, NaI, and KI, were chemically active and participated in the reaction and produced several by-products. Several catalysts were also examined under solvent-free conditions using alumina as the solid support, and Fe(III) chloride, Fe(III) oxide hydroxide, and Fe_3O_4 were found to be the most effective catalysts.

6.3.17 Reduction of carbonyl compounds with sodium borohydride mediated by microwave

A fast and efficient method for the reduction of varieties of carbonyl compounds such as aldehydes, ketones, conjugated aldehydes and ketones, α-diketones, and acyloins to their corresponding alcohols with sodium borohydride ($NaBH_4$) in water under microwave irradiation was presented [140].

A wide range of structurally different aliphatic and aromatic aldehydes was considered for the reduction reaction with sodium borohydride (Scheme 38). Six selected aldehydes are shown in Table 7. The use of 0.5–1 molar equivalents of sodium borohydride was sufficient for the complete microwave-assisted reduction of aldehydes. The corresponding primary alcohols were achieved in excellent yields (82%–98%) within 30–60 s. All the reduction reactions were irradiated in water with 300 W energy of a microwave oven.

As ketones are less reactive relative to aldehydes, reduction of ketones needs the use of molar equivalents of $NaBH_4$ under microwave irradiation (700 W). Several kinds of aromatic and aliphatic ketones were subjected to reduction with $NaBH_4$ in water under microwave irradiation. Table 8 shows six ketones and the corresponding secondary alcohols obtained in excellent yields within 15–60 s. With benzophenone, reduction reaction required 4 molar equivalents of $NaBH_4$, and benzhydrol was obtained in 90% yield after 5 min irradiation. This was due to steric hindrance.

The utility of this method was also examined for the reduction of α-diketones and acyloins. The reduction reactions were carried out with 2 molar equivalents of $NaBH_4$ in water under microwave irradiation with a power of 700 W. The reduction reactions were realized within 1–3 min, and the

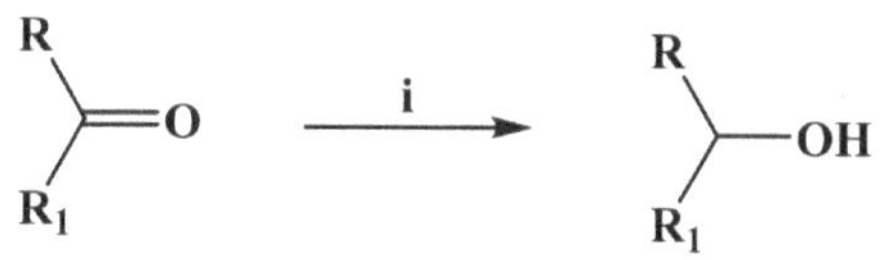

110 **111**

SCHEME 38 Reduction of carbonyl compounds with $NaBH_4$ in water under microwave irradiation. Reagents and conditions: (i) $NaBH_4/H_2O$, MW, 15 s–5 min.

TABLE 7 Six selected aldehydes and corresponding alcohols.

Substrate	Product	Time (s)	Yield (%)
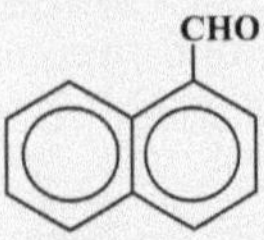	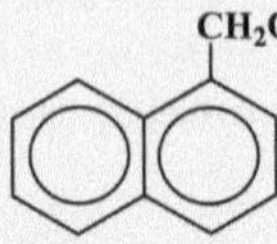	30	98
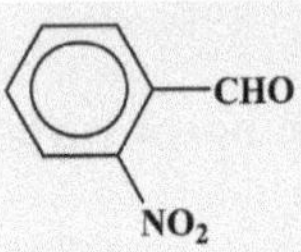	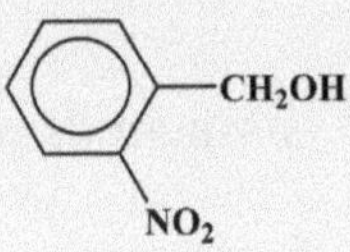	45	95
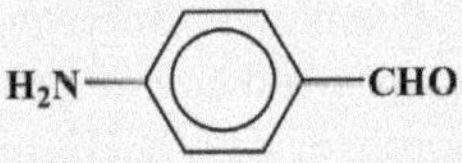	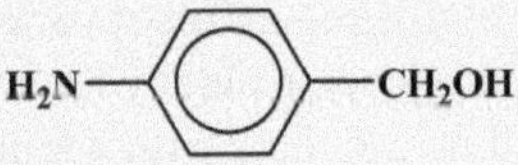	45	92
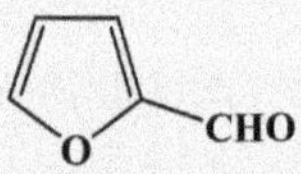	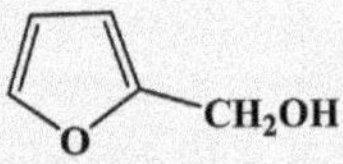	30	94
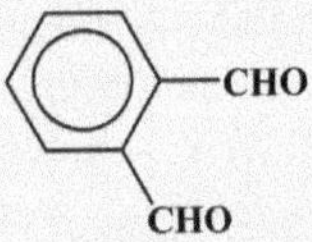	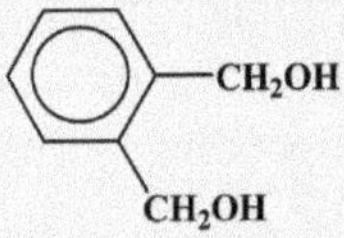	30	95
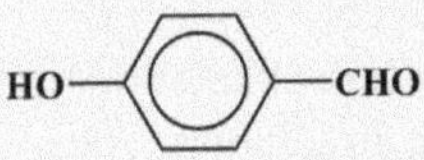	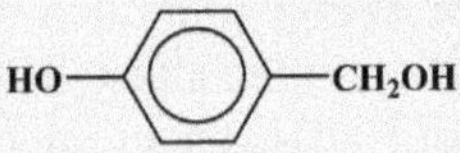	30	93

corresponding vicinal diols were obtained in excellent yields (91%–99%). Six selected compounds and corresponding products are shown in Table 9.

The possibility of 1,2-reduction of α, β-unsaturated aldehydes, and ketones with $NaBH_4$ in water under microwave irradiation was also investigated. The result showed that the microwave irradiation of cinnamaldehyde in the presence of one molar equivalent of $NaBH_4$ in water was accomplished exclusively in a 1,2-reduction manner within 50 s. And the cinnamyl alcohol was obtained in excellent yield (98%). Under microwave irradiation, the compound citral also showed the best efficiency and regioselectivity. Besides, the reduction of conjugated ketones like benzalacetone and β-ionone was achieved quickly and efficiently with 2 molar equivalents of $NaBH_4$ in water under microwave irradiation. The microwave irradiation of benzalacetophenone in 5 min gave only 70% yield. All the results are summarized in Table 10.

TABLE 8 Six selected ketones and corresponding alcohols.

Substrate	Product	Time (s)	Yield (%)
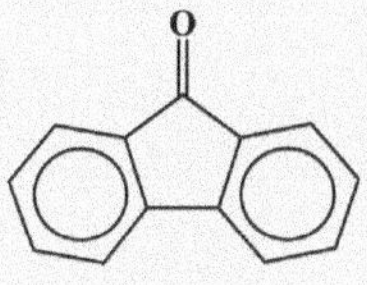	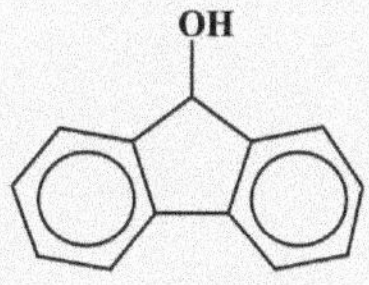	45	97
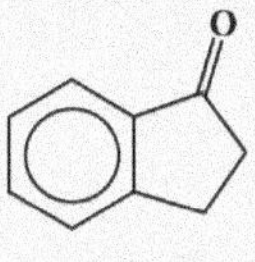	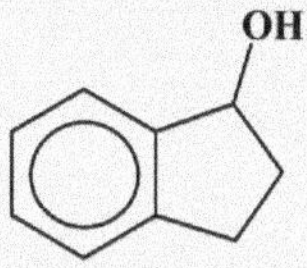	45	94
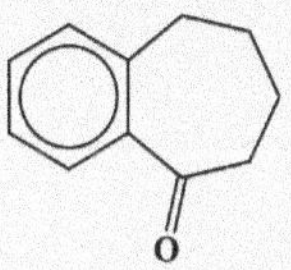	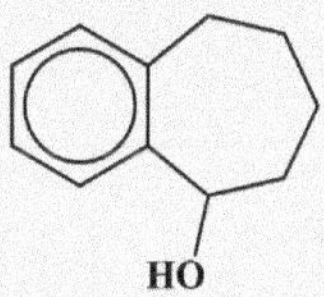	50	93
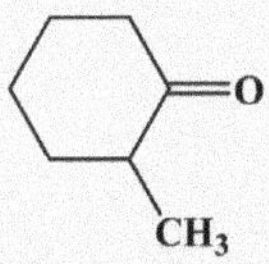	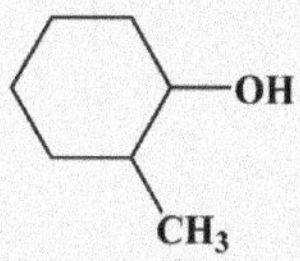	25	91
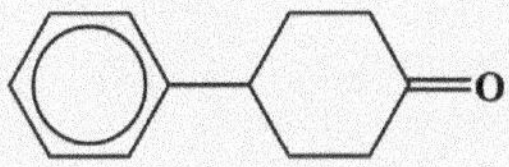	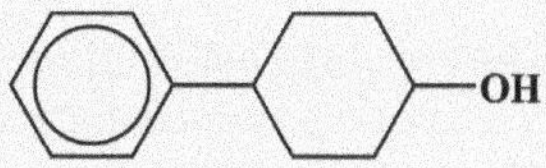	25	95
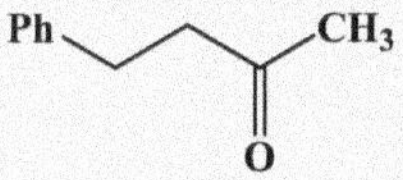	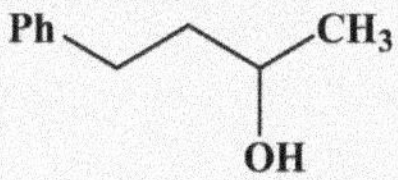	60	98

6.3.18 Reduction of alkene groups in β-lactams induced by microwave

Microwave-assisted catalytic transfer hydrogenation technique was applied to several of β-lactams using commercially available Raney nickel (RaNi) catalyst, and the reduction products were obtained in good yield in a few minutes [141]. For example, the β-lactam with an *N*-allyl group was reduced to a β-lactam with an *N*-propyl group in the presence of microwave irradiation (Scheme 39).

TABLE 9 Six selected α-diketones and acyloins.

Substrate	Product	Time (min)	Yield (%)
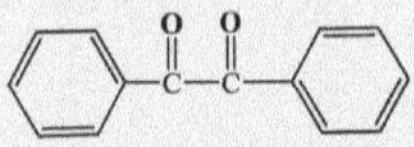	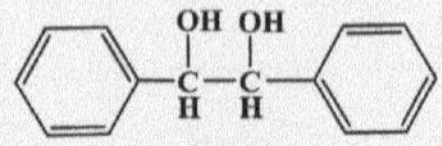	1.5	99
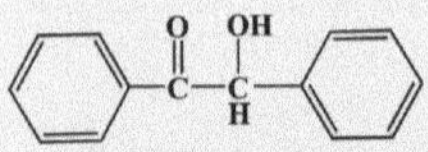	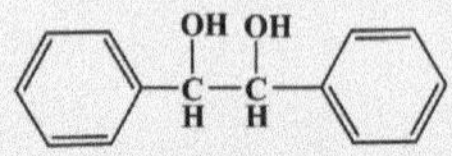	1	96
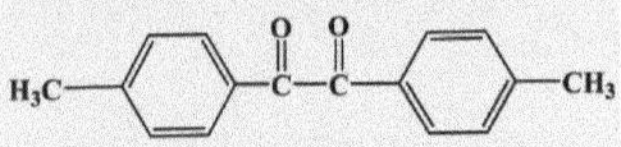	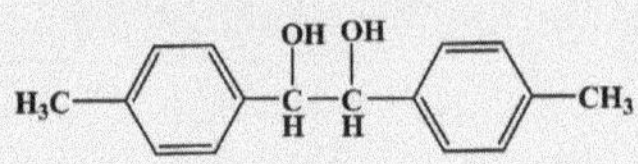	2	96
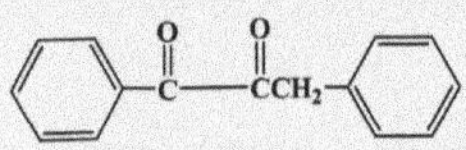	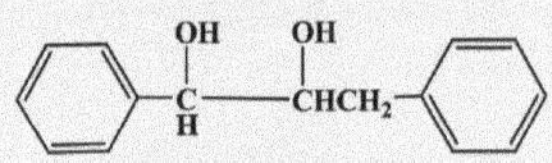	3	91
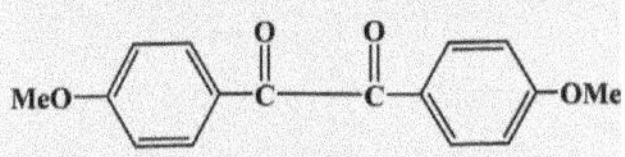	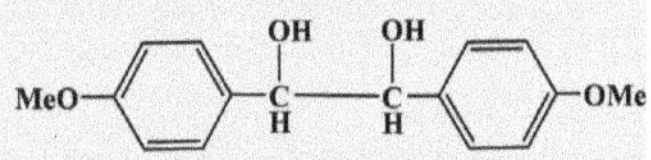	3	98
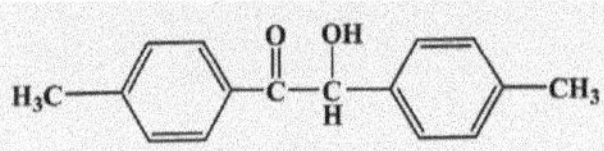	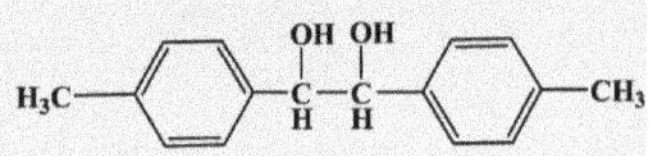	2	94

TABLE 10 List of conjugated carbonyl compounds.

Substrate	Product	Time (s)	Yield (%)
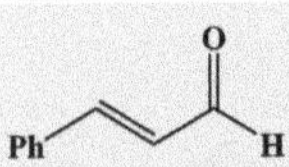	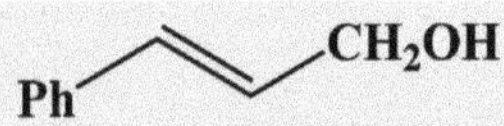	50	98
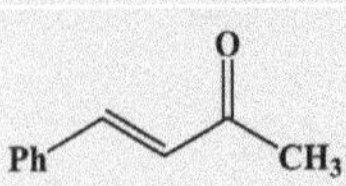	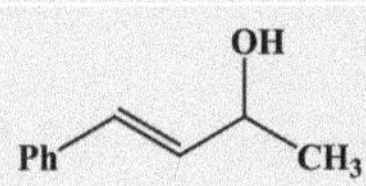	50	99
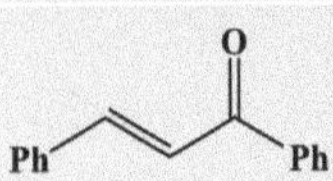	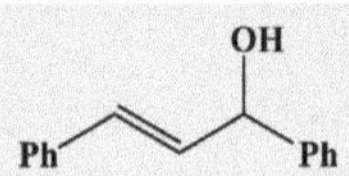	300	70
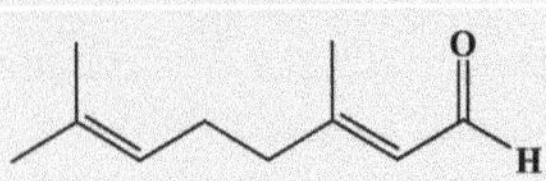	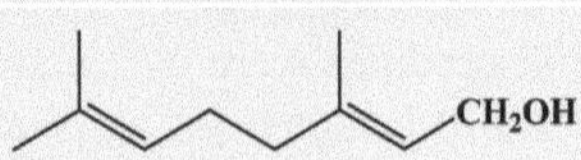	50	93
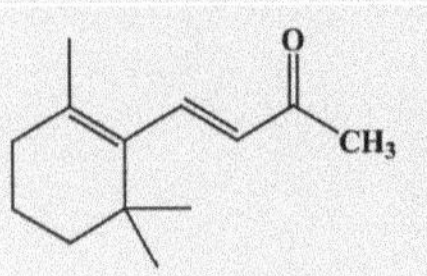	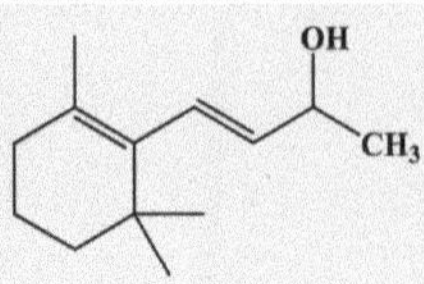	60	95

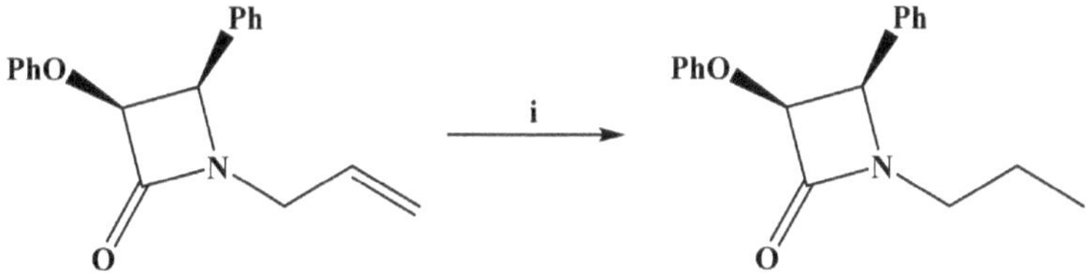

112 113

SCHEME 39 Reduction of β-lactam compounds under microwave irradiation. Reagents and conditions: (i) HCO_2NH_4, RaNi, ethylene glycol, MW, 2–3 min.

The lactams with vinyl group at C-3 position were converted to lactams with ethyl group at C-3 position using the microwave-assisted reduction (Scheme 40).

Scheme 41 shows the microwave-assisted Pd/C-catalyzed reduction of the another β-lactams, and during the reaction, the unsaturated ester is converted to a saturated ester side chain.

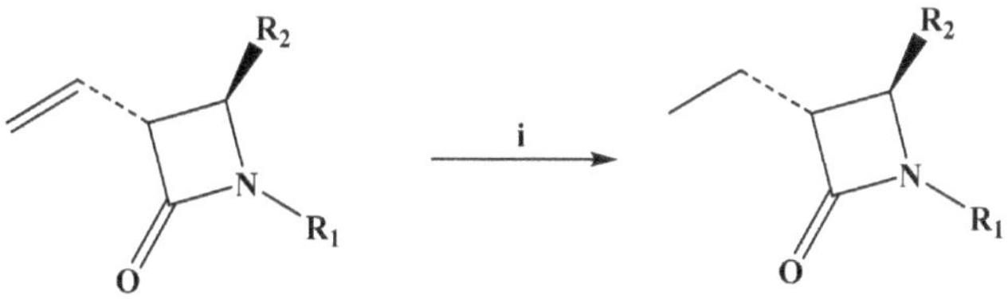

114 115

R_1= Ph, *p*-methoxyphenyl; R_2=Ph, 2-furyl.

SCHEME 40 Reduction of β-lactam compounds under microwave irradiation. Reagents and conditions: (i) HCO_2NH_4, RaNi, Ethylene glycol, MW, 2–3 min.

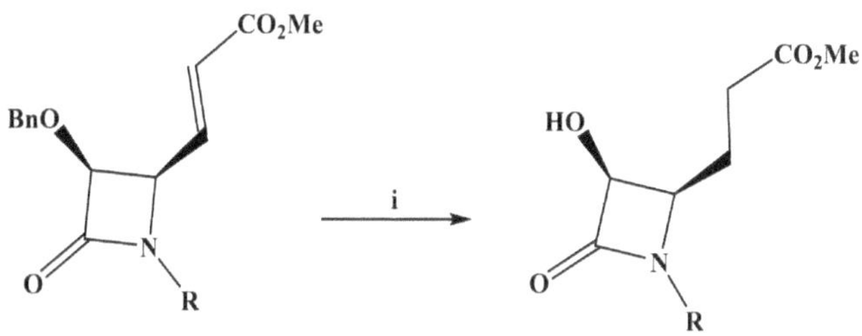

116 117

R= *p*-methoxyphenyl, Bn; Bn=$CH_2C_6H_5$

SCHEME 41 Reduction of β-lactam compounds under microwave irradiation. Reagents and conditions: (i) HCO_2NH_4, Pd/C (10%), ethylene glycol, MW, 2–3 min.

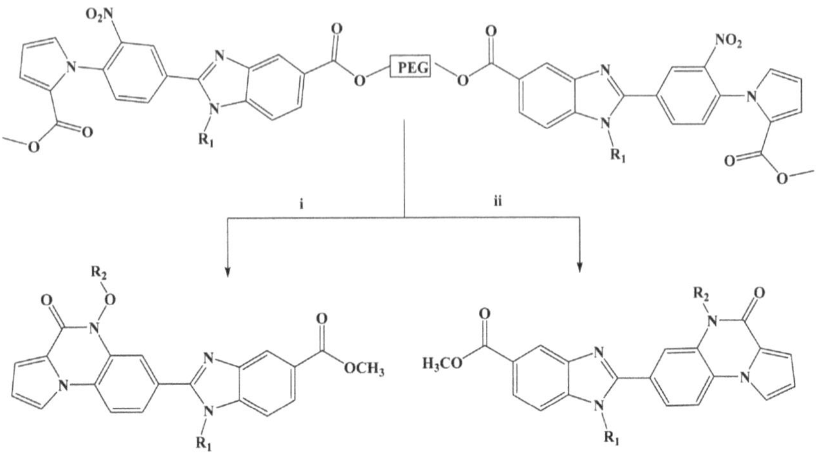

SCHEME 42 Synthesis of novel benzimidazole-alkyloxypyrrolo[1,2-a]quinoxalinones. Reagents and conditions: (i) $HCOONH_4$, Pd/C, MW, 60°C, 7 min (for partial reductive cyclization). (ii) $HCOONH_4$, Pd/C, MW, 85°C, 12 min (for full reductive cyclization).

6.3.19 Reductive cyclization induced by microwave

Microwave controlled reductive cyclization is important for the synthesis of organic compounds. For example, microwave-assisted synthesis of novel benzimidazole linked alkyloxypyrrolo[1,2-a]quinoxalinones was investigated on soluble polymer support [142]. By controlling the microwave energy, two methods were exploded for the partial and full reductive cyclization (Scheme 42).

Starting from the same substrate, ortho nitro pyrrol carboxylates, *N*-hydroxy pyrroloquinoxalinones were accomplished by partial reductive cyclization (60° C, 7 min). The synthesis of pyrroloquinoxalinones was obtained by full reductive cyclization (85°C, 12 min). In this method, Pd/C and ammonium formate were used as reducing agents.

6.4 Conclusion

In this chapter, we have described a highly efficient microwave-assisted modification of conventional heating procedure for the oxidation and reduction reactions of organic compounds. The advantages of this environmentally benign and safe protocol include a simple reaction setup, application of commercially available catalysts, high product yields, short reaction times, elimination of side products, and low chemical waste.

Acknowledgments

AD and BKB are grateful to Prince Mohammad Bin Fahd University for support. BKB is also grateful to US NIH, US NCI, and Kleberg Foundation of Texas for finance support.

References

[1] E. Adler, Lignin chemistry—past, present and future, Wood Sci. Technol. 11 (1977) 169–218.

[2] C.V. Stevens, in: C.V. Stevens, R.G. Verhe (Eds.), Renewable Resource Scope and Modification for Non-Food Applications, John Wiley and Sons, 2004, p. 160.

[3] C.N. Chang, Y.S. Ma, G.C. Fang, A.C. Chao, M.C. Tsai, H.F. Sung, Decolorizing of lignin wastewater using the photochemical UV/TiO_2 process, Chemosphere 56 (2004) 1011–1017.

[4] K. Tanaka, R.C.R. Calanag, T. Hisanaga, Photocatalyzed degradation of lignin on TiO_2, J. Mol. Catal. A Chem. 138 (1999) 287–294.

[5] C. Bonini, M.D. Auria, Degradation and recovery of fine chemicals through singlet oxygen treatment of lignin, Ind. Crop. Prod. 20 (2004) 243–259.

[6] S.J. Zhang, H.Q. Yu, L.X. Wu, Degradation of calcium lignosulfonate using gamma-ray irradiation, Chemosphere 57 (2004) 1181–1187.

[7] C. Crestini, M.D. Auria, Singlet oxygen in the photodegradation of lignin models, Tetrahedron 53 (1997) 7877–7888.

[8] I.A. Weinstock, R.H. Atalla, R.S. Reiner, M.A. Moen, K.E. Hammel, C.J. Houtman, C.L. Hill, M.K. Harrup, A new environmentally benign technology for transforming wood pulp into paper. Engineering polyoxometalates as catalysts for multiple processes, J. Mol. Catal. A Chem. 116 (1997) 59–84.

[9] S.K. Badamali, R. Luque, J.H. Clark, S.W. Breeden, Microwave-assisted oxidation of a lignin model phenolic monomer using Co(salen)/SBA-15, Catal. Commun. 10 (2009) 1010–1013.

[10] C. Crestini, A. Pastoni, P. Tagliatesta, Metalloporphyrins immobilized on motmorillonite as biomimetic catalysts in the oxidation of lignin model compounds, J. Mol. Catal. A Chem. 208 (2004) 195–202.

[11] C.W. Bailley, Reactions of alkaline hydrogen peroxide with softwood lignin model compounds, spruce milled-groundwood lignin and spruce groundwood, Tappi 52 (1969) 491–500.

[12] S.K. Badamali, R. Luque, J.H. Clark, S.W. Breeden, Unprecedented oxidative properties of mesoporous silica materials: towards microwave-assisted oxidation of lignin model compounds, Catal. Commun. 31 (2013) 1–4.

[13] R. Gažák, D. Walterová, V. Křen, Silybin and silymarin—new and emerging applications in medicine, Curr. Med. Chem. 14 (2007) 315–338.

[14] W. Lu, C. Lin, T.D. King, H. Chen, R.C. Reynolds, Y. Li, Silibinin inhibits Wnt/β-catenin signaling by suppressing Wnt co-receptor LRP6 expression in human prostate and breast cancer cells, Cell. Signal. 24 (2012) 2291–2296.

[15] T. Zhan, M. Digel, E.-M. Küch, W. Stremmel, J. Füllekrug, Silybin and dehydrosilybin decrease glucose uptake by inhibiting GLUT proteins, J. Cell. Biochem. 112 (2011) 849–859.

[16] O.B. Karim, K.-J. Rhee, G. Liu, D. Zheng, D.L. Huso, Chemoprevention utility of silibinin and Cdk4 pathway inhibition in Apc −/+ ice, BMC Cancer 13 (2013) 157.

[17] B.A. Payer, T. Reiberger, K. Rutter, S. Beinhardt, A.F. Staettermayer, M. Peck-Radosavljevic, P. Ferenci, Successful HCV eradication and inhibition of HIV replication by intravenous silibinin in an HIV–HCV coinfected patient, J. Clin. Virol. 49 (2010) 131–133.

[18] H. Kauntz, S. Bousserouel, F. Gossé, F. Raul, Silibinin triggers apoptotic signaling pathways and autophagic survival response in human colon adenocarcinoma cells and their derived metastatic cells, Apoptosis 16 (2011) 1042.

[19] A. Huber, P. Thongphasuk, G. Erben, W.D. Lehmann, S. Tuma, W. Stremmel, W. Chamulitrat, Significantly greater antioxidant anticancer activities of 2,3-dehydrosilybin than silybin, Biochim. Biophys. Acta 1780 (2008) 837–847.

[20] Z. Dvorˇák, R. Vrzal, J. Ulrichová, Silybin and dehydrosilybin inhibit cytochrome P450 1A1 catalytic activity: a study in human keratinocytes and human hepatoma cells, Cell Biol. Toxicol. 22 (2006) 81.

[21] G. Halbach, W. Trost, Chemistry and pharmacology of silymarin. Studies on various metabolites of silybin, Arzneim. Forsch. 24 (1974) 866.

[22] R. Gažák, A. Svobodová, J. Psotová, P. Sedmera, V. Přikrylová, D. Walterová, V. Křen, Oxidised derivatives of silybin and their antiradical and antioxidant activity, Bioorg. Med. Chem. 12 (2004) 5677–5687.

[23] P. Džubák, M. Hajdúch, R. Gažák, A. Svobodová, J. Psotová, D. Walterová, P. Sedmera, V. Křen, New derivatives of silybin and 2,3-dehydrosilybin and their cytotoxic and P-glycoprotein modulatory activity, Bioorg. Med. Chem. 14 (2006) 3793–3810.

[24] A. Pelter, R. Hänsel, Struktur des silybins: i. abbauversuche, Chem. Ber. 108 (1975) 790–802.

[25] L.X. Yang, K.X. Huang, H.B. Li, J.X. Gong, Y.B. Feng, Q.F. Tao, Y.H. Wu, X.K. Li, X.M. Wu, S. Zeng, S. Spencer, Y. Zhao, J. Qu, Design, synthesis, and examination of neuron protective properties of alkenylated and amidated dehydro-silybin derivatives†, J. Med. Chem. 52 (2009) 7732–7752.

[26] A. Zarrelli, A. Sgambato, V. Petito, L. De Napoli, L. Previtera, G. Di Fabio, New C-23 modified of silybin and 2, 3-dehydrosilybin: synthesis and preliminary evaluation of antioxidant properties, Bioorg. Med. Chem. Lett. 21 (2011) 4389–4392.

[27] G.D. Fabio, V. Romanucci, M.D. Nisco, S. Pedatella, C.D. Marino, A. Zarrelli, Microwave-assisted oxidation of silibinin: a simple and preparative method for the synthesis of improved radical scavengers, Tetrahedron Lett. 54 (2013) 6279–6282.

[28] C. Parmeggiani, C. Matassini, F. Cardona, A step forward towards sustainable aerobic alcohol oxidation: new and revised catalysts based on transition metals on solid supports†, Green Chem. 19 (2017) 2030–2050.

[29] R.A. Sheldon, I.W.C.E. Arends, A. Dijksman, New developments in catalytic alcohol oxidations for fine chemicals synthesis, Catal. Today 57 (2000) 157–166.

[30] R.A. Sheldon, Fundamentals of green chemistry: efficiency in reaction design†‡, Chem. Soc. Rev. 41 (2012) 1437–1451.

[31] M. Oszajca, A. Franke, M. Brindell, G. Stochel, R. van Eldik, Redox cycling in the activation of peroxides by iron porphyrin and manganese complexes. 'Catching' catalytic active intermediates, Coord. Chem. Rev. 306 (2016) 483–509.

[32] S. Sahu, D.P. Goldberg, Activation of dioxygen by iron and manganese complexes: a heme and nonheme perspective, J. Am. Chem. Soc. 138 (2016) 11410–11428.

[33] X. Engelmann, I. Monte-Perez, K. Ray, Oxidation reactions with bioinspired mononuclear non-heme metal–oxo complexes, Angew. Chem. Int. Ed. 55 (2016) 7632–7649.

[34] B. Biswas, A. Al-Hunaiti, M.T. Räisänen, S. Ansalone, M. Leskelä, T. Repo, Y.-T. Chen, H.-L. Tsai, A.D. Naik, A.P. Railliet, Y. Garcia, R. Ghosh, N. Kole, Efficient and selective oxidation of primary and secondary alcohols using an iron(iii)/phenanthroline complex: structural studies and catalytic activity, Eur. J. Inorg. Chem. 2012 (2012) 4479–4485.

[35] C. Crotti, E. Farnetti, Selective oxidation of glycerol catalyzed by iron complexes, J. Mol. Catal. A Chem. 396 (2015) 353–359.

[36] I.S. Cozzi, C. Crotti, E. Farnetti, Microwave-assisted green oxidation of alcohols with hydrogen peroxide catalyzed by iron complexes with nitrogen ligands, J. Organomet. Chem. 878 (2018) 38–47.

[37] A. Karmakar, L.M.D.R.S. Martins, M.F.C.G. da Silva, S. Hazra, A.J.L. Pombeiro, Solvent-free microwave-assisted peroxidative oxidation of alcohols catalyzed by iron(III)-tempo catalytic systems, Catal. Lett. 145 (2015) 2066–2076.

[38] R.R. Fernandes, J. Lasri, M.F.C.G. da Silva, J.A.L. da Silva, J.J.R.F. Silva, A.J.L. Pombeiro, Bis- and tris-pyridyl amino and iminothioether Cu and Fe complexes. Thermal and microwave-assisted peroxidative oxidations of 1-phenylethanol and cyclohexane in the presence of various N-based additives, J. Mol. Catal. A Chem. 351 (2011) 100–111.

[39] M. Sutradhar, E.C.B.A. Alegria, K.T. Mahmudov, M.F.C.G. da Silva, A.J.L. Pombeiro, Iron (III) and cobalt(III) complexes with both tautomeric (keto and enol) forms of aroylhydrazone ligands: catalysts for the microwave assisted oxidation of alcohols†, RSC Adv. 6 (2016) 8079–8088.

[40] N.M.R. Martins, K.T. Mahmudov, M.F.C. Guedes da Silva, L.M.D.R.S. Martins, A.J.L. Pombeiro, Copper(ii) and iron(iii) complexes with arylhydrazone of ethyl 2-cyanoacetate or formazan ligands as catalysts for oxidation of alcohols†, New J. Chem. 40 (2016) 10071–10083.

[41] P.J. Figiel, M.N. Kopylovich, J. Lasri, M.F.C.G. da Silva, J.J.R.F. da Silva, A.J.L. Pombeiro, Solvent-free microwave-assisted peroxidative oxidation of secondary alcohols to the corresponding ketones catalyzed by copper(II) 2,4-alkoxy-1,3,5-triazapentadienato complexes, Chem. Commun. 46 (2010) 2766–2768.

[42] F. Brühne, E. Wright, Benzaldehyde, in: Ullmann's Encyclopellmann's Encyclopedia of Industrial Chemistry, vol. 5, Wiley-VCH Verlag GmbH &Co. KGaA, 2011, pp. 223–233.

[43] C.C. Guo, Q. Liu, X.T. Wang, H.Y. Hu, Selective liquid phase oxidation of toluene with air, Appl. Catal. A 282 (2005) 55–59.

[44] L. Kesavan, R. Tiruvalam, M.H. Ab Rahim, M.I.B. Saiman, D.I. Enache, R.L. Jenkins, N. Dimitratos, J.A.L. Sanchez, S.H. Taylor, D.W. Knight, C.J. Kiely, G.J. Hutchings, Solvent-free oxidation of primary carbon-hydrogen bonds in toluene using Au-Pd alloy nanoparticles, Science 331 (2011) 195–199.

[45] G. Huang, A.P. Wang, S.Y. Liu, Y.A. Guo, H. Zhou, S.K. Zhao, An efficient oxidation of toluene over Co(II)TPP supported on chitosan using air, Catal. Lett. 114 (2007) 174–177.

[46] F. Wang, J. Xu, X. Li, J. Gao, L. Zhou, R. Ohnishi, Liquid phase oxidation of toluene to benzaldehyde with molecular oxygen over copper-based heterogeneous catalysts, Adv. Synth. Catal. 347 (2005) 1987–1992.

[47] M. Sutradhar, E.C.B.A. Alegria, T.R. Barman, F. Scorcelletti, M.F.C.G. da Silva, A.J.L. Pombeiro, Microwave-assisted peroxidative oxidation of toluene and 1-phenylethanol with monomeric keto and polymeric enol aroylhydrazone Cu(II) complexes, Mol. Catal. 439 (2017) 224–232.

[48] M. Hudlicky, Oxidations in organic chemistry, Am. Chem. Soc. Wash. (1990) 99.

[49] X. Wang, J. Wu, M. Zhao, Y. Lv, G. Li, C. Hu, Partial oxidation of toluene in CH_3COOH by H_2O_2 in the presence of VO (acac)2 catalyst, J. Phys. Chem. C 113 (2009) 14270–14278.

[50] Y. Wang, P. Prinsen, F. Mangin, A. Yepez, A. Pineda, E.R. Castellón, M.R.H.S. Gilani, G. Xu, C. Len, R. Luque, Mechanistic insights into the microwave-assisted cinnamyl alcohol oxidation using supported iron and palladium catalysts, Mol. Catal. 474 (2019) 110409.

[51] A.P.C. Ribeiro, L.M.D.R.S. Martins, S.A.C. Carabineiro, J.G. Buijnsters, J.L. Figueiredo, A. J.L. Pombeiro, Heterogenized C-scorpionate iron(ii) complex on nanostructured carbon materials as recyclable catalysts for microwave-assisted oxidation reactions, ChemCatChem 10 (2018) 1821–1828.

[52] R. Jlassi, A.P.C. Ribeiro, E.C.B.A. Alegria, H. Naïli, G.A.O. Tiago, T. Rüffer, H. Lang, F.I. Zubkov, A.J.L. Pombeiro, W. Rekik, Copper(II) complexes with an arylhydrazone of methyl 2-cyanoacetate as effective catalysts in the microwave-assisted oxidation of cyclohexane, Inorg. Chim. Acta 471 (2018) 658–663.

[53] H.Y. Lee, A.R. Kim, M. Park, J.M. Jo, D.H. Lee, J.W. Be, Combined steam and CO_2 reforming of CH_4 using coke oven gas onnickel-based catalyst: effects of organic acids to nickel dispersion and activity, Chem. Eng. 280 (2015) 771–781.

[54] H. Quitmann, R. Fan, P. Czermak, Acidic organic compounds in beverage, food, and feed production, in: H. Zorn, P. Czermak (Eds.), Biotechnology of Food and Feed Additives, Springer-Verlag, Berlin, 2014, pp. 91–141.
[55] C. Perego, D. Bianchi, Biomass upgrading through acid-base catalysis, Chem. Eng. J. 161 (2010) 314–322.
[56] S. Veeravali, A.P. Mathews, Continuous fermentation of xylose to short chain fatty acids by Lactobacillus buchneri under low pH conditions, Chem. Eng. J. 337 (2018) 764–771.
[57] H. Zhang, G. Liu, J. Zhang, J. Bao, Fermentative production of high titer gluconic and xylonic acids from corn stover feedstock by gluconobacteroxydans and technoeconomic analysis, Bioresour. Technol. 219 (2016) 123–131.
[58] A.M.C. Rodríguez, I.M.S. Dueñas, M.J.T. Martínez, A. Mas, J.E.J. Hornero, I.G. García, Preparation of a pure inoculum of acetic acid bacteria for the selective conversion of glucose in strawberry purée into gluconic acid, Food Bioprod. Process. 96 (2015) 35–42.
[59] Z. Zhang, G.W. Huber, Catalytic oxidation of carbohydrates into organic acids and furan chemicals, Chem. Soc. Rev. 47 (2018) 1351–1390.
[60] W. Deng, Q. Zhang, Y. Wang, Catalytic transformations of cellulose and cellulose derived carbohydrates into organic acids, Catal. Today 234 (2014) 31–41.
[61] H. Li, X. Kong, Z. Fang, R.L. Smith, Chapter 1 fundamentals of bifunctional catalysis for transforming biomass-related compounds into chemicals and biofuels, Biofuels Biorefineries 8 (2017) 3–30.
[62] H. Li, P.S. Bhadury, A. Riisager, S. Yang, One-pot transformation of polysaccharides via multi-catalytic processes, Catal. Sci. Technol. 4 (2014) 4138–4168.
[63] D. Chen, F. Liang, D. Feng, M. Xian, H. Zhang, H. Liu, F. Du, An efficient route from reproducible glucose to 5-hydroxymethylfurfuralcatalyzed by porous coordination polymer heterogeneous catalysts, Chem. Eng. J. 300 (2016) 177–184.
[64] X. Zhang, K. Wilson, A.F. Lee, Heterogeneously catalyzed hydrothermal processing of C5–C6 sugars, Chem. Rev. 116 (2016) 12328–12368.
[65] D. Bin, H. Wang, J. Li, H. Wang, Z. Yin, J. Kang, B. He, Z. Li, Controllable oxidation of glucose to gluconic acid and glucaric acid using an electrocatalytic reactor, Electrochim. Acta 130 (2014) 170–178.
[66] M. Besson, P. Gallezot, C. Pinel, Conversion of biomass into chemicals over metal catalysts, Chem. Rev. 114 (2014) 1827–1870.
[67] V.A. Timoshchuk, Uronic acids: synthesis and reactions, Russ. Chem. Rev. 64 (1995) 721–750.
[68] K.B.H. Finch, R.M. Richards, A. Richel, A.V. Medvedovici, N.G. Gheorghe, M. Verziu, S.M. Coman, V.I. Parvulescu, Catalytic hydroprocessing of lignin under thermal and ultrasound conditions, Catal. Today 196 (2012) 3–10.
[69] R.C. Sun, J.M. Lawther, W.B. Banks, Fractional and structural characterization of wheat straw hemicelluloses, Carbohydr. Polym. 29 (1996) 325–331.
[70] K. Khallouk, A. Solhy, N. Idrissi, V. Flaud, A. Kherbeche, A. Barakat, Microwave-assisted selective oxidation of sugars to carboxylic acids derivatives in water over zinc-vanadium mixed oxide, Chem. Eng. J. 385 (2020) 123914.
[71] V.B. Kumar, I.N. Pulidindi, A. Gedanken, Synergistic catalytic effect of the $ZnBr_2$–HCl system for levulinic acid production using microwave irradiation, RSC Adv. 5 (2015) 11043–11048.
[72] S. Rautiainen, P. Lehtinen, M. Vehkamäki, K. Niemelä, M. Kemell, M. Heikkilä, T. Repo, Microwave-assisted base-free oxidation of glucose on gold nanoparticle catalysts, Catal. Commun. 74 (2016) 115–118.

[73] S.J.H.F. Arts, E.J.M. Mombarg, H.V. Bekkum, R.A. Sheldon, Hydrogen peroxide and oxygen in catalytic oxidation of carbohydrates and related compounds, Synthesis 6 (1997) 597–613.

[74] M. Lukasiewicz, S. Bednarz, A. Ptaszek, Environmental friendly polysaccharide modification –microwave-assisted oxidation of starch, Starch-Starke 63 (2011) 268–273.

[75] S. Ruhemann, CXXXII.-cyclic di- and tri-ketones, J. Chem. Soc. Trans. 97 (1910) 1438–1449.

[76] M.M. Joullié, R.R. Hark, D.B. Hauze, O. Petrovskaia, Synthetic studies of novel ninhydrin analogs, Can. J. Chem. 79 (2001) 1632–1654.

[77] R.R. Hark, D.B. Hauze, O. Petrovskaia, D.B. Hansen, M.M. Joullié, The development of novel ninhydrin analogues, Chem. Soc. Rev. 34 (2005) 408–417.

[78] Y.S. Jung, C.K. Lee, H.S. Kim, H.C. Jeong, P.H. Kim, S.B. Han, J.S. Shin, J. Neyts, H.J. Thibaut, J. Eur. Pat. 2,722,042, 2014.

[79] Y.S. Jung, C.K. Lee, I.Y. Choi, H.S. Kim, P.H. Kim, S.B. Han, J. Neyts, H.J. Thibaut, J. Eur. Pat. 2,722,322, 2014.

[80] K. Bowden, S.J. Rumpal, J. Chem. Res. Miniprint (1997) 355–363.

[81] J. Tatsugi, Y. Izawa, A Convenient one-pot synthesis of indane-1,2,3-triones by oxidation of indan-1-ones with N-bromosuccinimide-dimethyl sulfoxide reagent, Synth. Commun. 28 (1998) 859–864.

[82] P. Maslak, A. Chopra, C.R. Moylan, R. Wortmann, S. Lebus, A.L. Rheingold, G.P.A. Yap, Optical properties of spiroconjugated charge-transfer dyes, J. Am. Chem. Soc. 118 (1996) 1471–1481.

[83] M.B. Smith, Organic Synthesis Int. Ed., McGraw-Hill Int. Ed., Singapore, 1994, p. 311.

[84] R. Frédérick, W. Dumont, F. Ooms, L. Aschenbach, C.J.V. da Schyf, N. Castagnoli, J. Wouters, A. Krief, Synthesis, structural reassignment, and biological activity of type B MAO inhibitors based on the 5H-indeno[1,2-c]pyridazin-5-one core, J. Med. Chem. 49 (2006) 3743–3747.

[85] J. Reniers, C. Meinguet, L. Moineaux, B. Masereel, S.P. Vincent, R. Frédérick, J. Wouters, Synthesis and inhibition study of monoamine oxidase, indoleamine 2,3-dioxygenase and tryptophan 2,3-dioxygenase by 3,8-substituted 5H-indeno[1,2-c]pyridazin-5-one derivatives, Eur. J. Med. Chem. 46 (2011) 6104–6111.

[86] C. Marminon, A. Nacereddine, Z. Bouaziz, P. Nebois, J. Jose, M.L. Borgne, Microwave-assisted oxidation of indan-1-ones into ninhydrins, Tetrahedron Lett. 56 (2015) 1840–1842.

[87] M.K. Mohammadi, Microwave-assisted oxidation of organic compounds with cetyltrimethylammonium chlorochromate, Open J. Synth. Theory Appl. 2 (2013) 87–90.

[88] W. Cao, H. Zhang, J. Chen, X. Zhou, M. Shao, M.C. McMills, Stereoselective synthesis of highly substituted trans-2,3-dihydrofuran and trans-1,2-cyclopropane derivatives containing sulfonyl groups, Tetrahedron 64 (2008) 163–167.

[89] G.J.L. Ruano, F. Bercial, G. Gonzalez, A.M.M. Castro, M.R. Martın, Asymmetric 1,3-dipolar cycloadditions of diazoalkanes to (5S,SS)-5-[(1R)-menthyloxy]-4-phenylsulfinyl (and phenylsulfonyl)furan-2(5H)-ones, Tetrahedron Asymmetry 13 (2002) 1993–2002.

[90] J.L.G. Ruano, F. Bercial, A. Fraile, A.M.M. Castro, M.R. Martın, Stereoselectivity control in Diels–Alder reactions of 4-thiosubstituted 5-alkoxyfuranones: synthesis and reactivity of enantiopure 4-sulfinyl and sulfonyl 5-(l-menthyloxy)furan-2(5H)-ones, Tetrahedron Asymmetry 11 (2000) 4737–4752.

[91] P. Shanmugam, K.H. Kumar, P.T. Perumal, Mn(III) acetate mediated diasteroselective [3+2] oxidative addition of 1,3-dicarbonyl compounds with chalcones, J. Heterocyclic Chem. 44 (2007) 827.

[92] C.-Y. Qian, H. Nishino, K. Kurosawa, Manganese(III)-mediated free-radical cyclization of active methylene compounds, alkenes and molecular oxygen. Syntheses of sulfur- and phosphorus-containing 1,2-dioxan-3-ols, J. Heterocycl. Chem. 30 (1993) 209.

[93] B.B. Snider, Y.F.B. Wan, B.O. Buckman, M.F. Bruce, Manganese(III)-based asymmetric oxidative free-radical cyclization of unsaturated. beta.-keto sulfoxides, J. Org. Chem. 56 (1991) 328–334.

[94] C. Curti, M.D. Crozet, P. Vanelle, Microwave-assisted manganese(III) acetate based oxidative cyclizations of alkenes with β-ketosulfones, Tetrahedron 65 (2009) 200–205.

[95] C.F. Gromboni, M.Y. Kamogawa, A.G. Ferreira, J.A. Nóbrega, A.R.A. Nogueira, Microwave-assisted photo-Fenton decomposition of chlorfenvinphos and cypermethrin in residual water, J. Photochem. Photobiol. A 185 (2007) 32–37.

[96] Y. Yang, P. Wang, S. Shi, Y. Liu, Microwave enhanced Fenton-like process for the treatment of high concentration pharmaceutical wastewater, J. Hazard. Mater. 168 (2009) 238–245.

[97] N. Remya, J.-G. Lin, Current status of microwave application in wastewater treatment–a review, Chem. Eng. J. 166 (2011) 797–813.

[98] V. Homem, A. Alves, L. Santos, Microwave-assisted Fenton's oxidation of amoxicillin, Chem. Eng. J. 220 (2013) 35–44.

[99] A.S. Amarasekara, M.A. Hasan, Pd/C catalyzed conversion of levulinic acid to γ-valerolactone using alcohol as a hydrogen donor under microwave conditions, Catal. Commun. 60 (2015) 5–7.

[100] R.C. Larock, Comprehensive Organic Transformations: A Guide to Functional Group Preparation, second ed., Wiley–VCH, Weinheim, 1999, pp. 821–828.

[101] M. Hudlicky, Reductions in Organic Chemistry, second ed., ACS Monograph, American Chemical Society, Washington, DC, 1996.

[102] G.W. Kabalka, R.S. Verma, in: B.M. Trost, I. Fleming (Eds.), Comprehensive Organic Synthesis, vol. 8, Pergamon, Oxford, 1991, pp. 363–379.

[103] C.A. Marlic, S. Motamed, B. Quinn, Structure determination and synthesis of fluoro nissl green: an RNA-binding fluorochrome, J. Org. Chem. 60 (1995) 3365–3369.

[104] A. Burawoy, J.P. Critchley, Electronic spectra of organic molecules and their interpretation—V: effect of terminal groups containing multiple bonds on the K-bands of conjugated systems, Tetrahedron 5 (1959) 340–351.

[105] K.M. Doxsee, M. Figel, K.D. Stewart, J.W. Canary, C.B. Knobler, D.J. Cram, Host-guest complexation. 42. Preorganization strongly enhances the tendancy of hemispherands to form hemispheraplexes, J. Am. Chem. Soc. 109 (1987) 3098–3107.

[106] B.K. Banik, C. Mukhopadhyay, M.S. Venkattman, F.F. Becker, A facile reduction of aromatic nitro compounds to aromatic amines by samarium and iodine, Tetrahedron Lett. 39 (1998) 7243.

[107] M.K. Basu, F.F. Becker, B.K. Banik, Ultrasound-promoted highly efficient reduction of aromatic nitro compounds to the aromatic amines by samarium/ammonium chloride, Tetrahedron Lett. 41 (2000) 5603–5606.

[108] L. Pehlivan, E. Metay, S. Laval, W. Dayoub, P. Demonchaux, G. Mignani, M. Lemaire, Iron-catalyzed selective reduction of nitro compounds to amines, Tetrahedron Lett. 51 (2010) 1939–1941.

[109] G.D. Yadav, S.V. Lande, Liquid-liquid-liquid phase transfer catalysis: a novel and green concept for selective reduction of substituted nitroaromatics, Adv. Synth. Catal. 347 (2005) 1235–1241.

[110] H.Y. Lee, M. An, Selective reduction of the nitro-group using $Co_2(CO)_8$-H_2O, Bull. Kor. Chem. Soc. 25 (2004) 1717–1719.

[111] S. Farhadi, M. Kazem, F. Siadatnasab, NiO nanoparticles prepared via thermal decomposition of the bis(dimethylglyoximato)nickel(II) complex: a novel reusable heterogeneous catalyst for fast and efficient microwave-assisted reduction of nitroarenes with ethanol, Polyhedron 30 (2011) 606–613.
[112] D.S. Soriano, Example of the Wolff-Kishner reduction procedure suitable for an undergraduate organic lab experiment: Preparation of oxindole, J. Chem. Educ. 70 (1993) 332.
[113] E. Parquet, Q. Lin, Microwave-assisted Wolff–Kishner reduction reaction, J. Chem. Educ. 74 (1997) 1225.
[114] S. Niwa, K. Soai, Catalytic asymmetric synthesis of optically active alkynyl alcohols by enantioselective alkynylation of aldehydes and by enantioselective alkylation of alkynyl aldehydes, J. Chem. Soc. Perkin Trans. 1 (1990) 937.
[115] A.A. Sinkula, S.H. Yalkowsky, Rationale for design of biologically reversible drug derivatives: prodrugs, J. Pharm. Sci. 64 (1975) 181–210.
[116] K.K. Wang, B. Liu, Y. Lu, Facile synthesis of [3]cumulenes via 1,4-elimination of hydroxytrimethylsilane from 4-(trimethylsilyl)-2-butyn-1-ols, J. Org. Chem. 60 (1995) 1885–1887.
[117] H. Firouzabadi, M. Seddighi, E. Mottaghinejad, M. Bolourtchian, Barium permanganate, Ba $(MnO_4)_2$, a versatile and mild oxidizing agent for use under aprotic and non-aqueous conditions, Tetrahedron 46 (1990) 6869–6878.
[118] J. Dunogues, R. Callas, M. Bolourtchian, C. Biran, N. Duffaut, B. Barbe, C-silylation et duplication réductrice observées lors de l'action de Me_3SiCl Mg HMPT sur des composés carbonylés α-éthyléniques.: application à la préparation d'εdicétones à partir de cétones α-éthyléniques, J. Organometallic Chem. 57 (1973) 55–69.
[119] E.N. Banfi, R. Riva, in: L. Paquette (Ed.), Reagents, for Organic Synthesis, vol. 7, Wiley, New York, 1995, p. 4522.
[120] R. Zadmard, M.R. Saidi, M. Bolourtchian, L. Nakhshab, Microwave-assisted reduction of P-trimethylsilyl carbonylcompounds by sodium borohydride, Phosphorus Sulfur Silicon 143 (1998) 63–66.
[121] A. Wolfson, C. Dlugy, Glycerol as an alternative green medium for carbonyl compound reductions, Org. Commun. 2 (2) (2009) 34–41.
[122] C.G. Swain, A.L. Powell, W.A. Sheppard, C.R. Morgan, Mechanism of the Cannizzaro reaction, J. Am. Chem. Soc. 101 (1979) 3576.
[123] R.S. Varma, G.W. Kabalka, L.T. Evans, R.M. Pagni, Aldol condensations on basic alumina: the facile syntheses of chalcones and enones in a solvent-free medium, Synth. Commun. 15 (1985) 279–284.
[124] R.S. Varma, K.P. Naicker, P.J. Liesen, Microwave-accelerated crossed Cannizzaro reaction using barium hydroxide under solvent-free conditions, Tetrahedron Lett. 39 (1998) 8437–8440.
[125] M.G. Al-Shaal, M. Calin, I. Delidovich, R. Palkovits, Microwave-assisted reduction of levulinic acid with alcohols producing γ-valerolactone in the presence of a Ru/C catalyst, Catal. Commun. 75 (2016) 65–68.
[126] M. Akira, T. Masashi, P.G.K. Surya, O.A. George, Microwave-assisted reduction of acetophenones using Ni–Al alloy in water, Bull. Chem. Soc. Jpn. 79 (2006) 791–792.
[127] N. Shankaraiah, N. Markandeya, M.E. Moraga, C. Arancibia, A. Kamal, L.S. Santos, One-pot microwave-assisted selective azido reduction/tandem cyclization in condensed and solid phase with nickel boride, Synthesis 13 (2009) 2163–2170.
[128] (a) S. Kobayashi, H. Ishitani, Catalytic enantioselective addition to imines, Chem. Rev. 99 (1999) 1069–1094. (b) N. Shankaraiah, L.S. Santos, Enantioselective total synthesis of pyrroloquinolone as a potent PDE5 inhibitor, Tetrahedron Lett. 50 (2009) 520–523. (c) N.

Shankaraiah, L.S. Santos, Enantioselective total synthesis of pyrroloquinolone as a potent PDE5 inhibitor, Tetrahedron Lett. 50 (2009) 2700.

[129] M.E. Moraga, A.G. Caceres, L.S. Santos, Palladium asymmetric reduction of b-carboline imines mediated by chiral auxiliaries assisted by microwave irradiation, Tetrahedron Lett. 50 (2009) 7059–7061.

[130] W.A. de Silva, M.T. Rodrigues Jr., N. Shankaraiah, R.B. Ferreira, C.K.Z. Andrade, R.A. Pilli, L.S. Santos, Novel supramolecular palladium catalyst for the asymmetric reduction of imines in aqueous media, Org. Lett. 11 (2009) 3238–3241.

[131] (a) M. Sakaitani, Y. Ohfune, Syntheses and reactions of silyl carbamates. 1. Chemoselective transformation of amino protecting groups via tert-butyldimethylsilyl carbamates, J. Org. Chem. 55 (1990) 870–876. (b) R.S. Coleman, A.J. Carpenter, Synthesis of the aziridino [1,2-a]pyrrolidine substructure of the antitumor agents azinomycin A and B, J. Org. Chem. 57 (1992) 5813–5815.

[132] M.J. Gronnow, R.J. White, J.H. Clark, D. Macquarrie, Energy efficiency in chemical reactions: a comparative study of different reaction techniques, Org. Process Res. Dev. 9 (2005) 516–518.

[133] H. Abbastabar Ahangar, K. Marjani, G.H. Mahdavinia, Microwave-assisted reduction of α, β-unsaturated carbonyl compounds in solid state using sodium borohydride supported on magnesium sulfate ($NaBH_4/MgSO_4$-$7H_2O$), Synth. Commun. 38 (2008) 3414–3421.

[134] R.S. Varma, R.K. Saini, Microwave-assisted reduction of carbonyl compounds in solid state using sodium borohydride supported on alumina, Tetrahedron Lett. 38 (1997) 4337–4338.

[135] E. Leyva, E. Moctezuma, M.D.S.S. Díaz, S.E.L. Carrillo, O.H. González, Fast microwave assisted bioreduction of aromatic aldehydes using aloe vera: a green chemistry reaction, Rev. Latinoamer. Quím. 40 (2012) 140–147.

[136] J.M. Abdullah, M.M. Suliman, Microwave-assisted reduction of aromatic nitro compounds with ammonium chloride and zinc dust, Iraqi Nat. J. Chem. 43 (2011) 418–423.

[137] J. Spencer, N. Anjum, H. Patel, R.P. Rathnam, J. Verma, Molybdenum hexacarbonyl and DBU reduction of nitro compounds under microwave irradiation, Synlett 16 (2007) 2557–2558.

[138] R.C. Larock, Comprehensive Organic Transformations, VHC Publishers, New York, 1987.

[139] A. Vass, J. Dudas, J. Toth, R.S. Varma, Solvent-free reduction of aromatic nitro compounds with alumina-supported hydrazine under microwave irradiation, Tetrahedron Lett. 42 (2001) 5347–5349.

[140] B. Zeynizadeh, D. Setamdideh, Water as a green solvent for fast and efficient reduction of carbonyl compounds with $NaBH_4$ under microwave irradiation, J. Chin. Chem. Soc. 52 (2005) 1179–1184.

[141] B.K. Banik, K.J. Barakat, D.R. Wagle, M.S. Manhas, A.K. Bose, Microwave-assisted rapid and simplified hydrogenation, J. Org. Chem. 64 (1999) 5746–5753.

[142] S. Dhole, M. Selvaraju, B. Maiti, K. Chanda, C.-M. Sun, Microwave controlled reductive cyclization: a selective synthesis of novel benzimidazole-alkyloxypyrrolo[1,2-a] quinoxalinones, ACS Comb. Sci. 17 (2015) 310–316.

Chapter 7

Microwave-assisted enzymatic reactions

Aparna Das* and Bimal Krishna Banik*
Department of Mathematics and Natural Sciences, College of Sciences and Human Studies, Prince Mohammad Bin Fahd University, Al Khobar, Kingdom of Saudi Arabia
Corresponding authors: E-mails: aparnadasam@gmail.com (Aparna Das); bimalbanik10@gmail.com, bbanik@pmu.edu.sa (Bimal Krishna Banik)

7.1 Introduction

Microwave irradiation and enzymatic catalysis synergistically raise the reaction rate significantly. Several reports have explained the biodegradation of toxic organic pollutants using different enzymes from bacteria, fungi, and plants. Bioremediation is a cost-effective and environmentally friendly biotechnology that is controlled by microbial enzymes. The studies in this area contribute to reduce the toxicity of the pollutants and also to obtain novel useful products. Thus, microwave technology, together with enzymatic catalysis and solvent-free chemical synthesis, is a nature-friendly method with low wastage of solvent and good yield of the products. This chapter provides information on the enzymes from various microorganisms involved in the biochemical processes of a wide range of compounds assisted by microwave irradiation.

7.2 Lipases

Lipase (triacylglycerol acylhydrolases) is an enzyme and it catalyzes the hydrolysis of fats (lipids) [1]. Lipase is included in the esterase family. Basically, lipases hydrolyze fats into their components of fatty acid and glycerol molecules. Lipases are found in the blood, pancreatic secretions, gastric juices, intestinal juices, and adipose tissues. There are different types of lipases and each has individual functions. Types of lipases include hepatic lipases (in the liver), hormone-sensitive lipases (in adipocytes), lipoprotein lipase (in the vascular endothelial surface), and pancreatic lipase (in the small intestine). The crystal structure of human gastric lipase is shown in Fig. 1. Lipase is used as a potent biocatalyst owing to its properties like high specificity, biodegradability, nontoxic nature, high catalytic efficiency, pH dependency, activity in organic

Microwaves in Chemistry Applications. https://doi.org/10.1016/B978-0-12-822895-1.00009-6

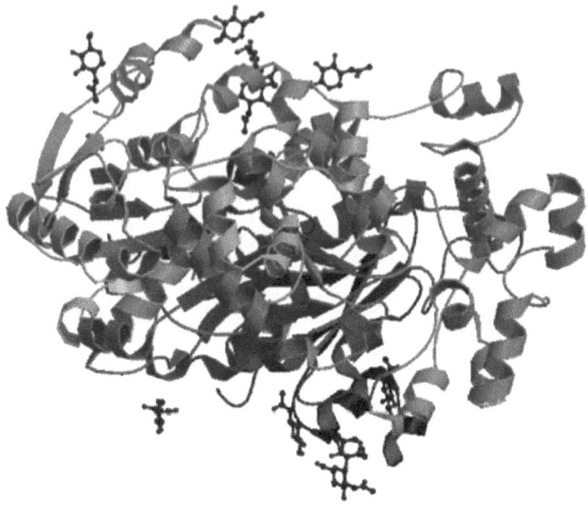

FIG. 1 Crystal structure of human gastric lipase. PDB ID: 1HLG [2].

solvents, low product inhibition, high yield in a nonaqueous medium, low reaction time, resistance to altered temperature, pH, alcohol, and reusability. Besides, lipases are playing a vital role in a wide range of industrial applications, such as in detergents, food, leather, textiles, cosmetics, paper, and pharmaceuticals [3–6].

Lipases can do a major role in the synthesis of organic compounds, polymer synthesis [7], functionalization reaction, transformations reaction [8, 9], hydrolysis [10], epoxidation and oxidations reaction [11, 12], esterification reaction [13–15] and transesterification reaction [16, 17].

Studies demonstrated that the microwave effect can enhance reactions such as hydrolysis, esterification, and transesterification using lipases [4, 18–28]. For example, Bradoo et al. reported microwave-assisted triolein hydrolysis and esterification of sucrose/methanol/ascorbic acid with different fatty acids using various commercial lipases from porcine pancreas, *Mucor miehei*, *Candida rugosa*, *Pseudomonas cepacia* as well as lipases from laboratory isolates (*Bacillus stearothermophilus* and *Burkholderia cepacia*). The reaction rate increased significantly compared to conventional methods. The hydrolysis rates increased by 7–12-fold and esterification were achieved within 30 s [29]. Similarly, in the hydrolysis of triolein using *Aspergillus carneus* lipase, both under normal methods and microwave-assisted methods at low (175 W) and high (800 W) power levels were carried out. The reaction took 24 h for complete hydrolysis of triolein under normal conditions and 160 s was enough for the completion of the reaction at the low power microwave level and 75 s at the high power level [4]. Thus, the microwave-assisted methods dynamically accelerated the enzyme catalytic reactions without negatively affecting the enzyme properties such as stability and selectivity [4, 29]. Moreover, several other studies confirmed that enzyme activity and product yields are markedly higher under microwave heating than classical thermal heating methods [19, 21–25, 27, 28]. Most of the studies indicated that microwaves can exert nonthermal effects [28, 30].

Modafinil (2-[(diphenylmethyl)sulfinyl]acetamide) is a CNS stimulant drug. It is successfully used for the treatment of hypersomnia and narcolepsy with fewer side effects [31]. Several methods are available for the synthesis of modafinil [32]. One of the key intermediates in the biotransformation process for the modafinil synthesis is *n*-butyl diphenyl methyl mercapto acetate (BDMMA). Yadav et al. synthesized diphenyl methyl mercapto butyrate by employing different lipases in the presence of microwave irradiation [33]. The esterification reaction of diphenyl methyl mercapto acid (DPMMA), with *n*-butanol to prepare *n*-butyl diphenyl methyl mercapto acetate (BDMMA), in the presence of immobilized lipases is shown in Scheme 1. Several reaction parameters such as catalyst loading, the mole ratio of reactants, reaction temperature, different lipases (Novozym 435, Lipozyme RM IM, and Lipozyme TL IM), water concentration, and reusability of catalysts under the microwave irradiation were considered to optimize the reaction conditions. Among the lipases, Novozym 435 was found to be the most active catalyst in the presence of microwave irradiation with toluene as the solvent and provided the conversion of 34% in 24 h at 60°C. And, the enzyme was reusable after the synthesis.

Starch or amylum is one of the most abundant polysaccharides found in nature. It comprises numerous glucose units joined by glycosidic bonds. Mainly it consists of two polymeric units: amylose and amylopectin, which are present in specific ratios depending on the source. Starch is used for multifarious applications. However, the native starch has many shortcomings for industrial applications [34, 35]. Therefore, it is important to modify starch to overcome these hurdles. One of the important and predominant ways of modification of starch is by esterification. The esterification process involves the formation of starch esters by acylation of hydroxyl groups of anhydrous glucose units in starch.

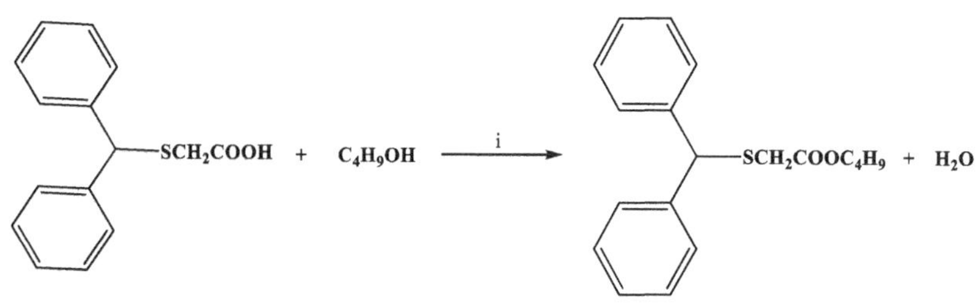

1 2 3

SCHEME 1 Microwave-assisted enzyme catalysis for synthesis of *n*-butyl diphenyl methyl mercapto acetate in nonaqueous media.

A green approach for starch modification was reported by Adak et al. [36]. In this study, esterification of cornstarch was performed by employing *Rhizopus oryzae* lipase, microwave irradiation, and novel imidazolium surfactants. Different reaction conditions that are affecting the process were analyzed for the esterification of cornstarch with oleic acid. The maximum degree of substitution (DS = 2.7) was obtained at 80% irradiation, 1:3 starch/oleic acid molar ratio, 150 IU enzyme, and 50 μmol of $[C_{16}\text{-}3\text{-}C_{16}im]Br_2$. Fig. 2 shows the scanning electron microscopic (SEM) images of the starch before and after esterification. The microwave irradiation and the imidazolium surfactants had enhanced the esterification process. Compared to esterification by conventional heating, high DS was observed in the microwave treatment owing to be the result of higher collision efficiency and the faster diffusion rates due to molecular vibrations along with dielectric heating effect [37, 38]. The modified starch had shown significant properties including better water resistance and oil holding capacity. These additional properties of modified starch may have potential applications in the food and biopolymer sectors.

Lukasiewicz et al. reported another method for starch esterification employing both biocatalysis and low power microwaves [39]. A series of starch esters were obtained using a microwave-assisted enzymatic transformation of starch. In this protocol, hog pancreas lipase was used as a biocatalyst as well as followed acyl donors: acetic, lauric, and stearic acid. DMF and DMSO were used as reaction media (Scheme 2). Low power-level microwave irradiation was used and compared with conventional heating of the reaction mixture. The results showed that the microwave-assisted reaction in DMF and lauric acid as an acyl donor gave the highest degree of substitution (0.5). And a significant reduction in the reaction time (around 2.5 times) for microwave-assisted reactions was observed. The protocol was useful for obtaining starch ester with higher efficiency in comparison with those known as conventional.

Ketorolac [(±)-5-benzoyl-1,2-dihydro-3*H*-pyrrolo[1,2-a]pyrrole-1-carboxylic acid] is a nonsteroidal antiinflammatory and analgesic drug, having one stereogenic carbon center [40]. Studies showed that a single-dose efficacy of ketorolac is greater than that of many other antiinflammatory drugs including aspirin, indomethacin, naproxen, ibuprofen, meperidine, morphine, and pentazocine [41, 42]. Therefore, it has gained significant therapeutic importance. Among the enantiomer of ketorolac, the (−)- or (*S*)-form is therapeutically active while (*R*)-isomer is inactive and shows adverse side effects [43]. Therefore, the pure synthesis of (*S*)-enantiomers is important. Different methods were reported for the stereoselective synthesis of (*S*)-enantiomers including hydrolysis in aqueous media, or esterification in organic media catalyzed by enzymes or by microorganisms [43–46]. However, enzymatic biotransformation needs a longer reaction time to achieve the desired enantiomeric excess form of ketorolac.

Microwave-assisted biocatalytic transformation is an important and green process for chiral drug resolution. Shinde et al. reported the synergistic effect of microwave irradiation in the kinetic resolution of RS-(±)-ketorolac via esterification using a lipase enzyme [47]. Under microwave conditions, different

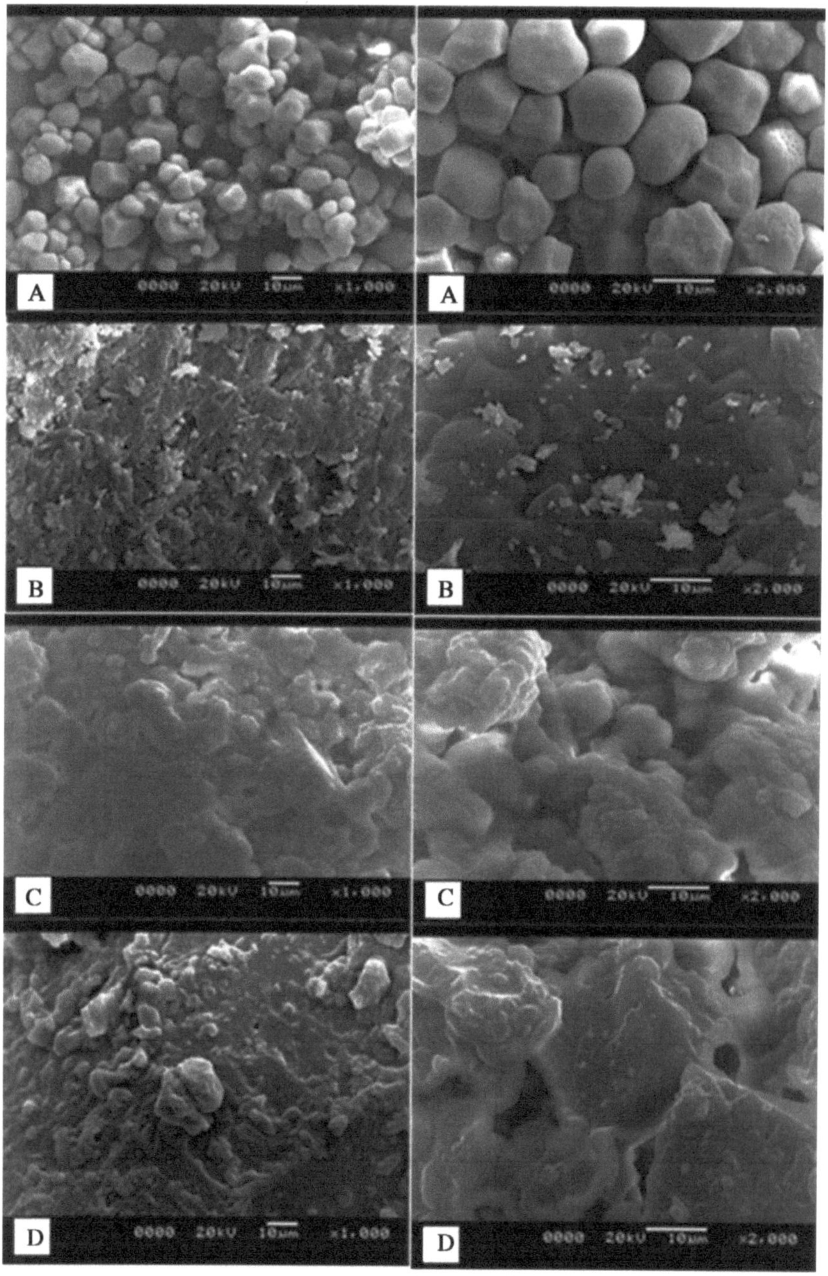

FIG. 2 SEM images of starch samples (A) maize starch, (B) control starch, (C) starch ester (1.8 DS), and (D) starch ester (2.7 DS). *(Reproduced with permission from S. Adak, R. Banerjee, A green approach for starch modification: esterification by lipase and novel imidazolium surfactant. Carbohydr. Polym. 150 (2016) 359–368.)*

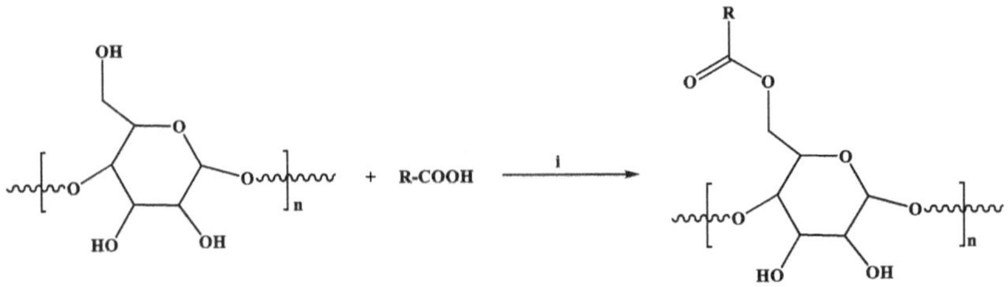

4 5 6

R=CH_3, $C_{11}H_{23}$, $C_{17}H_{35}$

SCHEME 2 Starch esterification process. Reagents and conditions: (i) Hog pancreas lipase, DMF/DMSO, MW.

immobilized enzymes, namely Novozym 435, Lipozyme TL IM, Lipozyme RM IM, Lipase Amano AS, and Lipase AYS amino, were tested for the kinetic resolution of RS-(±)-ketorolac. Effects of different parameters such as the speed of agitation, various solvents, temperature, and the concentration of substrate and catalyst were considered during the study. Under microwave conditions, an increase in initial rates up to 1.5-fold was observed compared to conventional methods. Compared to other lipases, Novozym 435 effectively catalyzed the enantioselective esterification of RS-(±)-ketorolac (Scheme 3). A conversion of 50% and enantiomeric excess >99% was obtained in 3 h at 50°C and 300 rpm. The results showed that the reaction followed the ping-pong bi-bi mechanism with inhibition by *n*-octanol.

Cinnamates are the esters of cinnamic acid and widely found in plants. Cinnamate esters have found wide-ranging applications in the food and pharmaceutical industry owing to their antioxidant, flavor, and fragrance properties. Compared to chemical synthesis, enzymatic synthesis of natural, semisynthetic, and synthetic cinnamate esters exhibit numerous advantages.

Microwave-assisted lipase-catalyzed synthesis of geranyl cinnamate was reported by Shinde et al. [48]. The transesterification reaction of ethyl cinnamate with geraniol under microwave irradiation for the synthesis of geranyl cinnamate is shown in Scheme 4.

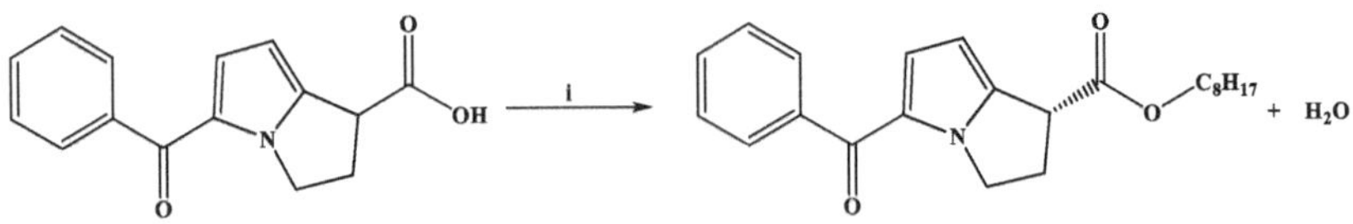

7 8

SCHEME 3 Microwave-assisted Novozym 435 catalyzed the kinetic resolution of ketorolac. Reagents and conditions: (i) Novozym 435, *n*-octanol, MW.

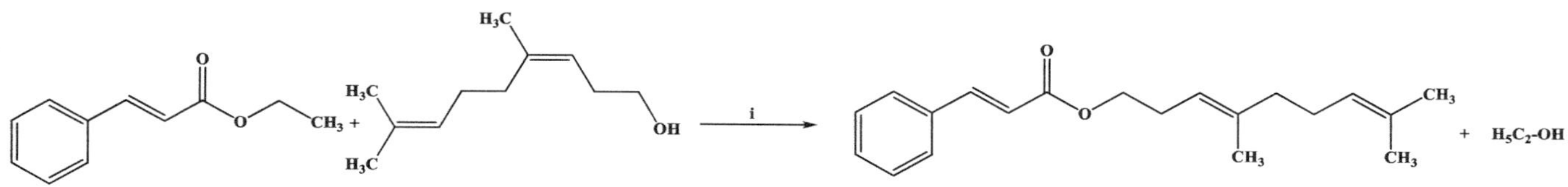

9 10 11

SCHEME 4 Microwave-assisted Novozym 435 catalyzed transesterification of ethyl cinnamate with geraniol. Reagents and conditions: (i) Novozym 435, MW.

The effects of different parameters such as biocatalyst, solvent, and temperature were analyzed in detail. Three lipases, Lipozyme RM IM (*Rhizomucor miehei*), Lipozyme TL IM (*Thermomyces lanuginosus*), and Novozym 435 (*Candida antarctica* lipase B), were screened for the study. The enzyme Novozym 435 showed the maximum conversion among all the studied enzymes; Novozym 435 (83%), Lipozyme RMIM (8%), and insignificant conversion were obtained with Lipozyme TL IM. Compared to conventional heating, an increase in initial rates up to 4.2-fold was observed under microwave irradiation. The optimal conditions obtained were as follows: enzyme loading, 60 mg; temperature, 65°C; the speed of agitation, 300 rpm; and substrate mole ratio, 1:2.

Triolein is a symmetrical triglyceride. It is also known as glyceryl trioleate and derived from glycerol and three units of the unsaturated fatty acid oleic acid. Triolein is one of the two components of Lorenzo's oil [49]. It is the main content (4%–30%) of olive oil [50]. Triolein acts as a plant metabolite and a Caenorhabditis elegans metabolite. Hydrolysis of triglyceride is an important and convenient method for the synthesis of fatty acid and glycerol. These products are used in a wide range of applications such as in nutrients, pharmaceuticals, and emulsifiers. Several researchers had investigated the hydrolysis of triolein. Among those methods, enzymatic hydrolysis is a useful method for the generation of fatty acids [4, 51–53]. The proposed mechanism for the triolein hydrolysis by *Candida rugosa* lipase in the biphasic oil-water system is shown in Fig. 3 [54]. Enhancement in the reaction rate is observed with microwave irradiation.

Chena et al. investigated the effect of microwave heating on the lipase-catalyzed hydrolysis of triolein [55]. The study was aimed to explore whether the reaction rate enhancements are purely due to thermal/heating effects or

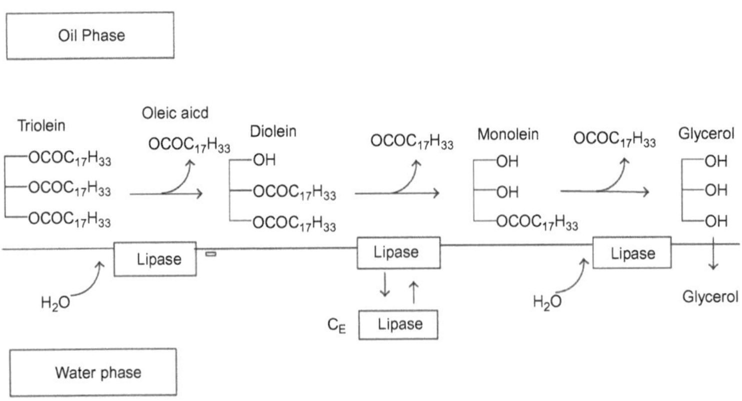

FIG. 3 Triolein hydrolysis by *Candida rugosa* lipase in the biphasic oil-water system, the proposed mechanism. *(Reproduced with permission from C.S. Karigar, S.S. Rao, Role of microbial enzymes in the bioremediation of pollutants: a review. Enzyme Res. 2011 (2011).)*

nonthermal effects. Hence, both the microwave-assisted method and conventional heating method were considered. Quantitative mass spectrometry was used to conduct an accurate kinetic analysis of lipase-catalyzed hydrolysis of triolein. For the analyses, commercial lipases from *Candida rugosa* (CRL), *Burkholderia cepacia* (BCL), and *porcine pancreas* were considered. No significant difference in the yield was observed between microwave and conventional heating conditions when the hydrolysis reactions were carried out at the same temperature. The study showed that CRL had optimum catalytic activity at 37°C, while the other two enzymes had better activities at 50°C. The results also depicted that a nonthermal effect does not exist in microwave-assisted lipase hydrolysis of triolein. Hence, conventional heating at higher temperatures, for example, at 50°C, can be also used to accelerate hydrolysis reactions.

n-Butyl propionate is a moderately fast evaporating solvent and it has a wide range of applications as precursors, food, fruit aroma, and also as a solvent for resin, nitrocellulose, paint, shellac, slow drying agent of banana water, and varnish. *n*-Butyl propionate is mainly synthesized by the esterification of *n*-butanol and propionic acid using catalyst like sulfuric acid. The acid-catalyzed reactions have their own limitation and other problems related to the yield of product, equipment erosion, and pollution of sulfuric acid waste [56]. Researchers had made efforts to synthesize *n*-butyl propionate by using microwave technology, enzymatic catalysis, and solvent-free conditions, as this synthesis is environment-friendly with low wastage of solvent and good yield of the product.

Synthesis of *n*-butyl propionate by esterification of propionic acid with *n*-butanol using immobilized lipase as catalyst under microwave irradiation was reported [57]. The reaction is shown in Scheme 5.

Different commercially available lipases such as Novozym 435, *Thermomyces lanuginosus* (Lipozyme-TLIM), and *Rhizomucor miehei* (Lipozyme-RMIM) were considered and tested for esterification reaction in the solvent-free synthesis of *n*-butyl propionate under the microwave. Among them, Novozym 435 exhibited the highest activity with a conversion of 92% in 8 h. The conversion efficiencies of Lipozyme-TLIM and Lipozyme-RMIM were 9% and 14%, respectively, after 8 h owing to steric hindrance toward substrate binding at active sites of the enzyme. But in the case of Novozym 435, the microenvironment around the active site pocket eased the interaction of substrate even in nonaqueous solvent-free conditions. Hence, conversion of 92% was achieved with

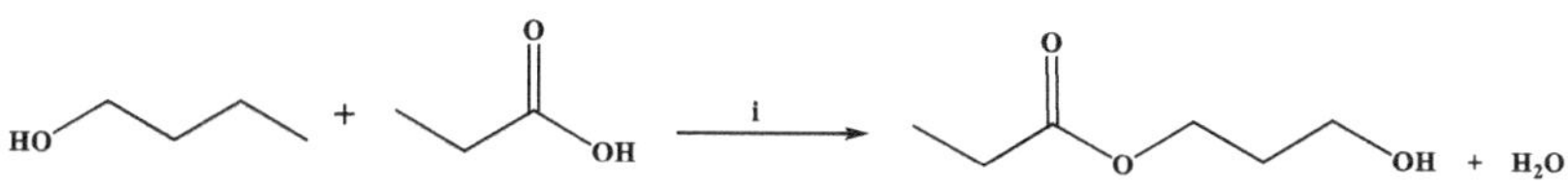

12 13 14

SCHEME 5 Microwave-assisted esterification reaction of *n*-butanol with propionic acid for the synthesis of *n*-butyl propionate by using the lipase catalyst. Reagents and conditions: (i) lipase, MW, 60°C, 50 W.

1:12 mol ratio of propionic acid to *n*-butanol and 120 mg catalyst at 60°C and 50 W of microwave power. The same reaction under conventional heating at 60°C with Novozym 435 afforded 72% conversion in 8 h. The studies on Lineweaver-Burk plots, to understand the mechanism and kinetics of the reaction, revealed that the reaction follows a ternary complex mechanism with inhibition by *n*-butanol.

Isoamyl butyrate also known as isopentyl butanoate is the butanoate ester of isoamylol. It belongs to the class of organic compounds known as fatty acid esters, and it has a role as a metabolite. It also serves as a flavoring agent, for example, in the preparation of a variety of fruit juice flavors, such as apricots, bananas, pears, and apples. And as a flavor and fragrance compounds, it is widely used in the food, cosmetic, beverage, and pharmaceutical industries. It is mainly synthesized by the lipase-catalyzed esterification of isoamyl alcohol and butyrate. It is also possible to carry out the enzymatic synthesis of isoamyl butyrate under microwave irradiation.

Bansode et al. investigated the impact of microwave irradiation on the synthesis of isoamyl butyrate ester in the presence of Novozym 435 as a biocatalyst in a solvent-free system [58]. Various reaction parameters were considered to optimize the reaction under the solvent-free condition to achieve maximum yield. The higher conversion (95%) was obtained at 60°C, 1:2 mole ratio of butyric acid: isoamyl alcohol, 2% (w/v) of lipase loading with 700 W microwave power, in 120 min. The thermodynamic parameters of the reaction were also analyzed and compared with the synthesis executed in ultrasonic and conventional systems. The enhanced reaction rate was observed in the microwave method, due to the lower activation energy with the microwave system. Thus, the lipase-catalyzed solvent-free esterification was intensified using microwave energy by 5- and 1.5-fold compared to conventional stirring reaction technique and ultrasonic system, respectively.

Ethyl valerate (ethyl pentanoate) is an organic compound used in flavors. It also possesses other applications such as biofuels, food, cosmetics, and pharmaceutical industries. Different synthetic routes are available for the production of ethyl valerate. An enzymatic route using valeric acid and ethanol with commercially available lipases, under the solvent-free condition and microwave irradiation was also examined [59]. Synthesis using three lipases, namely Lipozyme RM IM, Lipozyme TL IM, and Novozym 435 (*Candida antarctica* lipase B) was analyzed. A better result was obtained with the Novozym 435. A conversion of 69.2% of valeric acid was obtained at 50°C using the equimolar ratio of valeric acid and ethanol at an enzyme loading of 10.06 g/L under microwave irradiation in 40 min (Scheme 6). The other lipases, Lipozyme-RMIM and Lipozyme-TLIM, afforded only 17.2% and 6.3% conversion, respectively.

Biodiesel is a renewable, biodegradable fuel derived from plants or animals. It basically consists of long-chain fatty acid esters. Biodiesel is typically made by chemically reacting lipids such as animal fat or vegetable oil with alcohol that produces a methyl, ethyl, or propyl ester.

15 **16** **17**

SCHEME 6 Lipase-catalyzed esterification reaction of valeric acid with ethanol to produce ethyl valerate under solvent-free condition and microwave irradiation. Reagents and conditions: (i) Novozym 435, MW, 50°C, 50 W.

Conversion of *M. circinelloides* lipid into biodiesel by enzymatic synthesis assisted by microwave irradiations was explored [60]. Lipase, Novozym 435 was used as a catalyst and microwave irradiations were used to execute both the lipids extraction and transesterification heating. Three different approaches were considered for the synthesis of biodiesel. In two-step method, which combined the lipids extraction from microbial biomass followed by enzymatic transesterification reaction both carried out under microwave irradiation, produced ethyl esters (98.5%) in 10 h. The same reaction took 30 h under conventional heating. So, the microwave-assisted enzymatic reactions are a potential procedure for biodiesel production.

7.3 Cellulase

Cellulase comprises enzymes that catalyze cellulolysis, the decomposition of cellulose, and some related polysaccharides. Cellulase mostly hydrolyzes β-1,4-glucoside linkages of the cellulose chain. Cellulase is produced mainly by fungi, bacteria, and protozoans. Several kinds of cellulases are known that differ structurally and mechanistically. Fig. 4 shows the crystal structure of the cellulase Cel9M of *Clostridium cellulolyticum* bacteria. Cellulases are widely used in the textile industry, laundry detergents, food processing, brewery and wine, animal feed, pulp and paper industries, pharmaceutical applications, and agriculture. The microwave effect on cellulose activities is discussed later.

Enhancement in the catalytic efficiency of Cellulase by using microwave pretreatment of their substrates was reported [62]. Microwave pretreatment of carboxymethylcellulose was found to improve the catalytic efficiency of cellulase by 1.6-fold. Scanning electron microscopy images showed significant morphological changes in the substrate because of microwave pretreatment. The time of enzymatic hydrolysis depicted that the use of microwave pretreated substrates gave higher catalytic rates. Besides, a higher degree of bioconversion was found when the microwave pretreated substrates were used.

Li et al. reported enhanced cellulase production of the *Trichoderma viride* mutated by microwave and ultraviolet [63]. During the study, the cellulase-producing fungi *Trichoderma viride* were cultured and fermented on the solid-state wheat bran medium. After the compound mutagenesis by microwave

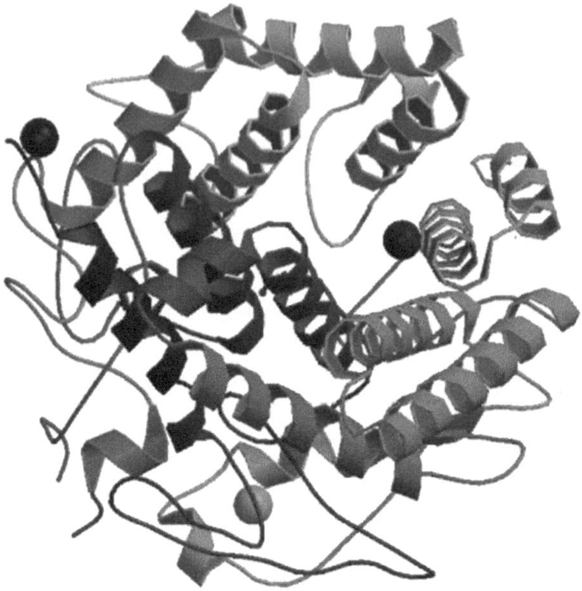

FIG. 4 Crystal structure of the cellulase Cel9M of *Clostridium cellulolyticum*. PDB ID: 1IA6 [61].

and ultraviolet, a significantly stronger ability to produce enzymes than the normal wild type was observed. Along with that, they were also very stable for a long period of up to nine generations to produce cellulose.

Usually, agricultural residues, wheat bran, and rice hulls were used as substrates for cellulase production with *Trichoderma* sp. by solid-state fermentation [64]. Microwave irradiation was employed to pretreat the substrates to increase the susceptibility. The highest cellulase yield was observed by the substrates pretreated by microwave at a power of 450 W for 3 min. It was also observed that the pretreatment time and microwave power have no significant effect on cellulase production. At the beginning of the treatment, the reducing sugar content (RSC) of substrates was decreased by microwave irradiation. However, more reducing sugars were produced in later fermentation. Alkali pretreatment combined with microwave pretreatment (APCMP) can make significant changes. For instance, APCMP of rice hulls significantly increased the cellulase yields and reducing sugar. The data revealed that maximum filter paper activity, carboxymethylcellulose, and RSC were increased by 35.2%, 21.4%, and 13%, respectively, in comparison with untreated rice hulls. Also, the fermented residues made more cellulase enzymes and reducing sugars than fresh rice hulls after they were treated with APCMP.

The Astragalus polysaccharide (APS) is an important water-soluble heteropolysaccharide having bioactive effects. This can be extracted from *Astragali Radix*, the dry root of *Astragalus membranaceus*. The pharmacological properties of Astragalus polysaccharide include antiaging, immunomodulation, antitumor, antioxidant, antiinflammatory, antiviral, and cardiovascular protection activities [65–67]. The main components of APS are glucose, rhamnose, galactose, arabinose, xylose, mannose, glucuronic acid, and galacturonic acid. Extraction of APS is important, as its chemical composition and structure play an important role in pharmacological activities. Numerous studies have reported different extraction methods of APS [68–76].

The microwave-assisted hydrolysis of cellulose for the extraction of Astragalus polysaccharides was also investigated [77]. The optimized enzymatic-microwave extraction conditions obtained were as follows: the liquid-to-solid ratio, 10:1; enzyme ratio, 57.6 U/g; and cellulase reaction time, 60 min, followed by microwave irradiation for 8 min (480 W). The maximum extraction rate (16%) and the purity (88%) were obtained under these conditions, which were higher than those achieved by other extraction methods. So the enzymatic-microwave-assisted method is a faster and efficient extraction method for extracting the Astragalus polysaccharides from *Astragali Radix*.

Lignocellulosic materials are the most abundant biopolymer in nature and are promising alternative feedstocks for bioethanol production. Examples of lignocellulosic include agricultural by-products, agricultural residues, and dedicated energy crops [78]. However, efficient pretreatment steps are needed to improve the yield of fermentable sugars and maximizing the enzymatic hydrolysis efficiency owing to the recalcitrant nature of lignocellulosic biomass. Microwave-assisted pretreatment is a good option as it can reduce the pretreatment time and improve the enzymatic activity during hydrolysis.

Nomanbhay et al. reported the microwave-assisted alkaline pretreatment and microwave-assisted enzymatic saccharification of oil palm empty fruit bunch fiber for enhanced fermentable sugar yield [79]. Empty fruit bunch is a complex lignocellulosic material and the major components are cellulose (41%–46%), hemicelluloses (25%–33%), and lignin (27%–32%) [80]. The result showed that microwave-assisted alkali pretreatment of empty fruit bunch using NaOH significantly improved the enzymatic saccharification of empty fruit bunch by removing more lignin and hemicellulose and increased its accessibility to hydrolytic enzymes. It was observed that the most effective way of utilizing microwave radiation as a pretreatment process was at lower power levels in combination with dilute NaOH. The optimum pretreatment condition was obtained at 3% (w/v) NaOH, 180 W for 12 min.

At this optimum condition, the component loss of lignin and holocellulose was about 74% and 24.5%, respectively (Fig. 5). The enzymatic saccharification of empty fruit bunch resulted in 411 mg of reducing sugar per gram EFB at a cellulase enzyme dosage of 20 FPU assisted by microwave. The overall enhancement in pretreatment and microwave-assisted enzymatic hydrolysis was fivefold (Fig. 6). The effect of microwave irradiation strongly depends on the power of the microwave, as the higher power levels of a microwave cause the denaturation of the enzyme. The study confirmed the importance of a well-controlled microwave-assisted enzymatic reaction to enhance the overall reaction rate of the enzymatic hydrolysis of lignocellulosic material.

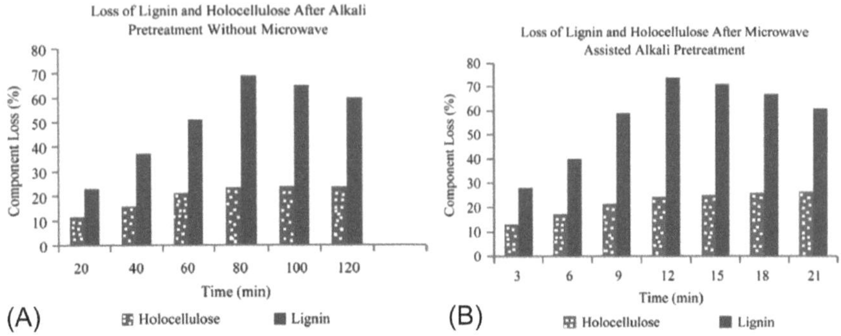

FIG. 5 Lignin and holocellulose reduction (A) caused by conventional heating at 50°C with 3% NaOH pretreatment of EFB, (B) caused by microwave-alkaline (3% NaOH) pretreatment of EFB. *(Reproduced with permission from S.M. Nomanbhay, R. Hussain, K. Palanisamy, Microwave-assisted alkaline pretreatment and microwave assisted enzymatic saccharification of oil palm empty fruit bunch fiber for enhanced fermentable sugar yield. J. Sustain. Bioenergy Syst. 3 (2013) 7–17.)*

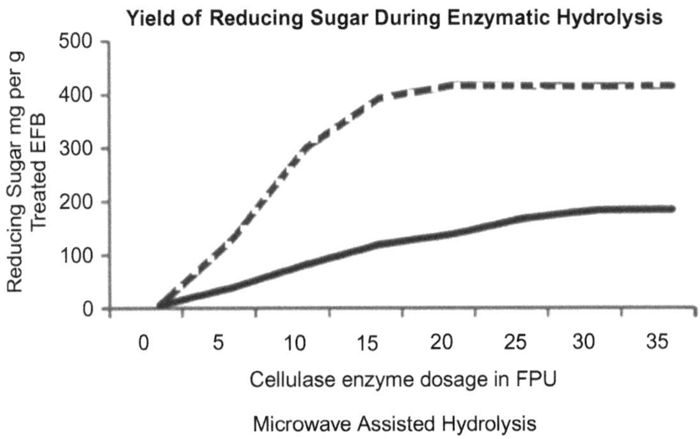

FIG. 6 Reducing sugar yield at various cellulose enzyme dosages for microwave-assisted hydrolysis and nonmicrowave hydrolysis. *(Reproduced with permission from S.M. Nomanbhay, R. Hussain, K. Palanisamy, Microwave-assisted alkaline pretreatment and microwave assisted enzymatic saccharification of oil palm empty fruit bunch fiber for enhanced fermentable sugar yield. J. Sustain. Bioenergy Syst. 3 (2013) 7–17.)*

The microwave-assisted pretreatment method was also adopted to treat rice straw as an alternative to conventional pretreatment methods for the production of cellulosic ethanol [81]. The influences of acid concentration, solid-liquid ratio, microwave intensity, irradiating time, and catalyst concentration on the removal ratio of lignin were investigated in detail. The optimal conditions of pretreatment were obtained at 25% acid concentration, 1:15 solid-liquid ratio, 230 W microwave intensity, and 5 min irradiating time. The removal ratios of lignin were 46% and 51% and the sugar yields were 71% and 80% when acetic acid and propionic acid were used as solvents, respectively, at the optimum condition. The results indicated that microwave-assisted organic acid pretreatment is an effective method in improving the enzymatic hydrolysis sugar yield of rice straw.

Eichhornia crassipes, commonly known as water hyacinth, is one of the worst weeds. As waste biomass, water hyacinth can potentially replace fossil fuels and handle urgent issues of energy deficit and serious environmental pollution. Several studies were reported for effectively converting water hyacinth biomass into clean renewable ethanol, hydrogen, and methane biofuels. Xia et al. reported the enhanced enzymatic saccharification of water hyacinth through microwave heating with dilute acid pretreatment for biomass energy utilization [82]. This microwave-assisted pretreatment method produced remarkable effects in promoting enzymatic saccharification of water hyacinth. The pretreatment resulted in hydrolysis of hemicellulose into monosaccharide and at the same time in the release of a large amount of cellulose from the degraded lignocellulose matrix to react with cellulase. Because of that, glucose yield increased after enzymatic hydrolysis. The highest reducing sugar yield achieved after enzymatic hydrolysis using cellulase was 48 g/100 g hyacinth and 20 g/L water hyacinth feedstock was pretreated with microwave at 140° C for 15 min with 1% H_2SO_4, for obtaining the mentioned yield.

7.4 Glycosidases

Glycosidases (glycoside hydrolases or glycosyl hydrolases) are important enzymes of carbohydrate metabolism. They are mostly responsible for in vivo degradation of oligo- and polysaccharides. Glycosidases normally catalyze the hydrolysis of glycosidic bonds in carbohydrates and related molecules. They are involved in several important processes such as in degradation of biomass such as cellulose, hemicellulose, and starch, in antibacterial defense strategies, in pathogenesis mechanisms, intestinal digestion, posttranslational modification of glycoproteins, lysosomal catabolism of glycoconjugates, and in normal cellular function. Fig. 7 shows the crystal structure of *Spodoptera frugiperda* β-glycosidase.

Pujic et al. reported the glycosidase-catalyzed reversed hydrolysis and transglycosidation in dry media under focused microwave irradiation (Scheme 7) [84].

Oligosaccharides and their derivatives play an important role in many biochemical reactions and are used in wide range of applications such as in therapeutics, diagnostic tools, cosmetics, and food industry [85]. Galacto-oligosaccharides stimulate the growth of Bifidobacteria in the human body, [86–88]. It is also known as Bifidus growth factors. In humans, the intestinal Bifidobacteria are useful to the maintenance of a normal intestinal balance, to the improvement of lactose tolerance and digestibility of milk products, for antitumorigenic activity, for reduction of serum cholesterol levels, to the synthesis of B-complex vitamins, and enhanced absorption of dietary calcium [89]. For this reason, the synthesis of galactooligosaccharides is important.

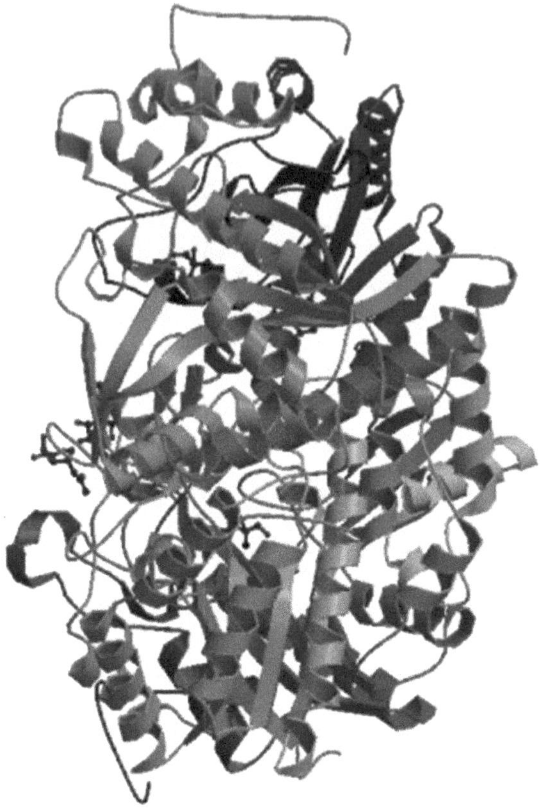

FIG. 7 Crystal structure of *Spodoptera frugiperda* β-glycosidase. PDB ID: 5CG0 [83].

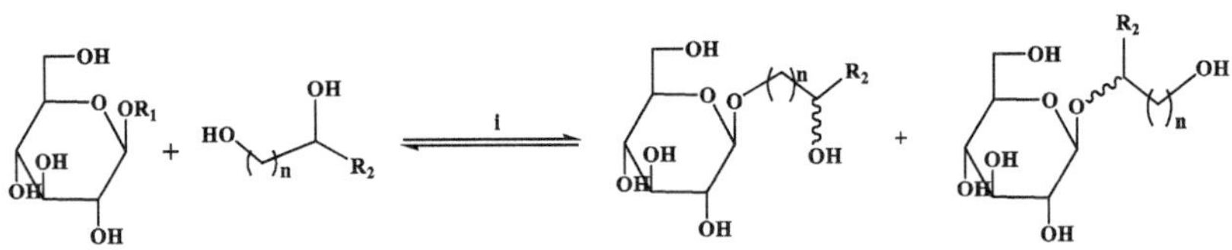

18 19 20 21

R1=H, Ph, 4-OGlc; R2=H, CH3.

SCHEME 7 The influence of microwave on transglycosylation reactions catalyzed by supported glycosidases. Reagents and conditions: (i) glycosidase support, MW.

Different chemical synthesis methods are available for the synthesis of oligosaccharides [90–94]. But the chemical synthesis of oligosaccharides is usually complicated. It produces unwanted enantiomers, requires long protection-deprotection steps for selectivity control, and normally gives low yields [95]. Enzyme-assisted methods of synthesis of oligosaccharides and their derivatives are promising alternatives.

Galacto-oligosaccharides were synthesized from lactose by immobilized β-galactosidase from *Kluyveromyces lactis* using either focused microwave irradiation [96]. Galacto-oligosaccharides synthesis was performed under microwave irradiation (12 W, 40°C) in phosphate buffer (50 mM, pH 6.5) supplemented with lactose (159 g/L), $MgCl_2$ (0.01 M), and immobilized β-galactosidase (80 U/mL) (Scheme 8). The reactions were executed until lactose depletion and stopped by heating at 100°C for 10 min. Galacto-oligosaccharides selectivity was increased by decreasing the water activity of media, increasing the initial lactose concentration, and using cosolvents in the media. The selectivity for galacto-oligosaccharides synthesis was increased 217-fold by exposing immobilized enzyme to microwave irradiation and with the addition of cosolvents such as hexanol, compared to the conventional method (in water and with free enzyme).

The synthesis of nucleotide-activated oligosaccharides by transglycosylation with β-galactosidase from *Bacillus circulans* under microwave irradiation was reported [97]. The schematic of the synthesis is shown in Scheme 9.

Glycosylation is the process by which saccharides are covalently attached to proteins and lipids or other organic molecules. It is an enzyme-directed site-specific process. Both the glycosylation and deglycosylation process play an important role such as in protein folding [98], protein stabilization [99], protein, and cellular trafficking [100], protease protection [101], and quaternary structure [102]. It also has significant effects on receptor binding, inflammation, and recovery from many diseases [103–106].

A standard and widely used method for deglycosylation is overnight digestion with their respective deglycosylating enzymes/chemicals. Several published techniques are available for the improvement of this method, such as optimization of conditions using anhydrous trifluoromethane sulfonic acid [107], on-chip deglycosylation using hydrophobic and hydrophilic chip technology [108], PVDF-immobilization of a glycosylated protein followed by

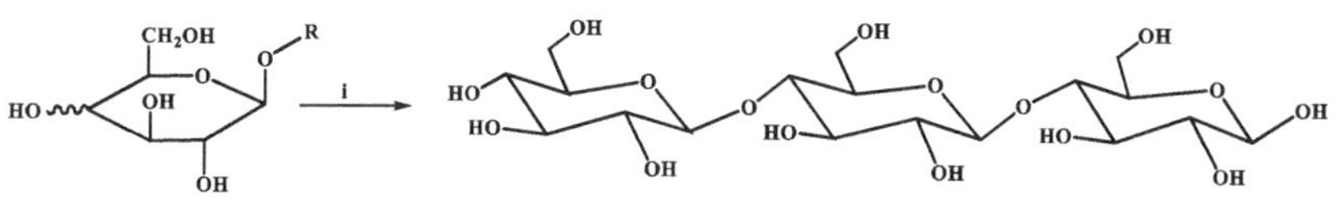

22

23

SCHEME 8 Microwave influence on the transglycosylation yield of lactose with a supported galactosidase. Reagents and conditions: (i) β-galactosidase, MW.

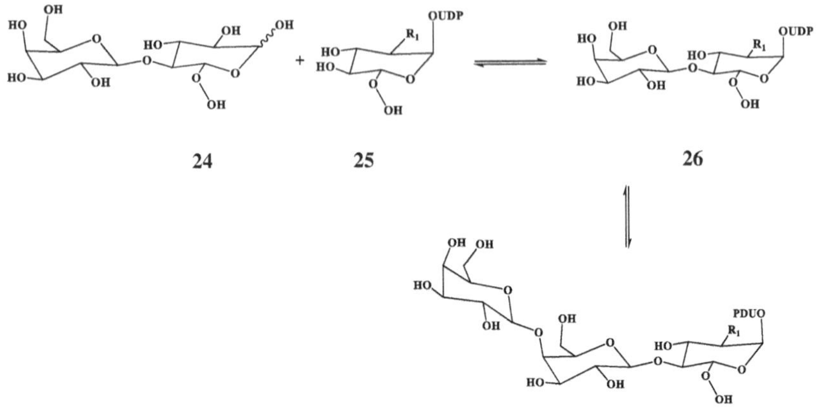

SCHEME 9 Kinetically controlled transglycosylation reactions for the synthesis of UDP-Lac (NAc) by β-galactosidase from *Bacillus circulans*.

incubation with a deglycosylating enzyme [109], incubation of glycoproteins with enzyme peptide *N*-glycosidase F (PNGase F) in the presence of enzyme-friendly surfactants [110], and engineering of hybrid de-glycosylation enzymes for facile immobilization on cellulose [111].

Sandoval et al. reported a method for the accelerated removal of N-linked oligosaccharides using PNGase F assisted by microwave irradiation [112]. Complete deglycosylation was achieved, without compromising the integrity of protein samples, in less than 30 min for most proteins, which was significantly faster than conventional deglycosylation protocols. This method was tested on a variety of glycoproteins, including antibodies, at the microgram level, and no detrimental effects such as poor recovery, deamidation, or cleavage at the protein backbone were observed.

7.5 Protease

Proteases (peptidases or proteinases) are enzymes that catalyze proteolysis, the breakdown of the peptide bonds of proteins. Proteases are divided into acid, neutral, and alkaline proteases. Protease enzymes are widely found in nearly all plants, animals, and microorganisms. The important animal proteases include pancreatic trypsin, pepsin, chymotrypsin, and renin. Proteases are involved in many biological activities, such as digestion of ingested proteins, protein catabolism, cell signaling, and so on [113, 114]. Fig. 8 shows the crystal structure of human uropepsin.

Normal proteolytic digestion methods involve several hours or overnight incubations. Researchers had reported several methods of improving digestion efficiency and reducing incubation time, including thermal denaturation [116], chemical denaturation [117, 118], using immobilized enzymes [119], and using

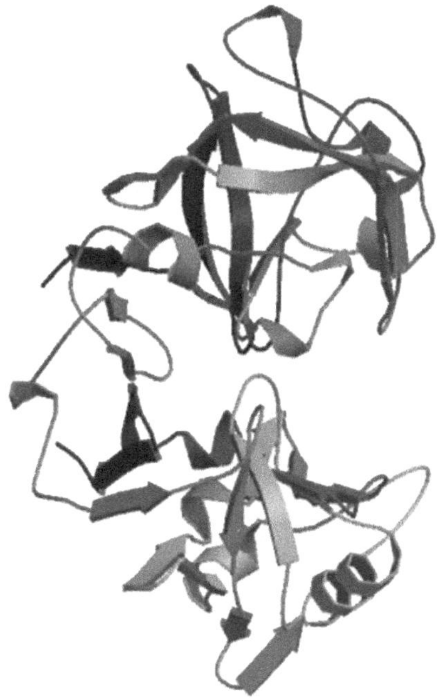

FIG. 8 Crystal structure of human uropepsin. PDB ID: 1FLH [115].

microwave irradiation [120]. In recent years, more studies were performed on microwave-assisted proteolytic digestion procedures, using tryptic [121–123] and acid-mediated proteolysis [124–127].

Because of the incomplete hydrolysis reactions between enzymes and substrates, the microwave-assisted proteolytic digestion often produces misscleaved peptides. Reddy et al. investigated the factors that affect microwave-assisted trypsin-catalyzed proteolytic digestion processes [128]. Optimum conditions for digestion were at 50 mM Tris buffer, 30 min, 60°C, and enzyme-to-protein molar ratio of 1:5. All protein molecules in a sample digested into peptides within a few minutes under microwave irradiation at optimum conditions. The results indicated that the factors that affect the number of misscleaved peptides and incomplete digestion percentages are the digestion temperature, reaction time, enzyme to substrate ratio, and digestion buffer.

Roy et al. investigated the microwave irradiation effects on protease-catalyzed esterification and transesterification reactions [26]. Two proteases, subtilisin, and chymotrypsin were chosen and explored the effect of microwaves at fixed temperatures on the rates of esterification [129] and transesterification [130] reactions catalyzed by these enzymes. To evaluate the nonthermal effects of microwave irradiation on enzyme-catalyzed reactions, the reaction temperature was kept constant during irradiation. And also solvents of differing polarities and at three different temperatures were considered for the study. The results depicted that, in all instances, microwave irradiation was observed to increase the initial reaction rates by 2.1–4.7 times at all hydration levels.

β-Lactoglobulin (LGB, β-Lg) is the major whey protein of bovine milk (~3 g/L) and is also present in many other mammalian species except in humans. Thus, in humans, this protein seems to be resistant to gastric digestion and remains intact after passing through the stomach [131]. Studies showed that β-lactoglobulin is one of the main allergens in bovine milk [132–135].

Enzyme-assisted proteolysis and conventional heat methods are the normally used methods for the production of partially or extensively hydrolyzed formulas to lower the content of β-lactoglobulin and other intact proteins.

Microwave-assisted digestion of β-lactoglobulin by pronase, α-chymotrypsin, and pepsin was reported by Izquierdo [136]. The effects of microwave irradiation on kinetic parameters for pronase, α-chymotrypsin, and pepsin hydrolysis of bovine β-lactoglobulin were evaluated in detail. The experiments were performed under microwave irradiation at 40°C. The results evidenced that the microwave irradiation treatment applied during the enzyme reaction enhanced the catalytic efficiency for pronase and chymotrypsin, mainly because of the reduction in Km, which indicates the highest substrate-enzyme affinity. Apart from that, pepsin showed very low activity on β-lactoglobulin at pH 4.0 regardless of the heating methods used.

Microwave treatment enhanced the enzymatic rates of hydrolysis of pure β-lactoglobulin and bovine whey proteins as compared to conventional heating [137]. The high extent of proteolysis of β-lactoglobulin and WPI by trypsin and chymotrypsin under microwave did not affect the immunoreactivity of digested proteins. At the same time, the application of microwave treatment at 200 W elevates the hydrolysis of β-lactoglobulin by pepsin in 3 min and reduces significantly its immune reactivity.

A fast and efficient proteolysis approach of microwave-assisted protein digestion using trypsin-immobilized magnetic silica microspheres was reported by Lin et al. [138]. The immobilization of the enzyme onto magnetic silica microspheres was attained through a one-step reaction with 3-glycidoxypropyltrimethoxysilane that provides the epoxy group as a reactive spacer. This protein digestion method aggregated the advantages of immobilized trypsin and microwave-assisted digestion, which ensued in high digestion efficiency. Bovine serum albumin and myoglobin were used to optimize the reaction conditions. The analysis showed that peptide fragments were produced in 15 s and equivalent or better digestion efficiency was observed compared to current in-solution digestion. Due to the unique magnetic responsivity, the immobilized trypsin can be easily isolated and reusable.

The influence of microwave radiation on trypsin activity using a CEM Cool-Mate apparatus at a constant bulk temperature was investigated [139]. Digestion of Nα(±)-benzoyl-D/L-arginine-4-nitroanilide hydrochloride, azocasein, and casein catalyzed by trypsin from the bovine pancreas were considered for the study. Through an external cooling system, the bulk temperature was maintained constant during the microwave radiation. Trypsin enzymatic activity was observed to be accelerated significantly when the reaction mixture was irradiated with microwave at a constant bulk temperature.

7.6 Baker's yeast

Yeast is a single-celled fungus, and hundreds of species are recognized in that family. *Saccharomyces cerevisiae* is one of the most renowned species of yeast and commonly known as baker's or brewer's yeast. Fig. 9 shows the crystal structure of ATP sulfurylase from *S. cerevisiae*. Baker's yeast is widely used for several biotechnological applications. Baker's yeast is available in different forms, depending on the moisture contents: cream yeast, compressed yeast, active dry yeast, instant yeast, rapid-rise yeast, deactivated yeast. Baker's yeast can be used in organic synthesis. It is widely used as a biocatalyst in various reactions such as oxidation, hydrogenation or reduction, condensation, hydrolysis, and cyclization. For example, it can reduce a carbonyl group into a hydroxyl group in reasonably high yield. During various reactions, the baker's yeast is used in two forms, immobilized form, and free form. Baker's yeast is

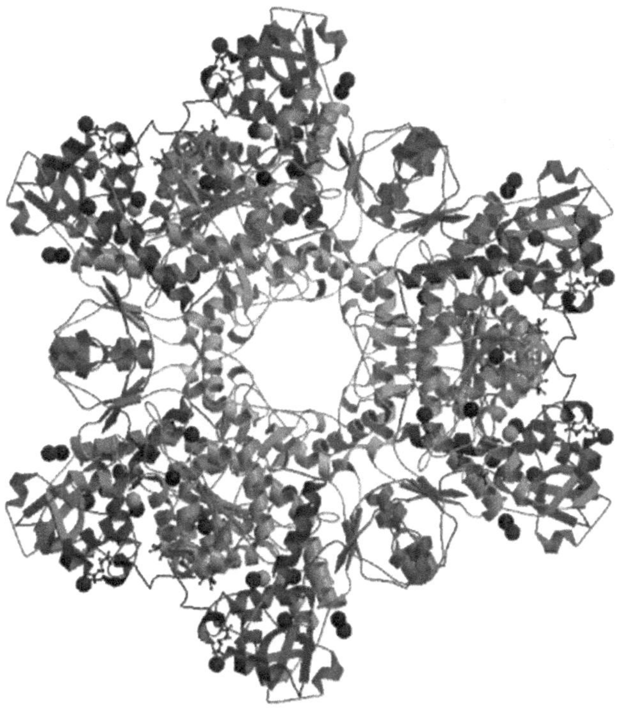

FIG. 9 Crystal structure of ATP sulfurylase from *Saccharomyces cerevisiae*. PDB ID: 1G8F [140].

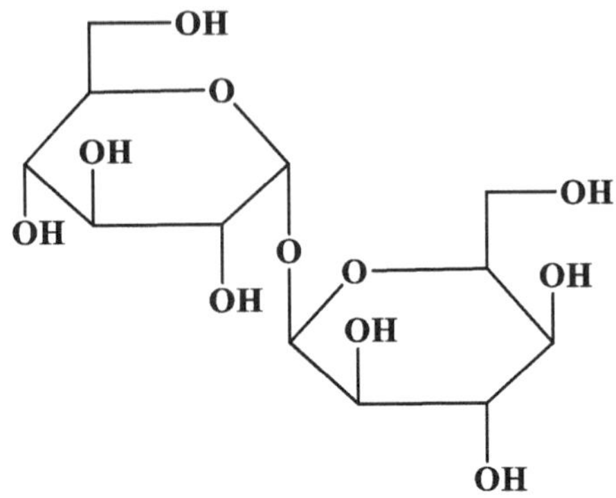

FIG. 10 Molecular structure of trehalose.

also useful to produce ethanol via fermentation. The following section describes the changes in the activity of baker's yeast due to microwave effects.

Trehalose (α-d-glucopyranosyl α-d-glucopyranoside) is a nonreducing disaccharide. It consists of two alpha, alpha-linked glucose moieties (Fig. 10). This glucose derivative is considered to function as storehouses of glucose. Trehalose is found in a variety of organisms including bacteria, fungi, yeast, insects, and plants, and absent in vertebrates.

Other than reserve carbohydrate, this diverse molecule has many other roles including protecting proteins and membranes against stresses such as desiccation, high osmolarity, frost, and heat; acting as an allosteric inhibitor of carbohydrate metabolism and a transcriptional regulator; and serving as a crucial component of mycobacterial cell walls and a donor of mycolic acid. Studies showed that trehalose is also used to preserve various unstable products such as enzymes, foods, pharmaceuticals, and cosmetics during the dry stage and is turning a preservative of practical importance [141]. Even though five known biosynthesis routes are available for the synthesis of trehalose, only three are more common [142]. The first methods involve the condensation of UDP-glucose and glucose-6-phosphate by the glycosyltransferase trehalose-6-phosphate synthase to afford trehalose-6-phosphate and the hydrolysis of the phosphoric acid ester by trehalose-6-phosphate phosphatase. The second route involves the rearrangement of maltose to trehalose catalyzed by the transglycosidase, trehalose synthase. The third pathway involves the isomerization of the terminal residue of maltooligosaccharides at the reducing end to maltooligosaccharyltrehalose by the action of maltooligosaccharyltrehalose synthase and hydrolysis of this product by maltooligosaccharyltrehalose trehalohydrolase to yield the product trehalose. Although most organisms possess only one of these biological pathways, mycobacteria and corynebacteria own all three.

For large-scale production, trehalose is mainly prepared from yeast (*S. cerevisiae*). Studies pointed out that the extraction of trehalose from the baker's yeast was affected by the activity of trehalase, which is a hydrolytic enzyme of trehalose in yeast cells [143]. Thus, inactivation of trehalase before extraction is necessary. Yoshikawa et al. demonstrated that activities of trehalase can be inhibited by adjusting the ethanol concentration, pH, and extraction temperature [143]. A high extraction ratio of trehalose is also achievable from thermally treated yeast [144]. Chuanbin et al. attempted to inactivate trehalase by treating fresh *S. cerevisiae* in the microwave field [145]. The study showed that after only 60 s of microwave treatment, yeast cell was disrupted and the activity of the trehalase enzyme was completely lost. It was practicable to extract trehalose easily and rapidly from microwave treated yeast by water in 10 min at room temperature, as trehalase was inactivated and yeast cell was lysed. Microwave treatment is appeared to be an effective pretreatment method for trehalose extraction from yeast. This microwave treatment is also applicable to other materials for intracellular product extraction.

Effect of 905 MHz microwave radiation on colony growth of the yeast Saccharomyces cerevisiae strains (FF18733, FF1481, and D7) was reported [146]. The study showed that three yeast strains show different patterns of colony growth after 15-, 30-, and 60-min exposure to a mobile phone radiofrequency of 905 MHz at SAR 0.12 W/kg. For instance, strains FF1481 (DNA repair mutant) and D7 (relatively genetically unstable) manifested an increased sensitivity to microwave radiation in comparison with strain FF18733 (wild-type). The analysis pointed out that pulsed radio frequency/microwave radiation at a low SAR level could cause DNA damage in yeast cells.

Fatty acid methyl esters (FAME) are a type of fatty acid ester. Normally, the transesterification of fats with methanol in the presence of a base such as sodium hydroxide, sodium methoxide, or potassium hydroxide produces FAME [147]. It commonly is known as biodiesel and is a renewable and environmental-friendly source of energy [148]. It can be derived from vegetable oils, animal fats, or recycled greases by transesterification. The main reasons for use of FAME in biodiesel instead of free fatty acids are to neutralize any corrosion that free fatty acids would induce the metals of engines, production facilities, and so on. FAMEs generally have almost 12–15 units higher cetane number than their unesterified counterparts [149].

The most commonly used method for the synthesis of FAMEs is acid-catalyzed transesterifications [150–154]. Even though the preparation of FAMEs using this method delivers good results, the reaction time to complete the conversion of lipids to FAMEs using the conventional heating method is very time-consuming, usually 1–4 h [155–157]. Studies had been made to prepare FAMEs from fatty acids using microwave-assisted methods. And various types of biological samples such as edible oil, bacterial cells, animal tissue, and plasma samples were used for the preparation [158–162]. However, there has been no attempt to establish a proper method for FAME preparation in yeast samples.

Khoomrung et al. established a proper, fast, and accurate method for the preparation of FAMEs using microwave-assisted derivatization of fatty acids

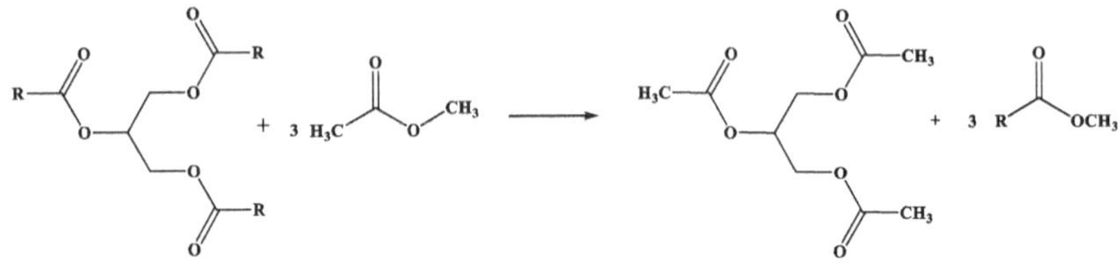

28 29 30 31

SCHEME 10 Synthesis of FAME and triacetin by interesterification of triglycerides with methyl acetate under strongly acidic conditions.

present in yeast (*S. cerevisiae*) samples [163]. By employing this microwave-assisted method, esterification of free/bound fatty acids to FAMEs was completed within 5 min, which is 24 times faster compared to conventional heating methods (Scheme 10). To investigate the yields of formation of FAMEs, the developed method was validated with a conventional method (hot plate) and with the standard reference material (SRM) 3275-2 omega-3 and omega-6 fatty acids in fish oil, and no significant differences were observed in yields of FAMEs with both validations.

Effects of nonthermal microwave exposures on the biological activity (proliferation rate) of *S. cerevisiae* yeast were investigated [164]. Yeast strains type II (Sigma) were exposed to the microwaves at 900 MHz with selected powers of 13, 3, and −7 dBm using the transverse electromagnetic (TEM) cell. The average specific absorption rate (SAR) observed for a single cell was 0.12 W/kg, and during the experiments, yeast cells were continuously exposed to microwave radiation.

The analysis showed that microwave radiation at 900 MHz with a power of −7 dBm induced no effect on yeast growth (Fig. 11B). Nevertheless, microwave radiation at 900 MHz with powers of 13 dB (Fig. 11A) and 3 dB affected significantly (14% and 11%, respectively) the biological activity of yeast cells. These data indicated that the external nonthermal microwave radiation at the selected frequency and powers can modulate the proliferation of the yeast cells.

Studies were also performed to modify magnetically the Baker's yeast (*S. cerevisiae*) cells with magnetic iron oxide particles prepared by microwave irradiation of iron(II) sulfate at high pH [165]. Owing to the presence of active intracellular invertase, both the noncross-linked and glutaraldehyde cross-linked magnetic cells enabled efficient and fast sucrose conversion into glucose and fructose. Because of the magnetic properties, separation from the reaction mixture was easier and simple. The prepared magnetic whole-cell biocatalyst was observed stable even after 1-month storage at 4°C and was reusable. Simple magnetic separation and stability of the developed biocatalyst enabled its reusability without significant loss of enzyme activity.

Microwave-assisted methods are available for fast screening of yeast lipids in *S. cerevisiae* [166]. Microwave-assisted methods allowed the extraction of

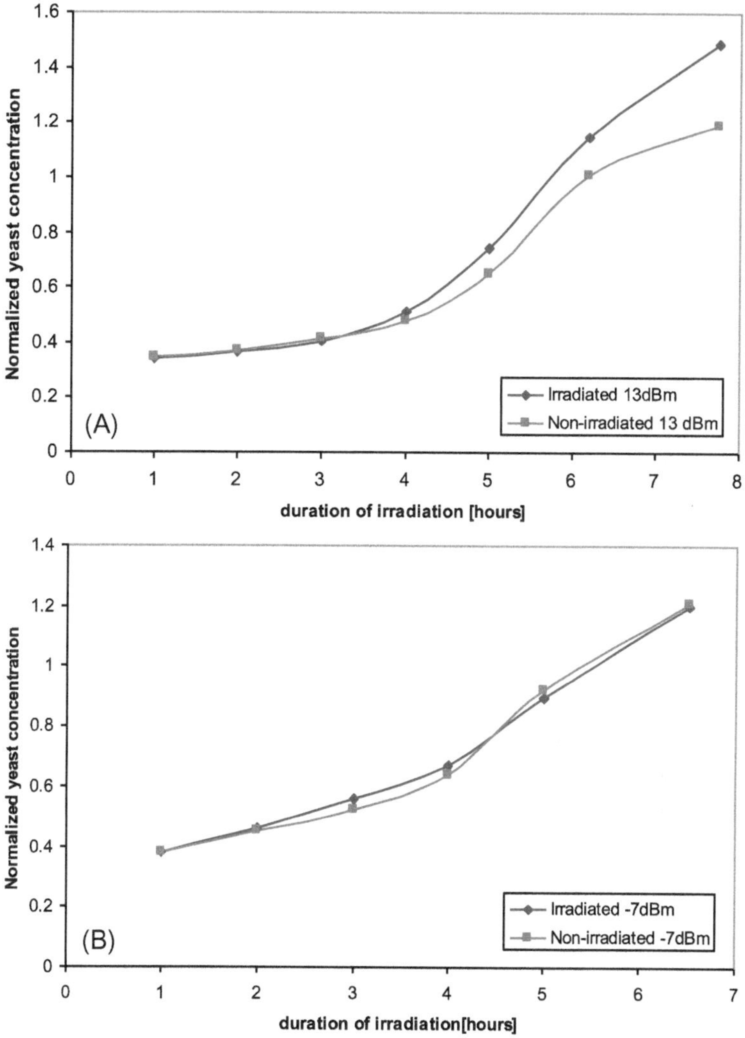

FIG. 11 Changes in the normalized concentration of yeast cells in time (nonirradiated vs irradiated) at 900 MHz and the powers of (A) 13 dBm, (B) −7 dBm. *(Reproduced with permission from H. Alsuhaim, V. Vojisavljevic, E. Pirogova, World congress on medical physics and biomedical engineering, effects of non-thermal microwave exposures on the proliferation rate of Saccharomyces cerevisiae yeast. IFMBE Proc. (2012) 48–51.)*

lipids within 10 min and enhanced the extraction rate by 27 times while maintaining product yields comparable to conventional methods. Furthermore, the microwave-assisted extraction method combines cell disruption and extraction in one step, and the approach, consequently, not only greatly simplifies sample handling but also scales down analysis time and minimizes sample loss during sample preparation.

Juice pasteurization using conventional method causes adverse effects on nutrients and nutritional value of food because of high temperature. Several studies were conducted by researchers to overcome this issue. One of the accomplishable methods is the use of microwave-assisted methods instead of conventional heating methods.

Samani et al. examined the effect of microwave output power, temperature, ultrasound power, and ultrasonic exposure time on *S. cerevisiae* in orange juice [167]. The investigation revealed that microwave output power, ultrasound power, ultrasonic exposure time, and orange juice temperature were the most effective factors to reduce *S. cerevisiae*. The quadratic model demonstrated that concerning the decrease of *S. cerevisiae,* the microwave-induced temperature was more effective than the microwave output power. The optimum processing condition was 350 W microwave power, 35°C temperature, and 11 min of exposure.

Microwave-assisted immobilization of yeast in cellulose biopolymer was practicable [168]. As an example, the baker's yeast (Saccharomyces cerevisiae) was immobilized by the microwave-assisted method in glutaraldehyde cross-linked cellulose and used as a novel and green adsorbent for the removal of toxic chromium Cr(VI). In microwave-assisted preparation, 200 s was enough for the incorporation of yeast in the biopolymer matrix. The analysis evidenced that yeast immobilized biopolymer acted as a good host to accommodate the Cr (VI) oxyanion and the functional groups present in the microbe and cellulose play a dominant role in the interaction with the hexavalent chromium. Using sodium hydroxide, the regeneration of yeast-impregnated cellulose adsorbent was achievable.

The effect of continuous 2.45 GHz microwave radiation on glucose uptake by baker's yeast cells under constant temperature is important to investigate as it can give a clear picture of microwave effects on cell cultures and living organisms [169]. A *S. cerevisiae* suspension of 2.9×10^8 cells/mL was used for the investigation under well-defined experimental conditions. In detail, all the investigated samples during glucose uptake were kept at well-defined bulk temperature (within 1°C) with the same energy flow through the system. The statistical analysis of data revealed insignificant changes in glucose uptake during the initial stages of the experiment (up to 10 min), in comparison with MW effects with control experiments. But statistically remarkable differences were observed during the next 20 min of the irradiation corresponding to thermal overheating of 2°C. The results showed that microwave radiation behaved as a slight nonchemical stressor, which may influence cell states. The consequences could be attributed to specific thermal microwave effects caused by either a local increase of temperature inside cells or a large flow of energy through the system. Thus, a longer exposure of cells to microwave radiation or any other electromagnetic irradiation may have impacts on biochemical applications and production of valuable biotechnological products.

Regarding the microwave interactions with living cells, two distinct actions were identified. In destructive action, microwave inactivates the microorganisms [170–173], and in nondestructive action, microwave can affect both functionality and development of living cells [146, 174–176]. The most important difference between these two effects is given by the microwave used to power, expressed as specific absorption ratio (SAR). Many researchers had reported the mechanism of action of microwaves on living cells [177–181]. The effect of microwave on living cells is observed normally at low temperatures and in almost adiabatic environmental conditions; hence, the microwaves are used mostly at very low power densities (10^{-19}–10^{-3} W/cm^2). Researchers also had described in great detail the nonthermal effects of low-intensity microwaves on living systems [182, 183] and numerous parameters that affect the nonthermal microwave effect on living cells [181].

Bioethanol is a biofuel, and it is made by fermentation from glucose feedstocks. Studies are going on to increase the rate of the reaction. Calinescu et al. attempted to identify the domain for microwave specific absorption ratio (SAR) that leads to an increase in fermentation rate of *S. cerevisiae* without affecting its viability [184]. For this purpose, an experimental setup was installed which allows for glucose fermentation to be implemented in the presence of microwave, with more effective control of both SAR and the temperature of the fermentative medium. In all the analyses, a positive effect of microwave exposure on yeast metabolic activity was detected, and ethanol concentrations obtained through microwave-assisted fermentation were higher than those from conventional processes. The study presented an optimum SAR domain, between 15 and 25 W/kg, assuring a significant effect on glucose fermentation by *S. cerevisiae*, with an increase in fermentation rates of up to 20%. Electron and optical microscopy studies on cell growth and viability revealed that microwave irradiation favors ethanol production of the cells, against their growth. Temperature-dependent calculations and analysis confirm that the positive effect of microwave exposure on the fermentation process is not due to the increase of the temperature of the yeast cells, it might be due to an increase of the cell membrane permeability and/or the activity of the enzymes catalyzing the reactions of the cell's metabolic pathways. However, further investigations are needed to clarify these complex processes.

7.7 Conclusion

In this chapter, we have discussed the importance of microwave technology in enzymatic reactions. Studies have showed that the activity, stability, and selectivity of the enzyme can be improved by microwave heating. However, the use of microwave irradiation in enzymatic synthesis remains still limited due to the high temperatures associated with microwave heating as enzymes are temperature-sensitive macromolecules. Several studies are going with new technology, by maintaining the temperature as low as 40°C and by precise power inputs.

Acknowledgments

AD and BKB are grateful to Prince Mohammad Bin Fahd University for support. BKB is also grateful to US NIH, US NCI, and Kleberg Foundation of Texas for financial support.

References

[1] A. Svendsen, Lipase protein engineering, Biochim. Biophys. Acta 1543 (2000) 223–228.

[2] A. Roussel, S. Canaan, M.P. Egloff, M. Riviere, L. Dupuis, R. Verger, C. Cambillau, Crystal structure of human gastric lipase and model of lysosomal acid lipase, two lipolytic enzymes of medical interest, J. Biol. Chem. 274 (1999) 16995–17002.

[3] I. Roy, M.N. Gupta, Applications of microwaves in biological sciences, Curr. Sci. 85 (2003) 1685.

[4] R.K. Saxena, J. Isar, S. Saran, R. Kaushik, W.S. Davidson, Efficient microwave-assisted hydrolysis of triolein and synthesis of bioester, bio-surfactant and glycerides using Aspergillus carneus lipase, Curr. Sci. 89 (2005) 1000–1003.

[5] A. Houde, A. Kademi, D. Leblanc, Lipases and their industrial applications, Appl. Biochem. Biotechnol. 118 (2004) 155.

[6] K.-E. Jaeger, M.T. Reetz, Microbial lipases form versatile tools for biotechnology, Trends Biotechnol. 16 (1998) 396.

[7] T.D. Matos, N. King, L. Simmons, C. Walker, A.R. McClain, A. Mahapatro, F.J. Rispoli, K.T. McDonnell, V. Shah, Microwave assisted lipase catalyzed solvent-free poly-ε-caprolactone synthesis, Green, Chem. Lett. Rev. 4 (2011) 73–79.

[8] G.D. Yadav, I.V. Borkar, Lipase-catalyzed hydrazinolysis of phenyl benzoate: kinetic modeling approach, Process Biochem. 45 (2010) 586–592.

[9] G.D. Yadav, I.V. Borkar, Synthesis of n-butyl acetamide over immobilized lipase, J. Chem. Technol. Biotechnol. 84 (2009) 420–426.

[10] G.D. Yadav, K.M. Devi, Kinetics of hydrolysis of tetrahydrofurfuryl butyrate in a three phase system containing immobilized lipase from Candida Antarctica, Biochem. Eng. J. 17 (2004) 57–63.

[11] C. Sanfilippo, N. D'Antona, G. Nicolosi, Chloroperoxidase from Caldariomyces fumago is active in the presence of an ionic liquid as co-solvent, Biotechnol. Lett. 26 (2004) 1815–1819.

[12] G.D. Yadav, K.M. Devi, A kinetic model for the enzyme-catalyzed self epoxidation of oleic acid, J. Am. Oil Chem. Soc. 78 (2001) 347–351.

[13] A. Buthe, T. Recker, M. Heinemann, W. Hartmeier, J. Büchs, M.B.A. Schumacher, pH-optima in lipase-catalysed esterification, Biocatal. Biotransformation 23 (2005) 307–314.

[14] L.N. Mutua, C.C. Akoh, Synthesis of alkyl glycoside fatty acid esters in non-aqueous media by Candida sp. Lipase, J. Am. Oil Chem. Soc. 70 (1993) 43–46.

[15] G.D. Yadav, I.V. Borkar, Kinetic modeling of immobilized lipase catalysis in synthesis of n-butyl levulinate, Ind. Eng. Chem. Res. 47 (2008) 3358–3363.

[16] A. Bajaj, P. Lohan, P.N. Jha, R. Mehrotra, Biodiesel production through lipase catalyzed transesterification: an overview, J. Mol. Catal. B Enzym. 62 (2010) 9–14.

[17] G.D. Yadav, S.R. Jadhav, Synthesis of reusable lipases by immobilization on hexagonal mesoporous silica and encapsulation in calcium alginate: transesterification in non-aqueous medium, Microporous Mesoporous Mater. 86 (2005) 215–222.

[18] I.C.R. Costa, S.G.F. Leite, I.C.R. Leal, L.S.M. Miranda, R. de Souza, Thermal effect on the microwave assisted biodiesel synthesis catalyzed by lipases, J. Braz. Chem. Soc. 22 (2011) 1993.

[19] G. Lin, W.-Y. Lin, Microwave-promoted lipase-catalyzed reactions, Tetrahedron Lett. 39 (1998) 4333.

[20] R.O.M.A. de Souza, O.A.C. Antunes, W. Kroutil, C.O. Kappe, Kinetic resolution ofrac-1-phenylethanol with immobilized lipases: a critical comparison of microwave and conventional heating protocols, J. Org. Chem. 74 (2009) 6157.

[21] M.C. Parker, T. Besson, S. Lamare, M.D. Legoy, Microwave radiation can increase the rate of enzyme-catalysed reactions in organic media, Tetrahedron Lett. 37 (1996) 8383.

[22] B. Rejasse, T. Besson, M.D. Legoy, S. Lamare, Influence of microwave radiation on free Candida antarctica lipase B activity and stability, Org. Biomol. Chem. 4 (2006) 3703.

[23] N.E. Leadbeater, L.M. Stencel, E.C. Wood, Probing the effects of microwave irradiation on enzyme-catalysed organic transformations: the case of lipase-catalysed transesterification reactions, Org. Biomol. Chem. 5 (2007) 1052.

[24] B. Rejasse, S. Lamare, M.-D. Legoy, T. Besson, Stability improvement of immobilized Candida antarctica lipase B in an organic medium under microwave radiation, Org. Biomol. Chem. 2 (2004) 1086.

[25] M. Vacek, M. Zarevúcka, Z. Wimmer, K. Stránský, K. Demnerová, M.-D. Legoy, Selective enzymic esterification of free fatty acids with n-butanol under microwave irradiation and under classical heating, Biotechnol. Lett. 22 (2000) 1565.

[26] I. Roy, M.N. Gupta, Non-thermal effects of microwaves on protease-catalyzed esterification and transesterification, Tetrahedron 59 (2003) 5431–5436.

[27] G.D. Yadav, I.V. Borkar, Kinetic and mechanistic investigation of microwave-assisted lipase catalyzed synthesis of citronellyl acetate, Ind. Eng. Chem. Res. 48 (2008) 7915.

[28] B. Rejasse, S. Lamare, M.-D. Legoy, T. Besson, Influence of microwave irradiation on enzymatic properties: applications in enzyme chemistry, J. Enzyme Inhib. Med. Chem. 22 (2007) 519.

[29] S. Bradoo, P. Rathi, R.K. Saxena, R. Gupta, Microwave-assisted rapid characterization of lipase selectivities, J. Biochem. Biophys. Methods 51 (2002) 115.

[30] M.A. Herrero, J.M. Kremsner, C.O. Kappe, Nonthermal microwave effects revisited: on the importance of internal temperature monitoring and agitation in microwave chemistry, J. Org. Chem. 73 (2008) 36.

[31] G. Carrea, S. Riva, Properties and synthetic applications of enzymes in organic solvents, Angew. Chem. Int. Ed. 39 (2000) 2226–2254.

[32] V. Gotor, R. Brieva, C. Gonzalez, F. Rebolledo, Enzymatic aminolysis and transamidation reactions, Tetrahedron 47 (1991) 9207–9214.

[33] G.D. Yadav, P.S. Lathi, Microwave assisted enzyme catalysis for synthesis of n-butyl dipheyl methyl mercapto acetate in non-aqueous media, Clean Techn. Environ. Policy 9 (2007) 281–287.

[34] K.A. Abbas, S.K. Khalil, A.S.M. Hussin, Modified starches and their usages in selected food products: a review study, J. Agric. Sci. 2 (2010) 90–100.

[35] S.X. Xie, Q. Liu, S.W. Cui, Starch modification and application, in: S.W. Cui (Ed.), Food Carbohydrates: Chemistry, Physical Properties, and Applications, CRC Press, Taylor & Francis Group, Boca Raton, FL, 2005, pp. 357–406.

[36] S. Adak, R. Banerjee, A green approach for starch modification: esterification by lipase and novel imidazolium surfactant, Carbohydr. Polym. 150 (2016) 359–368.

[37] L. Perreux, A. Loupy, A tentative rationalization of microwave effects inorganic synthesis according to the reaction medium, and mechanistic considerations, Tetrahedron 57 (2001) 9199–9223.

[38] V. Singh, A. Tiwari, Microwave-accelerated methylation of starch, Carbohydr. Res. 343 (2008) 15–154.

[39] M. Lukasiewicz, S. Kowalski, Low power microwave-assisted enzymatic esterification of starch, Starch/Stärke 64 (2012) 188–197.

[40] Y.H. Kim, C.S. Cheong, S.H. Lee, K.S. Kim, Enzymatic kinetic resolution of ketorolac, Tetrahedron Asymmetry 12 (2001) 1865–1869.
[41] B.Y. Kim, H.J. Doh, T.N. Le, W.J. Cho, C.S. Yong, H.G. Choi, J.S. Kim, C.-H. Lee, D.-D. Kim, Ketorolac amide prodrugs for transdermal delivery: stability and in vitro rat skin permeation studies, Int. J. Pharm. 293 (2005) 193–202.
[42] D.R. Brocks, F. Jamali, Clinical pharmacokinetics of ketorolac tromethamine, Clin. Pharmacokinet. 23 (1992) 415–427.
[43] J. Guo, Y. Zhuang, L. Chen, J. Liu, D. Li, N. Ye, Process optimization for microwave-assisted direct liquefaction of Sargassum polycystum C. Agardh using response surface methodology, Bioresour. Technol. 120 (2012) 19–25.
[44] G. Fulling, C.J. Sih, Enzymatic second-order asymmetric hydrolysis of ketorolac esters: in situ racemization, J. Am. Chem. Soc. 109 (1987) 2846–2848.
[45] A. Palomer, M. Cabrg, J. Ginesta, D. Mauleon, G. Carganico, Resolution of rac-ketoprofen esters by enzymatic reactions in organic media, Chirality 5 (1993) 320–328.
[46] G. Kraai, J.G.M. Winkelman, J.G. de Vries, H.J. Heeres, Kinetic studies on the Rhizomucor miehei lipase catalyzed esterification reaction of oleic acid with 1-butanolin a biphasic system, Biochem. Eng. J. 41 (2008) 87–94.
[47] S.D. Shinde, G.D. Yadav, Insight into microwave assisted immobilized Candida antarctica lipase B catalyzed kinetic resolution of RS-(±)-ketorolac, Process Biochem. 50 (2015) 230–236.
[48] S.D. Shinde, G.D. Yadav, Insight into microwave-assisted lipase catalyzed synthesis of geranyl cinnamate: optimization and kinetic modeling, Appl. Biochem. Biotechnol. 175 (2015) 2035–2049.
[49] B.H. Lerner, Complicated lessons: Lorenzo Odone and medical miracles, Lancet 373 (2009) 888–889.
[50] A. Thomas, Fats and fatty oils, in: Ullmann's Encyclopedia of Industrial Chemistry, Wiley-VCH, Weinheim, 2002, ISBN: 3527306730, https://doi.org/10.1002/14356007.a10_173.
[51] M. Goto, M. Goto, F. Nakashio, K. Yoshlzuka, K. Inoue, Hydrolysis of triolein by lipase in a hollow fiber reactor, J. Membr. Sci. 14 (1992) 207–214.
[52] F.J. Plou, M. Barandiarán, M.V. Calvo, A. Ballesteros, E. Pastor, High-yield production of mono- and di-oleylglycerol by lipase-catalyzed hydrolysis of triolein, Enzym. Microb. Technol. 18 (1996) 66–71.
[53] F. Jing, X. An, W. Shen, The characteristics of hydrolysis of triolein catalyzed by wheat germ lipase in water-in-oil microemulsions, J. Mol. Catal. B Enzym. 24–25 (2003) 53–60.
[54] (a) H. Hermansyah, A. Wijanarko, M. Gozan, R. Arbianti, T.S. Utami, M. Kubo, N.S. Kitakawa, T. Yonemoto, Consecutive reaction model for triglyceride hydrolysis using lipase, J. Teknol. 2 (2007) 151–157. (b) C.S. Karigar, S.S. Rao, Role of microbial enzymes in the bioremediation of pollutants: a review, Enzyme Res. 2011 (2011).
[55] C.-C. Chena, P.M. Reddy, C.S. Devia, P.-C. Changa, Y.-P. Hoa, Study of microwave effects on the lipase-catalyzed hydrolysis, Enzym. Microb. Technol. 82 (2016) 164–172.
[56] X.L. Yang, W.H. Zhang, Study on catalytic synthesis of N-butyl propionate using environmentally friendly catalyst, Adv. Mater. Res. 791–793 (2013) 89–92.
[57] K.V. Bhavsar, G.D. Yadav, Microwave assisted solvent-free synthesis of n-butyl propionate by immobilized lipase as catalyst, Biocatal. Agric. Biotechnol. 14 (2018) 264–269.
[58] S.R. Bansode, V.K. Rathod, Enzymatic synthesis of isoamyl butyrate under microwave irradiation, Chem. Eng. Process. Process Intensif. 129 (2018) 71–76.

[59] K.V. Bhavsar, G.D. Yadav, Process intensification by microwave irradiation in immobilized-lipase catalysis in solvent-free synthesis of ethyl valerate, Mole. Catal. 461 (2018) 34–39.
[60] A.K.F. Carvalho, H.B.S. Bento, H.J.I. Filho, H.F. de Castro, Approaches to convert Mucor circinelloides lipid into biodiesel by enzymatic synthesis assisted by microwave irradiations, Renew. Energy 125 (2018) 747–754.
[61] G. Parsiegla, A. Belaich, J.P. Belaich, R. Haser, Crystal structure of the cellulase Cel9M enlightens structure/function relationships of the variable catalytic modules in glycoside hydrolases, Biochemistry 41 (2002) 11134–11142.
[62] K. Mondal, I. Roy, M.N. Gupta, Enhancement of catalytic efficiencies of xylanase, pectinase and cellulase by microwave pretreatment of their substrates, Biocatal. Biotransformation 22 (2004) 9–16.
[63] X.-H. Li, H.-J. Yang, B. Roy, E.Y. Park, L.-J. Jiang, D. Wang, Y.-G. Miao, Enhanced cellulase production of the Trichoderma viride mutated by microwave and ultraviolet, Microbiol. Res. 165 (2010) 190–198.
[64] X. Zhao, Y. Zhou, G. Zheng, D. Liu, Microwave pretreatment of substrates for cellulase production by solid-state fermentation, Appl. Biochem. Biotechnol. 160 (2010) 1557–1571.
[65] Q. Li, J.H. Hu, G. Bo, H.D. Li, S. Song, L.X. Yang, Advances on immunoregulation effect of Astragalus polysaccharides, Chin. J. Exp. Tradit. Med. Formulae 2 (2017) 199–206.
[66] Q.Z. Yang, D.W. Liu, Y.X. Tian, L.F. Huang, Research progress on chemical structure and pharmacological activity of Astragalus polysaccharide, North. Hortic. 7 (2015) 168–175.
[67] D.Z. Bai, F. Dong, W.T. Tang, X. Chen, Study on process of pharmacology of Astragalus polysaccharide, Heilongjiang Med. J. 1 (2014) 103–106.
[68] Z.Y. Zhu, R.Q. Liu, F. Zhou, S.F. Li, J. Yuan, Extraction and purification of Astragalus polysaccharide and its anti-tumor activity, Mod. Food Sci. Technol. 4 (2011) 376–379.
[69] H.M. Li, R.Q. Huang, Y.Z. Wang, A technological study on enhancing the extraction rate of Astragalus polysaccharids, J. Northwest. Univ. 6 (2000) 509–510.
[70] S.Z. Gong, Z.R. Yang, Investigation into the microwave-assisted extraction technology for Astragalus polysaccharide, J. South China Univ. Technol. 8 (2004) 93–96.
[71] Y.X. Chen, F. Lin, J. Mo, X.D. Du, Comparison of two extraction methods of Astragalus polysaccharide, Res. Explor. Lab. 3 (2015) 20–22.
[72] R.C. Jin, S.T. Zhou, D.B. Zhang, Study on extraction of Astragalus membranaceus polysaccharides by optimized ultrasonic method with uniform design, J. Anhui Agric. Sci. 12 (2009) 5498–5499.
[73] G.F. Du, Z.H. Cai, G. Wang, B. Dai, L. He, Study on extraction of Astragalus polysaccharide by ultrasonic-microwave synergistically assisted technique, J. Nat. Prod. Res. Dev. 24 (2012) 114–117.
[74] Y.G. Bi, Z.C. Wu, Factors analysis of ultrasound combined with cellulose enzymatic extraction of total Astragalus polysaccharides, J. Guangdong Pharm. College 2 (2010) 134–137.
[75] X.W. Chen, S.L. Ma, Extraction of Astragalus iepsensis by cellulase degradation, Shanghai J. Tradit. Chin. Med. 1 (2005) 56–58.
[76] L.Y. Zheng, Y.M. Wei, L. Chen, Extraction of effective component from Radix astragli with cellulose, J. Gansu Agric. Univ. 1 (2005) 94–96.
[77] L.L. Dong, X. Huang, Y.G. Qi, H. Feng, Study on the enzymatic-microwave extraction of Astragalus polysaccharides, J. Zhejiang Univ. Technol. 3 (2011) 220–232.
[78] N. Mosier, C. Wyman, B. Dale, R. Elander, Y.Y. Lee, M. Holtzapple, M. Ladisch, Features of promising technologies for pretreatment of lignocellulosic biomass, Bioresour. Technol. 96 (2005) 673–686.

[79] S.M. Nomanbhay, R. Hussain, K. Palanisamy, Microwave-assisted alkaline pretreatment and microwave assisted enzymatic saccharification of oil palm empty fruit bunch fiber for enhanced fermentable sugar yield, J. Sustain. Bioenergy Syst. 3 (2013) 7–17.

[80] Y. Syafwina, T. Honda, M. Watanabe, M. Kuwahara, Pretreatment of oil palm empty fruit bunch by white-rot fungi for enzymatic saccharification, Wood Res. 89 (2002) 19–20.

[81] G. Gong, D. Liu, Y. Huang, Microwave-assisted organic acid pretreatment for enzymatic hydrolysis of rice straw, Biosyst. Eng. 107 (2010) 67–73.

[82] A. Xia, J. Cheng, W. Song, C. Yu, J. Zhou, K. Cen, Enhancing enzymatic saccharification of water hyacinth through microwave heating with dilute acid pretreatment for biomass energy utilization, Energy 61 (2013) 158–166.

[83] F.K. Tamaki, D.P. Souza, V.P. Souza, C.M. Ikegami, C.S. Farah, S.R. Marana, Using the amino acid network to modulate the hydrolytic activity of beta-glycosidases, PLoS One 11 (2016) 0167978.

[84] M.G. Pujic, E.G. Jampel, A. Loupy, A. Trincone, Enzymatic glycosidation in dry media under microwave irradiation, J. Chem. Soc. Perkin Trans. 1 (1997) 1001–1002.

[85] P. Monsan, P. Paul, Enzymatic synthesis of oligosaccharides, FEMS Microbiol. Rev. 16 (1995) 187–192.

[86] M.J. Kullen, J. Khil, F. Busta, D. Gallaher, L.J. Brady, Carbohydrate source and bifidobacteria influence the growth of clostridium perfringens in vivo and in vitro, Nutr. Res. 18 (1998) 1889–1897.

[87] R. Sharp, S. Fishbain, G.T. Macfarlane, Effect of short-chain carbohydrates on human intestinal bifidobacteria and Escherichia coli in vitro, J. Med. Microbiol. 50 (2001) 152–160.

[88] Y. Vandenplas, Oligosaccharides in infant formula, Br. J. Nutr. 87 (2002) 293–296.

[89] D.B. Hughes, D.G. Hoover, Viability and enzymatic activity of bifidobacteria in milk, J. Dairy Sci. 78 (1995) 268–276.

[90] R.A. Rastall, C. Bucke, Enzymatic synthesis of oligosaccharides, Biotechnol. Genet. Eng. Rev. 10 (1992) 253–281.

[91] Z. Guo, P.G. Wang, Utilization of glycosyltransferases to change oligosaccharide structures, Appl. Biochem. Biotechnol. 6 (1997) 1–20.

[92] G.M. Watt, P.A.S. Lowden, S.L. Flitsch, Enzyme-catalyzed formation of glycosidic linkages, Curr. Opin. Struct. Biol. 7 (1997) 652–660.

[93] D.H.G. Crout, G. Vic, Glycosidases and glycosyl transferases in glycoside and oligosaccharide synthesis, Curr. Opin. Chem. Biol. 2 (1998) 98–111.

[94] R. Öhrlein, Glycosyltransferase-catalyzed synthesis of non-natural oligosaccharides, Top. Curr. Chem. 200 (1999) 227–254.

[95] H.M. Flowers, Chemical synthesis of oligosaccharides, Methods Enzymol. 50 (1978) 93–121.

[96] T. Maugard, D. Gaunt, M.D. Legoy, T. Besson, Microwave-assisted synthesis of galacto-oligosaccharides from lactose with immobilized b-galactosidase from Kluyveromyces lactis, Biotechnol. Lett. 25 (2003) 623–629.

[97] C. Kamerke, M. Pattky, C. Huhn, L. Elling, Synthesis of UDP-activated oligosaccharides with commercial β-galactosidase from Bacillus circulans under microwave irradiation, J. Mol. Catal. B Enzym. 79 (2012) 27–34.

[98] J.F. Hare, Intracellular pathways of folded and misfolded amyloid precursor protein degradation, Arch. Biochem. Biophys. 451 (2006) 79–90.

[99] P. Hu, P. Berkowitz, V.J. Madden, D.S. Rubenstein, Stabilization of plakoglobin and enhanced keratinocyte cell-cell adhesion by intracellular O-glycosylation, J. Biol. Chem. 281 (2006) 12786.

[100] M.B. Olszewski, D. Trzaska, E.F. Knol, V. Adamczewska, J. Dastych, Efficient sorting of TNF-alpha to rodent mast cell granules is dependent on N-linked glycosylation, Eur. J. Immunol. 36 (2006) 997.

[101] L. Valmu, N. Kalkkinen, A. Husa, P.D. Rye, Differential susceptibility of transferrin glycoforms to chymotrypsin: a proteomics approach to the detection of carbohydrate-deficient transferrin, Biochemistry 44 (2005) 16007.

[102] X.L. Zhang, Roles of glycans and glycopeptides in immune system and immune-related diseases, Curr. Med. Chem. 13 (2006) 1141.

[103] R.A. Dwek, Biological importance of glycosylation, in: A.W. Coleman (Ed.), Molecular Recognition and Inclusion, Springer, Dordrecht, 1998.

[104] M.X. Silveyra, N. Cuadrado-Corrales, A. Marcos, M.A. Barquero, A. Rabano, M. Calero, J. Saez-Valero, Altered glycosylation of acetylcholinesterase in Creutzfeldt–Jakob disease, J. Neurochem. 96 (2005) 97–104.

[105] Y. Shi, E. Brandin, E. Vincic, M. Jansson, A. Blaxhult, K. Gyllensten, L. Moberg, C. Brostrom, E.M. Fenyo, J. Albert, Evolution of human immunodeficiency virus type 2 coreceptor usage, autologous neutralization, envelope sequence and glycosylation, J. Gen. Virol. 86 (2005) 3385.

[106] E.G. Sieskiewicz, P.A. Klimiuk, I. Domyslawska, S. Sierakowski, Defect of glycosylation of immunoglobulin G in rheumatoid arthritis patients, Postepy Hig. Med. Dosw. 59 (2005) 485–489.

[107] T.S. Raju, E.A. Davidson, New approach towards deglycosylation of sialoglycoproteins and mucins, Biochem. Mol. Biol. Int. 34 (1994) 943–954.

[108] Y. Ge, B.F. Gibbs, R. Masse, Complete chemical and enzymatic treatment of phosphorylated and glycosylated proteins on protein chip arrays, Anal. Chem. 77 (2005) 3644–3650.

[109] D.L. Papac, J.B. Briggs, E.T. Chin, A.T.S. Jones, A high-throughput microscale method to release N-linked oligosaccharides from glycoproteins for matrix-assisted laser desorption/ ionization time-of-flight mass spectrometric analysis, Glycobiology 8 (1998) 445–454.

[110] Y.Q. Yu, M. Gilar, J. Kaska, J.C. Gebler, A rapid sample preparation method for mass spectrometric characterization of N-linked glycans, Rapid Commun. Mass Spectrom. 19 (2005) 2331–2336.

[111] E.M. Kwan, A.B. Boraston, B.W. McLean, D.G. Kilburn, R.A. Warren, N-Glycosidase-carbohydrate-binding module fusion proteins as immobilized enzymes for protein deglycosylation, Protein Eng. Des. Sel. 18 (2005) 497–501.

[112] W.N. Sandoval, F. Arellano, D. Arnott, H. Raab, R. Vandlen, J.R. Lill, Rapid removal of N-linked oligosaccharides using microwave assisted enzyme catalyzed deglycosylation, Int. J. Mass Spectrom. 259 (2007) 117–123.

[113] J.V. King, W.G. Liang, K.P. Scherpelz, A.B. Schilling, S.C. Meredith, W.J. Tang, Molecular basis of substrate recognition and degradation by human presequence protease, Structure 22 (2014) 996–1007.

[114] Y. Shen, A. Joachimiak, M.R. Rosner, W.J. Tang, Structures of human insulin-degrading enzyme reveal a new substrate recognition mechanism, Nature 443 (2006) 870–874.

[115] F. Canduri, L.G. Teodoro, V. Fadel, C.C. Lorenzi, V. Hial, R.A. Gomes, J.R. Neto, W.F. de Azevedo, Structure of human uropepsin at 2.45 A resolution, Acta Crystallogr. D Biol. Crystallogr. 57 (2001) 1560–1570.

[116] Z.Y. Park, D.H. Russell, Thermal denaturation: a useful technique in peptide mass mapping, Anal. Chem. 72 (2000) 2667–2670.

[117] Y.Q. Yu, M. Gilar, P.J. Lee, E.S.P. Bouvier, J.C. Gebler, Enzyme-friendly, mass spectrometry-compatible surfactant for in-solution enzymatic digestion of proteins, Anal. Chem. 75 (2003) 6023–6028.
[118] W.K. Russell, Z.Y. Park, D.H. Russell, Proteolysis in mixed organic-aqueous solvent systems: applications for peptide mass mapping using mass spectrometry, Anal. Chem. 73 (2001) 2682–2685.
[119] G. Massolini, E. Calleri, Immobilized trypsin systems coupled on-line to separation methods: recent developments and analytical applications, J. Sep. Sci. 28 (2005) 7–21.
[120] J.R. Lill, E.S. Ingle, P.S. Liu, V. Pham, W.N. Sandoval, Microwave-assisted proteomics, Mass Spectrom. Rev. 26 (2007) 657–671.
[121] W.-Y. Chen, Y.-C. Chen, Acceleration of microwave-assisted enzymatic digestion reactions by magnetite beads, Anal. Chem. 79 (2007) 2394–2401.
[122] B.N. Pramanik, U.A. Mirza, Y.H. Ing, Y.H. Liu, P.L. Bartner, P.C. Weber, M.K. Bose, Microwave-enhanced enzyme reaction for protein mapping by mass spectrometry: a new approach to protein digestion in minutes, Protein Sci. 11 (2002) 2676–2687.
[123] H.F. Juan, S.C. Chang, H.C. Huang, S.T. Chen, A new application of microwave technology to proteomics, Proteomics 5 (2005) 840–842.
[124] H. Zhong, Y. Zhang, Z. Wen, L. Li, Protein sequencing by mass analysis of polypeptide ladders after controlled protein hydrolysis, Nat. Biotechnol. 22 (2004) 1291–1296.
[125] L. Hua, T.Y. Low, S.K. Sze, Microwave-assisted specific chemical digestion for rapid protein identification, Proteomics 6 (2006) 586–591.
[126] S. Swatkoski, P. Gutierrez, C. Wynne, A. Petrov, J.D. Dinman, N. Edwards, C. Fenselau, Evaluation of microwave-accelerated residue-specific acid cleavage for proteomic applications, J. Proteome Res. 7 (2008) 579–586.
[127] N.J. Hauser, F. Basile, On-line microwave D-cleavage LC-ESI-MS/MS of intact proteins: site-specific cleavages at aspartic acid residues and disulfide bonds, J. Proteome Res. 7 (2008) 1012–1026.
[128] P.M. Reddy, W.-Y. Hsu, J.-F. Hu, Y.-P. Ho, Digestion completeness of microwave-assisted and conventional trypsin-catalyzed reactions, J. Am. Soc. Mass Spectrom. 21 (2010) 421–424.
[129] M. Reslow, P. Adlercreutz, B. Mattiasson, On the importance of the support material for bioorganic synthesis, Eur. J. Biochem. 172 (1988) 573–578.
[130] A. Zaks, A.M. Klibanov, Enzymatic catalysis in nonaqueous solvents, J. Biol. Chem. 263 (1988) 3194–3201.
[131] M. Yvon, I.V. Hille, J.P. Pelissier, P. Guilloteau, R. Toullec, In vivo milk digestion in the calf abomasum, vol. II. Milk and whey proteolysis, Reprod. Nutr. Develop. 24 (1984) 835–843.
[132] Q. Huang, J.W. Coleman, D.R. Stanworth, Investigation of the allergenicity of b-lactoglobulin and its cleavage fragments, Int. Arch. Allergy Appl. Immunol. 78 (1985) 337–344.
[133] H. Otani, Antigenically reactive regions of bovine milk proteins, Agric. Res. Quart. 21 (1987) 135–142.
[134] M. Okamoto, R. Hayashi, A. Enomoto, S. Kaminogawa, K. Yamauchi, High-pressure proteolytic digestion of food proteins: selective elimination of b-lactoglobulin in bovine milk whey concentrate, Agric. Biol. Chem. 55 (1991) 1253–1257.
[135] I. Sélo, L. Negroni, C. Creminon, M. Yvon, G. Peltre, J.M. Wal, Allergy to bovine b-lactoglobulin specificity of human IgE using cyanogen bromide-derived peptides, Int. Arch. Allergy Appl. Immunol. 117 (1998) 20–28.
[136] F.J. Izquierdo, I. Alli, V. Yaylayan, R. Gomez, Microwave-assisted digestion of b-lactoglobulin by pronase, a-chymotrypsin and pepsin, Int. Dairy J. 17 (2007) 465–470.

[137] K.E.E. Mecherfi, D. Saidi, O. Kheroua, G. Boudraa, M. Touhami, O. Rouaud, S. Curet, Y. Choiset, H. Rabesona, J.-M. Chobert, T. Haertle, Combined microwave and enzymatic treatments for b-lactoglobulin and bovine whey proteins and their effect on the IgE immunoreactivity, Eur. Food Res. Technol. 233 (2011) 859–867.

[138] S. Lin, G. Yao, D. Qi, Y. Li, C. Deng, P. Yang, X. Zhang, Fast and efficient proteolysis by microwave-assisted protein digestion using trypsin-immobilized magnetic silica microspheres, Anal. Chem. 80 (2008) 3655–3665.

[139] S.A. Mazinani, B. DeLong, H. Yan, Microwave radiation accelerates trypsin-catalyzed peptide hydrolysis at constant bulk temperature, Tetrahedron Lett. 56 (2015) 5804–5807.

[140] T.C. Ullrich, M. Blaesse, R. Huber, Crystal structure of ATP sulfurylase from Saccharomyces cerevisiae, a key enzyme in sulfate activation, EMBO J. 20 (2001) 316–329.

[141] Y.M. Newman, S.G. Ring, G. Colaco, The role of trehalose and other carbohydrates in biopreservation, Biotech. Genetic Eng. Rev. 11 (1993) 263–294.

[142] A.D. Elbein, Y.T. Pan, I. Pastuszak, D. Carroll, New insights on trehalose: a multifunctional molecule, Glycobiology 13 (2003) 17R–27.

[143] Y. Yoshikawa, K. Nagata, K. Matsumoto, et al., Extraction of trehalose from bakers-yeast, Kagaku Kougaku Rronbunshu 17 (1991) 601–606.

[144] Y. Yoshikawa, K. Matsumoto, K. Nagata, T. Sato, Extraction of trehalose from thermally-treated bakers' yeast, Biosci. Biotechnol. Biochem. 58 (1994) 1226–1230.

[145] L. Chuanbin, X. Jian, B. Fengwu, S. Zhiguo, Trehalose extraction from Saccharomyces cerevisiae after microwave treatment, Biotechnol. Tech. 12 (1998) 941–943.

[146] I. Vrhovac, R. Hrascan, J. Franekic, Effect of 905 MHz microwave radiation on colony growth of the yeast Saccharomyces cerevisiae strains FF18733, FF1481 and D7, Radiol. Oncol. 44 (2010) 131–134.

[147] A.P. Vyas, J.L. Verma, N. Subrahmanyam, A review on FAME production processes, Fuel 89 (2010) 1–9.

[148] D.J. Anneken, S. Both, R. Christoph, G. Fieg, U. Steinberner, A. Westfechtel, Fatty acids, in: Ullmann's Encyclopedia of Industrial Chemistry, Wiley-VCH, Weinheim, 2006.

[149] H.H. Schobert, Chemistry of Fossil Fuels and Biofuels, Cambridge University Press, Cambridge, NY, 2013, pp. 62–64.

[150] M. Lamacka, J. Sajbidor, P. Bohov, Lipid isolation and fatty acid analysis in Saccharomyces cerevisiae. Comparison of different methods, Biotechnol. Tech. 12 (1998) 621–625.

[151] S. Abdulkadir, M. Tsuchiya, One-step method for quantitative and qualitative analysis of fatty acids in marine animal samples, J. Exp. Mar. Biol. Ecol. 354 (2008) 1–8.

[152] S. Khoomrung, K. Laoteng, S. Jitsue, S. Cheevadhanarak, Significance of fatty acid supplementation on profiles of cell growth, fatty acid, and gene expression of three desaturases in Mucor rouxii, Appl. Microbiol. Biotechnol. 80 (2008) 499–506.

[153] K. Ichihara, Y. Fukubayashi, Preparation of fatty acid methyl esters for gas-liquid chromatography, J. Lipid Res. 51 (2010) 635–640.

[154] C. Glaser, H. Demmelmair, B. Koletzko, High-throughput analysis of total plasma fatty acid composition with direct in situ transesterification, PLoS One 5 (2010) 1–6.

[155] F. Ulberth, M. Henninger, One-step extraction methylation method for determining the fatty-acid composition of processed foods, J. Am. Oil Chem. Soc. 69 (1992) 174–177.

[156] H. Shimasaki, F.C. Phillips, O.S. Privett, Direct transesterification of lipids in mammalian tissue for fatty acid analysis via dehydration with 2,2′-dimethoxypropane, J. Lipid Res. 18 (1977) 540–543.

[157] A. Masood, K.D. Stark, N. Salem Jr., A simplified and efficient method for the analysis of fatty acid methyl esters suitable for large clinical studies, J. Lipid Res. 46 (2005) 2299–2305.

[158] Z.X. Zhang, G.H. Xiong, G.K. Lie, X.Q. He, Sample pretreatment with microwave-assisted techniques, Anal. Sci. 16 (2000) 221–224.

[159] A. Tomas, M. Tor, G. Villorbina, R. Canela, M. Balcells, J. Eras, A rapid and reliable direct method for quantifying meat acylglycerides with monomode microwave irradiation, J. Chromatogr. A1216 (2009) 3290–3295.

[160] S. Itonori, M. Takahashi, T. Kitamura, K. Aoki, J.T. Dulaney, M. Sugita, Microwave-mediated analysis for sugar, fatty acid, and sphingoid compositions of glycosphingolipids, J. Lipid Res. 45 (2004) 574–581.

[161] N. Jeyashoke, K. Krisnangkura, S.T. Chen, Microwave induced rapid transmethylation of fatty acids for analysis of food oil, J. Chromatogr. A 818 (1998) 133–137.

[162] M. Liebeke, A. Wunder, M. Lalk, A rapid microwave-assisted derivatization of bacterial metabolome samples for gas chromatography/mass spectrometry analysis, Anal. Biochem. 401 (2010) 312–314.

[163] S. Khoomrung, P. Chumnanpuen, S. Jansa-ard, I. Nookaew, J. Nielsen, Fast and accurate preparation fatty acid methyl esters by microwave-assisted derivatization in the yeast Saccharomyces cerevisiae, Appl. Microbiol. Biotechnol. 94 (2012) 1637–1646.

[164] H. Alsuhaim, V. Vojisavljevic, E. Pirogova, World congress on medical physics and biomedical engineering, effects of non-thermal microwave exposures on the proliferation rate of Saccharomyces cerevisiae yeast, IFMBE Proc. (2012) 48–51.

[165] K. Pospiskova, G. Prochazkova, I. Safarik, One-step magnetic modification of yeast cells by microwave-synthesized iron oxide microparticles, Lett. Appl. Microbiol. 56 (2013) 456–461.

[166] S. Khoomrung, P. Chumnanpuen, S. Jansa-Ard, M. Ståhlman, I. Nookaew, J. Borén, J. Nielsen, Rapid quantification of yeast lipid using microwave-assisted total lipid extraction and HPLC-CAD, Anal. Chem. 85 (2013) 4912–4919.

[167] B.H. Samani, M.H. Khoshtaghaza, Z. Lorigooini, S. Minaei, H. Zareiforoush, Analysis of the combinative effect of ultrasound and microwave power on Saccharomyces cerevisiae in orange juice processing, Innov. Food Sci. Emerg. Technol. 32 (2015) 110–115.

[168] T. Sathvika, Manasi, V. Rajesh, N. Rajesh, Microwave assisted immobilization of yeast in cellulose biopolymer as a green adsorbent for the sequestration of chromium, Chem. Eng. 2791 (2015) 38–46.

[169] D. Stanisavljev, G.G. Cvijovic, I.N. Bubanja, Scrutinizing microwave effects on glucose uptake in yeast cells, Eur. Biophys. J. 46 (2017) 25–31.

[170] Y. Pan, D.-W. Sun, Z. Han, Applications of electromagnetic fields for nonthermal inactivation of microorganisms in foods: an overview, Trends Food Sci. Technol. 64 (2017) 13–22.

[171] Y. Wu, M. Yao, Inactivation of bacteria and fungus aerosols using microwave irradiation, J. Aerosol Sci. 41 (7) (2010) 682–693.

[172] L.R. Tonuci, C.F. Paschoalatto, R. Pisani Jr., Microwave inactivation of Escherichia coli in healthcare waste, Waste Manag. 28 (2008) 840–848.

[173] E.A. Oliveira, et al., Microwave inactivation of Bacillus atrophaeus spores in healthcare waste, Waste Manag. 30 (11) (2010) 2327–2335.

[174] Y. Shamis, et al., Review of the specific effects of microwave radiation on bacterial cells, Appl. Microbiol. Biotechnol. 96 (2012) 319–325.

[175] Y. Fang, et al., Effect of low-dose microwave radiation on Aspergillus parasiticus, Food Control 22 (2011) 1078–1084.

[176] H. Alsuhaim, et al., Effects of low power microwave radiation on biological activity of collagenase enzyme and growth rate of S. Cerevisiae yeast, SPIE–Int. Soc. Opt. Photon. 8923 (2013) 892357.

[177] L. Taylor, The mechanism of a thermal microwave biological effects, Bioelectromagnetics 2 (1981) 259–267.

[178] D.J. Panagopoulos, A. Karabarbounis, L.H. Margaritis, Mechanism for action of electromagnetic fields on cells, Biochem. Biophys. Res. Commun. 298 (2002) 95–102.

[179] W. Grundler, et al., Mechanisms of electromagnetic interaction with cellular systems, Naturwissenschaften 79 (1992) 551–559.

[180] A.V. Drozdov, et al., Quantum mechanical aspects of the effects of weak magnetic fields on biological objects, Biophysics 55 (4) (2010) 652–660.

[181] I.Y. Belyaev, Nonthermal biological effects of microwaves: current knowledge, further perspective, and urgent needs, Electromagn. Biol. Med. 24 (3) (2005) 375–403.

[182] H. Fröhlich, The biological effects of microwaves and related questions, Adv. Electron. Electron. Phys. 53 (1980) 85–152.

[183] G.J. Hyland, Non-thermal bioeffects induced by low-intensity microwave irradiation of living systems, Eng. Sci. Educ. J. 7 (1998) 261–268.

[184] I. Calinescu, A. Vlaicu, P. Chipurici, D. Ighigeanu, V. Lavric, Alcoholic fermentation in the presence of microwaves, Chem. Eng. Process. Process Intensif. 126 (2018) 16–22.

Part III

Applications of microwave chemistry

Chapter 8

Microwave-assisted sterilization

Aparna Das* and Bimal Krishna Banik*
Department of Mathematics and Natural Sciences, College of Sciences and Human Studies, Prince Mohammad Bin Fahd University, Al Khobar, Kingdom of Saudi Arabia
**Corresponding authors: e-mails: aparnadasam@gmail.com (Aparna Das); bimalbanik10@gmail.com, bbanik@pmu.edu.sa (Bimal Krishna Banik)*

8.1 Introduction

Infections caused by the pathogenic Gram-positive bacteria, such as *Bacillus subtilis* and *Methicillin-resistant Staphylococcus aureus*, as well as Gram-negative bacteria, including *Escherichia coli*, Salmonella, and Vibrio species, and *Pseudomonas aeruginosa*, remain as the significant public health-threatening issues in the world [1–3]. The viruses and fungi cause similar problems. Since the viruses are airborne, a good air purification system is required at places where people gather, such as in offices, schools, hospitals, theaters, assembly halls, and restaurants.

Sterilization is the process of inactivation of microorganisms. It is used essentially in several areas such as in the healthcare industry and hospital [4], food industry, and material manufacturing industry. Several methods are available for the inactivation of harmful microorganisms. These are performed using physical or chemical methods, such as heat, radiation, and chemical solutions or gases.

The most widely used sterilization processes include ozone gas sterilization [5], heat sterilization [6], plasma sterilization [7, 8], UV light sterilization [9], and microwave (MW) sterilization [10–15]. The full interior of the system is heated in the heat sterilization process. Thus, the system needs much energy and time. Ozone gas sterilization requires a few chemical or physical reactions. It poses problems, such as toxicity. To stabilize the electric discharge of the airflow system in the plasma sterilization is difficult. UV sterilization is compact. It possesses a high sterilization effect and is very easy to use. This is widely employed in hospitals and laboratories. But, UV irradiation can sterilize only the irradiated area (shadow effect). Among the sterilization methods, microwave sterilization has a lot of advantages and it is the most feasible method for several applications. In the microwave sterilization process, the material can be heated rapidly, directly, and selectively, leading to a reduction of power consumption. Besides, microwave sterilization is effective not only for the

Microwaves in Chemistry Applications. https://doi.org/10.1016/B978-0-12-822895-1.00011-4

surface sterilization of material but also for the interior of the material. As microwaves are irradiated from all angles, microorganisms can be sterilized homogeneously and effectively. Owing to these advantages, microwave sterilization is widely used in many fields.

In this chapter, we discuss microwave-assisted sterilization methods. The application of this method in the medical field, food industry, and material processing field is also included.

8.2 Microwave sterilization in medical industry

In the medical field, the sterilization process at low temperatures with no burden to the environment as well as with low cost and high safety is needed [16]. In the last several years, the use of microwave radiation has acquired much attention in the medical field [17, 18]. Microwave technology is used widely for sterilization of both laboratory items (e.g., plastic Petri dishes, media, towels, and contaminated plates) [18, 19] as well as biomedical waste [11, 20].

Biomedical waste management is an essential part of a healthcare facility [21]. Various biomedical waste management methods are currently available [22–25] and most of these methods disinfect the wastes at the time of generation. Although most of these methods are economical and straightforward, it has many disadvantages.

For example, sputum from pulmonary tuberculosis cases is dense, viscid, and sticky and contains live *Mycobacterium tuberculosis* bacilli in a large amount. Consequently, handling and disposal of such waste cause an administrative and environmental challenge. The available physical (autoclaving) and chemical (5% phenol) methods for sputum disinfection have many drawbacks [25–28]. Some disinfection methods require a great deal of human manipulation and the effectiveness to kill *M. tuberculosis* was poor [29].

Studies showed that microwave is effective in killing common nonmycobacterial organisms. Considering this efficacy of microwave, the effectiveness of microwave in sterilizing large volumes of *M. tuberculosis*-positive sputum samples was evaluated [30]. For sterilization of clinical biomedical waste, OptiMazer 30 was the most popular used commercial microwave technology. The sputum samples were treated in a microwave with 2450 MHz, 1.5 kW microwave radiation for 45 min. After the irradiation, the samples were checked for the presence of *M. tuberculosis* bacilli by Ziehl-Neelsen (ZN) staining. In addition, the viability was checked by inoculating in highly sensitive MGIT liquid culture media. The analysis showed that the percentage of sputum samples containing live *M. tuberculosis* bacilli descended from 95% in nonmicrowaved samples to $<1\%$ in postmicrowaved samples. Likewise, before microwaving, 93.8% of sputum samples were positive for acid-fast bacilli on Ziehl-Neelsen staining; but only 14.2% was found in postmicrowave samples (Figs. 1 and 2).

Similar investigations were performed for assessing the microwave effect on nonmycobacterial organisms such as *Clostridium perfringens* [31], *E. coli*

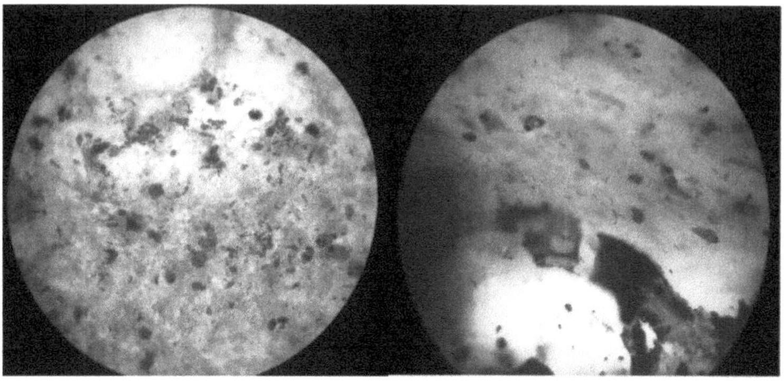

FIG. 1 Premicrowave ZN staining images showing acid-fast bacilli in clusters with large multilobed pus cells in the background. *(Reproduced with permission from V.P. Myneedu, A. Aggarwal, Disposal of the large volume of sputum positive for Mycobacterium tuberculosis by using microwave sterilisation technology as an alternative to traditional autoclaving in a tertiary respiratory care hospital in Delhi, India. Infect. Prev. Pract. 2 (2020) 100072.)*

[32, 33], *Salmonella* [34, 35], *Staphylococcus aureus* [36, 37], and *Listeria* spp. [38]. The analyses reported a 99.99% reduction in nonmycobacterial organism count.

The microwave radiation induced apoptosis in the cell [39, 40]. The decrease in the percentage of live bacilli in sputum samples was because of apoptotic changes occurring at the cellular level. This was noticeable by altered size and increased fragmentation of acid-fast bacilli and the surrounding pus cells. Studies on microwaved *Lactobacillus* phage PL-1 particle also presented similar results and proved that the effect of microwave irradiation is more intense at

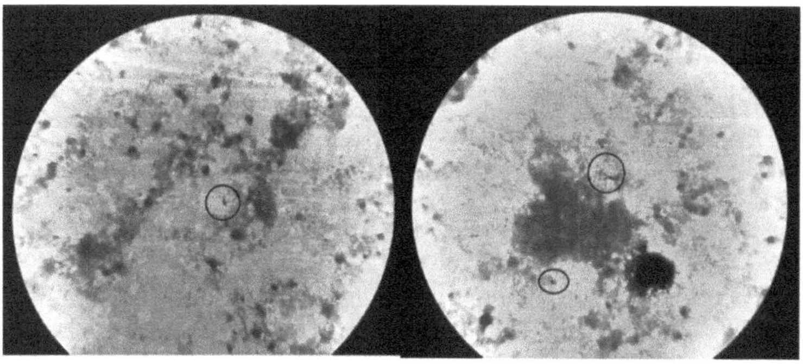

FIG. 2 Postmicrowave ZN staining images showing fragmented acid-fast bacilli of small size, and not present in clusters. In the background are the small pus cells without multilobed nuclei. *(Reproduced with permission from V.P. Myneedu, A. Aggarwal, Disposal of the large volume of sputum positive for Mycobacterium tuberculosis by using microwave sterilisation technology as an alternative to traditional autoclaving in a tertiary respiratory care hospital in Delhi, India. Infect. Prev. Pract. 2 (2020) 100072.)*

the level of the DNA chain [41]. Besides, the radiation appeared to fragment the nuclear material of cells. This was observed from the altered nuclear material of cells.

Microwaving at 650 W for 3 min was proved to remove Candida species from infected dentures in an equal or more effective way than chemical methods (e.g., sterilization for 8 h in 0.02% solution of sodium hypochlorite) [42, 43]. Wemken et al. reported the influence of hydrothermal aging and microwave sterilization on the trueness of milled, additively manufactured and injection-molded denture bases [44].

Rohrer et al. reported the sterilization of hydrophilic contact lenses contaminated with a variety of fungal, bacterial, and viral corneal pathogens using standard 2450-MHz microwave irradiation [45]. To overcome the problem of cold spots, a three-dimensional rotisserie was used within the microwave oven. The contact lenses became dehydrated in two minutes and rehydration with normal saline restored their original shape and appearance. The time required to prohibit the growth of the bacterial and fungal organisms was in the range of 45 s to 8 min. Regarding the viral contaminants, all the contaminants were wholly inactivated after 4 min of microwave exposure. Even after 101 exposures to microwaves for 10 min, the refractive properties were unaffected.

For reconstructing segmental skeletal disorders obtained from the tumor, trauma, or total joint arthroplasties, the usage of bone grafts is one of the most important and common treatment options [46, 47]. At the same time, the major problem related to bone allografts is the chance for disease transmission. Sterilization methods are necessary to prevent infection via allografts. Singh et al. explored the use of microwave radiation for sterilization of bone allografts and compared it with gamma radiation sterilization [48]. The study depicted that bone allografts contaminated with bacteria are sterilized by short exposure to microwave radiation for 2 min (2450 MHz, 900 W). The inactivation patterns of the microwave-radiated Gram-positive bacteria (*B. subtilis*, *Corynebacterium*, and *Staphylococcus aureus*) and Gram-positive bacteria (*E. coli*, *Klebsiella pneumoniae*, and *Pseudomonas aeruginosa*) are presented in Fig. 3.

8.3 Microwave sterilization in food industry

Several studies reported changing the microwave heating characteristics by changing food components like fat, salt, and water content [49, 50]. These changes were attained by altering the polarity of the food components, thereby varying the electromagnetic characteristics of the microwaves. Chen et al. investigated the synergistic effect of basic amino acids and microwaves [51].

The study showed that basic amino acids lysine and arginine coupled with microwave irradiation at 2450 MHz could promote a sterilizing effect on *E. coli*. The effect was mainly associated with the destruction of the cell membrane and the leakage of electrolytes. The bactericidal effect was proportional to the amino acid concentration, in the amino acid concentration range of 2.5%. At

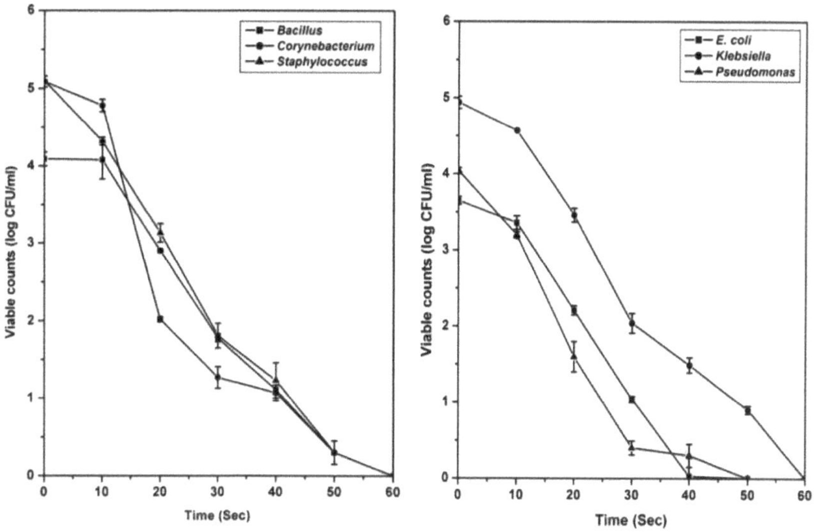

FIG. 3 (A) Effect of microwave radiation on the survival of Gram-positive bacterial cells. (B) Effect of microwave radiation on the survival of Gram-negative bacterial cells. *(Reproduced with permission from R. Singh, D. Singh, Sterilization of bone allografts by microwave and gamma radiation. Int. J. Radiat. Biol. 88 (2012) 661–666.)*

the same concentration and the same microwave treatment, the effect of lysine was significantly lower than that of arginine. On the contrary, histidine, another basic amino acid, had no promotion effect on microwave sterilization.

Scanning electron microscope (SEM) images of *E. coli* with different treatment conditions are shown in Fig. 4. Fig. 4A and (a) shows the untreated *E. coli* bacterial suspension. After the microwave treatment for 105 s at 3 W/g, the bacterial cell surface had experienced depression, swelling, and roughness (Fig. 4B and (b)). After treatment with both lysine and microwave irradiation, the *E. coli* cells were damaged and adhered in big pieces (Fig. 4C and (c)). Besides, after the addition of arginine, most of the *E. coli* cells went through significant adhesion and no longer had a complete short rod shape as shown in Fig. 4D and (d).

Tajchakavit et al. reported the destruction kinetics of two spoilage microorganisms, *Saccharomyces cerevisiae* and *Lactobacillus plantarum* in apple juice. These were evaluated under continuous flow microwave heating conditions [52]. The result was compared with conventional batch heating in a water bath. The inoculated apple juice was heated in a microwave oven (2450 MHz, 700 W) to find exit temperatures (52–65°C) under continuous-flow conditions. Fig. 5 shows the schematic diagram of the system. For the comparison study, inoculated juice was also subjected to conventional batch heating (50–80°C) in a well-stirred water bath.

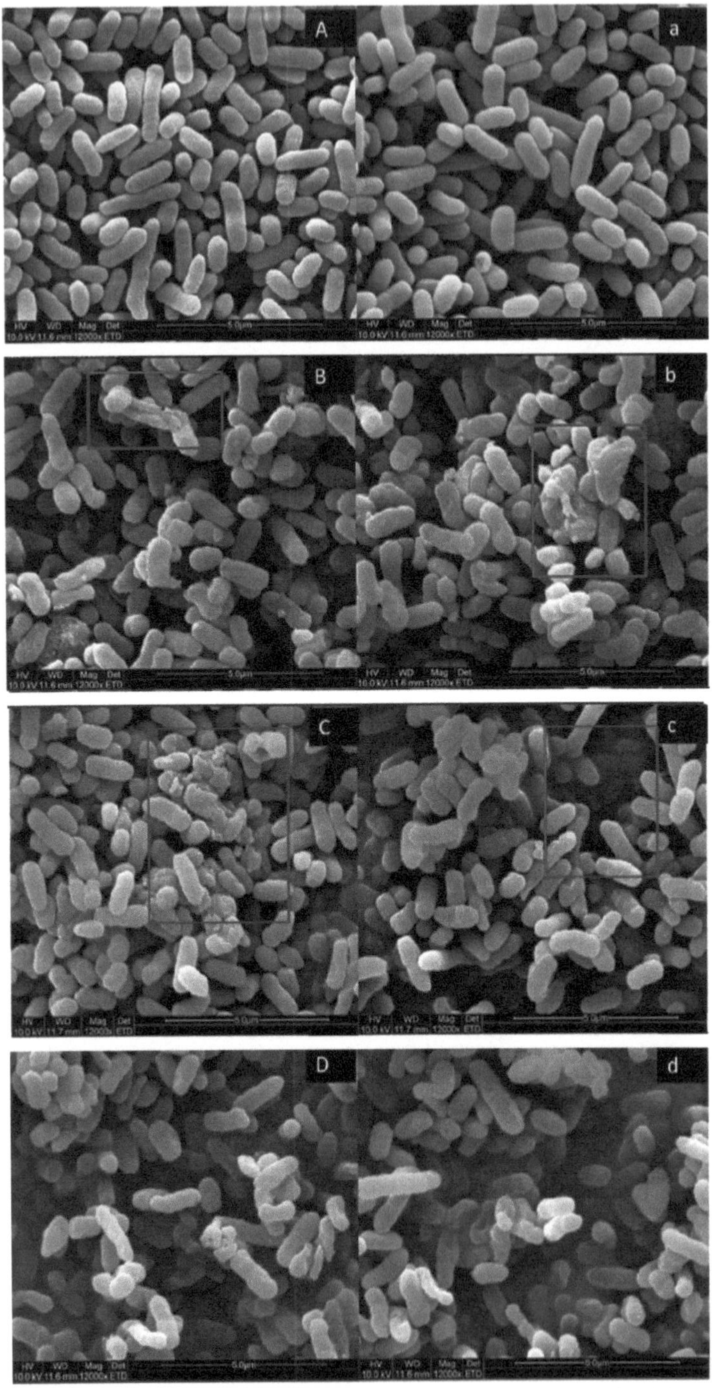

FIG. 4 SEM picture of the 3 W/g microwave dose treatment with the addition of arginine or lysine. (A) and (a) Untreated bacterial suspension, (B) and (b) microwave only, (C) and (c) microwave plus lysine, (D) and (d) microwave plus arginine. *(Reproduced with permission from M. Chen, D. Fan, T. Li, B. Yan, Y. Gao, J. Zhao, H. Zhang, Synergistic bactericidal effects of basic amino acids and microwave treatment on Escherichia coli. LWT 84 (2017) 99–105. https://doi.org/10.1016/j.lwt.2017.05.038.)*

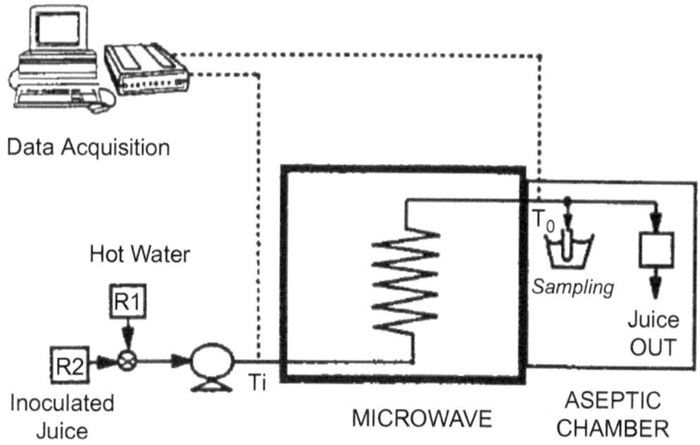

FIG. 5 Continuous-flow microwave heating system. *(Reproduced with permission from S. Tajchakavit, H.S. Ramaswamy, P. Fustier, Enhanced destruction of spoilage microorganisms in apple juice during continuous flow microwave heating. Food Res. Int. 31 (1998) 713–722. https://doi.org/10.1016/S0963-9969(99)00050-2.)*

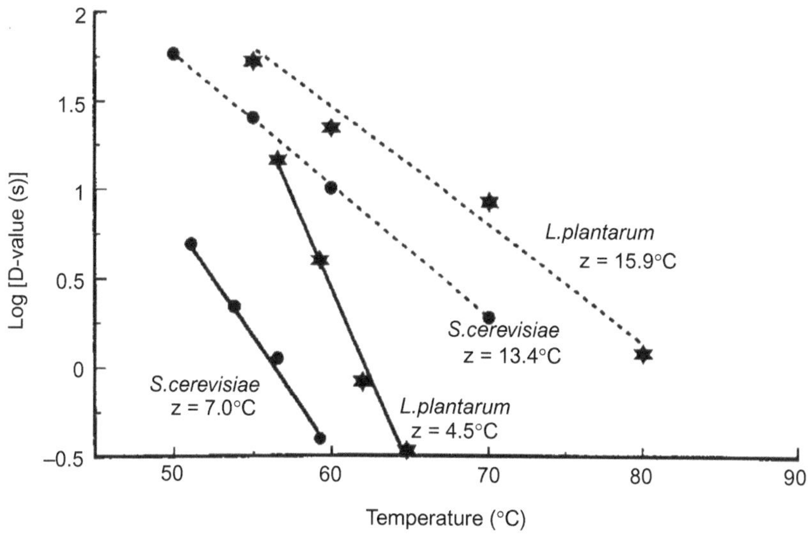

FIG. 6 Temperature sensitivity curves for continuous microwave heating and conventional batch heating destruction rates of *Saccharomyces cerevisiae* and *Lactobacillus plantarum* in apple juice. *(Reproduced with permission from S. Tajchakavit, H.S. Ramaswamy, P. Fustier, Enhanced destruction of spoilage microorganisms in apple juice during continuous flow microwave heating. Food Res. Int. 31 (1998) 713–722. https://doi.org/10.1016/S0963-9969(99)00050-2.)*

Fig. 6 shows the temperature sensitivity of the inactivation rates (z-values) of the yeast and bacteria obtained during continuous microwave heating *(solid lines)* and conventional batch heating *(dotted lines)* conditions.

The time-temperature profiles were used to correct both comeup and come-down times (Fig. 7). Under microwave heating, the time-corrected D-values

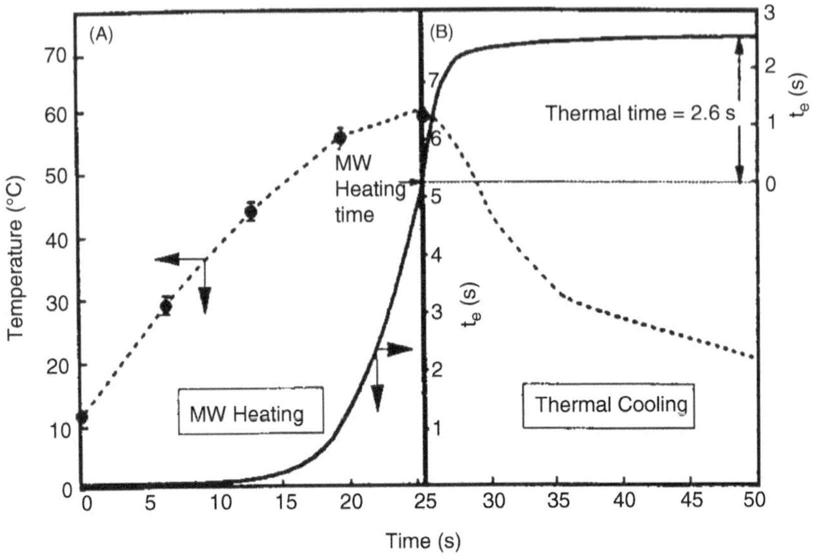

FIG. 7 Time-temperature profiles during (a) microwave heating and (b) thermal cooling (implying no microwave), and computed effective times. *(Reproduced with permission from S. Tajchakavit, H.S. Ramaswamy, P. Fustier, Enhanced destruction of spoilage microorganisms in apple juice during continuous flow microwave heating. Food Res. Int. 31 (1998) 713–722. https://doi.org/10.1016/S0963-9969(99)00050-2.)*

were 14, 3.8, and 0.79 s at 57.5°C, 60°C, and 62.5°C, respectively, for *Lactobacillus plantarum* and 4.8, 2.1, and 1.1 s at 52.5°C, 55°C, and 57.5°C, respectively, for *Saccharomyces cerevisiae* with corresponding z-values of 7°C and 4.5°C. The study revealed that the continuous microwave heating method is an appropriate method for the destruction of microorganisms. This was able to destroy the microorganisms by an order of magnitude faster than the conventional heating method.

Even though fresh vegetables and fruits are rich in nutrients, they have a short shelf life and are spoilable [53]. Some studies showed that thermal processes can control the safety of foods and significantly increase their shelf life. However, during thermal processes, all the important vitamins in vegetables were decreased [53–56]. For instance, during conventional thermal processes, around 90% of the loss was observed in heat-sensitive vitamins like vitamin C. At the same time, studies showed that storage retention of vitamins is greater in thermally processed vegetables than in either fresh or frozen vegetables [53, 55]. Besides, thermally stabilized vegetable products had other applications [57].

Microwave-assisted thermal sterilization (MATS) is established as the important food processing method to make low-acid, shelf-stable packaged foods [58]. In addition, MATS is found to retain the color, texture, and sensory attributes better than conventional methods [58, 59]. Adaption of polymer-based packaging that is transparent to microwaves is one of the important factors regarding shelf life for MATS-processed foods [59, 60]. High-barrier polymer pouches basically prohibited the oxidation, Millard browning, and other quality decaying reactions [61]. Zhang et al. investigated the effects of package barrier properties on shelf-life of vitamin C-fortified sweet potato puree (SPP) processed with MATS [62]. The analysis found a variation in SPP color during processing and storage. Besides, a change in vitamin C was also observed. It decreased from 201.7 ± 4.7 to 185.8 ± 15.6 mg/100 g during MATS. At the end of 9 or 18 months, depending on the oxygen barrier and temperature, a further decrease in vitamin C was also observed (as low as 13.6 ± 4.1 and 10.0 ± 0.3 mg/100 g). After processing and storage, the total β-carotene content was slightly more than before processing.

For the efficient development of commercial microwave sterilization processes that assure the treated foods are safe for use. It is essential to find the location of cold and hot spots in packaged foods [63, 64].

At the same time, during microwave sterilization, it is hard to evaluate the distribution of temperature and to locate the hot and cold spots within packaged foods. The dielectric heating using microwaves depends on direct interactions between the food and the electromagnetic fields. Furthermore, direct temperature measurement devices like thermocouples, infrared sensors, and fiber optic sensors have lots of limitations in these types of heating [64].

To estimate the heating patterns and to find the locations of the hot and cold spot in developing MATS processes, model foods as chemical marker carriers were required. Auksornsri et al. determined the dielectric constants, loss factors, and the penetration depths of rice grain and rice flour gel model foods with moisture contents between 50% and 60% and of cooked rice over a frequency range of 300–3000 MHz at temperatures range of 20–121°C [65]. The study depicted that the dielectric properties of rice models with 0% salt and 0.5% D-ribose closely match with the properties of cooked rice (Fig. 8). Thus, those rice model foods can be employed to emulate cooked rice for heating pattern and cold and hot spot identification at 915 and 2450 MHz in the development of MATS processes.

The mesocarp and the kernel of the palm fruits are the source of palm oil and palm kernel oil, respectively [66]. The palm oil extraction process starts with the sterilization of the fresh fruit bunches [67, 68]. By deactivating lipase activities on the fresh fruit bunch, the sterilization process protects oil quality from free fatty acid [69, 70]. Besides, the other main benefit of the sterilization process is to supply heat for the detachment of fruit from fresh fruit bunch stalks [71, 72]. Presently, the sterilization process is actioned by using a pressure vessel in a steam batch process (pressure: 15–45 psi; temperature: >100°C; time: 90 min) [73]. This process has several demerits. A large amount of wastewater was produced due to larger consumption of water for the preparation [74]. Furthermore, this steam process needs a large amount of steam for heating purposes. Combining greener technology, the alternatives for this steam process

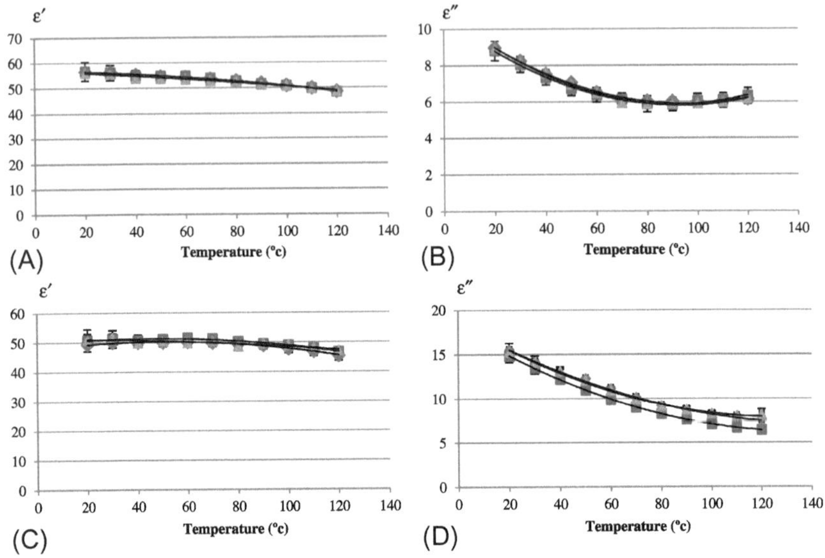

FIG. 8 (A and C) Comparison of dielectric constants. (B and D) dielectric loss factors of cooked rice (◆), rice flour gel (■), and rice grain (▲) with 0.5% D-ribose and no added salt at 915 MHz (A and B) and 2450 MHz (C and D) between 20°C and 121°C. *(Reproduced with permission from T. Auksornsri, J. Tang, Z. Tang, H. Lin, S. Songsermpong, Dielectric properties of rice model food systems relevant to microwave sterilization process. Innovative Food Sci. Emerg. Technol. 45 (2018) 98–105. https://doi.org/10.1016/j.ifset.2017.09.002.)*

include microwave heating, continuous sterilizer, and dry oven heating to provide the environment with [75, 76].

Among those methods, the microwave irradiation method has advantages in terms of the shorter sterilization process, lower operating temperature, uniform heating, and lower sintering temperature [77, 78]. To improve the sterilization process, Hock et al. introduced a hybrid steam-microwave method to perform the sterilization of palm at atmospheric pressure [79]. The hybrid system has advantages in terms of efficiency with less pollution and the whole sterilization duration can be reduced significantly, and the process can be done in an industry-scale system.

Mesocarp temperature during pure-microwave treatment and presteaming microwave treatment with a duration of 4 min is shown in Fig. 9. The data depicted that the mesocarp temperature during the presteaming system is generally greater than that in the pure-microwave system. It was observed that a moisture layer is created at the surface of the mesocarp with the presence of the presteaming process and it provided a platform for the microwave to heat up without drying out the palm oil fruit. In pure microwave treatment for the first 2 min, it was not easy to detach the kernel from the palm fruit compared to presteaming microwave treatment. Another important difference observed was in the temperature. The temperature of the presteaming microwave process

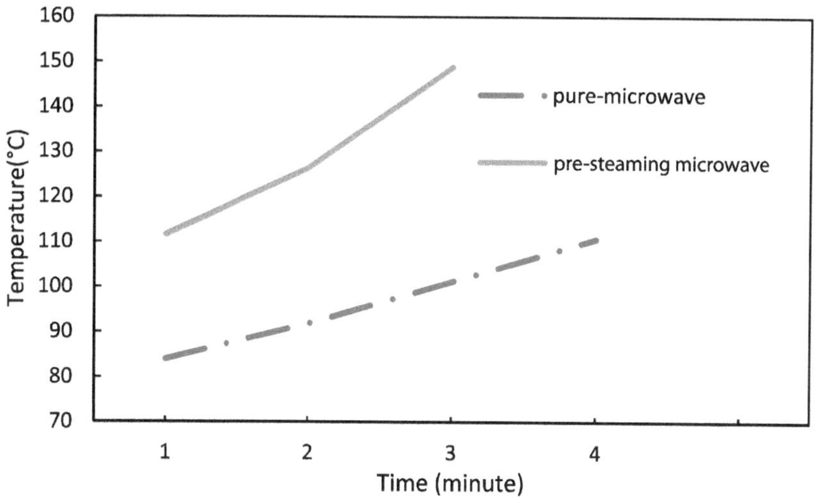

FIG. 9 Mesocarp temperature at pure-microwave and presteaming microwave. *(Reproduced with permission from T.K. Hock, G.T. Chala, H.H. Cheng, An innovative hybrid steam-microwave sterilization of palm oil fruits at atmospheric pressure. Innovative Food Sci. Emerg. Technol. 60 (2020) 102289.)*

was 111.73°C in the first minute; this is slightly higher than the temperature of the pure-microwave process for 4 min (110.73°C). This analysis pointed out that the hybrid method has a higher efficiency all over the heating process.

Changes in kernel temperature during pure-microwave and presteaming microwave treatment with the duration of 4 min are shown in Fig. 10. The result showed that kernel temperatures under presteaming microwave were generally greater than those in the pure-microwave process. However, the difference between the two conditions is smaller than the mesocarp.

The appearance of fruit under different heating methods like pure-microwave heating, presteam microwave heating, and poststeam hybrid heating is shown in Fig. 11. For the atmospheric steaming process, the temperature of the fruit was at around 100°C and the fruit detachment cannot be done under that condition. It can be seen that the palm fruits are generally drier as opposed to the poststeaming sample.

Microwave sterilization holds promise to reduce process time and improve product quality [80, 81]. Estimation of cold spot locations in foods during microwave sterilization is an important task as it can ensure that the processed foods are safe for consumers. Studies showed that computer simulation models can aid in understanding the sterilization process well [82, 83]. However, simulation models need validation and may not be always authentic owing to the complexity of the heat transfer coupling and heating of dielectric in complex microwave sterilization cavities [84–86]. During microwave sterilization processes, it was observed to be difficult to identify the cold spots in packaged

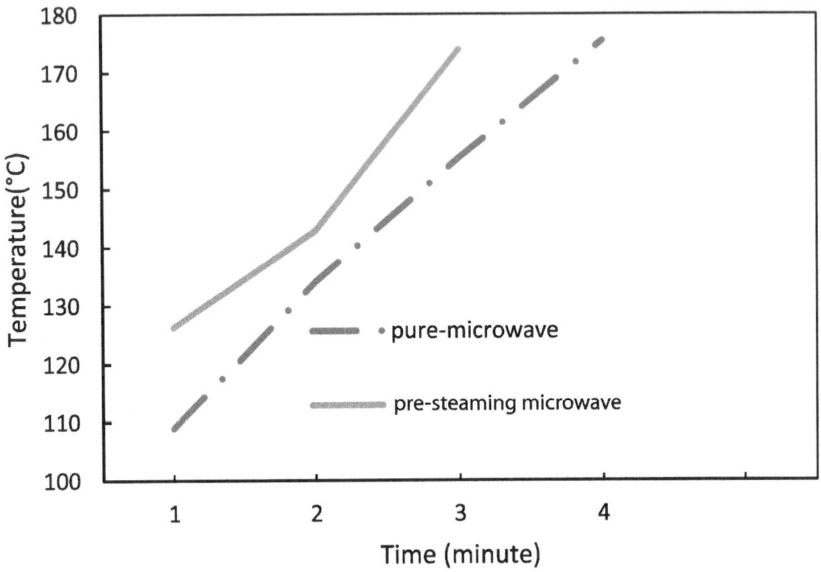

FIG. 10 Kernel temperature at pure-microwave and presteaming microwave. *(Reproduced with permission from T.K. Hock, G.T. Chala, H.H. Cheng, An innovative hybrid steam-microwave sterilization of palm oil fruits at atmospheric pressure. Innovative Food Sci. Emerg. Technol. 60 (2020) 102289.)*

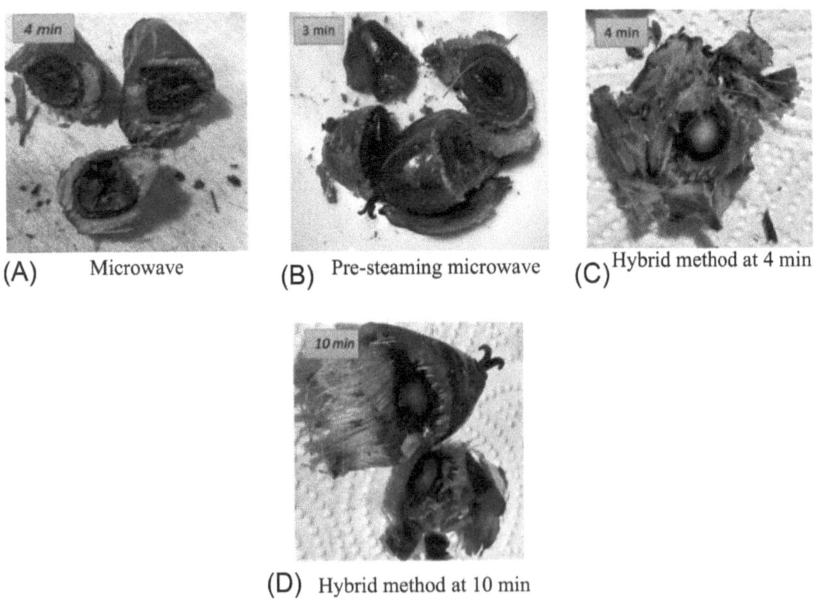

FIG. 11 Comparison of fruit appearance: (A) microwave, (B) presteam microwave, and (C) poststeam (hybrid) process at 4 min, and (D) hybrid method at 10 min. *(Reproduced with permission from T.K. Hock, G.T. Chala, H.H. Cheng, An innovative hybrid steam-microwave sterilization of palm oil fruits at atmospheric pressure. Innovative Food Sci. Emerg. Technol. 60 (2020) 102289.)*

foods using point temperature measurement methods. Also, the accuracy of the chemical marker methods was poor [64, 87, 88].

A combination of the chemical marker method and computer vision method was proposed as a suitable technology for evaluating the heating patterns of sterilized foods [89]. Computer vision has a wide range of applications in quality inspection, classification, evaluation of products, and processes in the food industry [90–93]. Pandit et al. developed a method that involved the application of chemical marker M-2 yield to a model food, using a computer vision system (Fig. 12) and an image processing software to capture and analyze color patterns after thermal processes [63].

Tang et al. investigated the feasibility to develop a short-time sterilization method for a highly inhomogeneous food [58]. The food was prepackaged in polymeric trays and the study was performed using microwave energy. The schematic diagram of the 915-MHz, single-mode, 10-kW pilot-scale microwave system developed for the study is shown in Fig. 13. The inhomogeneous food used consisted of sliced beef and gravy packaged in 7-oz trays.

Specially formulated whey protein gel (WPG) was chosen as a model food and it is used to emulate the real food for the analysis of heating patterns and cold spots inside food trays. Comparisons of dielectric properties of WPG with different compositions and pretreated beef samples are shown in Figs. 14 and 15.

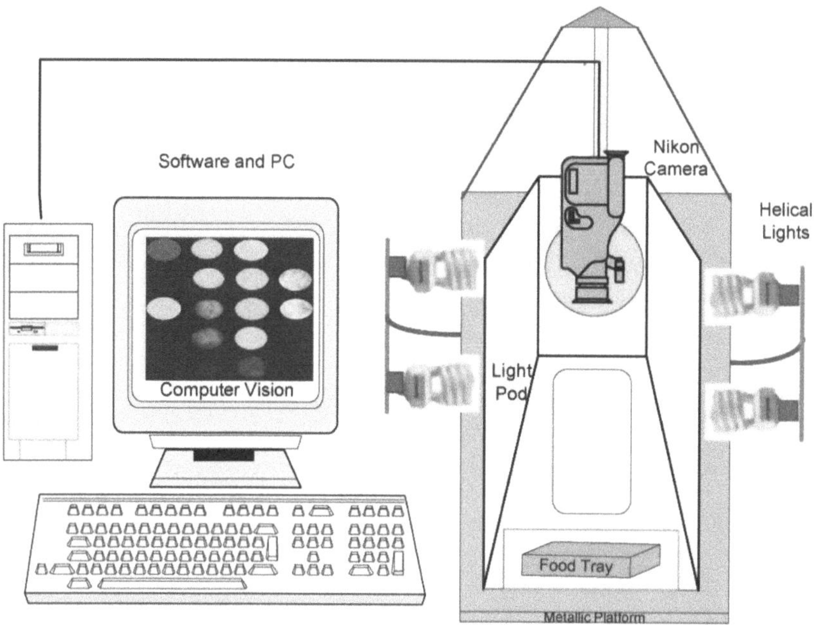

FIG. 12 Schematic of the computer vision system. *(Reproduced with permission from R.B. Pandit, J. Tang, F. Liu, G. Mikhaylenko, A computer vision method to locate cold spots in foods in microwave sterilization processes. Pattern Recogn. 40 (2007) 3667–3676.)*

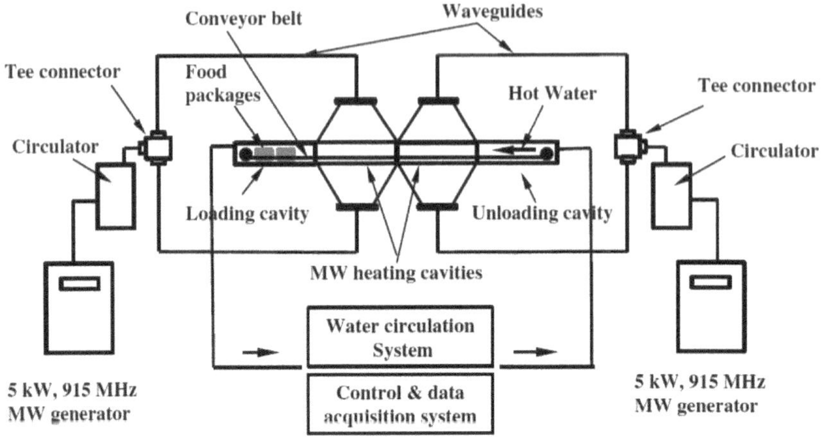

FIG. 13 Schematic diagram of pilot microwave sterilization system. *(Reproduced with permission from Z. Tang, G. Mikhaylenko, F. Liu, J.-H. Mah, R. Pandit, F. Younce, J. Tang, Microwave sterilization of sliced beef in gravy in 7-oz trays. J. Food Eng. 89 (2008) 375–383. https://doi.org/10.1016/j.jfoodeng.2008.04.025.)*

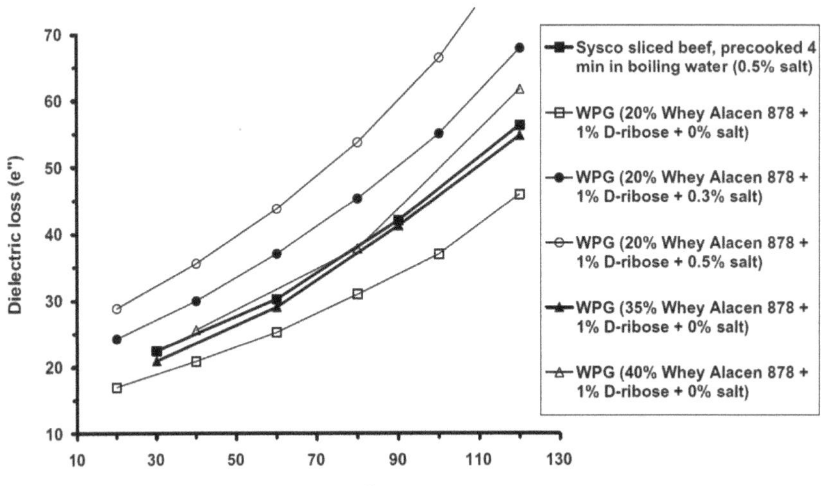

FIG. 14 Dielectric loss factors at 915 MHz for WPG with different compositions and precooked sliced beef. *(Reproduced with permission from Z. Tang, G. Mikhaylenko, F. Liu, J.-H. Mah, R. Pandit, F. Younce, J. Tang, Microwave sterilization of sliced beef in gravy in 7-oz trays. J. Food Eng. 89 (2008) 375–383. https://doi.org/10.1016/j.jfoodeng.2008.04.025.)*

The demonstrated processing schedules were confirmed by inoculated pack analysis using *C. sporogenes* spores. The investigation showed that the 915-MHz single-mode microwave sterilization technology was effective for processing the inhomogeneous food as well. The techniques are also useful for the

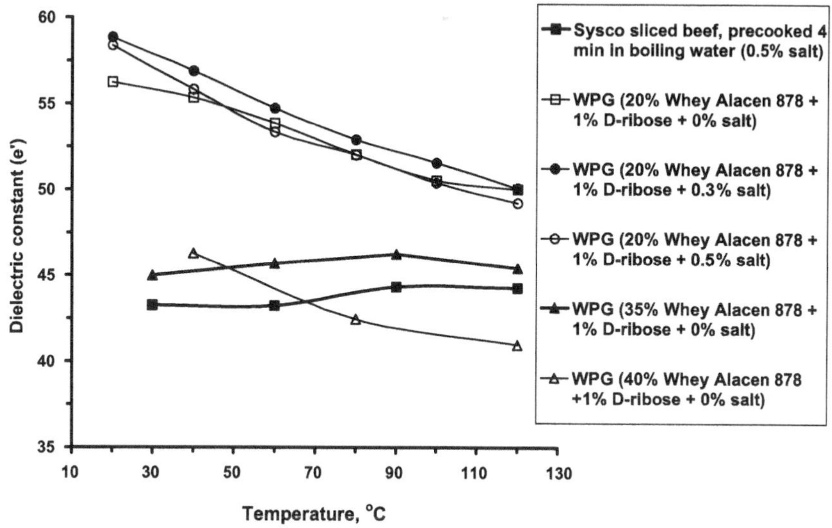

FIG. 15 Dielectric constants at 915 MHz for WPG with different compositions and precooked sliced beef. *(Reproduced with permission from Z. Tang, G. Mikhaylenko, F. Liu, J.-H. Mah, R. Pandit, F. Younce, J. Tang, Microwave sterilization of sliced beef in gravy in 7-oz trays. J. Food Eng. 89 (2008) 375–383. https://doi.org/10.1016/j.jfoodeng.2008.04.025.)*

sterilization of other packaged inhomogeneous foods, like chicken meat in gray in trays and fish in gravy in pouches.

Aggregatibacter (*Actinobacillus*) *actinomycetemcomitans* is naturally transformable [94, 95], Gram-negative, facultative anaerobe, a nonmotile bacterium. It belongs to the Pasteurellaceae family. *Aggregatibacter actinomycetemcomitans* is the main reason for localized aggressive periodontitis (LAP)/localized juvenile periodontitis (LJP), a severe form of periodontal disease, and several other types of extraoral infections [96–98]. However, studies also exhibited the presence of these bacteria in healthy oral cavities [96, 99, 100]. This bacterium grows best when incubated at 37°C in the presence of carbon dioxide (5%) [96, 101, 102]. To alleviate growth, normally sodium bicarbonate is added to both plates and broth. As it is a fastidious microorganism, the study of *Aggregatibacter actinomycetemcomitans* was limited. It multiplies slowly, grows in two distinct phenotypes and normally it takes more than two days on plates. It was reported that when grown in broth, fresh clinical isolates were rough and, autoaggregate [103]. The growth medium of *Aggregatibacter actinomycetemcomitans* includes tryptic soy broth, glucose, and yeast extract [101].

The slow growth and the decreased viability were due to the conventional method of making plates by autoclaving the growth medium [104]. Normally, microwave radiation effectively uses for disinfecting materials and food [105–108]. Bhattacharjee et al. demonstrated that the microwave method of sterilization is of special significance for *Aggregatibacter actinomycetemcomitans* [104]. Microwave irradiation is not just an alternative method of sterilization but also boosts the faster growth and viability of the bacteria. This investigation was carried out on a smooth variant, Y4 [109] of *Aggregatibacter actinomycetemcomitans* and nalidixic acid-resistant strains of two clinical isolates; CU1000 [101] and IDH781 [110].

Table 1 demonstrates the effectiveness of the microwave sterilization method. No changes in the growth of *E. coli* were observed with microwaving irradiation. But the growth of *Aggregatibacter actinomycetemcomitans* decreased when the medium was microwaved for more than 90 s after the initiation of boiling. Also, faster growth of *Aggregatibacter actinomycetemcomitans* was observed in microwaved broth than in autoclaved broth (Fig. 16).

It was demonstrated that growth media sterilized by microwave method promotes faster growth of *Aggregatibacter actinomycetemcomitans* in broth as well as on plates. Besides, the viability of the cells was also more when spread on microwaved plates compared to cells on autoclaved plates. The study evidenced that microwave irradiation is an effective method of sterilization of liquid as well as solid growth media.

Some studies showed that bacterial species can be inactivated microwave [111, 112]. For example, the *E. coli* destruction profile by microwave exposure is generally occurred by means of thermal effects [113]. Also, microwaves can destroy bacterial spores in a better way in the presence of water than in the dry state [13, 114]. It was reported that longer time of microwave exposure to the growth medium decreases the growth rate of bacteria [104]. Thus, the effect of microwave radiation is comparable to the thermal effects of autoclaving [115]. In the microwave method, the sterilization time can be better controlled, whereas autoclaving time cannot be controlled easily.

Wang et al. demonstrated a microwave sterilization system that can raise the temperature in quite a short time using a lower microwave power (100 W) [116]. In the system, filters built with Kao wool (Al_2O_3) and coated with

TABLE 1 Sterilization of *Aggregatibacter actinomycetemcomitans* growth medium (AAGM) by microwave method.

	A_{600}			
	Microwaving time (s)			
Cells	*60*	*90*	*120*	*150*
Aggregatibacter actinomycetemcomitans	0.907 ± 0.018	0.908 ± 0.008	0.773 ± 0.010	0.312 ± 0.017
E. coli	1.017 ± 0.003	1.012 ± 0.003	1.028 ± 0.003	1.025 ± 0.005

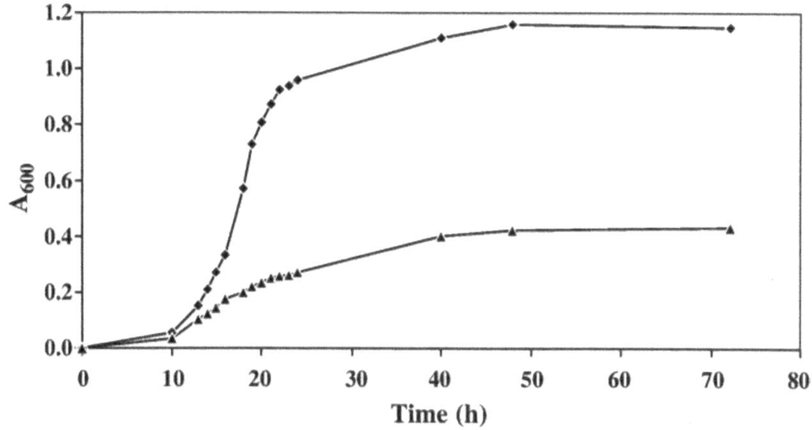

FIG. 16 Growth of *Aggregatibacter actinomycetemcomitans* in microwaved and autoclaved broth. Cells were grown in 10 mL of AAGM sterilized by the microwave (◆) or autoclave (▲) methods and A600 values were determined at indicated times. *(Reproduced with permission from M.K. Bhattacharjee, K. Sugawara, O.T. Ayandeji, Microwave sterilization of growth medium alleviates inhibition of Aggregatibacter actinomycetemcomitans by Maillard reaction products. J. Microbiol. Methods 78 (2009) 227–230.)*

TiO_2 were used to trap microorganisms. These filters were also coated with Pt or Ag. The microorganisms, *B. subtilis* and *B. stearothermophilus* were used in this investigation. After microwave irradiation, filters loaded with *B. subtilis* and *B. stearothermophilus* were incubated for 48 h in a TSB medium at 37°C and 56°C, respectively. The result showed that *B. subtilis* and *B. stearothermophilus* loaded onto Ag-impregnated filters were sterilized in 30 and 15 s, respectively. The results showed that the microwave sterilization system could sterilize *B. subtilis* and *B. stearothermophilus* in a short time and that microwave absorbable materials were effective as microwave sterilization filters.

The sterilization effects of various kinds of catalysts that were loaded on SiO_2-TiO_2 pellets were investigated. Among the tested catalyst (Ag, Pt, La, Pd, and Ni), *E. coli* were sterilized by Ag and/or Pt catalyst pellets in a very short time [117].

Takashima et al. developed a sterilization method in which low microwave power (~100 W) was irradiated at a pellet that was coated with a particular catalyst metal and loaded with targeting microorganisms [118]. The schematic of the sterilization system is shown in Fig. 17. Two microorganisms, *E. coli* and *B. subtilis,* were considered for the investigation. The result demonstrated that this system could sterilize *E. coli* and *B. subtilis* in a very short time and the metal films and TiO_2 play a significant role in this process.

The comparison of the sterilization effects of microwave irradiation for various kinds of pellets coated with Ag or Pt is shown in Table 2. The result showed that the pellets irradiated with microwaves fully sterilized *E. coli* within a short

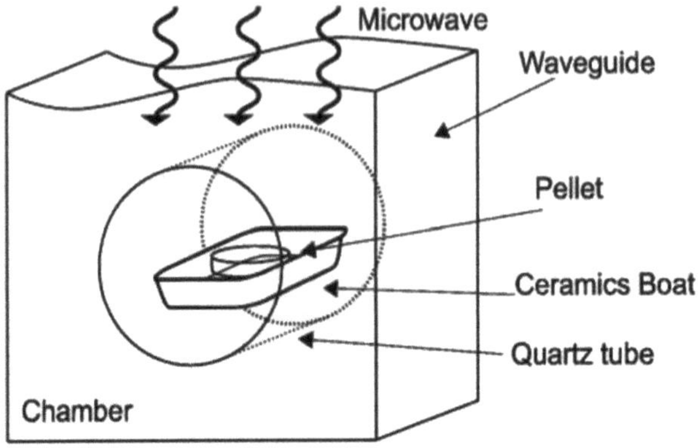

FIG. 17 Microwave irradiation system. *(Reproduced with permission from H. Takashima, K. Nakamura, Y. Kanno, Microwave sterilization by TiO_2 filter coated with Ag thin film, in: 2006 IEEE International Conference on Systems, Man and Cybernetics, IEEE, 2006, pp. 1413–1418.)*

TABLE 2 Microwave sterilization of *E. coli* test, + and – indicate the positive and negative growth, respectively.

		Irradiation time (s)				
Pellet	**Impregnated metal**	**0**	**5**	**15**	**30**	**60**
SiO_2	Ag	+	+	–	–	–
SiO_2	Pt	+	+	–	–	–
$SiO_2 + TiO_2$ (anatase)	Ag	+	–	–	–	–
$SiO_2 + TiO_2$ (anatase)	Pt	+	+	–	–	–
$SiO_2 + TiO_2$ (anatase)	None	+	+	–	–	–
Al_2O_3	Ag	+	+	–	–	–
Al_2O_3	Pt	+	+	–	–	–
$Al_2O_3 + TiO_2$ (rutile)	Ag	+	+	–	–	–
$Al_2O_3 + TiO_2$ (rutile)	Pt	+	–	–	–	–
$Al_2O_3 + TiO_2$ (rutile)	None	+	+	–	–	–

Reproduced with permission from H. Takashima, K. Nakamura, Y. Kanno, Microwave sterilization by TiO_2 filter coated with Ag thin film, in: 2006 IEEE International Conference on Systems, Man and Cybernetics, IEEE, 2006, pp. 1413–1418.

time (15 s). *E. coli* on the SiO_2-TiO_2 coated with Ag and on the Pt-coated Al_2O_3-TiO_2 pellet was completely sterilized within 5 s. Besides, these data depicted that the pellets coated with catalyst were better than uncoated pellets. Also, the pellets that involve TiO_2 with catalyst metal elements were efficient for the microwave sterilization of *E. coli*.

Microwave-assisted sterilization time of *B. subtilis* with various kinds of pellets is shown in Table 3. *B. subtilis* on SiO_2 pellet coated with Ag was sterilized within 5 s.

Novotny et al. reported the sterilization of wood-destroying insects by the application of microwave radiation [119]. Three species, namely *Hylotrupes bajulus*, *Anobium pertinax*, and *Anobium striatum*, were considered for the study. It was observed that for successful elimination of wood-destroying insects, temperature of around 60°C was sufficient. Moreover, it was found that at this temperature range, irreversible changes in the insects' tissues occur and subsequent death happens in all developmental stages. The microwave power was in the range of 400–700 W to attain these temperatures and irradiation was maintained for approximately 10 min.

TABLE 3 MW sterilization of B. subtilis test, + and − indicate the positive and negative growth, respectively.

Pellet	Impregnated metal	Irradiation time (s)					
		0	5	15	30	60	120
SiO_2	Ag	+	−	−	−	−	−
SiO_2	Pt	+	+	−	−	−	−
SiO_2 + TiO_2 (anatase)	Ag	+	+	+	−	−	−
SiO_2 + TiO_2 (anatase)	Pt	+	+	+	−	−	−
SiO_2 + TiO_2 (anatase)	None	+	+	+	+	+	−
Al_2O_3	Ag	+	+	+	−	−	−
Al_2O_3	Pt	+	+	+	−	−	−
Al_2O_3 + TiO_2 (rutile)	Ag	+	+	+	+	−	−
Al_2O_3 + TiO_2 (rutile)	Pt	+	+	+	+	−	−
Al_2O_3 + TiO_2 (rutile)	None	+	+	+	+	+	−

Reproduced with permission from H. Takashima, K. Nakamura, Y. Kanno, Microwave sterilization by TiO_2 filter coated with Ag thin film, in: 2006 IEEE International Conference on Systems, Man and Cybernetics, IEEE, 2006, pp. 1413–1418.

8.4 Sterilization using microwave plasmas

Some material sterilization methods produced some level of damage to the material or make only incomplete sterilization [120, 121]. Furthermore, removal of lipopolysaccharide (LPS) from a biomaterial is difficult, due to its ability to adhere strongly to biomaterials. LPS produces various harmful biological activities even in a small amount [122, 123].

So it is important to consider sterilization techniques with superior characteristics, such as nontoxic, short sterilization times, and medium preservation. Sterilization by plasma is a substitute for the conventional sterilization process. Besides, microwave-based plasma sterilization has several advantages. This sterilization is possible at a relatively low temperature with medium preservation and it is safe compared to other techniques [121]. Sterilization using microwave-induced plasma is not only efficient in killing bacteria and viruses but also efficient in removing the dead bacteria and viruses from the surface of the material [120, 124–127]. Several studies showed gram-negative and gram-positive bacterial strains could be sterilized and inactivated with plasma treatment within 20 s or less than that [125–127]. Plasma sterilization using oxygen, argon, helium and their mixture has been published.

Park et al. investigated the sterilization effect of the microwave-induced plasma system at atmospheric pressure against *E. coli* and the LPS elimination efficiency of the method from contaminated materials [128]. In this study, to generate plasma at atmospheric pressure, a 2.45 GHz, waveguide-based, microwave-induced Ar plasma system was used. The system mainly consists of a 1 kW magnetron power supply that is commonly used in a microwave oven, an applicator, and a nozzle section made of quartz. Fig. 18 proves that *E. coli*

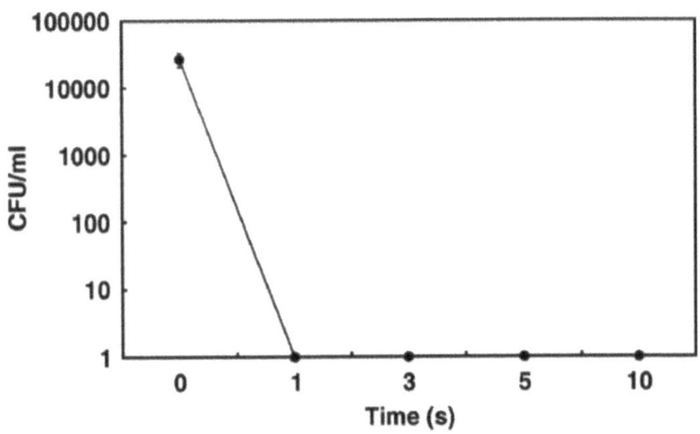

FIG. 18 Sterilization effects of microwave-induced argon plasma at atmospheric pressure against *E. coli* ATCC 8739. CFU: colony-forming unit. *(Reproduced with permission from B.J. Park, K. Takatori, M.H. Lee, D.-W. Han, Y.I. Woo, H.J. Son, J.K. Kim, K.-H. Chung, S.O. Hyun, J.-C. Park, Escherichia coli sterilization and lipopolysaccharide inactivation using microwave-induced argon plasma at atmospheric pressure. Surf. Coat. Technol. 201 (2007) 5738–5741.)*

ATCC 8739 used for the study was completely sterilized after 1 s by microwave-induced Ar plasma at atmospheric pressure. It was claimed that UV light and free radicals generated during the process by the plasma could diffuse into bacteria via chemically and physically robust membrane more quickly and then directly interact with the bacterial cell components.

The SEM observations (Fig. 19) showed severe morphological alterations during the plasma-treatment of *E. coli*. The normal untreated spores (Fig. 19A) had an ellipsoidal shape and the cells were damaged after the plasma treatment for 1–10 s and are clearly visible from Fig. 19B–F.

Plasma treatment completely inactivated the Limulus amoebocyte lysate (LAL) activity of LPS after 10 s (Fig. 20).

Itarashiki et al. reported the sterilization effect of nitrogen oxide radicals generated by microwave plasma using air [129]. In this case, plasma was produced by irradiating air with microwaves and the sterilization was caused by O-radicals and NO-radicals. Low-temperature sterilization process was possible using this method and can substitute for ethylene oxide gas (EOG) sterilization and conventional

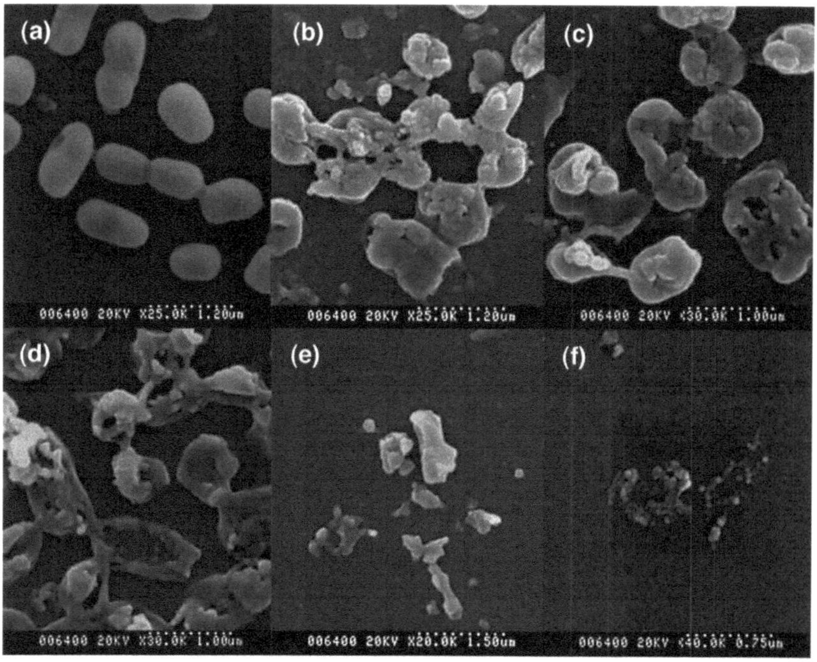

FIG. 19 SEM photographs of *E. coli*: untreated control (A) and spores exposed to microwave-induced argon plasma at atmospheric pressure for 1 s (B), 3 s (C), 5 s (D), 7 s (E), and 10 s (F). *(Reproduced with permission from B.J. Park, K. Takatori, M.H. Lee, D.-W. Han, Y.I. Woo, H.J. Son, J.K. Kim, K.-H. Chung, S.O. Hyun, J.-C. Park, Escherichia coli sterilization and lipopolysaccharide inactivation using microwave-induced argon plasma at atmospheric pressure. Surf. Coat. Technol. 201 (2007) 5738–5741.)*

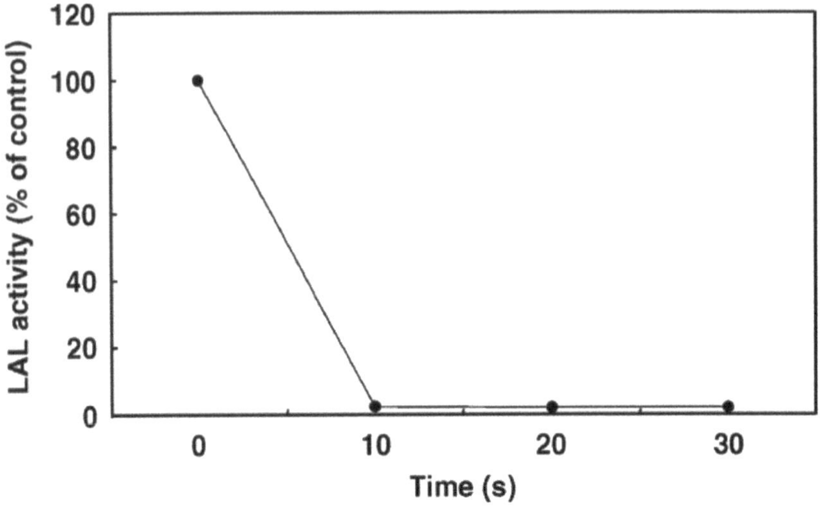

FIG. 20 LPS inactivation by microwave-induced argon plasma at atmospheric pressure. *(Reproduced with permission from B.J. Park, K. Takatori, M.H. Lee, D.-W. Han, Y.I. Woo, H.J. Son, J.K. Kim, K.-H. Chung, S.O. Hyun, J.-C. Park, Escherichia coli sterilization and lipopolysaccharide inactivation using microwave-induced argon plasma at atmospheric pressure, Surf. Coat. Technol. 201 (2007) 5738–5741.)*

hydrogen peroxide sterilization. Furthermore, chemical agents and sterilization gases were not needed for this method, and hence, it is a low-cost method. Along with that, the microwave plasma can produce high energy to a small area without creating a huge quantity of heat. These high-density and high-energy active species at low temperature were appropriate for medical sterilization application. High-energy electrons present in the microwave plasma produce active species like nitrogen oxides that cannot be obtainable in a conventional plasma sterilizer. The NO radical produced in the air plasma can deoxidize the organic compounds. When these radicals interact with the proteins on the surface of microorganisms, deoxidization and decomposition occur. Thus, the addition of NO radical to protein molecules on the surface of microorganisms modifies the original protein functions and the microorganisms become inactive [130, 131]. The biological indicator, *Geobacillus stearothermophilus* (ATCC7953), was used to examine the optimal conditions such as pressure and microwave output. The schematic view of the sterilization apparatus is shown in Fig. 21. It was found from the study that the sterilization was completed within 6 h with a microwave output power of 280 W and a pressure of 60 Pa in the vacuum chamber at the temperature of 60°C or less.

Changes in the number of bacilli (*B. subtilis* var. natto) by plasma exposure time are shown in Fig. 22. Within 10 min, the number decreased to $\frac{1}{10}^{\text{th}}$ of the original number. One hour was needed in order to decrease the number to $\frac{1}{100}^{\text{th}}$ of the original. This was mainly due to the overlapping of the bacilli and the radicals do not reach them well.

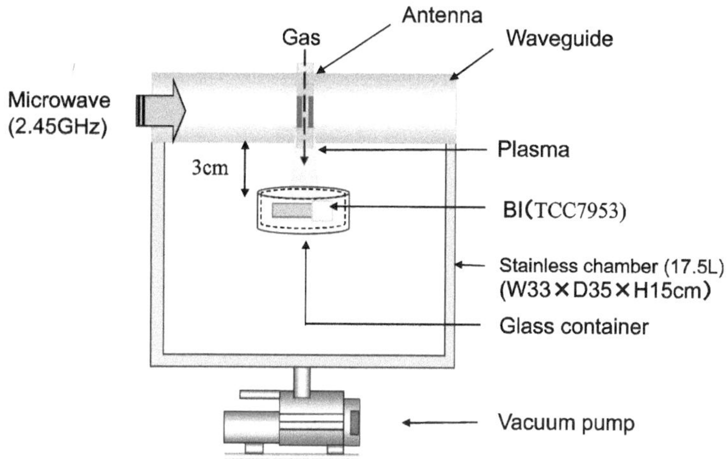

FIG. 21 Schematic view of the sterilization apparatus. *(Reproduced with permission from T. Itarashiki, N. Hayashi, A. Yonesu, Sterilization effect of nitrogen oxide radicals generated by microwave plasma using air, Vacuum 110 (2014) 213–216. https://doi.org/10.1016/j.vacuum.2014.06.018.)*

A method for dry sterilization employing Electron cyclotron resonance (ECR) microwave plasma was described [120]. ECR microwave plasma processing method has several advantages in terms of a high degree of ionization, low average electron temperature, and flexible operating temperature and pressure [132]. The schematic diagram of the ECR microwave plasma sterilization system is shown in Fig. 23.

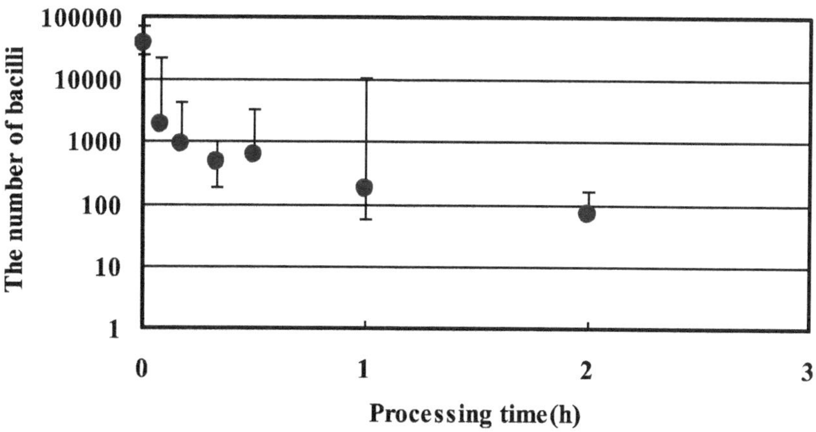

FIG. 22 Changes in the number of bacilli (*B. subtilis* var. natto) by plasma exposure time. *(Reproduced with permission from T. Itarashiki, N. Hayashi, A. Yonesu, Sterilization effect of nitrogen oxide radicals generated by microwave plasma using air, Vacuum 110 (2014) 213–216. https://doi.org/10.1016/j.vacuum.2014.06.018.)*

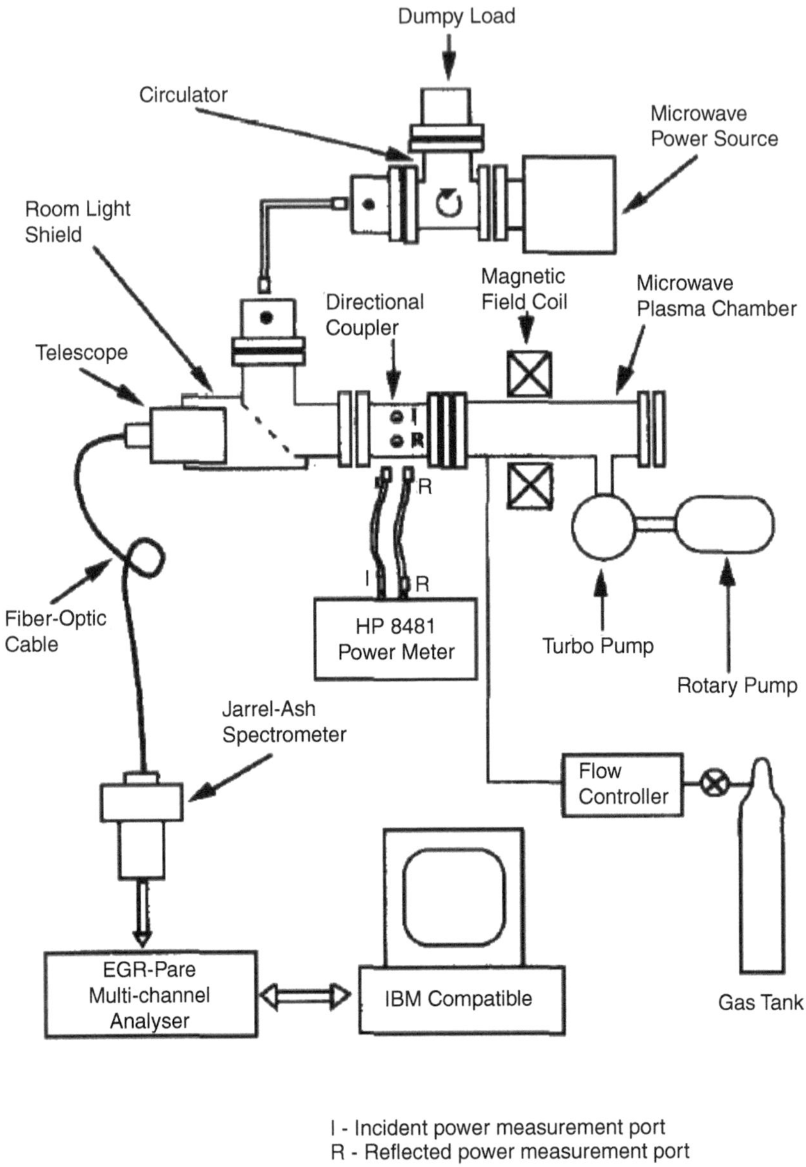

FIG. 23 Schematic diagram of the ECR microwave plasma sterilization system. *(Reproduced with permission from T.T. Chau, K.C. Kao, G. Blank, F. Madrid, Microwave plasmas for low-temperature dry sterilization. Biomaterials 17 (1996) 1273–1277.)*

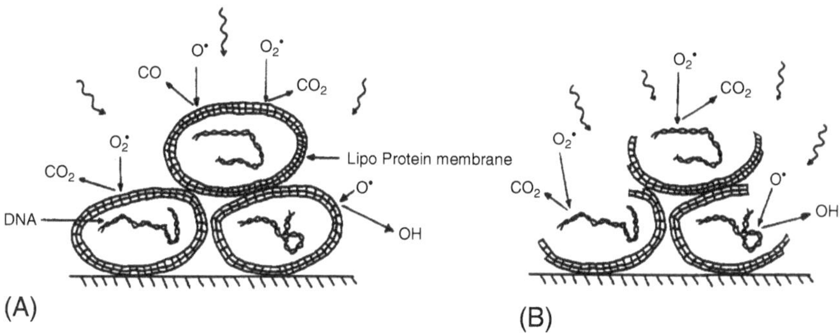

FIG. 24 Dry sterilization process in microwave N_2O plasma. (A) Prior to sterilization. (B) During sterilization. *(Reproduced with permission from T.T. Chau, K.C. Kao, G. Blank, F. Madrid, Microwave plasmas for low-temperature dry sterilization. Biomaterials 17 (1996) 1273–1277.)*

Fig. 24 shows the dry sterilization process in microwave N_2O plasma. Prior to sterilization, the activated oxygen species started to react with the hydrocarbon bonds of the cell wall of bacteria (Fig. 24A) and during the sterilization, the removal of the upper portion of the cell wall and the reaction of the plasma with the lower portion of the cell wall and DNA occurred (Fig. 24B). The bacteria, including *E. coli*, *Proteus vulgaris*, *Pseudomonas fluorescens*, and *B. stearothermophilus,* were considered in this study. The study showed that the low-temperature dry sterilization method is easy to use, nontoxic, and requires only less time compared to other methods for sterilization. It is practicable for application in the field of sterilization at medical and dental clinics.

Microwave-induced plasma treatment is one of the suitable methods for the sterilization of microorganisms in silk fabrics. The sterilization effects of microwave-induced argon plasma on the microorganisms in silk fabrics at atmospheric pressure were reported [133].

Fig. 25 shows the sterilization effects of microwave-induced argon plasma on bacterial and fungal strains in silk fabrics at atmospheric pressure. The report showed that *Staphylococcus aureus* was the most resistant bacterial strain during plasma treatment. However, all the strains were completely sterilized in less than 7 s. Several other studies also demonstrated that bacterial and fungal strains can be fully sterilized and inactivated by plasma treatment in 20 s or less [125, 126, 128].

The morphological changes in the strains of *Pseudomonas aeruginosa* by plasma treatment for 5 s are shown in SEM micrographs (Fig. 26). The morphology of the nontreated *Pseudomonas aeruginosa* spores was round in shape (Fig. 26A and B). After the plasma treatment, the spores were totally destroyed (Fig. 26C and D).

SEM images of *Aspergillus niger* before and after microwave-induced argon plasma treatment are shown in Fig. 27. The morphology of the nontreated *Aspergillus niger* was in a globular shape (Fig. 27A and B). After the plasma-treatment for 5 s, the spores were clearly damaged with holes in the cell walls (Fig. 27C and D).

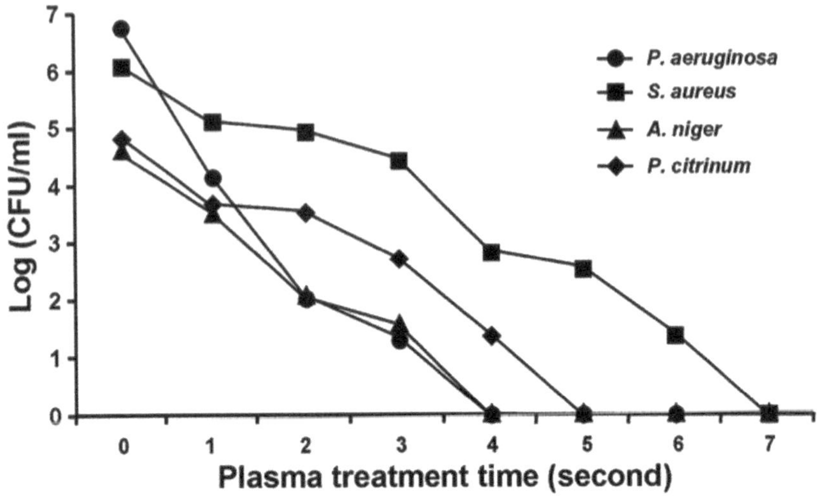

FIG. 25 Sterilization effects of microwave-induced argon plasma treatment on silk fabrics inoculated with different kinds of fungal and bacterial strains. *(Reproduced with permission from D.J. Park, M.H. Lee, Y.I. Woo, D.-W. Han, J.B. Choi, J.K. Kim, S.O. Hyun, K.-H. Chung, J.-C. Park, Sterilization of microorganisms in silk fabrics by microwave-induced argon plasma treatment at atmospheric pressure. Surf. Coat. Technol. 202 (2008) 5773–5778. https://doi.org/10.1016/j.surfcoat.2008.06.039.)*

Lee et al. investigated the sterilization effects of *E. coli* and *Methicillin-resistant Staphylococcus aureus* (MRSA) using microwave-induced argon plasma at atmospheric pressure [127]. Sterilization was performed using a self-designed, low-cost, and reliable, 2.45 GHz, waveguide-based applicator to generate plasma at atmospheric pressure (Fig. 28).

Only a much shorter plasma treatment time was needed for bacterial sterilization. Both the bacteria were wholly sterilized in less than one second. The SEM images of *E. coli* are shown in Fig. 29. The untreated normal spores were in an ellipsoidal shape (Fig. 29A), while the plasma-treated *E. coli* spores for 1, 2, 3, 4, and 5 s were completely damaged with holes in the cell walls (Fig. 29B–F).

The SEM images of *methicillin-resistant Staphylococcus aureus* are shown in Fig. 30. The untreated *methicillin-resistant Staphylococcus aureus* cells were in a globular shape (Fig. 30A). The plasma-treated *methicillin-resistant Staphylococcus aureus* spores were completely damaged (Fig. 30B–F).

The study confirmed three basic mechanisms for sterilization. The processes were DNA destruction by UV irradiation, the erosion of the microorganism through intrinsic photodesorption, and etching in the plasma [121, 125, 126, 134, 135].

Studies also showed that low-pressure microwave plasmas were appropriate for the sterilization of materials for different kinds of applications. For instance, Feichtinger et al. proved that low-pressure microwave plasmas were able to inactivate four different kinds of spores effectively [136]. They used different

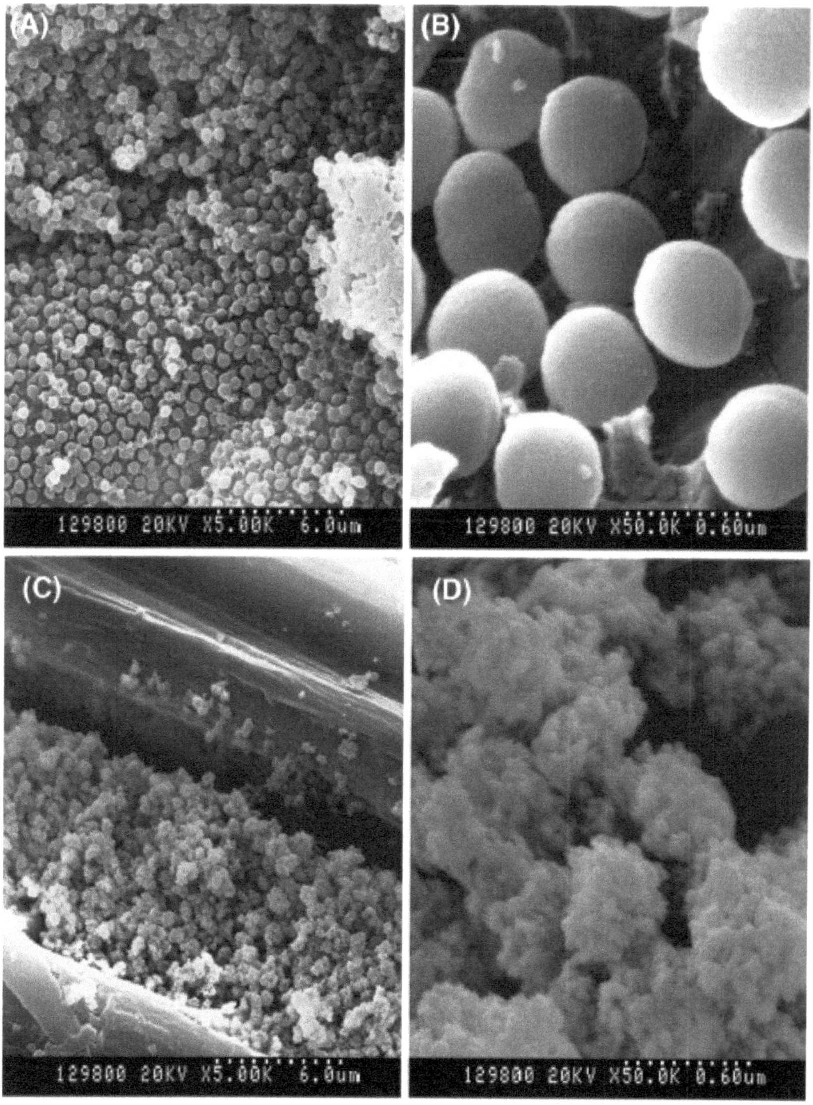

FIG. 26 (A) and (B) SEM micrographs of morphology of *Pseudomonas aeruginosa* in silk fabrics before microwave-induced argon plasma treatment. (C) and (D) after microwave-induced argon plasma treatment for 5 s. *(Reproduced with permission from D.J. Park, M.H. Lee, Y.I. Woo, D.-W. Han, J.B. Choi, J.K. Kim, S.O. Hyun, K.-H. Chung, J.-C. Park, Sterilization of microorganisms in silk fabrics by microwave-induced argon plasma treatment at atmospheric pressure. Surf. Coat. Technol. 202 (2008) 5773–5778. https://doi.org/10.1016/j.surfcoat.2008.06.039.)*

FIG. 27 (A) and (B) SEM micrographs of the morphology of *Aspergillus niger* in silk fabrics before microwave-induced argon plasma treatment. (C) and (D) after microwave-induced argon plasma treatment for 5 s. *(Reproduced with permission from D.J. Park, M.H. Lee, Y.I. Woo, D.-W. Han, J.B. Choi, J.K. Kim, S.O. Hyun, K.-H. Chung, J.-C. Park, Sterilization of microorganisms in silk fabrics by microwave-induced argon plasma treatment at atmospheric pressure, Surf. Coat. Technol. 202 (2008) 5773–5778. https://doi.org/10.1016/j.surfcoat.2008.06.039.)*

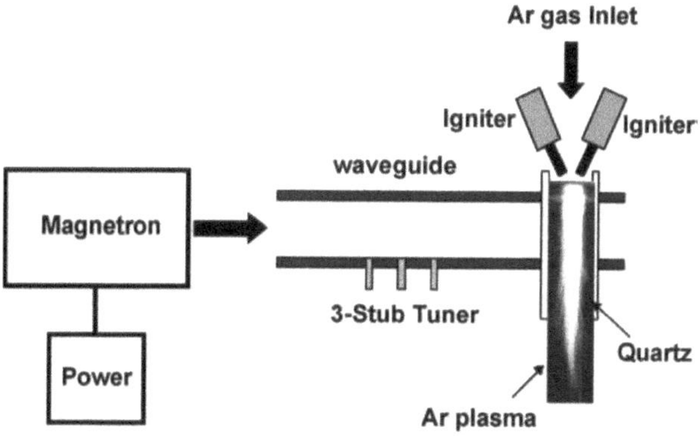

FIG. 28 Schematic diagram of the microwave-induced argon plasma system. *(Reproduced with permission from K.-Y. Lee, B. Joo Park, D. Hee Lee, I.-S. Lee, S. O. Hyun, K.-H. Chung, J.-C. Park, Sterilization of Escherichia coli and MRSA using microwave-induced argon plasma at atmospheric pressure. Surf. Coat. Technol. 193 (2005) 35–38. https://doi.org/10.1016/j.surfcoat.2004.07.034.)*

microwave plasma sources (Plasmodul and Planartron) to treat test substrates with defined initial contamination of test spores. The reduction kinetics of spores were reported for four different test spores found with different working gases and different plasma sources. Fig. 31 shows the reduction kinetics of *B. subtilis*; three different working gases (air, argon, and ammonia) were tested. The working pressure was 15 Pa. For all working gases, within the first 10 s of the plasma treatment, a very fast inactivation of the *B. subtilis* spores was observed.

Fig. 32 shows the very fast spore inactivation for all four test spores with the Planartron plasma source with air as the working gases. The working pressure was 20 Pa. After 1 s of plasma treatment, all the test spores were reduced by four orders of magnitude. Shorter treatment times were found for the Planartron source because of the direct contact of the test samples with the plasma.

After less than 1 s of plasma treatment, a fast spore inactivation of four orders of magnitude could be reached. The study proves that the fast inactivation is due to the effect of UV and VUV light that was produced in the plasma.

8.5 Sterilization using combined system

It is reported that the combined microwave-infrared (MW-IR) heating technology could be effectual for microbial inactivation. It was also useful for the restriction of moisture deteriorative physicochemical reactions [137–141]. The volumetric heating phenomena produced by the microwave-assisted heating aid to get the temperature of the product rapidly and uniformly within the entire volume [142]. Besides, spatial changes in the temperature distribution produced because of heating mechanisms of microwave (volumetric heating) and infrared (surface heating) radiations as well as the subsequent variation

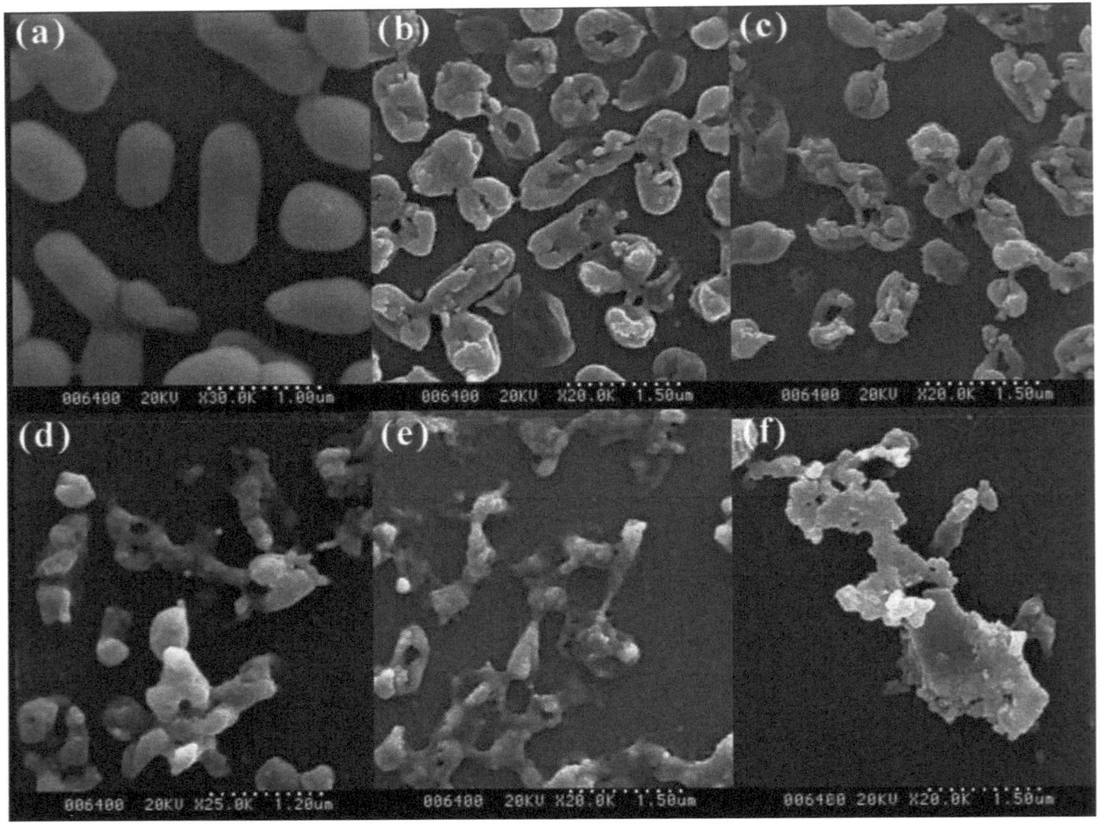

FIG. 29 SEM photographs of *E. coli*: untreated control (A) and spores exposed to the microwave-induced argon plasma at atmospheric pressure for 1 s (B), 2 s (C), 3 s (D), 4 s (E), and 5 s (F). *(Reproduced with permission from K.-Y. Lee, B. Joo Park, D. Hee Lee, I.-S. Lee, S. O. Hyun, K.-H. Chung, J.-C. Park, Sterilization of Escherichia coli and MRSA using microwave-induced argon plasma at atmospheric pressure. Surf. Coat. Technol. 193 (2005) 35–38. https://doi.org/10.1016/j.surfcoat.2004.07.034.)*

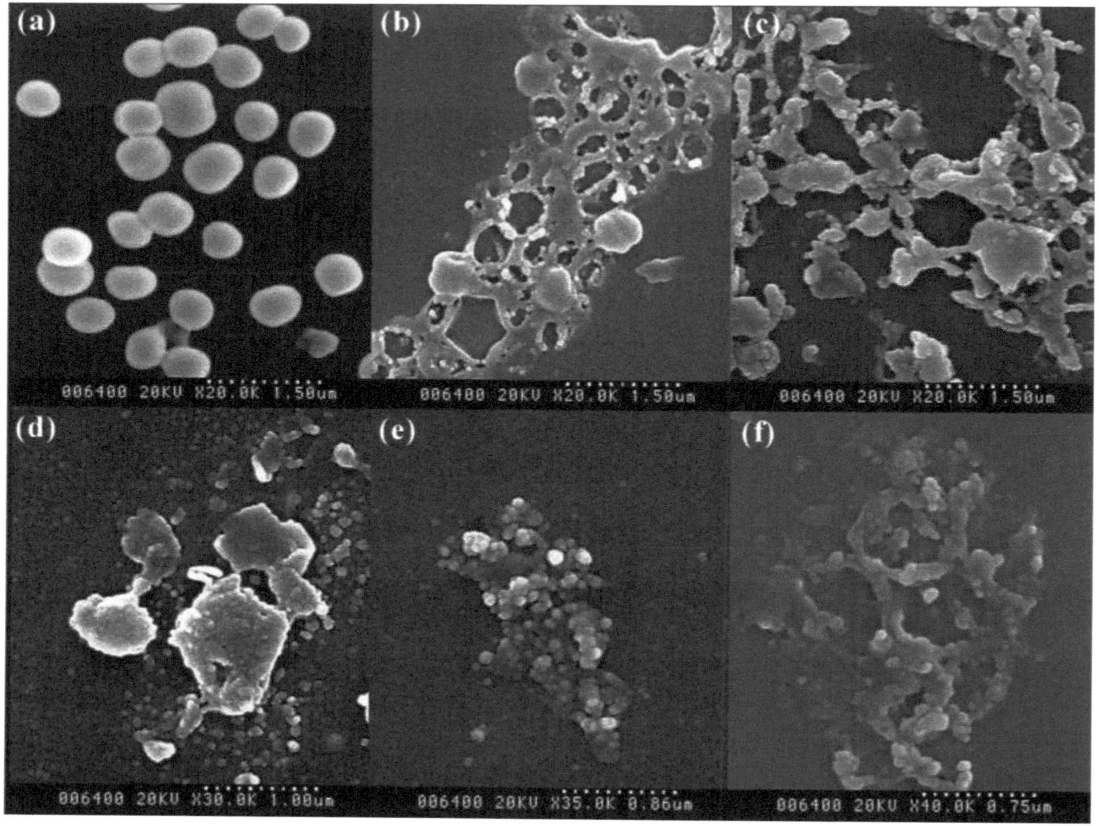

FIG. 30 SEM photographs of MRSA: untreated control (A) and spores exposed to the microwave-induced argon plasma at atmospheric pressure for 1 s (B), 2 s (C), 3 s (D), 4 s (E), and 5 s (F). *(Reproduced with permission from K.-Y. Lee, B. Joo Park, D. Hee Lee, I.-S. Lee, S. O. Hyun, K.-H. Chung, J.-C. Park, Sterilization of Escherichia coli and MRSA using microwave-induced argon plasma at atmospheric pressure. Surf. Coat. Technol. 193 (2005) 35–38. https://doi.org/10.1016/j.surfcoat.2004.07.034.)*

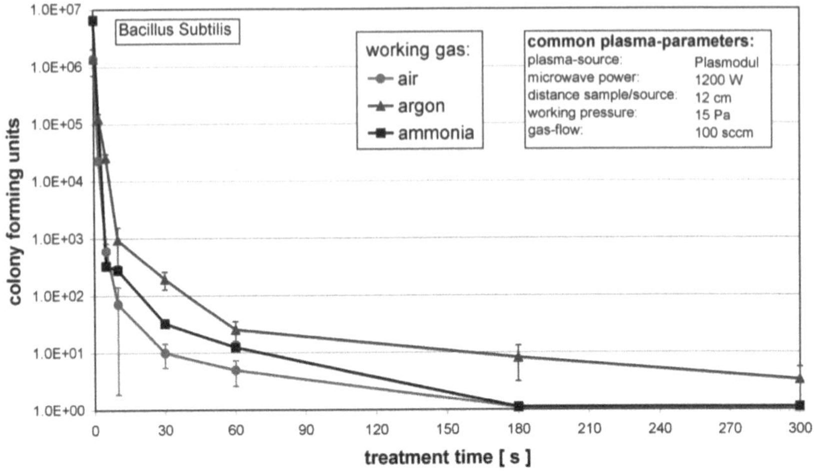

FIG. 31 *Bacillus subtilis* reduction kinetics for three different working gases. *(Reproduced with permission from J. Feichtinger, A. Schulz, M. Walker, U. Schumacher, Sterilisation with low-pressure microwave plasmas. Surf. Coat. Technol. 174–175 (2003) 564–569. https://doi.org/10.1016/S0257-8972(03)00404-3.)*

in water activity of the product are effective for inactivation of microbial and limiting the moisture-related deteriorative reactions.

Recently, Shirkole et al. reported the inactivation of *Salmonella typhimurium* and *Aspergillus flavus* in paprika by short-time intensive

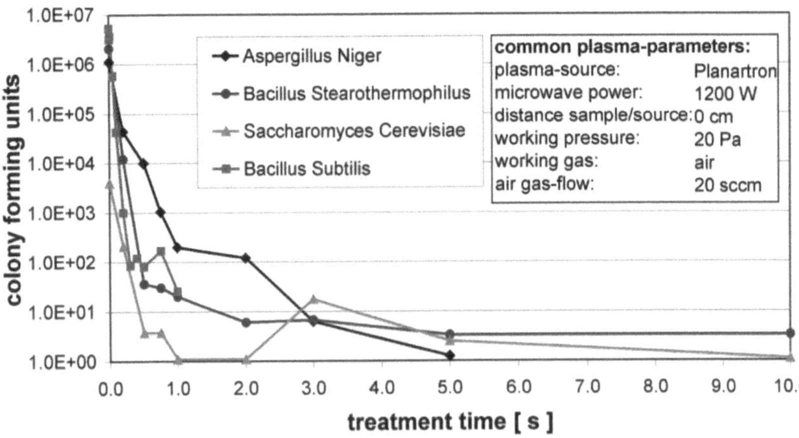

FIG. 32 Fast spore inactivation with the Planartron plasma source. *(Reproduced with permission from J. Feichtinger, A. Schulz, M. Walker, U. Schumacher, Sterilisation with low-pressure microwave plasmas. Surf. Coat. Technol. 174–175 (2003) 564–569. https://doi.org/10.1016/S0257-8972(03)00404-3.)*

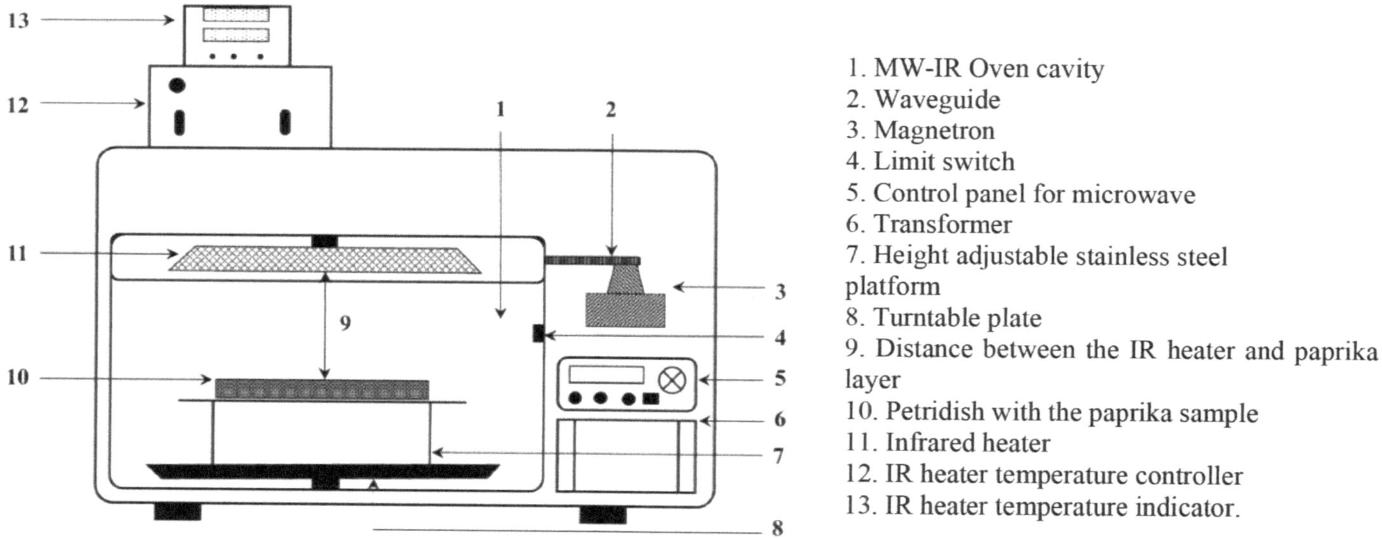

FIG. 33 Schematic diagram of the experimental setup used for combined microwave-infrared-assisted heating of paprika. *(Reproduced with permission from S.S. Shirkole, R. Jayabalan, P.P. Sutar, Dry sterilization of paprika (*Capsicum annuum *L.) by short time-intensive microwave-infrared radiation: establishment of process using glass transition, sorption, and quality degradation kinetic parameters, Innov. Food Sci. Emerg. Technol. 62 (2020) 102345. https://doi.org/10.1016/j.ifset.2020.102345.)*

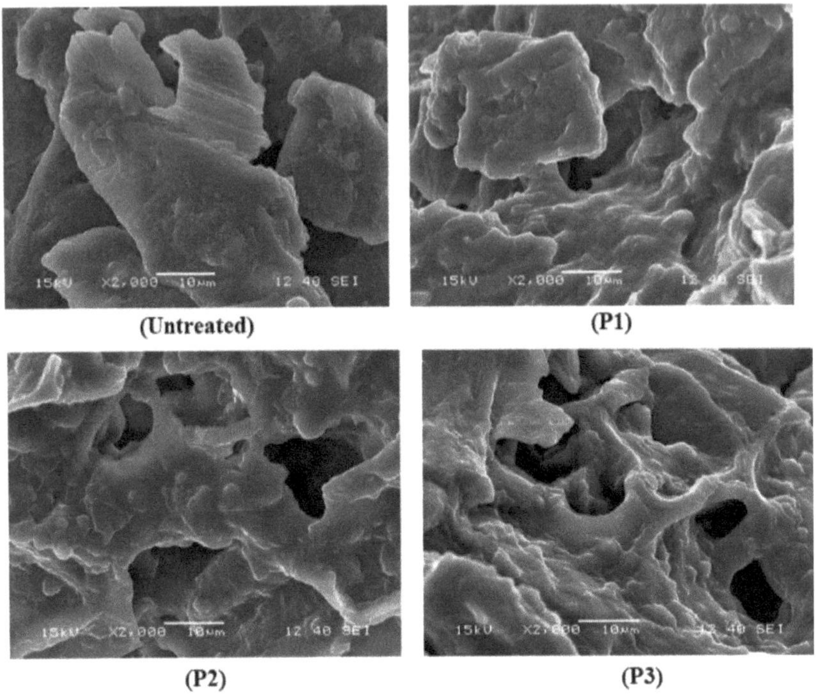

FIG. 34 SEM images of untreated and treated paprika at three power levels (P1, P2, and P3) of combined MW-IR-assisted heating for 120 s. *(Reproduced with permission from S.S. Shirkole, R. Jayabalan, P.P. Sutar, Dry sterilization of paprika (*Capsicum annuum *L.) by short time-intensive microwave-infrared radiation: establishment of process using glass transition, sorption, and quality degradation kinetic parameters, Innov. Food Sci. Emerg. Technol. 62 (2020) 102345. https://doi.org/10.1016/j.ifset.2020.102345.)*

microwave-infrared radiation [143]. Fig. 33 shows the experimental setup used for the combined microwave-infrared-assisted dry sterilization of paprika.

The SEM images of untreated and treated paprika at three power levels (P1, P2, and P3) are shown in Fig. 34. The microstructure of treated paprika was deeply influenced by the power levels of combined radiations. The increase in the average temperature of paprika from 70°C to 105°C because of the increase in power from P1 to P3 led to the enlarged porous structure. Besides, the use of high-power combined radiations resulted in a quicker rate of heat and mass transfer because of that expansion and intercellular spaces was arose [144].

The study evidenced that the short-time intensive Microwave-infrared-assisted heating process is an alternative process for microbial and enzyme inactivation, finish drying, and restriction of moisture deteriorative interactions in low moisture products.

A sterilization system using microwave and UV light was also demonstrated [9]. The Schematics diagram of the system is shown in Fig. 35.

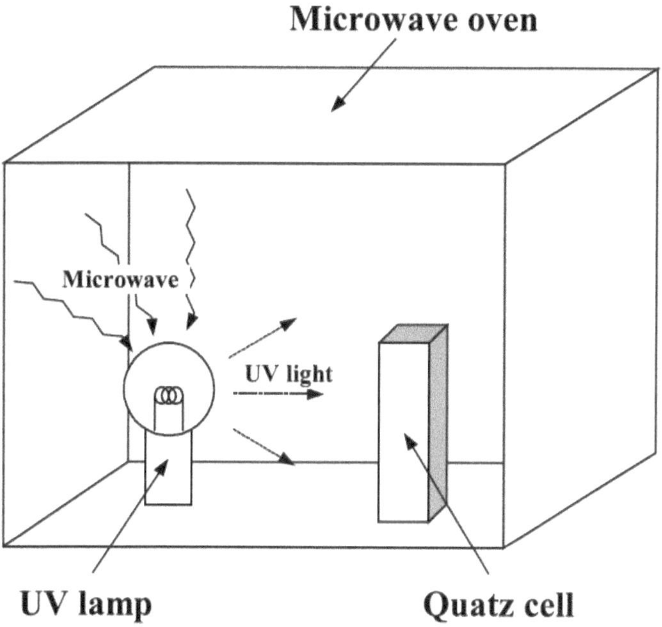

FIG. 35 Schematics diagram of microwave-UV light sterilization system. *(Reproduced with permission from S. Iwaguch, K. Matsumura, Y. Tokuoka, S. Wakui, N. Kawashima, Sterilization system using microwave and UV light. Colloids Surf. B: Biointerfaces 25 (2002) 299–304.)*

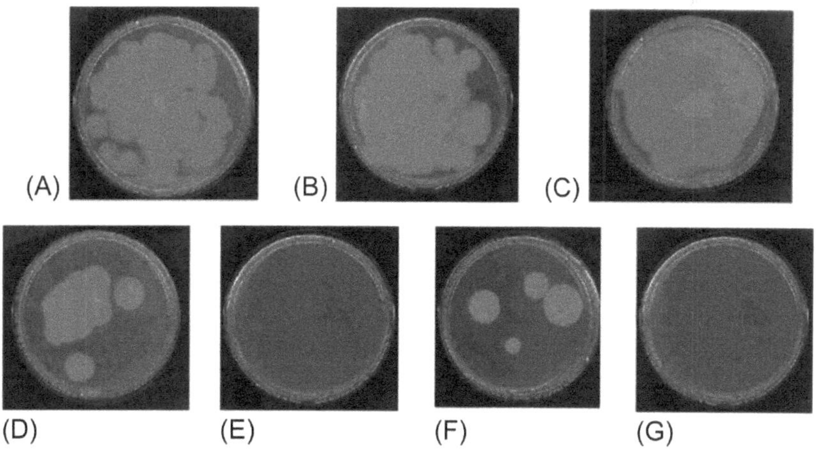

FIG. 36 Sterilization of *E. coli*. (A) Untreated, (B) irradiated with microwave alone for 5 s, (C) irradiated with microwave alone for 10 s, (D) irradiated with microwave and UV light for 5 s, (E) irradiated with microwave and UV light for 10 s, (F) irradiated with UV light alone for 5 s, and (G) irradiated with UV light alone for 10 s. *(Reproduced with permission from S. Iwaguch, K. Matsumura, Y. Tokuoka, S. Wakui, N. Kawashima, Sterilization system using microwave and UV light. Colloids Surf. B: Biointerfaces 25 (2002) 299–304.)*

The demonstrated sterilization system can emit UV light by irradiation of microwave without other power. The number of active oxygen species such as hydroxyl radical or superoxide generated on aqueous DMPO solution by irradiation of microwave and UV light was larger than that by irradiation of UV light alone. This was because of the promotion of emission of UV light photons by microwave and UV light irradiation.

Sterilization of *E. coli* under various conditions is shown in Fig. 36. No colony was seen after irradiation of microwave and UV light for 10 s (Fig. 36E) and after irradiation of microwave alone, lots of colonies were observed (Fig. 36B and C).

Thus, microwave-UV light sterilization is highly effective for the sterilization of microorganisms. The generated active oxygen species played an important role in the sterilization.

8.6 Conclusion

In this chapter, we have described the application of microwaves in the sterilization process. Mainly the application of microwave sterilization in the medical industry and food industry is described in detail. Microwave-assisted plasma sterilization methods and combined microwave systems are also included in this chapter.

Acknowledgments

AD and BKB are grateful to Prince Mohammad Bin Fahd University for support. BKB is also grateful to US NIH, US NCI, and Kleberg Foundation of Texas for financial support.

References

[1] S.C. Clarke, R.D. Haigh, P.P.E. Freestone, P.H. Williams, Virulence of enteropathogenic Escherichia coli, a global pathogen, Clin. Microbiol. Rev. 16 (2003) 365–378.

[2] K. Hiramatsu, L. Cui, M. Kuroda, T. Ito, The emergence and evolution of methicillin-resistant Staphylococcus aureus, Trends Microbiol. 9 (2001) 486–493.

[3] D. Metry, R. Katta, New and emerging pediatric infections, Dermatol. Clin. 21 (2003) 269.

[4] A.G. Whittaker, E.M. Graham, R.L. Baxter, A.C. Jones, P.R. Richardson, G. Meek, G.A. Campbell, A. Aitken, H.C. Baxter, Plasma cleaning of dental instruments, J. Hosp. Infect. 56 (2004) 37–41.

[5] M.-W. Byun, H.-S. Yook, I.-J. Kang, C.-K. Chung, J.-H. Kwon, K.-J. Choi, Comparative effects of gamma irradiation and ozone treatment on hygienic quality of Korean red ginseng powder, Radiat. Phys. Chem. 52 (1998) 95–99.

[6] H. Honda, M. Sugiyama, A. Kakamu, A. Mizoguchi, Y. Ishikawa, T. Kobayashi, Characterization on heat sterilization of a novel autoclave with a turntable for rotation, J. Ferment. Bioeng. 81 (1996) 570–572.

[7] S. Lerouge, A.C. Fozza, M.R. Wertheimer, R. Marchand, L. Yahia, Sterilization by low-pressure plasma: the role of vacuum-ultraviolet radiation, Plasma. Polym. 5 (2000) 31–46.

[8] T.C. Montie, K. Kelly-Wintenberg, J.R. Roth, An overview of research using the one atmosphere uniform glow discharge plasma (OAUGDP) for sterilization of surfaces and materials, IEEE Trans. Plasma Sci. 28 (2000) 41–50.
[9] S. Iwaguch, K. Matsumura, Y. Tokuoka, S. Wakui, N. Kawashima, Sterilization system using microwave and UV light, Colloids Surf. B: Biointerfaces 25 (2002) 299–304.
[10] M.R. Sanborn, S.K. Wan, R. Bulard, Microwave sterilization of plastic tissue culture vessels for reuse, Appl. Environ. Microbiol. 44 (1982) 960–964.
[11] R.A. Dunsmuir, G. Gallacher, Microwave sterilization of femoral head allograft, J. Clin. Microbiol. 41 (2003) 4755–4757.
[12] F. Celandroni, I. Longo, N. Tosoratti, F. Giannessi, E. Ghelardi, S. Salvetti, A. Baggiani, S. Senesi, Effect of microwave radiation on Bacillus subtilis spores, J. Appl. Microbiol. 97 (2004) 1220–1227.
[13] D.K. Jeng, K.A. Kaczmarek, A.G. Woodworth, G. Balasky, Mechanism of microwave sterilization in the dry state, Appl. Environ. Microbiol. 53 (1987) 2133–2137.
[14] J.M. Latimer, J.M. Matsen, Microwave oven irradiation as a method for bacterial decontamination in a clinical microbiology laboratory, J. Clin. Microbiol. 6 (1977) 340–342.
[15] D.I. Martin, I. Margaritescu, E. Cirstea, I. Togoe, D. Ighigeanu, M.R. Nemtanu, C. Oproiu, N. Iacob, Application of accelerated electron beam and microwave irradiation to biological waste treatment, Vacuum 77 (2005) 501–506.
[16] N. Hayashi, S. Tsutsui, T. Tomari, W. Guan, Sterilization of medical equipment using oxygen radicals produced by water vapor RF plasma, IEEE Trans. Plasma Sci. 36 (2008) 1302–1303, https://doi.org/10.1109/TPS.2008.924453.
[17] M.A. Brondani, A.R. Siqueira, A critical review of protocols for conventional microwave oven use for denture disinfection, Community Dent. Health (2018).
[18] A. Yezdani, K. Mahalakshmi, K. Padmavathy, Orthodontic instrument sterilization with microwave irradiation, J. Pharm. Bioallied Sci. 7 (2015) S111.
[19] G.V. Barbosa-Cánovas, I. Medina-Meza, K. Candoğan, D. Bermúdez-Aguirre, Advanced retorting, microwave assisted thermal sterilization (MATS), and pressure assisted thermal sterilization (PATS) to process meat products, Meat Sci. 98 (2014) 420–434.
[20] R.B. Sabnis, A. Bhattu, V. Mohankumar, Sterilization of endoscopic instruments, Curr. Opin. Urol. 24 (2014) 195–202.
[21] G.S. Bhalla, K. Bandyopadhyay, K. Sahai, Keeping in pace with the new biomedical waste management rules: what we need to know! Med. J. Armed Forces India 75 (2019) 240–245.
[22] N. Ciplak, S. Kaskun, Healthcare waste management practice in the West Black Sea Region, Turkey: a comparative analysis with the developed and developing countries, J. Air Waste Manage. Assoc. 65 (2015) 1387–1394.
[23] W.-C. Chao, C.-L. Wu, P.-Y. Liu, C.-C. Shieh, Regular sputum check-up for early diagnosis of tuberculosis after exposure in healthcare facilities, PLoS One 11 (2016), e0157054.
[24] T. Singh, T.R. Ghimire, S.K. Agrawal, Awareness of biomedical waste management in dental students in different dental colleges in Nepal, Biomed. Res. Int. 2018 (2018).
[25] A.C. Abreu, R.R. Tavares, A. Borges, F. Mergulhão, M. Simões, Current and emergent strategies for disinfection of hospital environments, J. Antimicrob. Chemother. 68 (2013) 2718–2732.
[26] B.T. Garibaldi, M. Reimers, N. Ernst, G. Bova, E. Nowakowski, J. Bukowski, B.C. Ellis, C. Smith, L. Sauer, K. Dionne, Validation of autoclave protocols for successful decontamination of category a medical waste generated from care of patients with serious communicable diseases, J. Clin. Microbiol. 55 (2017) 545–551.
[27] W.A. Rutala, APIC guideline for selection and use of disinfectants, Am. J. Infect. Control 24 (1996) 313–342.

[28] N. Wurtz, A. Papa, M. Hukic, A. Di Caro, I. Leparc-Goffart, E. Leroy, M.P. Landini, Z. Sekeyova, J.S. Dumler, D. Bădescu, Survey of laboratory-acquired infections around the world in biosafety level 3 and 4 laboratories, Eur. J. Clin. Microbiol. Infect. Dis. 35 (2016) 1247–1258.
[29] C. Doig, A.L. Seagar, B. Watt, K.J. Forbes, The efficacy of the heat killing of Mycobacterium tuberculosis, J. Clin. Pathol. 55 (2002) 778–779.
[30] V.P. Myneedu, A. Aggarwal, Disposal of the large volume of sputum positive for Mycobacterium tuberculosis by using microwave sterilisation technology as an alternative to traditional autoclaving in a tertiary respiratory care hospital in Delhi, India, Infect. Prev. Pract. 2 (2020) 100072.
[31] S.C. Ojha, S. Chankhamhaengdecha, S. Singhakaew, P. Ounjai, T. Janvilisri, Inactivation of Clostridium difficile spores by microwave irradiation, Anaerobe 38 (2016) 14–20.
[32] L. Huang, J. Sites, New automated microwave heating process for cooking and pasteurization of microwaveable foods containing raw meats, J. Food Sci. 75 (2010) E110–E115.
[33] E. Benjamin, A. Reznik, A.L. Williams, Mathematical models for conventional and microwave thermal deactivation of Enterococcus faecalis, Staphylococcus aureus and Escherichia coli, Cell. Mol. Biol. 53 (2007) 42–48 (Noisy-Le-Grand, France).
[34] S.-S. Kim, H.-J. Sung, H.-S. Kwak, I.-S. Joo, J.-S. Lee, G. Ko, D.-H. Kang, Effect of power levels on inactivation of Escherichia coli O157: H7, Salmonella typhimurium, and Listeria monocytogenes in tomato paste using 915-megahertz microwave and Ohmic heating, J. Food Prot. 79 (2016) 1616–1622.
[35] B. Köhler, H. Hübner, M. Krautschick, The use of ionizing radiation for the decontamination of salmonella-containing slaughtered broiler chickens and powdered eggs, Z. Gesamte Hyg. Ihre Grenzgeb. 35 (1989) 665.
[36] H. Khalil, R. Villota, Comparative study on injury and recovery of Staphylococcus aureus using microwaves and conventional heating, J. Food Prot. 51 (1988) 181–186.
[37] H. Khalil, R. Villota, The effect of microwave sublethal heating on the ribonucleic acids of Staphylococcus aureus, J. Food Prot. 52 (1989) 544–548.
[38] P.J. Coote, C.D. Holyoak, M.B. Cole, Thermal inactivation of Listeria monocytogenes during a process simulating temperatures achieved during microwave heating, J. Appl. Bacteriol. 70 (1991) 489–494.
[39] D. Eriksson, T. Stigbrand, Radiation-induced cell death mechanisms, Tumor Biol. 31 (2010) 363–372.
[40] L. Zhou, R. Yuan, S. Lanata, Molecular mechanisms of irradiation-induced apoptosis, Front Biosci 8 (2003) d9.
[41] Y. Kakita, N. Kashige, K. Murata, A. Kuroiwa, M. Funatsu, K. Watanabe, Inactivation of Lactobacillus bacteriophage PL-1 by microwave irradiation, Microbiol. Immunol. 39 (1995) 571–576.
[42] B.C. Webb, C.J. Thomas, D.W. Harty, M.D. Willcox, Effectiveness of two methods of denture sterilization, J. Oral Rehabil. 25 (1998) 416–423.
[43] T. Klironomos, A. Katsimpali, G. Polyzois, The effect of microwave disinfection on denture base polymers, liners and teeth: a basic overview, Acta Stomatol. Croat. 49 (2015) 242–253.
[44] G. Wemken, B.C. Spies, S. Pieralli, U. Adali, F. Beuer, C. Wesemann, Do hydrothermal aging and microwave sterilization affect the trueness of milled, additive manufactured and injection molded denture bases? J. Mech. Behav. Biomed. Mater. 111 (2020) 103975.
[45] M.D. Rohrer, M.A. Terry, R.A. Bulard, D.C. Graves, E.M. Taylor, Microwave sterilization of hydrophilic contact lenses, Am J. Ophthalmol. 101 (1986) 49–57, https://doi.org/10.1016/0002-9394(86)90464-2.

[46] D.R. Clohisy, H.J. Mankin, Osteoarticular allografts for reconstruction after resection of a musculoskeletal tumor in the proximal end of the tibia, J. Bone Joint Surg. Am. 76 (1994) 549–554.

[47] L. Vastel, C.T. Lemoine, M. Kerboull, J.P. Courpied, Structural allograft and cemented long-stem prosthesis for complex revision hip arthroplasty: use of a trochanteric claw plate improves final hip function, Int. Orthop. 31 (2007) 851–857.

[48] R. Singh, D. Singh, Sterilization of bone allografts by microwave and gamma radiation, Int. J. Radiat. Biol. 88 (2012) 661–666.

[49] J.G. Lyng, J.M. Arimi, M. Scully, F. Marra, The influence of compositional changes in reconstituted potato flakes on thermal and dielectric properties and temperatures following microwave heating, J. Food Eng. 124 (2014) 133–142, https://doi.org/10.1016/j.jfoodeng.2013.09.032.

[50] L. Zhang, J.G. Lyng, N.P. Brunton, The effect of fat, water and salt on the thermal and dielectric properties of meat batter and its temperature following microwave or radio frequency heating, J. Food Eng. 80 (2007) 142–151, https://doi.org/10.1016/j.jfoodeng.2006.05.016.

[51] M. Chen, D. Fan, T. Li, B. Yan, Y. Gao, J. Zhao, H. Zhang, Synergistic bactericidal effects of basic amino acids and microwave treatment on Escherichia coli, LWT 84 (2017) 99–105, https://doi.org/10.1016/j.lwt.2017.05.038.

[52] S. Tajchakavit, H.S. Ramaswamy, P. Fustier, Enhanced destruction of spoilage microorganisms in apple juice during continuous flow microwave heating, Food Res. Int. 31 (1998) 713–722, https://doi.org/10.1016/S0963-9969(99)00050-2.

[53] J.C. Rickman, D.M. Barrett, C.M. Bruhn, Nutritional comparison of fresh, frozen and canned fruits and vegetables. Part 1. Vitamins C and B and phenolic compounds, J. Sci. Food Agric. 87 (2007) 930–944.

[54] B.H. Chen, H.Y. Peng, H.E. Chen, Changes of carotenoids, color, and vitamin A contents during processing of carrot juice, J. Agric. Food Chem. 43 (1995) 1912–1918.

[55] L.A. Howard, A.D. Wong, A.K. Perry, B.P. Klein, β-Carotene and ascorbic acid retention in fresh and processed vegetables, J. Food Sci. 64 (1999) 929–936.

[56] J. Peng, J. Tang, D.M. Barrett, S.S. Sablani, N. Anderson, J.R. Powers, Thermal pasteurization of ready-to-eat foods and vegetables: critical factors for process design and effects on quality, Crit. Rev. Food Sci. Nutr. 57 (2017) 2970–2995.

[57] M. Cooper, G. Douglas, M. Perchonok, Developing the NASA food system for long-duration missions, J. Food Sci. 76 (2011) R40–R48.

[58] Z. Tang, G. Mikhaylenko, F. Liu, J.-H. Mah, R. Pandit, F. Younce, J. Tang, Microwave sterilization of sliced beef in gravy in 7-oz trays, J. Food Eng. 89 (2008) 375–383, https://doi.org/10.1016/j.jfoodeng.2008.04.025.

[59] H. Zhang, Z. Tang, B. Rasco, J. Tang, S.S. Sablani, Shelf-life modeling of microwave-assisted thermal sterilized mashed potato in polymeric pouches of different gas barrier properties, J. Food Eng. 183 (2016) 65–73.

[60] S. Dhawan, C. Varney, G.V. Barbosa-Cánovas, J. Tang, F. Selim, S.S. Sablani, The impact of microwave-assisted thermal sterilization on the morphology, free volume, and gas barrier properties of multilayer polymeric films, J. Appl. Polym. Sci. 131 (2014).

[61] H. Zhang, K. Bhunia, P. Kuang, J. Tang, B. Rasco, D.S. Mattinson, S.S. Sablani, Effects of oxygen and water vapor transmission rates of polymeric pouches on oxidative changes of microwave-sterilized mashed potato, Food Bioprocess Technol. 9 (2016) 341–351.

[62] H. Zhang, J. Patel, K. Bhunia, S. Al-Ghamdi, C.R. Sonar, C.F. Ross, J. Tang, S.S. Sablani, Color, vitamin C, β-carotene and sensory quality retention in microwave-assisted thermally

sterilized sweet potato puree: effects of polymeric package gas barrier during storage, Food Packag. Shelf Life 21 (2019) 100324.

[63] R.B. Pandit, J. Tang, F. Liu, G. Mikhaylenko, A computer vision method to locate cold spots in foods in microwave sterilization processes, Pattern Recogn. 40 (2007) 3667–3676.

[64] Y. Wang, M.H. Lau, J. Tang, R. Mao, Kinetics of chemical marker M-1 formation in whey protein gels for developing sterilization processes based on dielectric heating, J. Food Eng. 64 (2004) 111–118, https://doi.org/10.1016/j.jfoodeng.2003.09.019.

[65] T. Auksornsri, J. Tang, Z. Tang, H. Lin, S. Songsermpong, Dielectric properties of rice model food systems relevant to microwave sterilization process, Innovative Food Sci. Emerg. Technol. 45 (2018) 98–105, https://doi.org/10.1016/j.ifset.2017.09.002.

[66] G.T. Chala, Y.P. Lim, S.A. Sulaiman, C.L. Liew, Thermogravimetric analysis of empty fruit bunch, MATEC Web Conf. 225 (2018) 02002. EDP Sciences.

[67] M. Basyuni, N. Amri, L.A.P. Putri, I. Syahputra, D. Arifiyanto, Characteristics of fresh fruit bunch yield and the physicochemical qualities of palm oil during storage in North Sumatra, Indonesia, Indones. J. Chem. 17 (2017) 182–190.

[68] D. Kumaradevan, K.H. Chuah, L.K. Moey, V. Mohan, W.T. Wan, Optimising the operational parameters of a spherical steriliser for the treatment of oil palm fresh fruit bunch, IOP Conf. Ser. Mater. Sci. Eng. 88 (2015) 012031. IOP Publishing.

[69] A.M. Omar, T.T. Norsalwani, M.S. Asmah, Z.Y. Badrulhisham, A.M. Easa, F.M. Omar, M.S. Hossain, M.H. Zuknik, N.N. Norulaini, Implementation of the supercritical carbon dioxide technology in oil palm fresh fruits bunch sterilization: a review, J. CO_2 Util. 25 (2018) 205–215.

[70] R. Nokkaew, V. Punsuvon, Sterilization of oil palm fruits by microwave heating for replacing steam treatment in palm oil mill process, Adv. Mater. Res. 1025–1026 (2014) 470–475. Trans Tech Publ.

[71] V. Subramaniam, C.Y. May, H. Muhammad, Z. Hashim, Y.A. Tan, P.C. Wei, Life cycle assessment of the production of crude palm oil (part 3), J. Oil Palm Res. 22 (2010) 895–903.

[72] C.J. Vincent, R. Shamsudin, A.S. Baharuddin, Pre-treatment of oil palm fruits: a review, J. Food Eng. 143 (2014) 123–131.

[73] I. Umudee, M. Chongcheawchamnan, M. Kiatweerasakul, C. Tongurai, Sterilization of oil palm fresh fruit using microwave technique, Int. J. Chem. Eng. Appl. 4 (2013) 111.

[74] K. Sivasothy, R.M. Halim, Y. Basiron, A new system for continuous sterilization of oil palm fresh fruit bunches, J. Oil Palm Res. 17 (2005) 145.

[75] M.C. Chow, A.N. Ma, Microwave in the processing of fresh palm fruits, in: Cutting-Edge Technologies for Sustained Competitiveness: Proceedings of the 2001 PIPOC International Palm Oil Congress, Chemistry and Technology Conference, Kuala Lumpur, Malaysia, 20–22 August 2001, Malaysian Palm Oil Board (MPOB), 2001, pp. 3–8.

[76] N. Sukaribin, K. Khalid, Effectiveness of sterilisation of oil palm bunch using microwave technology, Ind. Crop. Prod. 30 (2009) 179–183.

[77] C.T. Ponne, P.V. Bartels, Interaction of electromagnetic energy with biological material-relation to food processing, Radiat. Phys. Chem. 45 (1995) 591–608.

[78] A.O. Valsechi, J. Horii, D.F. De Angelis, The effect of microwaves on microorganisms, Arq. Inst. Biol. São Paulo 71 (2004) 399–404.

[79] T.K. Hock, G.T. Chala, H.H. Cheng, An innovative hybrid steam-microwave sterilization of palm oil fruits at atmospheric pressure, Innovative Food Sci. Emerg. Technol. 60 (2020) 102289.

[80] D. Guan, V.C. Plotka, S. Clark, J. Tang, Sensory evaluation of microwave treated macaroni and cheese, J. Food Process. Preserv. 26 (2002) 307–322.

[81] D. Guan, P. Gray, D.-H. Kang, J. Tang, B. Shafer, K. Ito, F. Younce, T.C.S. Yang, Microbiological validation of microwave-circulated water combination heating technology by inoculated pack studies, J. Food Sci. 68 (2003) 1428–1432.

[82] S.K. Pathak, F. Liu, J. Tang, Finite difference time domain (FDTD) characterization of a single mode applicator, J. Microw. Power Electromagn. Energy. 38 (2003) 37–48.

[83] H. Zhang, A.K. Datta, Coupled electromagnetic and thermal modeling of microwave oven heating of foods, J. Microw. Power Electromagn. Energy. 35 (2000) 71–85.

[84] K.G. Ayappa, H.T. Davis, E.A. Davis, J. Gordon, Two-dimensional finite element analysis of microwave heating, AICHE J. 38 (1992) 1577–1592, https://doi.org/10.1002/aic.690381009.

[85] V.R. Romano, F. Marra, U. Tammaro, Modelling of microwave heating of foodstuff: study on the influence of sample dimensions with a FEM approach, J. Food Eng. 71 (2005) 233–241, https://doi.org/10.1016/j.jfoodeng.2004.11.036.

[86] R.B. Pandit, S. Prasad, Finite element analysis of microwave heating of potato—transient temperature profiles, J. Food Eng. 60 (2003) 193–202.

[87] M.H. Lau, J. Tang, I.A. Taub, T.C.S. Yang, C.G. Edwards, R. Mao, Kinetics of chemical marker formation in whey protein gels for studying microwave sterilization, J. Food Eng. 60 (2003) 397–405.

[88] R.B. Pandit, J. Tang, G. Mikhaylenko, F. Liu, Kinetics of chemical marker M-2 formation in mashed potato—a tool to locate cold spots under microwave sterilization, J. Food Eng. 76 (2006) 353–361.

[89] R.B. Pandit, J. Tang, F. Liu, M. Pitts, Development of a novel approach to determine heating pattern using computer vision and chemical marker (M-2) yield, J. Food Eng. 78 (2007) 522–528.

[90] H.-H. Wang, D.-W. Sun, Melting characteristics of cheese: analysis of effect of cheese dimensions using computer vision techniques, J. Food Eng. 52 (2002) 279–284.

[91] K.L. Yam, S.E. Papadakis, A simple digital imaging method for measuring and analyzing color of food surfaces, J. Food Eng. 61 (2004) 137–142.

[92] M.Z. Abdullah, L.C. Guan, K.C. Lim, A.A. Karim, The applications of computer vision system and tomographic radar imaging for assessing physical properties of food, J. Food Eng. 61 (2004) 125–135.

[93] H. Hwang, B. Park, M. Nguyen, Y.-R. Chen, Hybrid image processing for robust extraction of lean tissue on beef cut surfaces, Comput. Electron. Agric. 17 (1997) 281–294.

[94] M.K. Bhattacharjee, D.H. Fine, D.H. Figurski, tfoX (sxy)-dependent transformation of Aggregatibacter (Actinobacillus) actinomycetemcomitans, Gene 399 (2007) 53–64.

[95] Y. Wang, S.D. Goodman, R.J. Redfield, C. Chen, Natural transformation and DNA uptake signal sequences in Actinobacillus actinomycetemcomitans, J. Bacteriol. 184 (2002) 3442–3449.

[96] B. Henderson, S.P. Nair, J.M. Ward, M. Wilson, Molecular pathogenicity of the oral opportunistic pathogen Actinobacillus actinomycetemcomitans, Annu. Rev. Microbiol. 57 (2003) 29–55.

[97] L. Paturel, J.P. Casalta, G. Habib, M. Nezri, D. Raoult, Actinobacillus actinomycetemcomitans endocarditis, Clin. Microbiol. Infect. 10 (2004) 98–118.

[98] A.J. van Winkelhoff, J. Slots, Actinobacillus actinomycetemcomitans and Porphyromonas gingivalis in nonoral infections, Periodontology 2000 20 (1999) 122–135.

[99] J. Slots, H.S. Reynolds, R.J. Genco, Actinobacillus actinomycetemcomitans in human periodontal disease: a cross-sectional microbiological investigation, Infect. Immun. 29 (1980) 1013–1020.

[100] J.J. Zambon, L.A. Christersson, J. Slots, Actinobacillus actinomycetemcomitans in human periodontal disease: prevalence in patient groups and distribution of biotypes and serotypes within families, J. Periodontol. 54 (1983) 707–711.

[101] D.H. Fine, D. Furgang, H.C. Schreiner, P. Goncharoff, J. Charlesworth, G. Ghazwan, P. - Fitzgerald-Bocarsly, D.H. Figurski, Phenotypic variation in Actinobacillus actionmycetemcomitans during laboratory growth: implications for virulence, Microbiology 145 (1999) 1335–1347.

[102] J. Slots, Selective medium for isolation of Actinobacillus actinomycetemcomitans, J. Clin. Microbiol. 15 (1982) 606–609.

[103] S.C. Kachlany, P.J. Planet, M.K. Bhattacharjee, E. Kollia, R. DeSalle, D.H. Fine, D.H. Figurski, Nonspecific adherence by Actinobacillus actinomycetemcomitans requires genes widespread in bacteria and archaea, J. Bacteriol. 182 (2000) 6169–6176.

[104] M.K. Bhattacharjee, K. Sugawara, O.T. Ayandeji, Microwave sterilization of growth medium alleviates inhibition of Aggregatibacter actinomycetemcomitans by Maillard reaction products, J. Microbiol. Methods 78 (2009) 227–230.

[105] V.H. Cardoso, D.L. Goncalves, E. Angioletto, F. Dal-Pizzol, E.L. Streck, Microwave disinfection of gauze contaminated with bacteria and fungi, Indian J. Med. Microbiol. 25 (2007) 428–429.

[106] Y. Kakita, M. Funatsu, F. Miake, K. Watanabe, Effects of microwave irradiation on bacteria attached to the hospital white coats, Int. J. Occup. Med. Environ. Health 12 (1999) 123–126.

[107] L.E. Steed, V.-D. Truong, J. Simunovic, K.P. Sandeep, P. Kumar, G.D. Cartwright, K.R. Swartzel, Continuous flow microwave-assisted processing and aseptic packaging of purple-fleshed sweetpotato purees, J. Food Sci. 73 (2008) E455–E462, https://doi.org/10.1111/j.1750-3841.2008.00950.x.

[108] M.P. Iacoviello, S.A. Rubin, Sterile preparation of antibiotic-selective LB agar plates using a microwave oven, BioTechniques 30 (2001) 963–965.

[109] V.J. Thomson, M.K. Bhattacharjee, D.H. Fine, K.M. Derbyshire, D.H. Figurski, Direct selection of IS903 transposon insertions by use of a broad-host-range vector: isolation of catalase-deficient mutants of Actinobacillus actinomycetemcomitans, J. Bacteriol. 181 (1999) 7298–7307.

[110] D. Haubek, K. Poulsen, S. Asikainen, M. Kilian, Evidence for absence in northern Europe of especially virulent clonal types of Actinobacillus actinomycetemcomitans, J. Clin. Microbiol. 33 (1995) 395–401.

[111] M. Hayashi, Y. Shimazaki, S. Kamata, N. Kakiichi, Effects of sodium chloride on destruction of microorganisms by microwave heating in potatoes, [Nihon Koshu Eisei Zasshi] Japanese, J. Public Health 38 (1991) 431–437.

[112] L. Humpheson, M.R. Adams, W.A. Anderson, M.B. Cole, Biphasic thermal inactivation kinetics in Salmonella enteritidis PT4, Appl. Environ. Microbiol. 64 (1998) 459–464.

[113] H. Fujikawa, H. Ushioda, Y. Kudo, Kinetics of Escherichia coli destruction by microwave irradiation, Appl. Environ. Microbiol. 58 (1992) 920–924.

[114] L. Najdovski, A.Z. Dragaš, V. Kotnik, The killing activity of microwaves on some non-sporogenic and sporogenic medically important bacterial strains, J. Hosp. Infect. 19 (1991) 239–247.

[115] K. Sasaki, Y. Mori, W. Honda, Y. Miyake, Selection of biological indicator for validating microwave heating sterilization, PDA J. Pharm. Sci. Technol. 52 (1998) 60–65.

[116] H. Wang, H. Takashima, Y. Miyakawa, Y. Kanno, Development of catalyst materials being effective for microwave sterilization, Sci. Technol. Adv. Mater. 6 (2005) 921–926, https://doi.org/10.1016/j.stam.2005.07.005.

[117] H. Takashima, K. Nakamura, Y. Kanno, Microwave sterilization by TiO_2 filter coated with Ag thin film, in: 2006 IEEE International Conference on Systems, Man and Cybernetics, IEEE, 2006, pp. 1413–1418.

[118] H. Takashima, Y. Miyakawa, Y. Kanno, Microwave sterilization with metal thin film coated catalyst in liquid phase, Mater. Sci. Eng. C 27 (2007) 898–903.

[119] M. Novotny, J. Skramlik, K. Suhajda, V. Tichomirov, Sterilization of biotic pests by microwave radiation, Procedia Eng. 57 (2013) 1094–1099.

[120] T.T. Chau, K.C. Kao, G. Blank, F. Madrid, Microwave plasmas for low-temperature dry sterilization, Biomaterials 17 (1996) 1273–1277.

[121] N. Philip, B. Saoudi, M.-C. Crevier, M. Moisan, J. Barbeau, J. Pelletier, The respective roles of UV photons and oxygen atoms in plasma sterilization at reduced gas pressure: The case of N/sub 2/-O/sub 2/mixtures, IEEE Trans. Plasma Sci. 30 (2002) 1429–1436.

[122] S. Sasaki, Biological activity of lipopolysaccharides isolated from bacteria in human periodontal lesions, Bull. Tokyo Dent. Coll. 20 (1979) 159.

[123] M.B. Gorbet, M.V. Sefton, Endotoxin: the uninvited guest, Biomaterials 26 (2005) 6811–6817.

[124] B.A. Campbell, Circular Waveguide Plasma Microwave Sterilizer Apparatus, 1993.

[125] J.-C. Park, B. PARK, D.-W. Han, D.H. Lee, I.-S. Lee, S.O. Hyun, M.-S. Chun, K.-H. Chung, M. Aihara, Fungal Sterilization Using Microwave-Induced Argon Plasma at Atmospheric, J. Microbiol. Biotechnol. 14 (2004) 188–192.

[126] B.J. Park, D.H. Lee, J.-C. Park, I.-S. Lee, K.-Y. Lee, S.O. Hyun, M.-S. Chun, K.-H. Chung, Sterilization using a microwave-induced argon plasma system at atmospheric pressure, Phys. Plasmas 10 (2003) 4539–4544.

[127] K.-Y. Lee, B. Joo Park, D. Hee Lee, I.-S. Lee, S.O. Hyun, K.-H. Chung, J.-C. Park, Sterilization of Escherichia coli and MRSA using microwave-induced argon plasma at atmospheric pressure, Surf. Coat. Technol. 193 (2005) 35–38, https://doi.org/10.1016/j.surfcoat.2004.07.034.

[128] B.J. Park, K. Takatori, M.H. Lee, D.-W. Han, Y.I. Woo, H.J. Son, J.K. Kim, K.-H. Chung, S. O. Hyun, J.-C. Park, Escherichia coli sterilization and lipopolysaccharide inactivation using microwave-induced argon plasma at atmospheric pressure, Surf. Coat. Technol. 201 (2007) 5738–5741.

[129] T. Itarashiki, N. Hayashi, A. Yonesu, Sterilization effect of nitrogen oxide radicals generated by microwave plasma using air, Vacuum 110 (2014) 213–216, https://doi.org/10.1016/j.vacuum.2014.06.018.

[130] H. Liu, J. Chen, L. Yang, Y. Zhou, Long-distance oxygen plasma sterilization: effects and mechanisms, Appl. Surf. Sci. 254 (2008) 1815–1821, https://doi.org/10.1016/j.apsusc.2007.07.152.

[131] A. Bol'Shakov, B. Cruden, R. Mogul, M. Rao, S. Sharma, B. Khare, M. Meyyappan, Radio-frequency oxygen plasma as a sterilization source, AIAA J. 42 (2004) 823–832.

[132] C. Keqiang, Z. Erli, W. Jinfa, Z. Hansheng, G. Zuoyao, Z. Bangwei, Microwave electron cyclotron resonance plasma for chemical vapor deposition and etching, J. Vac. Sci. Technol. A 4 (1986) 828–831.

[133] D.J. Park, M.H. Lee, Y.I. Woo, D.-W. Han, J.B. Choi, J.K. Kim, S.O. Hyun, K.-H. Chung, J.-C. Park, Sterilization of microorganisms in silk fabrics by microwave-induced argon plasma treatment at atmospheric pressure, Surf. Coat. Technol. 202 (2008) 5773–5778, https://doi.org/10.1016/j.surfcoat.2008.06.039.

[134] M. Moisan, J. Barbeau, S. Moreau, J. Pelletier, M. Tabrizian, Y. L'H, Low-temperature sterilization using gas plasmas: a review of the experiments and an analysis of the inactivation mechanisms, Int. J. Pharm. 226 (2001) 1–21.

[135] M. Laroussi, Nonthermal decontamination of biological media by atmospheric-pressure plasmas: review, analysis, and prospects, IEEE Trans. Plasma Sci. 30 (2002) 1409–1415.

[136] J. Feichtinger, A. Schulz, M. Walker, U. Schumacher, Sterilisation with low-pressure microwave plasmas, Surf. Coat. Technol. 174–175 (2003) 564–569, https://doi.org/10.1016/S0257-8972(03)00404-3.

[137] L. Eliasson, S. Isaksson, M. Lövenklev, L. Ahrné, A comparative study of infrared and microwave heating for microbial decontamination of paprika powder, Front. Microbiol. 6 (2015), https://doi.org/10.3389/fmicb.2015.01071.

[138] S.S. Shirkole, P.P. Sutar, High power short time microwave finish drying of paprika (Capsicum annuum L.): Development of models for moisture diffusion and color degradation, Dry. Technol. 37 (2019) 253–267, https://doi.org/10.1080/07373937.2018.1454941.

[139] J. Wang, K. Sheng, Far-infrared and microwave drying of peach, LWT Food Sci. Technol. 39 (2006) 247–255, https://doi.org/10.1016/j.lwt.2005.02.001.

[140] A.K. Datta, H. Ni, Infrared and hot-air-assisted microwave heating of foods for control of surface moisture, J. Food Eng. 51 (2002) 355–364, https://doi.org/10.1016/S0260-8774(01)00079-6.

[141] G. Sumnu, S. Sahin, M. Sevimli, Microwave, infrared and infrared-microwave combination baking of cakes, J. Food Eng. 71 (2005) 150–155, https://doi.org/10.1016/j.jfoodeng.2004.10.027.

[142] Q. Guo, D.-W. Sun, J.-H. Cheng, Z. Han, Microwave processing techniques and their recent applications in the food industry, Trends Food Sci. Technol. 67 (2017) 236–247, https://doi.org/10.1016/j.tifs.2017.07.007.

[143] S.S. Shirkole, R. Jayabalan, P.P. Sutar, Dry sterilization of paprika (Capsicum annuum L.) by short time-intensive microwave-infrared radiation: establishment of process using glass transition, sorption, and quality degradation kinetic parameters, Innov. Food Sci. Emerg. Technol. 62 (2020) 102345, https://doi.org/10.1016/j.ifset.2020.102345.

[144] H.C.P. Karunasena, P. Hesami, W. Senadeera, Y.T. Gu, R.J. Brown, A. Oloyede, Scanning electron microscopic study of microstructure of gala apples during hot air drying, Dry. Technol. 32 (2014) 455–468, https://doi.org/10.1080/07373937.2013.837479.

Chapter 9

Microwave-assisted CVD processes for diamond synthesis

Aparna Das* and Bimal Krishna Banik*

Department of Mathematics and Natural Sciences, College of Sciences and Human Studies, Prince Mohammad Bin Fahd University, Al Khobar, Kingdom of Saudi Arabia

**Corresponding authors: e-mails: aparnadasam@gmail.com (Aparna Das); bimalbanik10@gmail.com, bbanik@pmu.edu.sa (Bimal Krishna Banik)*

9.1 Diamond: Structure, properties, and applications

Diamond is an allotrope of carbon. The other common allotropes of pure carbon include graphite, Lonsdaleite, C_{60} (Buckminsterfullerene or buckyball), C_{540}, C_{70}, amorphous carbon, and single-walled carbon nanotube, or buckytube.

Fig. 1 shows the structures of the allotropes of carbon. Due to the different structures, these allotropes exhibit different properties. For instance, the diamond structure is a lattice of tetrahedrally bonded carbon atoms [1]. The carbon atoms in diamond are sp^3-hybridized, and the bond lengths of the carbon-carbon atom are equal. Thus, diamond builds a three-dimensional network of strong covalent bonds. On the other hand, graphite is composed of planar sheets comprising relatively strongly bound six-membered carbon rings, and these sheets are only weakly bound to each other by van der Waals attractions.

Because of the stable network of covalent bonds and hexagonal rings, diamond is incredibly strong and exhibits the highest hardness and thermal conductivity compared with any bulk materials. On other hand, graphite is one of the softest materials.

Diamond exhibits many other excellent properties. It is chemically inert, thermally conductive, highly wear resistant, electrically insulate, and broadly optically transparent from ultraviolet to far-infrared [2]. At the same time, the diamond has the lowest thermal expansion coefficient. Table 1 shows some of the outstanding properties of diamond. On account of these unique, excellent, and extreme properties, diamond is a promising material for a wide range of applications. For example, diamond is used in cutting tools (because of the hardness), heat spreaders (because of the high thermal conductivity), electrical insulation layer for microcircuitry (because of the high resistivity), electronic

Microwaves in Chemistry Applications. https://doi.org/10.1016/B978-0-12-822895-1.00004-7

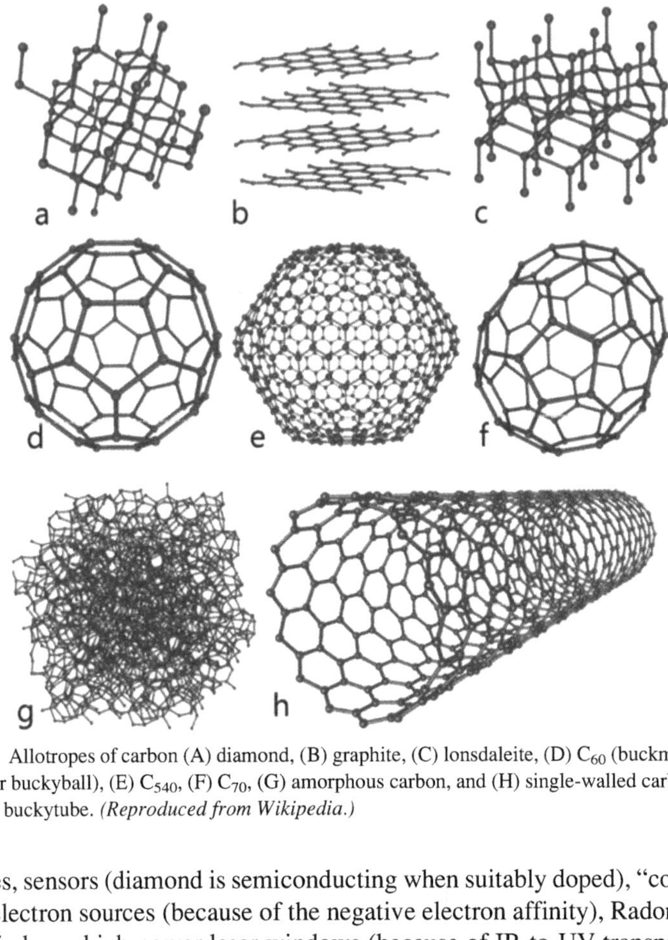

FIG. 1 Allotropes of carbon (A) diamond, (B) graphite, (C) lonsdaleite, (D) C_{60} (buckminsterfullerene or buckyball), (E) C_{540}, (F) C_{70}, (G) amorphous carbon, and (H) single-walled carbon nanotube, or buckytube. *(Reproduced from Wikipedia.)*

devices, sensors (diamond is semiconducting when suitably doped), "cold cathode" electron sources (because of the negative electron affinity), Radome/missile windows, high-power laser windows (because of IR-to-UV transparency), electrochemical sensors (because of the chemical inertness), in vitro applications (coatings/sensors) (because of the biological inertness), and robust particle detectors (because of the Radiation "hard") [3–11].

9.1.1 Natural diamonds

Natural diamonds are mainly classified into two types (I and II) based on the level of nitrogen impurities found within them and further subdivided in accordance with the arrangement of nitrogen atoms (isolated or aggregated) and the occurrence of boron impurities [12].

Type Ia: Most of the natural diamonds are of this type and contain up to 0.3% nitrogen.

Type Ib: This type of diamond is very rare in nature (only ~0.1%). But almost all synthetic diamonds are of this type and contain nitrogen at concentrations of up to 500 ppm.

TABLE 1 Selected properties of diamond.

Property	Value
Hardness	10,000 kg/mm^2
Tensile strength	>1.2 GPa
Compressive strength	>110 GPa
Density	3.52 g/cm^3
Thermal conductivity	20.0 W/cm K
Thermal expansion coefficient	0.0000011 K^{-1}
Thermal shock parameter	30000000 W/m
Resistivity	10^{13}–10^{16} Ω cm
Bandgap	5.45 eV
Dielectric strength	10,000,000 V/cm
Dielectric constant	5.7
Electron mobility	2200 cm^2/V s
Hole mobility	1600 cm^2/V s
Electron saturated velocity	27000000 cm/s
Hole saturated velocity	10000000 cm/s
Sound velocity	18,000 m/s
Optical refractive index	2.41 (at 591 nm)
Optical transmissivity	225
Young's modulus	1220 GPa
Poisson's ratio	0.2
Debye temperature	2200 K
Loss tangent at 40 Hz	0.0006

Type IIa: It is also very rare in nature. It contains very little nitrogen; hence, it can't be easily detected by the usual IR or UV absorption measurements.

Type IIb: This type of diamond is extremely rare in nature. It contains a very low concentration of nitrogen, even lower than type IIa. This diamond crystal is a p-type semiconductor, as it contains uncompensated boron acceptor impurities.

9.1.2 Synthetic diamond

The exceptional properties and applications of diamond led to the discovery of synthetic diamond. Studies were aimed to supply diamonds of regular dimensions at as low a price as possible.

The phase diagram for elemental carbon is shown in Fig. 2 [13]. At ambient conditions, the stable bonding configuration of carbon is graphite. Diamond is kinetically stable and thermodynamically unstable, and said to be metastable. Because of the high energetic barrier between the two phases of carbon, the transition from diamond to graphite at normal temperature and pressure is very slow. However, at high temperatures (above 4500 K) and at favorable pressure, diamond rapidly converts to graphite [13]. There are mainly two methods available for the synthesis of diamonds from graphite. These are the high-pressure high-temperature (HPHT) growth method and chemical vapor deposition (CVD) method.

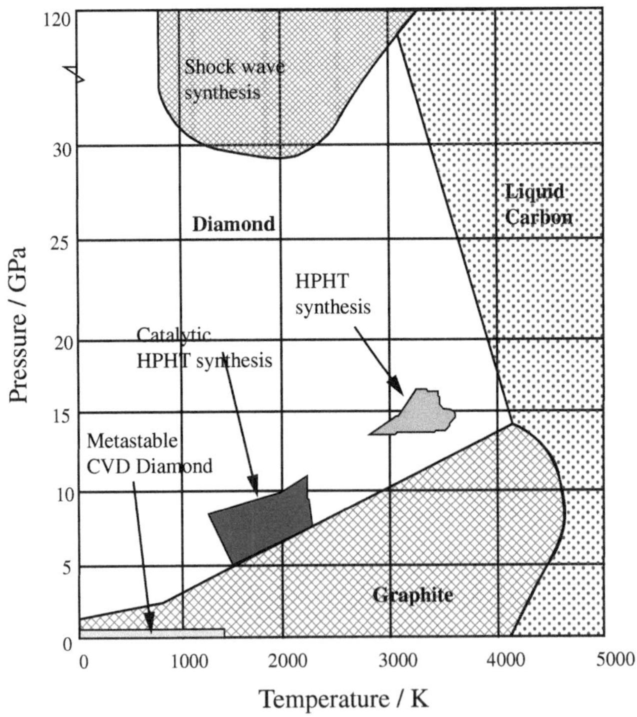

FIG. 2 The phase diagram for carbon. *(Reproduced with permission from F.P. Bundy, W.A. Bassett, M.S. Weathers, R.J. Hemley, H.U. Mao, A.F. Goncharov, The pressure-temperature phase and transformation diagram for carbon; updated through 1994. Carbon 34 (1996) 141–153.)*

9.1.3 High-pressure high-temperature method

The breakthrough in the synthesis came in 1954 when General Electric used a combination of high pressure and high temperature to obtain a diamond from graphite [14]. This high-pressure high-temperature (HPHT) growth method had been used to produce an "industrial" diamond for several decades [14–17]. In this method, graphite is put into a huge hydraulic press at high temperatures and pressures, and the conversion of graphite to diamond occurs over a period of a few hours with the addition of a metallic catalyst. In the method, there are three main press designs named the belt press, the cubic press, and the split-sphere (BARS) press. Schematic of a BARS system for the growth of HPHT diamonds is shown in Fig. 3. Studies showed that using this technique, monocrystalline diamonds with high purity can produce up to several millimeters.

9.1.4 Chemical vapor deposition method

The second method, named chemical vapor deposition (CVD), is a process that includes both gas-phase reactions and gas-solid surface reactions. The advantage of the CVD method is that it does not require high-pressure and high-temperature conditions to synthesize diamond crystallites. In this method, a carbon plasma on top of a substrate helps to form a diamond. This method allows depositing thin films, in the range of nm to mm, of polycrystalline diamond onto a wide range of substrate materials. Studies in this field were started in the 1950s.

In 1958, the first successful deposition of the diamond from the gas phase at low pressures was reported [18]. And later, Derjaguin et al. verified the result [19]. During the synthesis, thermal decomposition of carbon-containing gases (hydrocarbon gases) at less than 1-atmosphere pressure was used to grow diamond on diamond seed crystals heated nearly to 900°C. The drawback of this method was the co-deposition of graphite with the diamond. Later, several studies were reported to improve the growth method [20–24]. For instance, improvements in the method were observed when researchers used atomic hydrogen to etch graphite [20, 21], boron was incorporated into diamond during growth [22] and the diamond was deposited on nondiamond substrates [23, 24].

An important contribution in this field was occurred in 1981, when researchers were able to grow the good-quality films of diamond on nondiamond substrates (silicon substrate), at significant rates of around 1 μm/h using a hot filament reactor [25, 26]. The growth system was operated using 1% CH_4 in H_2 at a pressure of 20 Torr (0.026 atm). As a continuation study, the same group reported another method for diamond synthesis under similar conditions in a "microwave plasma" reactor [27] and later on others [28].

Later on, several methods for the growth of CVD diamond have been developed [2]. They include direct current (DC) plasma method [29–32], DC plasma

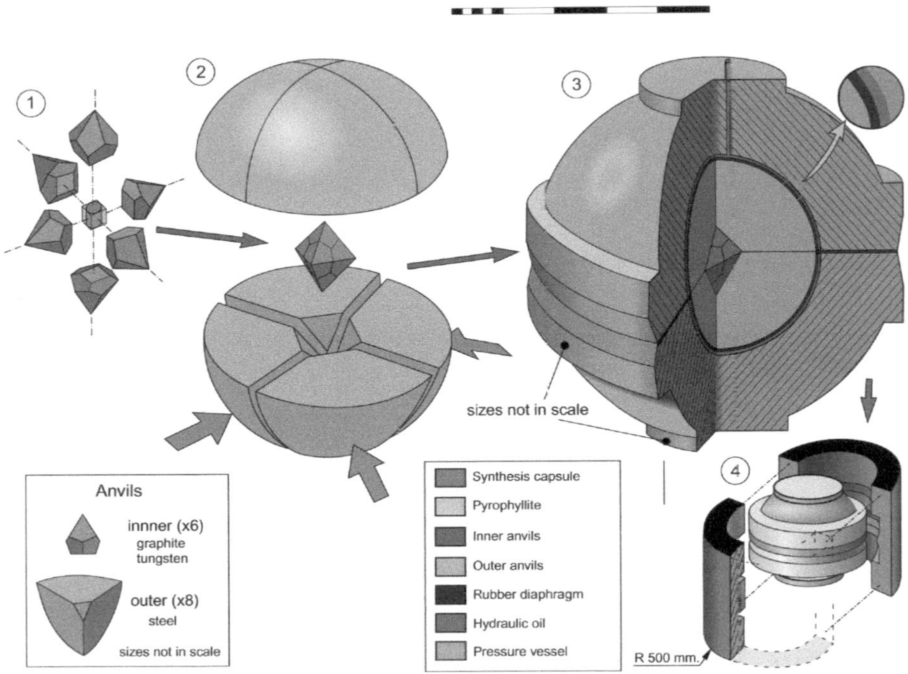

FIG. 3 Schematic setup of a BARS system for the growth of HPHT diamonds. *(Reproduced from Wikipedia.)*

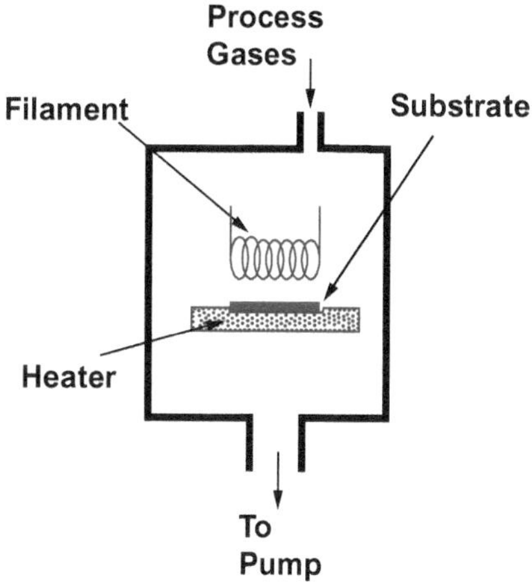

FIG. 4 Schematic of a hot filament CVD reactor. *(Reproduced with permission from S.H. Mortazavi, M. Ghoranneviss, M. Dadashbaba, R. Alipour, Synthesis and investigation of silicon carbide nanowires by HFCVD method. Bull. Mater. Sci. 39 (2016) 953–960.)*

jet method [33–35], radio frequency (RF) plasma method [36, 37], microwave plasma jet method [38, 39], electron cyclotron resonance (ECR) microwave plasma method [40–43], and combustion flame method [44–46]. It is interesting to note that most of these techniques are plasma-based.

Considering the hot filament CVD (HFCVD) technique, a schematic diagram of the HFCVD reactor is shown in Fig. 4. In the reactor, there are filaments of tungsten (W) or tantalum (Ta), maintained at a temperature of 2000–2200°C. The filaments are normally placed at 5–10 mm above the substrate (maintained at ~900°C). The gas pressure inside the reactor is usually 20–30 Torr. One can apply a bias voltage either to the filaments or to the substrate. During the process, the CH_4 and H_2 molecules that have interacted with the hot filament are thermally dissociated to be fragmented hydrocarbons and atomic hydrogen. They then diffuse onto the substrate to create diamonds and the exhaust gases are pumped from the chamber. This method is relatively simple and inexpensive. At the same time, the major drawbacks are its limited growth areas (~1 cm^2), contamination of diamond films from the filament, and a limited range of gases available for use in HFCVD.

The DC arc jet (DC plasma jet) is another CVD method to produce a diamond. In the reactor system, an anode and a cathode are connected by a DC power supply (Fig. 5), and a discharge region is formed between the electrodes. Ionization occurs and a jet of plasma is generated when gases such as an $Ar/H_2/CH_4$ mixture pass through this region. Then, it is accelerated by a pressure drop towards the substrate, where the diamond film is deposited. This technique offers a high growth rate (~1 mm/h) compared with other methods [47]. Like HFCVD, drawbacks are its limited growth areas and the contamination of diamond films from the cathode.

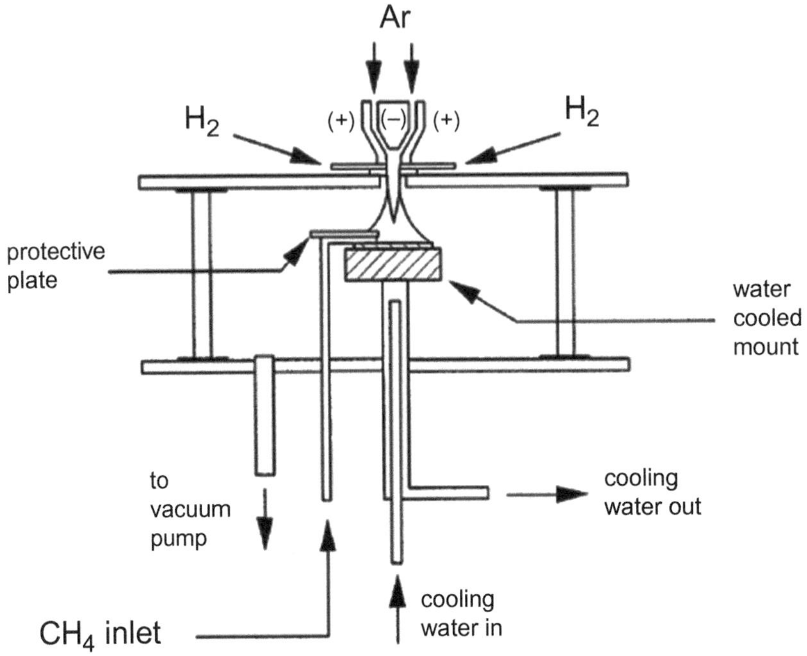

FIG. 5 Schematic of DC arc jet reactor incorporating the substrate CH_4 injection system. *(Reproduced with permission from S.W. Reeve, W.A. Weimer, D.S. Dandy, On the optimization of a dc arc-jet diamond chemical vapor deposition reactor. J. Mater. Res. 11 (1996) 694–702.)*

In a microwave plasma CVD (MWCVD) system, the substrate is placed on a heater at a position where a plasma is formed. The growth parameters such as pressure, gas mixture, and substrate temperature are identical to those used for HFCVD.

In the reactor, via a dielectric window (usually quartz), microwave power is coupled into the chamber to create a discharge. During the process, microwaves couple energy into gas-phase electrons, and through collisions, the electrons transfer their energy to the gas. This leads to heating and dissociation of the gas molecules and then the formation of active species. And finally, the active species react on the substrate surface and form the diamond film. The two most common types of MWCVD reactor are shown in Figs. 6 and 7.

In the NIRIM-type reactor [27], for the propagation of 2.45-GHz microwaves, a quartz discharge tube was introduced through the side of a fundamental mode rectangular waveguide (Fig. 6). In this configuration, the position of the plasma can be tuned by using a sliding short in the waveguide. To prevent the microwave leakage to the outside, the substrate was inserted from the bottom of the discharge tube using a dielectric rod. In the ASTEX-type reactor (Fig. 7), through a quartz window, microwaves are coupled into a water-cooled

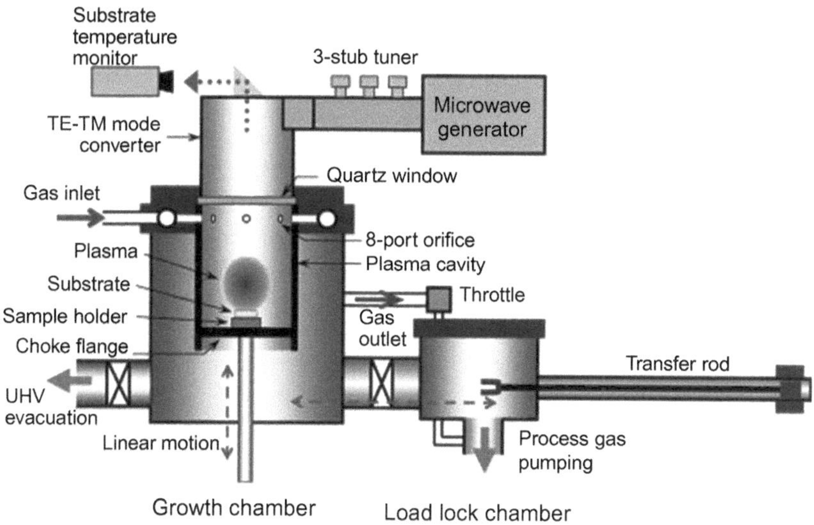

FIG. 6 Schematic of NIMS (NIRIM)-type microwave plasma reactor. *(Reproduced with permission from R. Ohtani, T. Yamamoto, S.D. Janssens, S. Yamasaki, S. Koizumi, Large improvement of phosphorus incorporation efficiency in n-type chemical vapor deposition of diamond. Appl. Phys. Lett. 105 (2014) 232106.)*

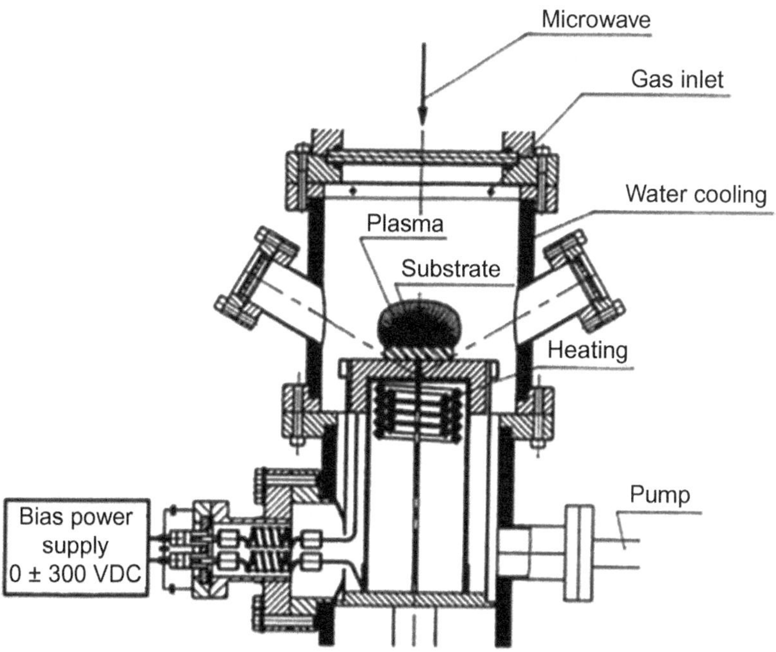

FIG. 7 Schematic of ASTEX-type microwave plasma reactor. *(Reproduced with permission from S.-T. Lee, Z. Lin, X. Jiang, CVD diamond films: nucleation and growth. Mater. Sci. Eng. R. Rep. 25 (1999) 123–154.)*

metal cavity. The reactor has a tuning antenna that converts the TE_{10} microwave mode in the waveguide to the TM_{01} mode in the cavity. Microwave powers of up to 5 kW can be used in such systems, and it can provide a growth rate of 10 μm/h.

There is no electrode or filament used in the reactor, so this method provides a clean environment for diamond growth. Thus, MWCVD has become the system of choice for electronic applications. Also, the growth rate in this method is relatively fast due to high input power and the immersion of the substrate into the plasma. The other merits of MWCVD system over other reactors are that it can use a broad variety of gas mixtures, for example, gas mixtures with high oxygen content, or ones containing chlorinated or fluorinated gases. Due to these advantages, the MWCVD method has become the most widely used technique for diamond growth. In the coming section, we discuss the detailed progress in the MWCVD diamond growth.

9.2 Synthesis of the single-crystal diamond

Single-crystal diamond (SCD) synthesized by the microwave plasma-enhanced CVD deposition is one of the centerpieces of CVD diamond science and technology. Using microwave technology, advances have been achieved in high growth rates and the production of large and high-quality diamonds [48–51].

Even though MWCVD is one of the most widely adopted methods to grow diamond films and bulk crystals [52], the commercialized size of SCD is still several millimeters in the edge length. Also, their prices are still much higher compared with other semiconductor materials such as Si, GaN, SiC. Thus, it is essential to establish a way to synthesize diamond crystal efficiently over a large area with sufficient quality and acceptable cost. One way to enhance the growth rate is by increasing the gas pressure [49, 53–55]. In this case, the increase of the gas pressure limits the discharge region, and the power density increases under a fixed input power. But several other studies exhibited that this method sacrifices the uniformity drastically [56–61]. Another way to enhance the growth rate is through the introduction of argon into the source gas mixture [38, 62–65]. A linear relation was observed between the growth rate and argon concentration [57, 58].

Yamada et al. reported the growth of single-crystal diamond over inch size area and uniformity by using MWCVD with the introduction of argon [66]. The enhanced growth rate was obtained by argon addition under various gas pressures, as shown in Fig. 8.

In addition, improvement of the uniformity of the substrate temperature and the growth rate was detected by introducing argon. At optimized conditions, over 40% nonuniformity was minimized to $<10\%$ without surrendering the absolute magnitude of the growth rate over 1.5 in. area. It was also observed that the temperature distributions do not entirely correspond to the growth rate.

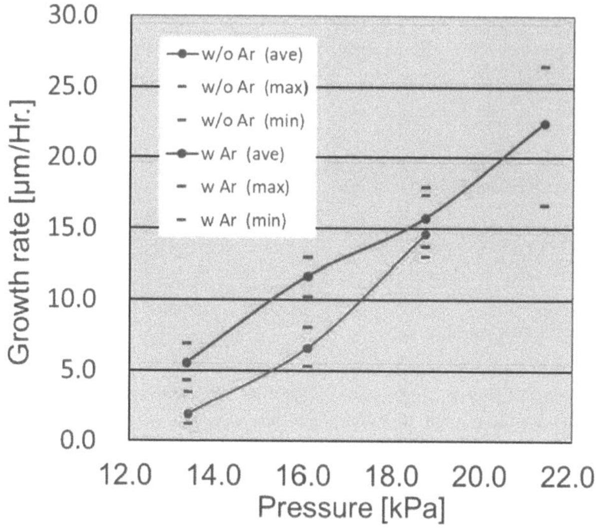

FIG. 8 Dependence of the growth rate on the gas pressure with and without Ar. Symbols ● and — represent the average and maximum/minimum growth rates, respectively. *(Reproduced with permission from H. Yamada, A. Chayahara, Y. Mokuno, Effect of Ar addition on uniformity of diamond growth by using microwave plasma chemical vapor deposition. Diam. Relat. Mater. 87 (2018) 143–148.)*

The theoretical analysis indicated that dissociation of CH_3 and C_2H_2 into CH_x is elevated by the argon addition.

In normal laboratory cases, to create energetic plasma at high gas pressures (150–300 Torr), MPCVD reactors use up to 6 kW of microwave power at 2.45 GHz. And the backside sample cooling is required to control the substrate growth temperature [67]. No difficulties were reported for the production of small-volume diamond using these types of reactors. However, the synthesis area is restrained by the plasma discharging area (normally <25 mm in diameter) [58]. Studies showed that a decrease in the growth pressure or an increase in the incident microwave power can increase the deposition area [49, 58]. But the normal tubular 2.45-GHz MPCVD rector systems can't handle the excessive heat generated under those conditions and can result in a catastrophic system failure. Reactor system operating at 915 MHz is considered as a cost-effective tool for scaling up high-quality CVD diamond production [68]. Studies were made in producing up to 70 single-crystal diamond crystals in a batch using a bell jar-type MPCVD system, and an average growth rate of 10–30 μm/h was reported for those samples with large amounts of nitrogen addition [69]. Studies of SCD growth in a 915-MHz MPCVD reactor operated at a microwave power of up to 70 kW and with a deposition area, up to 300 mm in diameter, were reported by Liang et al [55]. UV-visible absorption and PL spectra

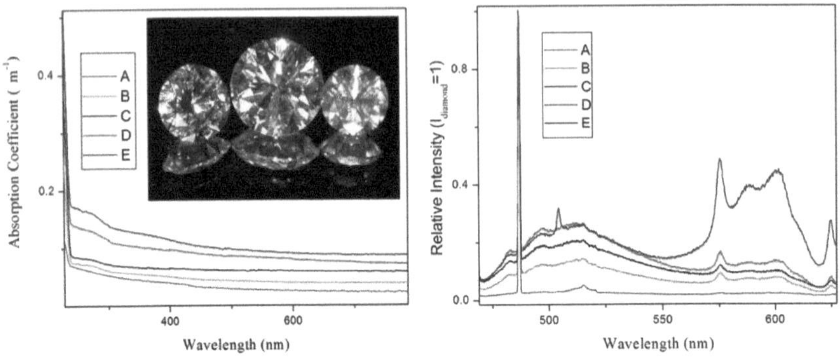

FIG. 9 (a) UV-visible absorption measured at room temperature for 5 samples (A–E). The inset is a photograph of representative single-crystal diamond anvils produced by 915-MHz systems. (b) Photoluminescence spectra of samples A–E. *(Reproduced with permission from Q. Liang, C. Yan, J. Lai, Y. Meng, S. Krasnicki, H. Shu, H. Mao, R.J. Hemley, Large area single-crystal diamond synthesis by 915 MHz microwave plasma-assisted chemical vapor deposition. Cryst. Growth Des. 14 (2014) 3234–3238.)*

(Fig. 9) confirmed a general relationship between the brown color of diamonds and the nitrogen addition. Unintentional air leakage was the major source of the incorporation of nitrogen.

One of the critical issues in diamond synthesis is the plasma-substrate interactions, particularly during the long-run deposition in an MPCVD reactor. To achieve a higher growth rate and a better diamond quality, the challenge of process control concerning the plasma-and-substrate environment was need to be conquered and optimized. For this purpose, the formation of hot spots noticed during the single-crystal diamond synthesis in 2.45-GHz cylindrical cavity reactors was examined after the long-run deposition [70]. A 2.45-GHz microwave cylindrical cavity reactor with TM_{01} mode excitation was used for analysis. The microwave plasma was generated at 150-torr operating pressure, 4-kW microwave input power, and 350- and 60-sccm flow rates of H_2 and CH_4, respectively, and the substrate was the single-crystal diamond.

Depending on growth parameters such as substrate temperature, plasma power density, partial pressure, and diffusion transport, the growth directions and growth rate can vary on the diamond surface during the synthesis. Also, studies showed that the growth rate in the different parts of the substrate was different; for instance, edge of the substrate tends to grow faster in (100) direction than the center of the surface. It was observed that microwave energy coupled more easily to the sharper edge due to a stark gradient growth surface profile between the edge and substrate center. The localized microwave coupling onto the edge might be the reason for hot spots. Besides, the microwave energy made electrical breakdown and thermal instabilities via ohmic heating on the substrate tip edge. Bright luminescence with the measured substrate temperature above the range of 2200°C was observed in such type of hot spot. As this temperature is beyond the diamond temperature deposition window, this can produce graphitic or diamond-like carbon. The hot spots observe normally if the diamond growth time exceeds 24 h and produces thermal runaway.

Often, SC diamond growth is carried out under high methane concentration in the hydrogen environment, as well as high power and high-pressure discharges. The latter factor result in an enhancement of absorbed microwave power density (MWPD) in MPCVD rectors, which governs the effectiveness of the gas decomposition, and active radical concentrations in the methane-hydrogen plasma [71]. Mostly MWPD in the SC diamond deposition process could be as high as 300–700 W/cm^3 as a result of shrinking the plasma ball [59, 62], promoting the growth rate. The reported growth rate for SC diamond in CH_4-H_2 and CH_4-H_2-Ar mixtures amounts to 70–165 μm/h [48, 51, 62, 72–74].

Recently, Vikharev et al. investigated the single-crystal diamond growth on (100)-oriented high-pressure high-temperature type 2a diamond substrate to optimize growth conditions, thus providing high growth rate and high quality of diamond simultaneously [75]. CVD operating regime for SCD synthesis was measured that allowed achieving the highest possible growth rate and at the same time the minimum number of defects in diamond at the minimum amount of nitrogen (<200 pbb) in the reactor. At the optimized microwave plasma (CVD growth condition at high power density), the formation of nonepitaxial diamond crystallites or hillocks or polycrystalline diamond rim around the SCD top surface was prevented. The high-power density was achieved with a relatively low gas pressure of 160 Torr due to the local electric field amplification above the substrate holder with projection. High-quality SCD grown with a relatively high growth rate in the range of 6 – 8 μm/h at a microwave power of 3 kW and a temperature of 900°C was obtained. It was found that the formation of surface defects could be effectively suppressed by choosing the misorientation angle of the substrate surface. At this angle, the step-flow growth regime was implemented. By choosing the misorientation angle, CH_4/H_2 ratio, and the power density or the concentration of atomic hydrogen, the high-quality SCDs of 1 mm thickness were successfully grown.

9.3 Synthesis of polycrystalline diamond film

Su et al. described a new type of MPCVD reactor possessing a novel dome-shaped resonant cavity [76]. Fig. 10 shows the schematics of the reactor. The experimental results showed that with the new MPCVD reactor, a uniform and transparent polycrystalline diamond film of high quality and purity could be deposited on 2-in.-diameter Si substrates at an input power of 8.5 kW. The growth rate was about 3.5 mm/h. Fig. 11 shows a diamond film sample that was removed from the Si substrate by acid etching. A microscopic image and a

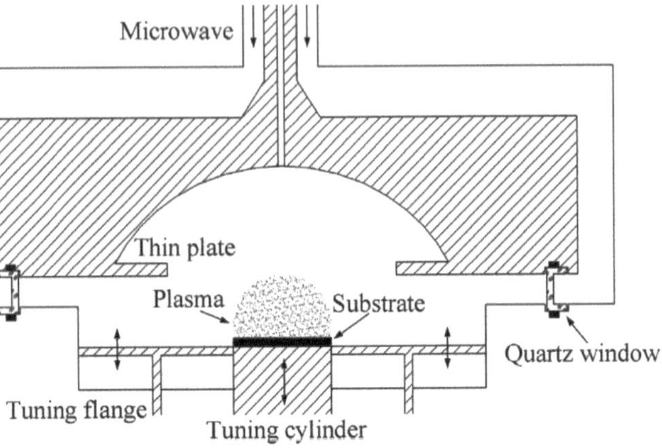

FIG. 10 Schematic cross-sectional views of the dome-shaped cavity-type reactor. *(Reproduced with permission from J.J. Su, Y.F. Li, M.H. Ding, X.L. Li, Y.Q. Liu, G. Wang, W.Z. Tang, A dome-shaped cavity type microwave plasma chemical vapor deposition reactor for diamond films deposition. Vacuum 107 (2014) 51–55. doi:10.1016/j.vacuum.2014.04.002.)*

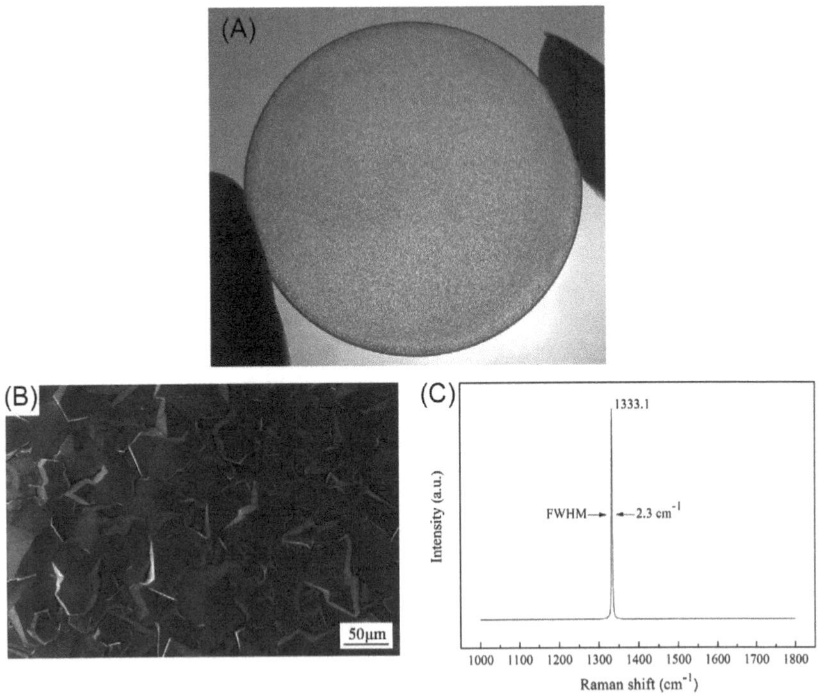

FIG. 11 (A) A diamond film sample, (B) scanning electron micrograph, and (C) Raman spectrum of the sample, respectively. *(Reproduced with permission from J.J. Su, Y.F. Li, M.H. Ding, X.L. Li, Y. Q. Liu, G. Wang, W.Z. Tang, A dome-shaped cavity type microwave plasma chemical vapor deposition reactor for diamond films deposition. Vacuum 107 (2014) 51–55. doi:10.1016/j.vacuum.2014.04.002.)*

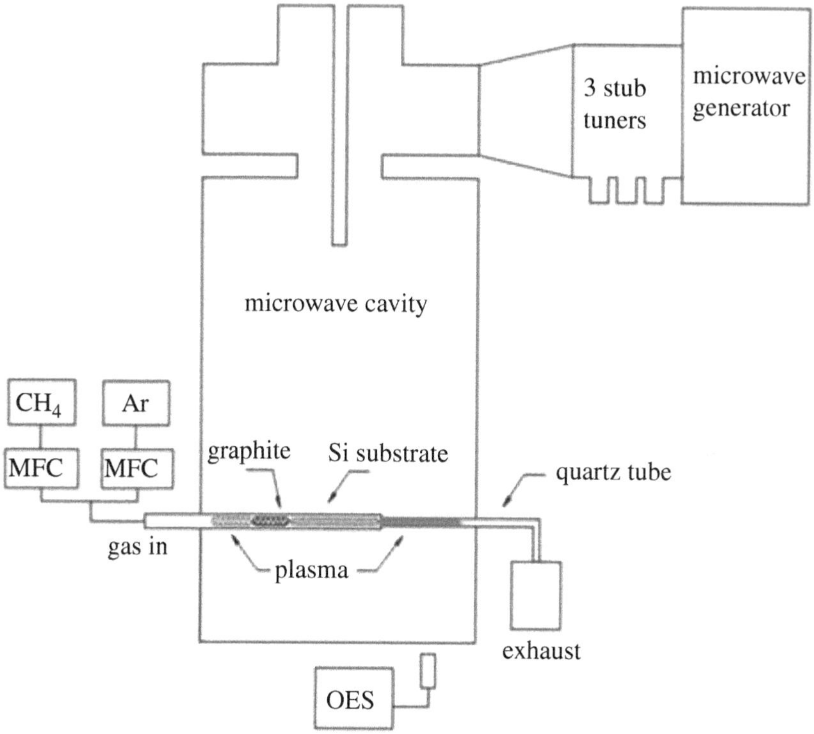

FIG. 12 Schematic view of the atmospheric pressure MPCVD. *(Reproduced with permission from K.W. Hemawan, H. Gou, R.J. Hemley, Diamond synthesis at atmospheric pressure by microwave capillary plasma chemical vapor deposition, Appl. Phys. Lett. 107 (2015) 181901. doi:10.1063/1.4934751.)*

Raman spectrum of the diamond film sample are also shown in Fig. 11. It was possible to apply an input microwave power of up to 10 kW at 2.45 GHz.

The synthesis of the high-quality diamond at high gas-phase pressures is challenging. Hemawan et al. reported the synthesis of the polycrystalline diamond by MPCVD at atmospheric pressure using a mixture of argon and methane gas without the addition of hydrogen [77]. The schematic of the reactor is shown in Fig. 12. The CH_4/Ar plasma was generated inside of quartz capillary tubes using 2.45-GHz microwave excitation. Raman measurements of deposited material indicated the formation of a well-crystallized diamond. Carbon microstructures of diamond films showed coral and cauliflower-like morphologies or well-facetted diamond crystals with grain sizes ranging from 100 nm to 10 μm depending on the growth conditions. SEM images are shown in Fig. 13.

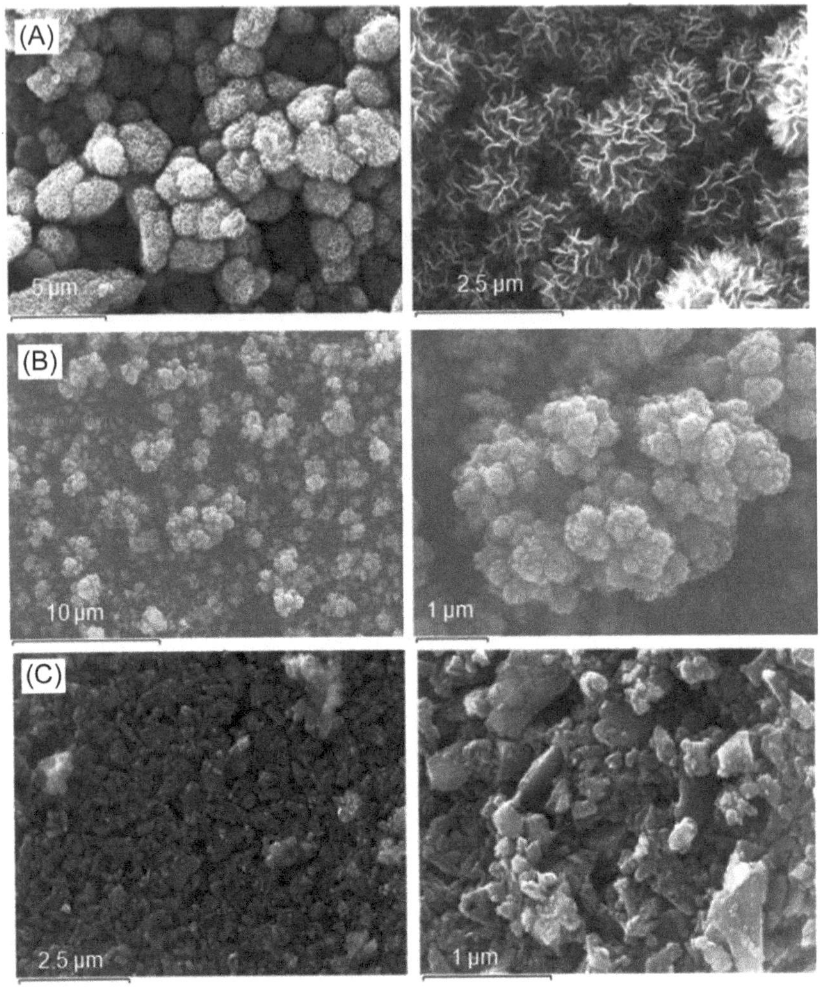

FIG. 13 SEM micrographs of microcrystalline diamond and other carbonaceous material grown on silicon. (A) Ar/CH_4 6%, (B) Ar/CH_4 4%, (C) Ar/CH_4 2%. *(Reproduced with permission from K. W. Hemawan, H. Gou, R.J. Hemley, Diamond synthesis at atmospheric pressure by microwave capillary plasma chemical vapor deposition. Appl. Phys. Lett. 107 (2015) 181901. doi:10.1063/1.4934751.)*

In comparison with SCD, the relatively low growth rates of PCD films were typically a few microns per hour for substrate sizes from 1 to 5 in. [53, 78–88]. It was basically because of the low microwave power density needed to get an enough-sized plasma cloud for homogeneous growth all over the large-sized substrates. The growth rate around 31 μm/h for PCD film on a very small PCD substrate (2.5 × 2.5mm^2 in dimension) was reported by Vikharev et al.

[89]. Under very high temperature (1450°C) and high methane concentration (16%), the diamond growth rate ~30 μm/h was achieved for 0.5-in.-diameter Mo substrate [90]. Growth rate as high as 21 μm/h on 1 in. Si substrate was demonstrated by Hemawan et al. [91].

Recently, Bolshakov et al. reported the growth of polycrystalline diamond by MPCVD under high absorbed microwave power density and high pressures [92]. Under the pressure of 320 Torr and microwave power density of ~700 W/cm^3, growth rates as high as 30–36 μm/h were achieved in CH_4-H_2 mixtures even at relatively low (3%) CH_4 concentration. Fig. 14 shows the SEM image of the diamond film grown on the black polycrystalline diamond substrate with dimensions 5 × 5 × 0.33mm^3.

The plasma shapes and spatial profiles of species were analyzed with spatially resolved optical emission spectroscopy at moderate and high microwave power density. The photographs of the plasma in low and high microwave power density regimes for large and small substrates are shown in Fig. 15.

Analysis of PCD growth rate vs substrate size in MPCVD reactors disclosed a clear trend of increased growth rate with decreasing substrate dimension. A strong increase in growth rate for small-sized substrates was observed owing to an enhanced absorbed power density in the plasma.

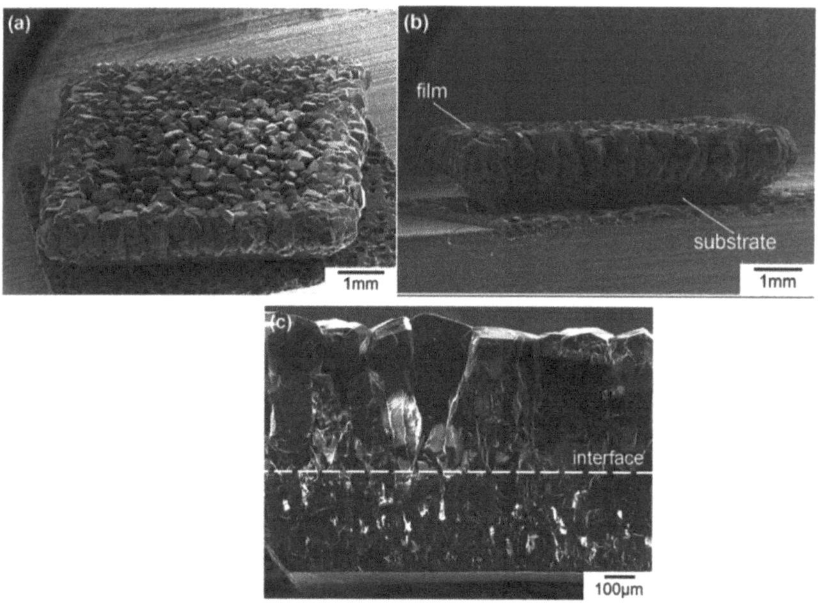

FIG. 14 Panoramic SEM images of the film deposited on black PCD substrate taken at oblique angle (A) and side view by tilt 90 degrees (B). The fracture surface of the film deposited on white PCD substrate (C). *(Reproduced with permission from A. Bolshakov, V. Ralchenko, V. Yurov, G. Shu, E. Bushuev, A. Khomich, E. Ashkinazi, D. Sovyk, I. Antonova, S. Savin, Enhanced deposition rate of polycrystalline CVD diamond at high microwave power densities. Diam. Relat. Mater. 97 (2019) 107466.)*

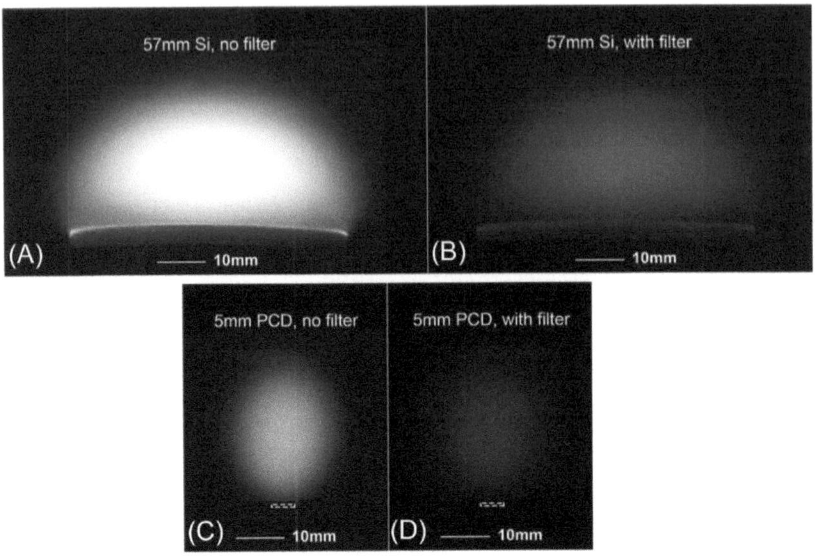

FIG. 15 Plasma photographs of the plasma for 57-mm-diameter Si substrate (A, B) and PCD substrate 5 × 5 mm^2 (C, D). *(Reproduced with permission from A. Bolshakov, V. Ralchenko, V. Yurov, G. Shu, E. Bushuev, A. Khomich, E. Ashkinazi, D. Sovyk, I. Antonova, S. Savin, Enhanced deposition rate of polycrystalline CVD diamond at high microwave power densities. Diam. Relat. Mater. 97 (2019) 107466.)*

9.4 Diamond doping

Doping of diamond films is important to improve the mechanical, electrical, and electrochemical properties of diamond films. The main doping elements are nitrogen [93], boron [94], sulfur [94], phosphorus [95], and some metal elements [96, 97].

Color centers in diamond, like nitrogen-vacancy, silicon-vacancy, chromium-, or nickel-related color centers, are important owing to their wide range of applications in photonics [98–102]. Among these color centers, the silicon-vacancy (SiV) defect produces strong luminescence with zero phonon line (ZPL) at 738 nm and also has other advantages in terms of narrow ZPL, very weak vibronic sidebands, and short lifetime at room temperature [101, 103]. There are mainly two methods to make color centers in diamond. One way is to use the ion implantation technique to embed the doping atoms into diamond [104, 105]. This method could precisely control the doping (e.g., Si) concentration. However, the collateral lattice damage can direct to a reduction in the efficiency of the PL. The second method is in situ doping with Si atoms during the CVD process to form the SiV centers.

In some cases, the Si doping was accomplished by diamond deposition on a Si-containing substrate, like Si [101, 103, 106, 107] or SiO_2 substrate [108].

Atomic hydrogen of the plasma could etch the Si substrate, SiO_2 substrate, or SiO_2 mask on the substrate [109], and the etched substrate served as the dopant source. This unintentional self-doping procedure was not a well-controlled one. Another drawback associated with this doping was the nonuniform concentration of imbedded Si. Another doping process was using stand-alone pieces of Si [110] or SiC [111] crystal, as a solid precursor, by placing it near the substrate and immersing in the plasma. However, the doping was not at a quantified rate. Some of these shortcomings could be avoided by adding Si-containing gases, such as silanes (SiH_4, Si_2H_6, CH_3SiH_3, etc.), $HSiCl_3$, SiF_4, and $Si(OC_2H_5)_4$, directly in the process gas [112–114]. For instance, SiH_4 was used to dope microcrystalline diamond films grown by HFCVD in CH_4/H_2 mixtures [114]. Later, Si doping was demonstrated using SiH_4 as a dopant source in MWCVD [112]. In both cases, it was observed that the diamond grown at high enough SiH_4 concentrations had a more defective microstructure and holds a significant amount of amorphous carbon phase. Similar observations were reported when polycrystalline diamond films deposited on Si substrates in a HFCVD reactor with the addition of tetraethoxysilane.

Sedov et al. used monosilane SiH_4 additions to H_2/CH_4 gas mixture to grow Si-doped continuous microcrystalline diamond films and nanocrystalline diamond films by MWCVD [115]. Fig. 16 shows the SEM images of undoped nanocrystalline diamond and microcrystalline diamond film surfaces.

The study also compared the Raman and PL spectra of doped microcrystalline diamond films and nanocrystalline diamond films. Raman and PL spectra for nanocrystalline diamond films and microcrystalline diamond films are shown in Figs. 17 and 18, respectively. The data showed that microcrystalline films exhibit a much brighter PL emission compared with nanocrystalline diamond films. This higher PL efficiency was related to their less defective structure. Besides, no signals of crystalline or amorphous silicon and SiC were detected for both nanocrystalline and microcrystalline diamond films. At higher

FIG. 16 SEM images of undoped 1-μm-thick (A) nanocrystalline and (B) microcrystalline diamond films. *(Reproduced with permission from V. Sedov, V. Ralchenko, A. Khomich, I. Vlasov, A. Vul, S. Savin, A. Goryachev, V. Konov, Si-doped nano-and microcrystalline diamond films with controlled bright photoluminescence of silicon-vacancy color centers. Diam. Relat. Mater. 56 (2015) 23–28.)*

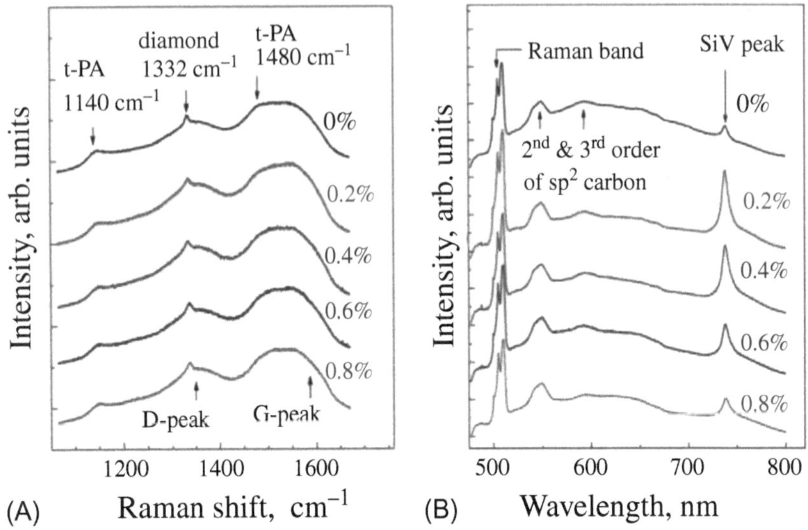

FIG. 17 Raman (A) and PL (B) spectra of nanocrystalline diamond films, grown in MWCVD at SiH_4/CH_4 ratios of 0%–0.8%. *(Reproduced with permission from V. Sedov, V. Ralchenko, A. Khomich, I. Vlasov, A. Vul, S. Savin, A. Goryachev, V. Konov, Si-doped nano and microcrystalline diamond films with controlled bright photoluminescence of silicon-vacancy color centers. Diam. Relat. Mater. 56 (2015) 23–28.)*

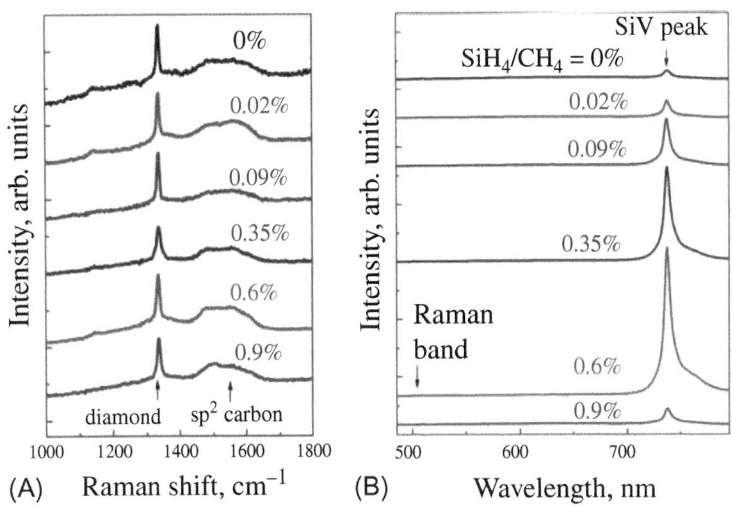

FIG. 18 Raman (A) and PL (B) spectra of microcrystalline diamond films, grown in MWCVD at different SiH_4/CH_4 ratios. *(Reproduced with permission from V. Sedov, V. Ralchenko, A. Khomich, I. Vlasov, A. Vul, S. Savin, A. Goryachev, V. Konov, Si-doped nano-and microcrystalline diamond films with controlled bright photoluminescence of silicon-vacancy color centers. Diam. Relat. Mater. 56 (2015) 23–28.)*

CH_4 addition, PL quenching occurred; at the same time, no significant degradation of the film structure was found within the SiH_4 concentration range of 0%–0.9%.

Even though diamond is a promising material for electronic applications, a major obstacle for the realization of its superior performance is associated with difficulties in the doping of diamond. The studies showed that all known dopants create deep-lying levels with high activation energy. For instance, doping of boron produces p-type conductivity in diamond but creates a level with the activation energy of 0.37 eV. The alternative approach to deal with this issue is the delta-doping technology [116].

Vikharev et al. reported a novel microwave plasma-assisted CVD reactor for diamond delta-doping [117]. The main features of the reactor were the use of a rapid gas switching system and the reactor design providing the laminar gas flow. The schematic of the reactor is shown in Fig. 19. Details: (1) represents quartz tube, (2) substrate holder, (3) plasma, (4) cavity, (5) 2.45-GHz magnetron, (6) rectangular waveguide, (7) gas feeding system, (8) gas pumping system, (9) optical fiber, (10) SOLAR TII monochromator, (11) PMT, (12) oscilloscope, and (13) PC.

The results showed ultrasharp interfaces between doped and undoped materials, which is important for an electronic application. Using the reactor, it was possible to synthesize nanometer-thin layers of the boron-doped diamond. The FWHM of the boron concentration profile obtained from SIMS measurements

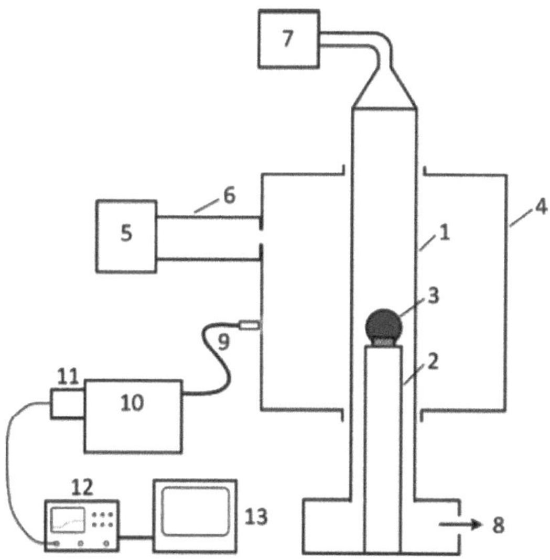

FIG. 19 Scheme of the novel 2.45-GHz MPACVD reactor for diamond delta-doping. *(Reproduced with permission from A. Vikharev, A. Gorbachev, M. Lobaev, A. Muchnikov, D. Radishev, V. Isaev, V. Chernov, S. Bogdanov, M. Drozdov, J. Butler, Novel microwave plasma-assisted CVD reactor for diamond delta doping. Phys. Status Solidi-Rapid Res. Lett. 10 (2016) 324–327.)*

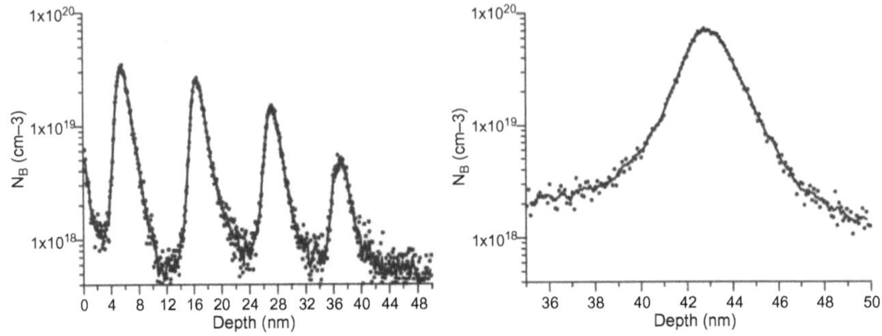

FIG. 20 (A) SIMS characterization of four boron delta-doped layers. (B) SIMS characterization of nanometric boron delta-doped layer. *(Reproduced with permission from A. Vikharev, A. Gorbachev, M. Lobaev, A. Muchnikov, D. Radishev, V. Isaev, V. Chernov, S. Bogdanov, M. Drozdov, J. Butler, Novel microwave plasma-assisted CVD reactor for diamond delta doping. Phys. Status Solidi-Rapid Res. Lett. 10 (2016) 324–327.)*

was about 2 nm (Fig. 20). Using this MWCVD reactor, both single delta layer and multiple delta layers could be grown effectively. Diamond delta doping with other dopants, like phosphorus and nitrogen, was also practicable with this reactor.

Nitrogen is a common impurity found in both natural and synthetic diamonds, and also, it is an n-type dopant in the diamond. Nitrogen-doped diamond has attracted an interest for electronic applications, especially as a potential high-electron-mobility semiconductor. Several studies had demonstrated that the presence of nitrogen in trace amounts increases the rate of diamond growth in a MWCVD process significantly [48, 52, 54, 56, 58, 85, 118–121]. It is also observed that small nitrogen additions affected the surface morphology of the film [52, 122, 123]. In particular, it promoted the formation of {100}-faceted surfaces instead of {111}-faceted surfaces. The {100}-faceted surfaces are typically less rough and attractive for mechanical applications [124]. On the other hand, the addition of a large amount of nitrogen in the source gas mixture led to smaller and less-well-oriented surface facets, along with a higher sp^2 fraction in the deposited material [118, 125]. Besides, a reduction in the thermal conductivity of the process gas mixture was also observed, with the addition of the N_2 gas in a larger amount, which was beneficial for power coupling efficiency [56].

Several nitrogen-containing species were used as participants in gas-surface reactions contributing to diamond growth, such as CN radicals, gas-phase NH_x and CNH_x species [52, 126–134]. Truscott et al. reported spatially resolved absorption and emission measurements of several gas-phase species, such as H ($n = 2$, 3) atoms, NH, CH, CN, and C_2 radicals, and triplet N_2 molecules in MW-activated $CH_4/N_2/H_2$ plasmas operating at pressures of around 150 Torr and powers of around 1.5 kW relevant to contemporary MWCVD processes [135]. The same group also reported the studies based on N_2/H_2 and NH_3/H_2 plasmas [136].

Several metal elements such as Ti, Cr, Ni, W, and Al were used for diamond doping [zha10-tsa08]. Among them, low-concentration Ti-doped films had shown high hardness and elastic moduli [137], thermal stability [138], low residual stress [139], high good biocompatibility [140], and excellent wear and corrosion resistance [141, 142]. Titanium doping is also advantageous due to several other reasons: Ti is remarkably adhesive to diamond [143], Ti doping effectively reduces the stress in the film and enhances the bonding strength of membrane base [144], and Ti doping can adjust the friction coefficient and wear resistance of diamond film [145]. Liu et al. reported for the first time the Ti-doped diamond films were synthesized by MPCVD [146]. They also investigated the effects of the addition of Ti on the morphology, microstructure, and quality of diamond films. Fig. 21 shows the SEM images of the diamond film and Ti-doped diamond films. A significant change in the film morphology and orientation of the crystal was observed when Ti is incorporated into the $CH_4 + H_2$ gas mixture. SEM as well as XRD results showed that Ti doping is beneficial to the growth of the (110)-faceted grains and the second nucleation.

FIG. 21 SEM images of the diamond film and Ti-doped diamond films with different Ti contents: (A) undoped diamond, (B) with 30 sccm, (C) with 45 sccm, and (D) with 60 sccm dopant flux. *(Reproduced with permission from X. Liu, P. Lu, H. Wang, Y. Ren, X. Tan, S. Sun, H. Jia, Morphology and structure of Ti-doped diamond films prepared by microwave plasma chemical vapor deposition. Appl. Surf. Sci. 442 (2018) 529–536.)*

9.4.1 Co-deposition (composites)

The application window is larger for composites consisting of two wide bandgap materials. For example, diamond/silicon carbide composites are of high interest to a number of important applications, such as heat spreaders for high-power electronic devices [147, 148], superhard cutting tools and protective layers [149–152], selective protein absorbent in medical applications [153], and field electron emitters [154]. Different approaches to produce the diamond/SiC composites are reported in the literature [148, 151, 154–158]. Several of these methods had problems with the residual porosity and produced carbon and/or Si inclusions on interfaces. To overcome these issues, the promising alternative approach is the synthesis of SiC/diamond composites by the MWCVD process [159–162]. The composite was produced via the co-deposition of diamond and cubic polytype β-SiC (3C-SiC) using H_2-CH_4-tetramethylsilane ($Si(CH_3)_4$, TMS) mixtures. When the film was grown at the low absorbed microwave power density, the composite structure was nanocrystalline. On the other hand, microcrystalline with a better quality of both phases was observed at high microwave power density (in the range of 30 W/cm^3) [160]. Using TMS as the Si (and also C) precursor with low concentration, the growth rates of the order of 100 nm/h for the composite films were reported [160, 162]. The use of the TMS as the precursor limits the SiC deposition process as the Si and C sources are combined in one molecule. Another appropriate Si-containing gas precursor for obtaining SiC films is silane (SiH_4), added to H_2-CH_4 mixtures in a small quantity (about <1%) [163]. In this case, CH_4 and SiH_4 were used as two separate precursors and allowed a much broader choice for the Si/C ratio, from zero to infinity.

The simultaneous growth of diamond and SiC phases with H_2-CH_4-SiH_4 mixtures at high silane contents and enhanced microwave power density using MWCVD was explored [164]. The study demonstrated the production of diamond-SiC composite films with two different buffer layers (single-phase diamond or single-phase SiC) on Si (100) substrates in one process at a microwave power density of up to 150 W/cm^3. The growth parameters were calibrated to get the almost equal deposition rates for both phases (~0.3 μm/h).

Prior to the deposition, the Si substrates were seeded with nanodiamond (ND) particles to provide diamond nucleation centers (Fig. 22). The Si substrate was then transferred to the MWCVD reactor and placed in a central zone of a Mo substrate holder. The deposition process was carried out at a substrate temperature of 750–875°C, and the deposition time varied from 5 to 100 h. The used microwave power was in the range of 2.8–4.5 kW and the pressures of the 60–80 Torr range. The SEM images of the film structure are shown in Fig. 23.

The synthesis of diamond nanorods (DNRs), the one-dimensional sp^3-configured analog of carbon nanotubes (CNTs), is still difficult, especially with few nanometers in the diameter. The first synthesis was realized in 2001 using MPCVD [165]. The bandgap of diamond nanowires is narrower than that

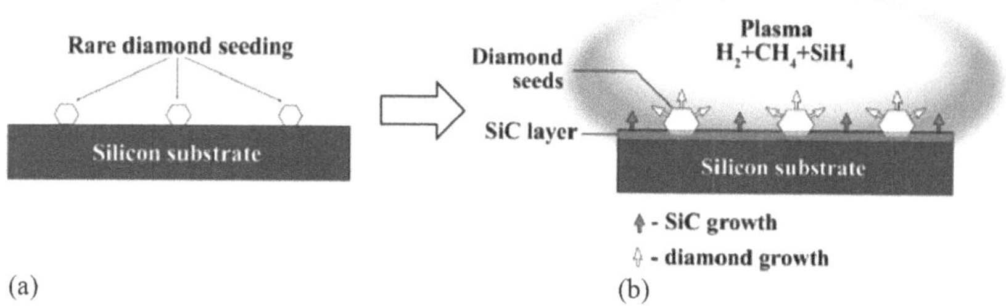

FIG. 22 Schematics for the Si substrate seeding with ND particles and the growth of the SiC-diamond composite: (A) rare diamond seeding on a Si substrate and (B) simultaneous deposition of diamond and SiC phases. *(Reproduced with permission from V. Sedov, A. Martyanov, A. Khomich, S. Savin, V. Voronov, R. Khmelnitskiy, A. Bolshakov, V. Ralchenko, Co-deposition of diamond and β-SiC by microwave plasma CVD in H_2-CH_4-SiH_4 gas mixtures. Diam. Relat. Mater. 98 (2019) 107520.)*

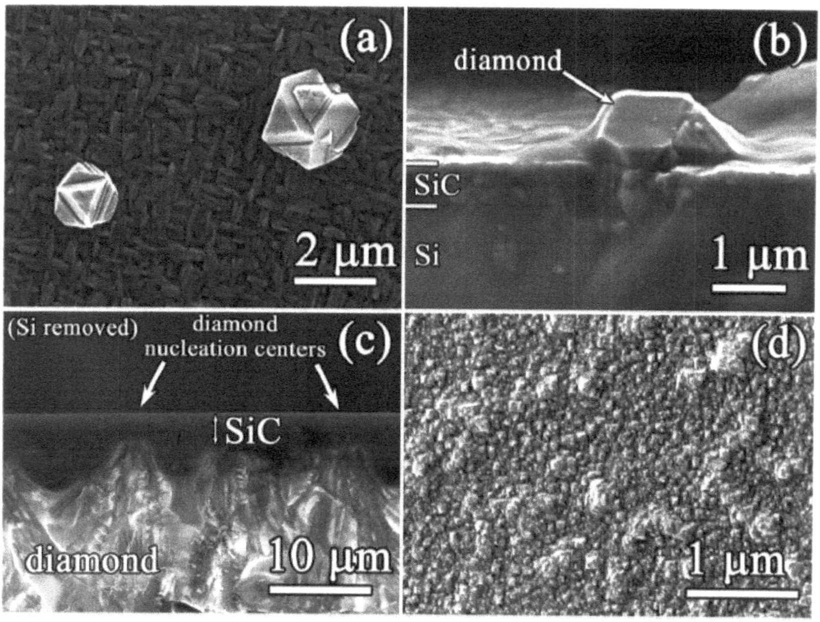

FIG. 23 SEM images of the textured SiC/diamond film deposited on the Si substrate. (A) plan-view and (B) cross section of 550-nm-thick SiC/diamond film. (C) Cross section of the SiC/diamond film grown. (D) Nucleation side of the free-standing composite film. *(Reproduced with permission from V. Sedov, A. Martyanov, A. Khomich, S. Savin, V. Voronov, R. Khmelnitskiy, A. Bolshakov, V. Ralchenko, Co-deposition of diamond and β-SiC by microwave plasma CVD in H_2-CH_4-SiH_4 gas mixtures. Diam. Relat. Mater. 98 (2019) 107520.)*

of the bulk diamond, and the bandgap varies with the diameter, surface morphology, and the orientation of the principal axis [166]. The diameter and crystallographic direction are the main factors affecting the structural stability of the diamond nanowires [166, 167]. Different methods for the synthesis of DNRs are reported by several researchers such as using hydrogen plasma treatment on the CNTs [168] and hydrogen plasma treatment on nanodiamond films, using Ar-rich MPCVD [169]. In most of these cases, the synthesized DNRs were covered with amorphous carbon. In the MPCVD method, when the methane percentage was increased (up to 10%) in the feeding gas, the diamond-graphite nanorods were also observed along with DNRs [170].

Rakha et al. reported the synthesis of diamond-graphite nanorods by MWCVD [170]. The synthesis was performed in a 2.45-GHz microwave plasma system on a (100) n-type Si substrate using $Ar/N_2/CH_4$ mixtures. The Si substrate temperature and the plasma pressure were 850°C and 80 Torr, respectively, and the microwave power was kept at 1800 W. The total gas flow was maintained at 40 sccm with 16 sccm N_2 (40%), 20 sccm Ar, and 4 sccm (10%) CH_4. Fig. 24 shows the low-magnification TEM micrograph demonstrating several nanorods in the matrix. Selected area electron diffraction (SAED) patterns were taken over a 100-nm region and clearly showed the presence of two crystalline structures: diamond and graphite.

Fig. 25A shows the XRD spectrum of the DNRs, and the results indicated the coexistence of diamond and graphite structures in the DNRs. The Raman spectrum (Fig. 25B) exhibited that the diamond feature peak is completely covered by the D band of graphite or amorphous carbon [171]. Also, the existence of ultrafine nanodiamond was observed (a weak band centered approximately at 1150 cm^{-1}) [172–175].

Even though nanocrystalline diamond (NCD) or ultrananocrystalline diamond (UNCD) film possesses a lot of advantages, these films still require to overcome several remaining problems. For instance, due to trapped high non-diamond carbons, the deposited NCD film with nanocrystals suffers from poor adhesion [176] and a great number of grain boundaries [153]. On the other hand, the crystallized microcrystalline diamond film showed superior adhesion characteristics [177]. Thus, a MCD and NCD dual-layer composite diamond coating are some of the best approaches to exploit the advantages of both layers. The high-quality and smooth micro- and nanocrystalline dual-layer composite diamond films were successfully prepared using $H_2/CH_4/Ar/CO_2$ microwave plasma [178]. The synthesis process includes nucleation/MCD growth/nucleation/NCD growth (Fig. 26).

The influences of CO_2 concentration on the morphology, microstructure, and growth rate of the deposited NCD layer were investigated (Fig. 27). The results showed that the additional carbonaceous sources supplied by CO_2 are supportive to enhance the growth rate. Besides, the appropriate addition of CO_2 can increase the quality and decrease the surface roughness. It is evidenced that the addition of CO_2 strongly affects the contents of various reaction species in the plasma.

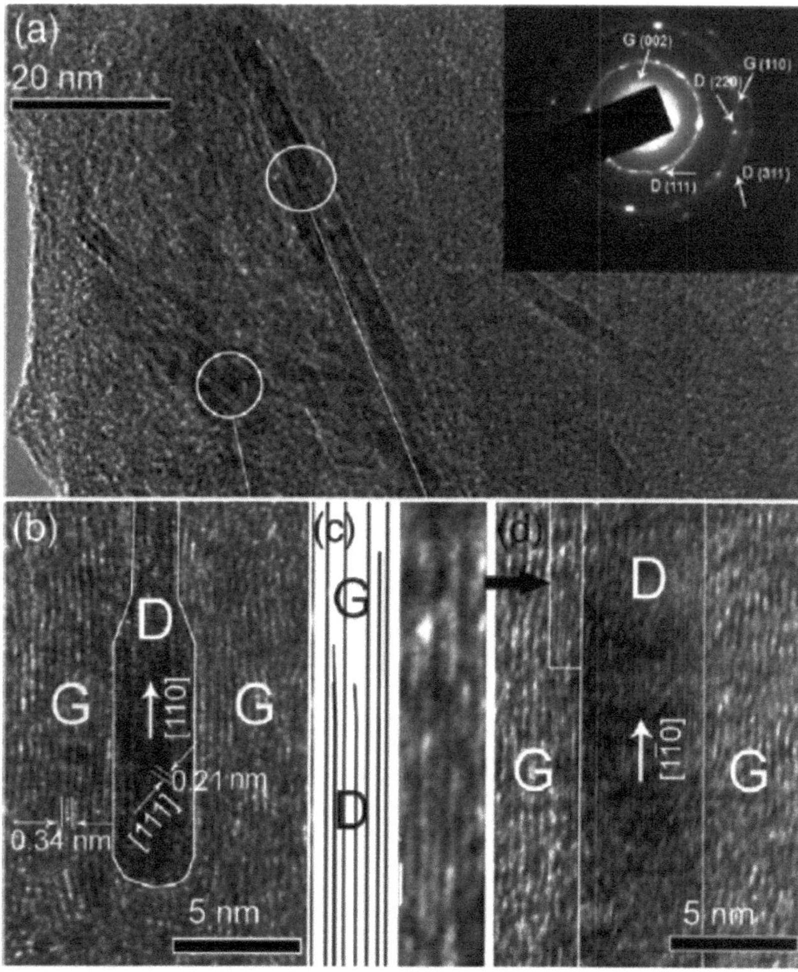

FIG. 24 (A) Low-magnification TEM images. The inset shows the corresponding SAED pattern (B, D). (C) Schematic indicating the lattice spacing of graphite and diamond structures. *(Reproduced with permission from S.A. Rakha, G. Yu, J. Cao, S. He, X. Zhou, Diamond-graphite nanorods produced by microwave plasma chemical vapor deposition. Diam. Relat. Mater. 19 (2010) 284–287.)*

9.5 Ultrananocrystalline diamond films

The ultrananocrystalline diamond (UNCD) film bears an unparalleled combination of physical and chemical properties, and several other important properties exceed those of single-crystal and microcrystalline diamond films. Beside, UNCD films possess excellent electron field emission [179, 180], far smoother surface [181], tribology [182], and unique ability to incorporate n-type dopants [183]. UNCD could be potentially used for the fabrication of important

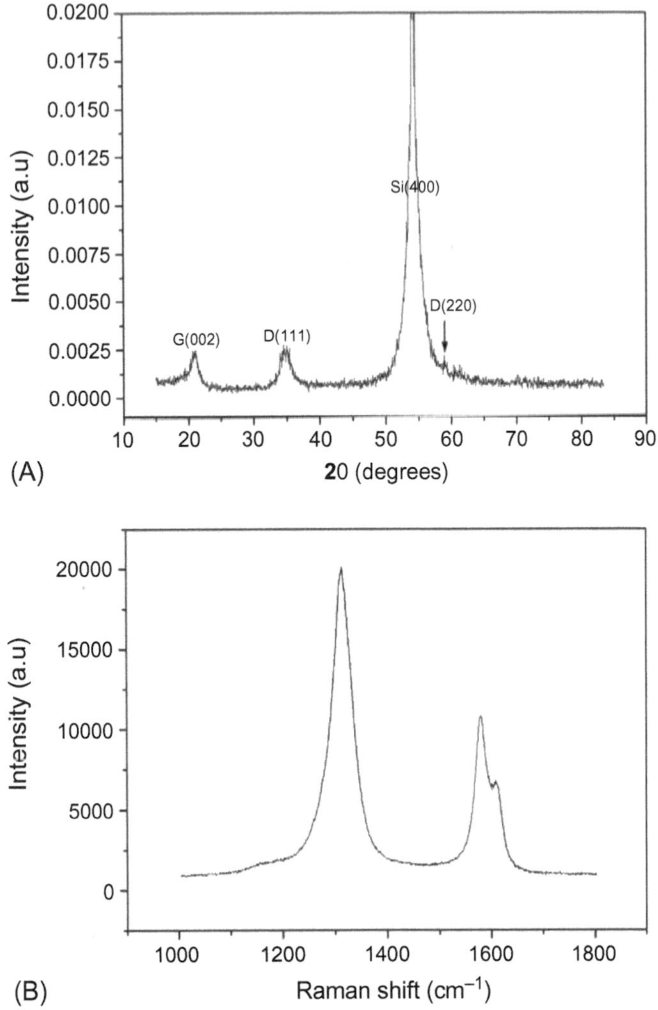

FIG. 25 (A) XRD and (B) Raman spectrum of DNRs indicating the characteristics peaks of diamond and graphite structures. *(Reproduced with permission from S.A. Rakha, G. Yu, J. Cao, S. He, X. Zhou, Diamond-graphite nanorods produced by microwave plasma chemical vapor deposition. Diam. Relat. Mater. 19 (2010) 284–287.)*

multifunctional devices [184–187]. Especially, the synthesis of highly smooth UNCD films is important in the area of micro-electromechanical systems (MEMS), surface acoustic wave (SAW) devices, and medical implants, as they can avoid the polishing step needed for microcrystalline diamond films. Even though several methods are available, MPCVD using Ar-rich/CH_4 growth chemistry was frequently used for achieving the UNCD films. Enhancement of species activity and diamond secondary nucleation to form nano-sized grains were observed during MPCVD deposition [183, 188–190].

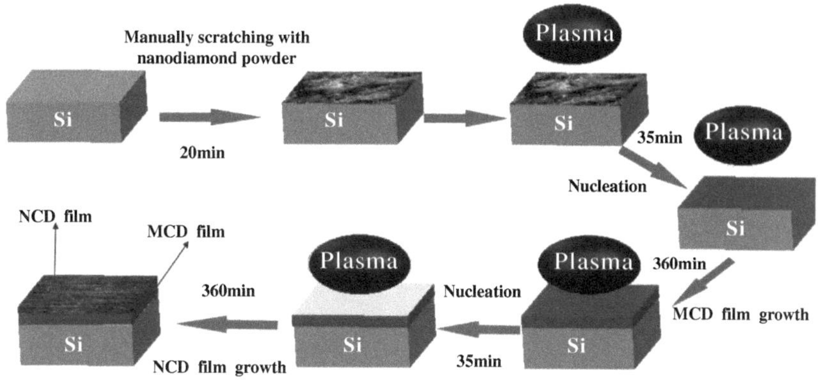

FIG. 26 Schematic diagram for growing micro- and nanocrystalline dual-layer composite diamond films. *(Reproduced with permission from C. Liu, J.-H. Wang, J. Weng, Growth of micro-and nanocrystalline dual layer composite diamond films by microwave plasma CVD: influence of CO_2 concentration on growth of nano-layer, J. Cryst. Growth 410 (2015) 30–34.)*

UNCD films prepared by MWCVD using Ar-rich $CH_4/H_2/Ar$ plasmas with varying argon concentration from 96% to 98% and negative bias voltage from 0 to 150 V was reported [191]. The influences of argon concentration as well as the negative bias voltage on the microstructure, morphology, and phase composition of the UNCD films were analyzed using scanning electron microscopy, X-ray diffraction, atom force microscopy, and visible and UV Raman spectroscopy.

The result showed that due to the introduction of Ar in the plasma, the grain size and surface roughness of the films were decreased (Fig. 28). When Ar content increased from 96% to 98%, the grain size decreased from 12.7 to 7.3 nm and the surface roughness decreased from 23.4 to 15.2 nm. Besides, very smooth films with a grain size of about 5.7 nm and a surface roughness of 9.6 nm were obtained at −125 V in $1\%CH_4/1\%H_2/98\%Ar$.

To enhance the formation of diamond nuclei on foreign substrates, ultrasonication in diamond powder suspended solutions was the commonly used procedure [190, 192–194]. Several studies are carried out to improve the syntheses and performance of UNCD films for functional applications. New techniques for the process, pretreatment, higher growth rate, higher doping efficiency, smoother surface, and lower cost were needed. Lin et al. used a unique synthesis system named microwave plasma jet chemical vapor deposition (MPJCVD) to improve the UNCD deposition compared with the conventional MPCVD method [195].

The schematic diagram of the MPJCVD system is shown in Fig. 29. The introduction of a conical antenna system into the deposition chamber provided the key function in the system. During the synthesis, the plasma generation was focused on the antenna tip to produce a plasma jet with energetic species. The

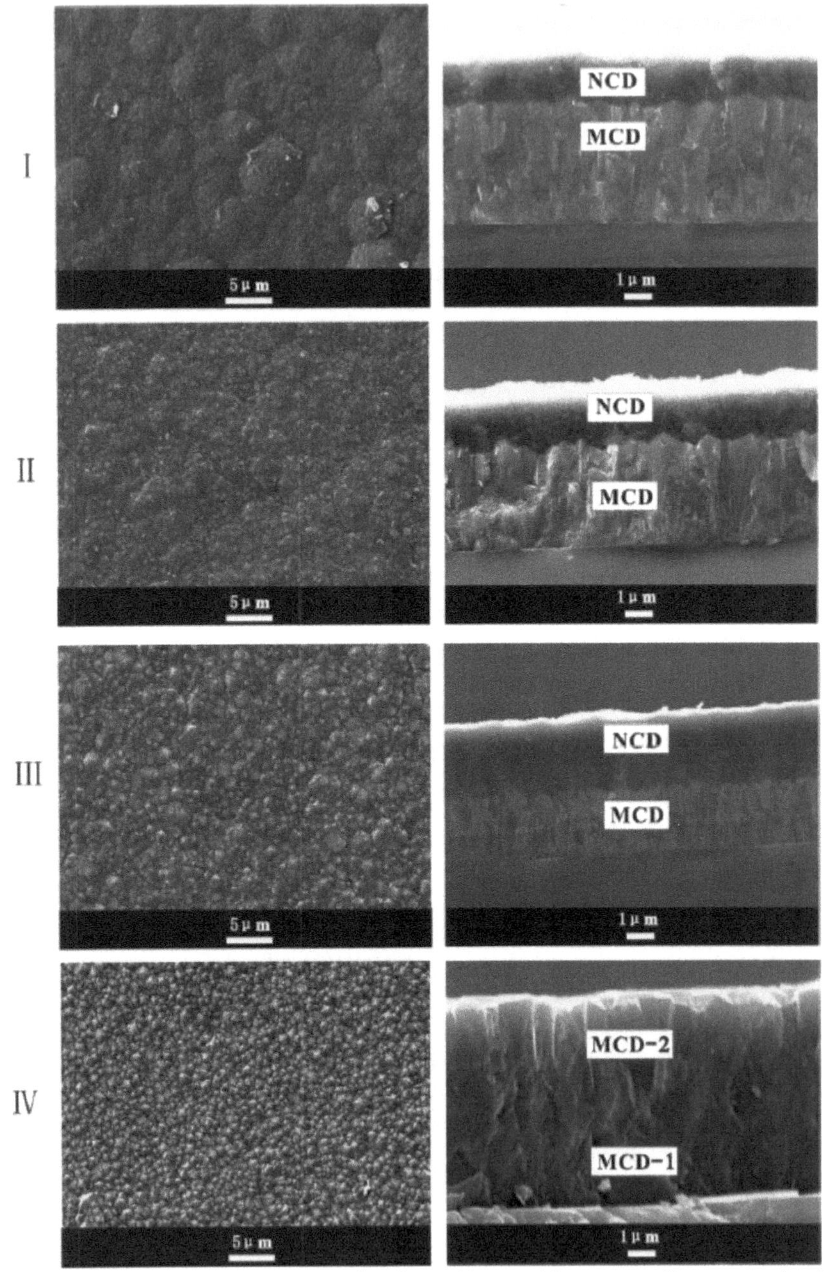

FIG. 27 The SEM images of nanocrystalline diamond layer and the corresponding cross-sectional SEM images of micro- and nanocrystalline dual-layer composite diamond films at different CO_2 concentration (I) 0.0% CO_2; (II) 0.8% CO_2; (III) 2.5% CO_2; (IV) 4.0% CO_2. *(Reproduced with permission from C. Liu, J.-H. Wang, J. Weng, Growth of micro-and nanocrystalline dual layer composite diamond films by microwave plasma CVD: influence of CO_2 concentration on growth of nano-layer. J. Cryst. Growth 410 (2015) 30–34.)*

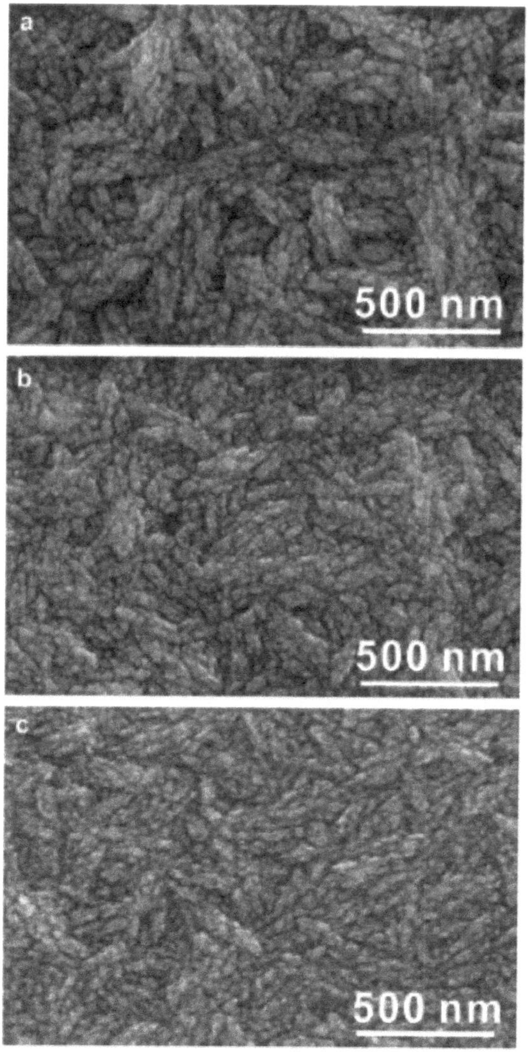

FIG. 28 The SEM image of ultrananocrystalline diamond films deposited under different Ar contents (A) 96% Ar, (B) 97% Ar, (C) 98% Ar. *(Reproduced with permission from Y. Zou, Z. Li, Y. Wu, Deposition and characterization of smooth ultra-nanocrystalline diamond film in CH4/H2/Ar by microwave plasma chemical vapor deposition. Vacuum 84 (2010) 1347–1352.)*

dissociation and energy of plasma species were improved with a focused plasma jet. It also created growth species with relatively high density and activity at the same growth parameters compared with the regular MPCVD.

Fig. 30 shows the evolution of the surface of diamond films with an increase in Ar gas to the CH_4/H_2 plasma during the synthesis by MPJCVD. The transition of the films from microcrystalline to ultrananocrystalline diamond film is very

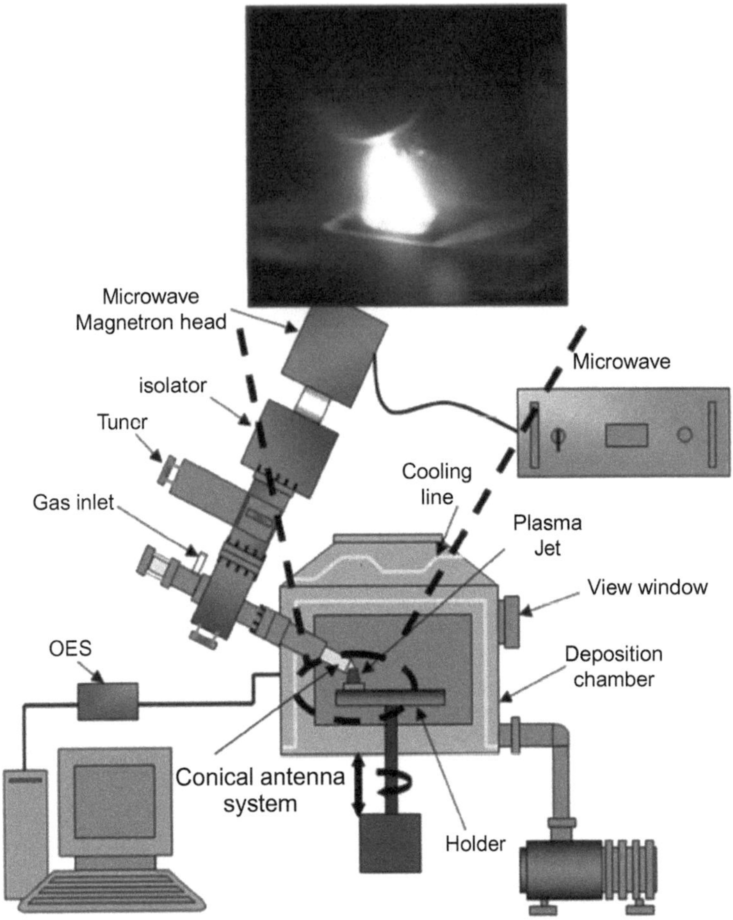

FIG. 29 Schematic diagram and construction are for home-made MPJCVD system. *(Reproduced with permission from C.-R. Lin, W.-H. Liao, D.-H. Wei, C.-K. Chang, W.-C. Fang, C.-L. Chen, C.-L. Dong, J.-L. Chen, J.-H. Guo, Improvement on the synthesis technique of ultrananocrystalline diamond films by using microwave plasma jet chemical vapor deposition. J. Cryst. Growth 326 (2011) 212–217.)*

clear from the SEM images. A schematic diagram of the transition is shown in Fig. 31. The Ar addition into plasma markedly changed the microstructure of diamond films, directing to the decrease in the microcrystalline grains size and surface roughness. Besides, with the increase in Ar concentration, the sp^3 bonding carbon concentration in the films also decreased. This was due to the great increment in renucleation during the growth.

Tang et al. investigated the influence of minute amounts of pure nitrogen addition into conventional methane/hydrogen mixtures during the growth of nanocrystalline diamond (NCD) films by MPCVD, under high power

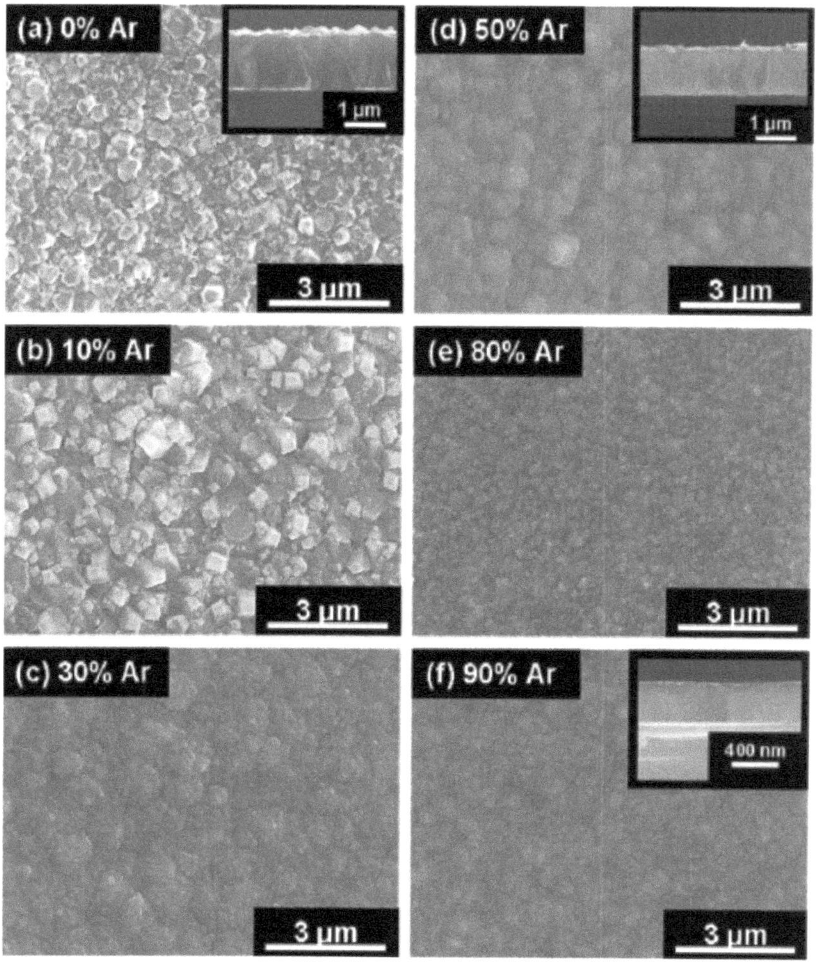

FIG. 30 SEM images are for the UNCD films grown at various Ar ($Ar/Ar+H_2$) concentrations: (A) 0%, (B) 10%, (C) 30%, (D) 50%, (E) 80%, and (F) 90%. *(Reproduced with permission from C.-R. Lin, W.-H. Liao, D.-H. Wei, C.-K. Chang, W.-C. Fang, C.-L. Chen, C.-L. Dong, J.-L. Chen, J.-H. Guo, Improvement on the synthesis technique of ultrananocrystalline diamond films by using microwave plasma jet chemical vapor deposition. J. Cryst. Growth 326 (2011) 212–217.)*

conditions [196]. The gas mixture of 4% CH_4/H_2 with two different concentrations of N_2 additive and microwave power ranging from 3.0 to 4.0 kW was used for the analysis. Under these conditions, growth rates ranging from 5.4 μm/h up to 9.6 μm/h were achieved. Nitrogen addition was found to be the main parameter responsible for the growth characteristics of NCD films.

The growth of diamond film on the Si substrate by MPCVD requires a nucleation step for the improvement of yield, quality, purity, and uniformity of the film at relatively low temperature and pressure. Thus, diamond nucleation on

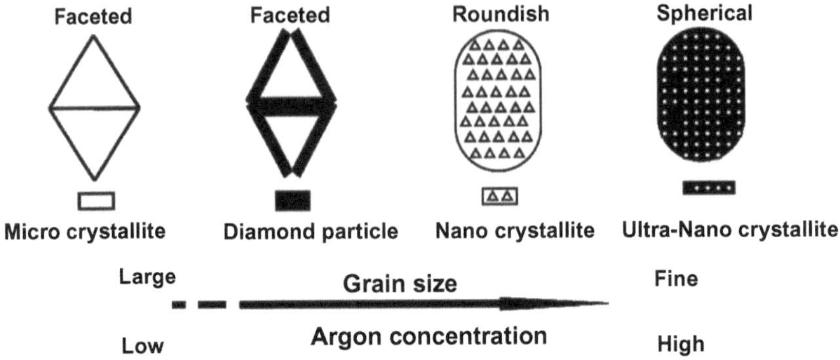

FIG. 31 Schematic diagram representing the transition from microcrystalline to ultrananocrystalline films during growth with Ar addition. *(Reproduced with permission from C.-R. Lin, W.-H. Liao, D.-H. Wei, C.-K. Chung, W.-C. Fang, C.-L. Chen, C.-L. Dong, J.-L. Chen, J.-H. Guo, Improvement on the synthesis technique of ultrananocrystalline diamond films by using microwave plasma jet chemical vapor deposition. J. Cryst. Growth 326 (2011) 212–217.)*

the unscratched Si surface is of great importance. The substrate pretreatments influence the initial growth period. Tiwari et al. reported the synthesis of diamond films on adamantane-coated crystalline Si {100} substrate by MPCVD from a gaseous mixture of methane and hydrogen gases without the application of a bias voltage to the substrates [197]. The substrate temperature was at 530°C during the synthesis. The schematic of diamond synthesis on the adamantane-coated Si substrate is shown in Fig. 32.

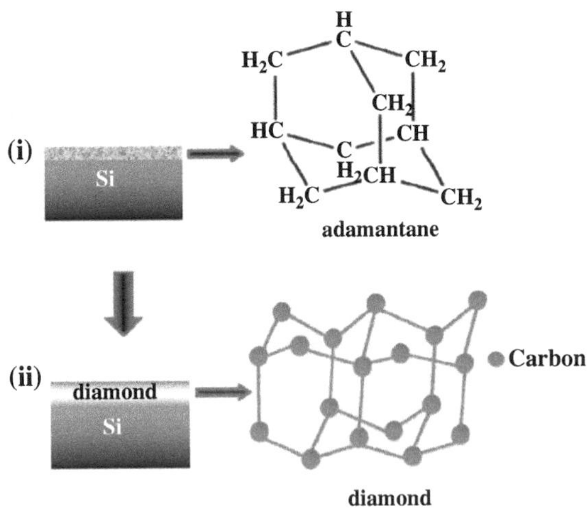

FIG. 32 Schematic diagram showing diamond synthesis in two steps; (i) adamantane deposited on silicon surface by hotplate method and (ii) diamond growth by MPCVD. *(Reproduced with permission from R.N. Tiwari, L. Chang, Growth, microstructure, and field-emission properties of synthesized diamond film on adamantane-coated silicon substrate by microwave plasma chemical vapor deposition. J. Appl. Phys. 107 (2010) 103305.)*

The deposited films were characterized by scanning electron microscopy, Raman spectrometry, X-ray diffraction, and X-ray photoelectron spectroscopy, and the analysis provided a definitive evidence for the high-crystalline-quality diamond film. The diamond nucleation from adamantane molecules at various deposition times (15–45 min) is shown in Fig. 33.

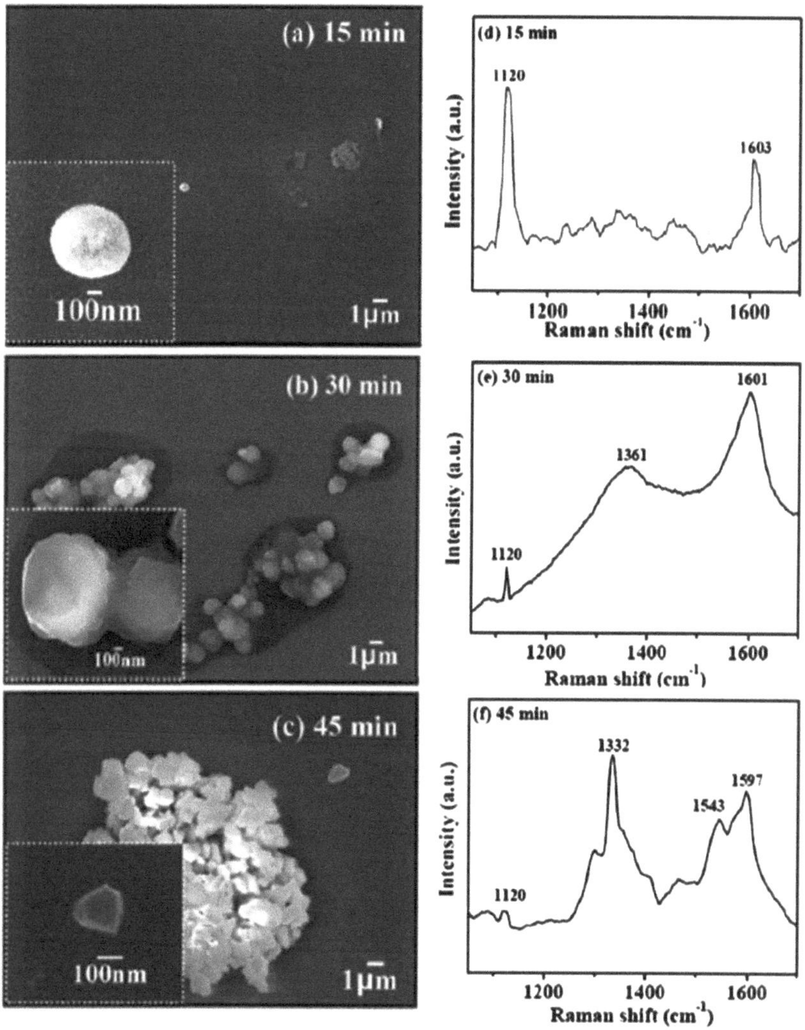

FIG. 33 Plan-view SEM images and micro-Raman spectra after growth for deposition times of (A) and (D) 15 min, (B) and (E) 30 min, and (C) and (F) 45 min. *(Reproduced with permission from R.N. Tiwari, L. Chang, Growth, microstructure, and field-emission properties of synthesized diamond film on adamantane-coated silicon substrate by microwave plasma chemical vapor deposition. J. Appl. Phys. 107 (2010) 103305.)*

9.6 Conclusion

This chapter has described the development of diamond synthesis by the microwave-enhanced chemical vapor deposition. Chemical vapor deposition gives diamonds in shapes and sizes and with composition and quality control unattainable using synthesis at high pressure. Consequently, it is an enabling technology that permits the exploitation of diamond's exceptional properties in a range of technologies outside its traditional role in cutting, grinding, and dressing operations.

Acknowledgments

AD and BKB are grateful to Prince Mohammad Bin Fahd University for support. BKB is also grateful to US NIH, US NCI, and Kleberg Foundation of Texas for financial support.

References

[1] H.O. Pierce, Handbook of Carbon, Graphite, Diamond and Fullerenes, Noyes Publications, Park Ridge, NJ, 1993.

[2] P.W. May, Diamond thin films: a 21st-century material, Philos. Trans. R. Soc. London, Ser. A 358 (2000) 473–495.

[3] M.N. Yoder, The vision of diamond as an engineered material, in: Synthetic Diamond—Emerging CVD Science and Technology, John Wiley & Sons, Inc, New York, 1994, pp. 3–17.

[4] J.E. Field, The Properties of Natural and Synthetic Diamond, Academic Press, 1992.

[5] K. Hayashi, Y. Yokota, T. Tachibana, K. Kobashi, T. Fukunaga, T. Takada, Fabrication and characterization of diamond gas sensors for detection of semiconductor doping gases, New Diamond Front. Carbon Technol. 11 (2001) 101–111.

[6] H. Shiomi, Y. Nishibayashi, N. Fujimori, Field-effect transistors using boron-doped diamond epitaxial films, Jpn. J. Appl. Phys. 28 (1989) L2153.

[7] D.R. Kania, M.I. Landstrass, M.A. Plano, L.S. Pan, S. Han, Diamond radiation detectors, Diam. Relat. Mater. 2 (1993) 1012–1019.

[8] C.E. Nebel, D. Shin, B. Rezek, N. Tokuda, H. Uetsuka, H. Watanabe, Diamond and biology, J. R. Soc. Interface 4 (2007) 439–461.

[9] C.E. Nebel, B. Rezek, D. Shin, H. Uetsuka, N. Yang, Diamond for bio-sensor applications, J. Phys. D. Appl. Phys. 40 (2007) 6443.

[10] S. Katsumata, Y. Oobuchi, T. Asano, Patterning of CVD diamond films by seeding and their field emission properties, Diam. Relat. Mater. 3 (1994) 1296–1300.

[11] M.W. Geis, N.N. Efremow, D.D. Rathman, Device Applications of Diamonds, (Ph.D. thesis), American Vacuum Society, 1988.

[12] C.M. Breeding, J.E. Shigley, The "type" classification system of diamonds and its importance in gemology, Gems Gemol. 45 (2009) 96–111.

[13] F.P. Bundy, W.A. Bassett, M.S. Weathers, R.J. Hemley, H.U. Mao, A.F. Goncharov, The pressure-temperature phase and transformation diagram for carbon; updated through 1994, Carbon 34 (1996) 141–153.

[14] F.P. Bundy, H.T. Hall, H.M. Strong, R.H. Wentorf, Man-made diamonds, Nature 176 (1955) 51–55.

[15] J.E. Field, The Properties of Diamond, in: J.E. Field (Ed.), Academic Press, London, New York, 1979.
[16] R.C. DeVries, Synthesis of diamond under metastable conditions, Annu. Rev. Mater. Sci. 17 (1987) 161–187.
[17] K. Nassau, J. Nassau, The history and present status of synthetic diamond, J. Cryst. Growth 46 (1979) 157–172.
[18] W.G. Eversole, US Patents 3,030,187 and 3,030,188, Synthesis of Diamond, 1962.
[19] B.V. Derjaguin, D.V. Fedoseev, V.M. Lukyanovich, B.V. Spitzin, V.A. Ryabov, A.V. Lavrentyev, Filamentary diamond crystals, J. Cryst. Growth 2 (1968) 380–384.
[20] J.C. Angus, H.A. Will, W.S. Stanko, Growth of diamond seed crystals by vapor deposition, J. Appl. Phys. 39 (1968) 2915–2922.
[21] J.C. Angus, N.C. Gardner, D.J. Poferl, S. Chauhan, T. Sung, Growing semiconductor diamonds at subatmospheric pressures, Sinteticheskie Almazy 3 (1971) 38–39.
[22] D.J. Poferl, N.C. Gardner, J.C. Angus, Growth of boron-doped diamond seed crystals by vapor deposition, J. Appl. Phys. 44 (1973) 1428–1434.
[23] B.V. Derjaguin, B.V. Spitsyn, L.L. Builov, A.A. Klochkov, A.E. Gorodetsky, A.V. Smolyaninov, Synthesis of diamond crystals on non-diamond substrates, Dokl. Acad. Nauk. SSSR (1976) 333–335.
[24] B.V. Spitsyn, L.L. Bouilov, B.V. Derjaguin, Vapor growth of diamond on diamond and other surfaces, J. Cryst. Growth 52 (1981) 219–226.
[25] S. Matsumoto, Y. Sato, M. Kamo, N. Setaka, Vapor deposition of diamond particles from methane, Jpn. J. Appl. Phys. 21 (1982) L183.
[26] S. Matsumoto, Y. Sato, M. Tsutsumi, N. Setaka, Growth of diamond particles from methane-hydrogen gas, J. Mater. Sci. 17 (1982) 3106–3112.
[27] M. Kamo, Y. Sato, S. Matsumoto, N. Setaka, Diamond synthesis from gas phase in microwave plasma, J. Cryst. Growth 62 (1983) 642–644.
[28] Y. Saito, S. Matsuda, S. Nogita, Synthesis of diamond by decomposition of methane in microwave plasma, J. Mater. Sci. Lett. 5 (1986) 565–568.
[29] K. Suzuki, A. Sawabe, H. Yasuda, T. Inuzuka, Growth of diamond thin films by dc plasma chemical vapor deposition, Appl. Phys. Lett. 50 (1987) 728–729.
[30] T. Chonan, M. Uenura, S. Futaki, S. Nishi, The preparation and characterization of diamond powder by DC arc plasma, Jpn. J. Appl. Phys. 28 (1989) L1058.
[31] F. Akatsuka, Y. Hirose, K. Komaki, Rapid growth of diamond films by arc discharge plasma CVD, Jpn. J. Appl. Phys. 27 (1988) L1600–L1602.
[32] K. Suzuki, A. Sawabe, T. Inuzuka, Growth of diamond thin films by DC plasma chemical vapor deposition and characteristics of the plasma, Jpn. J. Appl. Phys. 29 (1990) 153.
[33] K. Kurihara, K. Sasaki, M. Kawarada, N. Koshino, High rate synthesis of diamond by dc plasma jet chemical vapor deposition, Appl. Phys. Lett. 52 (1988) 437–438.
[34] A. Boudina, E. Fitzer, G. Wahl, Diamond film preparation by arc-discharge plasma-jet-CVD and thermodynamic calculation of the equilibrium gas composition, Diam. Relat. Mater. 1 (1992) 380–387.
[35] E. Lugscheider, W. Schlump, F. Deuerler, P. Remer, Extensive diamond film deposition by dc plasma jet chemical vapor deposition, Diam. Relat. Mater. 3 (1994) 325–327.
[36] S. Matsumoto, Chemical vapour deposition of diamond in RF glow discharge, J. Mater. Sci. Lett. 4 (1985) 600–602.
[37] J. Beckman, R.B. Jackman, J.S. Foord, Capacitively coupled rf plasma sources: a viable approach for CVD diamond growth? Diam. Relat. Mater. 3 (1994) 602–607.

[38] Y. Mitsuda, T. Yoshida, K. Akashi, Development of a new microwave plasma torch and its application to diamond synthesis, Rev. Sci. Instrum. 60 (1989) 249–252.

[39] D.K. Smith, E. Sevillano, M. Besen, V. Berkman, L. Bourget, Large area diamond reactor using a high flow velocity microwave plasma jet, Diam. Relat. Mater. 1 (1992) 814–817.

[40] J. Wei, H. Kawarada, J. Suzuki, A. Hiraki, Growth of diamond films at low pressure using magneto-microwave plasma CVD, J. Cryst. Growth 99 (1990) 1201–1205.

[41] H. Kawarada, K.S. Mar, A. Hiraki, Large area chemical vapour deposition of diamond particles and films using magneto-microwave plasma, Jpn. J. Appl. Phys. 26 (1987) L1032.

[42] W. Tsai, G.J. Reynolds, S. Hikido, C.B. Cooper III, Electron cyclotron resonance plasma-enhanced filament-assisted diamond growth, Appl. Phys. Lett. 60 (1992) 1444–1446.

[43] J. Suzuki, H. Kawarada, K.-S. Mar, J. Wei, Y. Yokota, A. Hiraki, The synthesis of diamond films at lower pressure and lower temperature using magneto-microwave plasma CVD, Jpn. J. Appl. Phys. 28 (1989) L281.

[44] W.A. Yarbrough, M.A. Stewart, J.A. Cooper Jr., Combustion synthesis of diamond, Surf. Coat. Technol. 39 (1989) 241–252.

[45] Y. Tzeng, C. Cutshaw, R. Phillips, T. Srivinyunon, A. Ibrahim, B.H. Loo, Growth of diamond films on silicon from an oxygen-acetylene flame, Appl. Phys. Lett. 56 (1990) 134–136.

[46] P.G. Kosky, D.S. McAtee, An experimental and theoretical investigation of flame-formed diamonds, Mater. Lett. 8 (1989) 369–374.

[47] N. Ohtake, M. Yoshikawa, Diamond film preparation by arc discharge plasma jet chemical vapor deposition in the methane atmosphere, J. Electrochem. Soc. 137 (1990) 717.

[48] C. Yan, Y.K. Vohra, H. Mao, R.J. Hemley, Very high growth rate chemical vapor deposition of single-crystal diamond, Proc. Natl. Acad. Sci. 99 (2002) 12523–12525.

[49] Q. Liang, C.Y. Chin, J. Lai, C. Yan, Y. Meng, H. Mao, R.J. Hemley, Enhanced growth of high quality single crystal diamond by microwave plasma assisted chemical vapor deposition at high gas pressures, Appl. Phys. Lett. 94 (2009), 024103.

[50] S. Ho, C. Yan, Z. Liu, H. Mao, R. Hemley, Prospects for large single crystal CVD diamonds, Ind. Diamond Rev. 66 (2006).

[51] Y. Meng, C. Yan, S. Krasnicki, Q. Liang, J. Lai, H. Shu, T. Yu, A. Steele, H. Mao, R.J. Hemley, High optical quality multicarat single crystal diamond produced by chemical vapor deposition, Phys. Status Solidi A 209 (2012) 101–104.

[52] H. Yamada, A. Chayahara, Y. Mokuno, Effects of intentionally introduced nitrogen and substrate temperature on growth of diamond bulk single crystals, Jpn J. Appl. Phys. 55 (2015) 01AC07.

[53] Y.F. Li, X.M. An, X.C. Liu, L. Jiang, P.W. Zhang, H. Guo, Z.L. Sun, H.Z. Zhao, W.Z. Tang, A 915 MHz/75 kW cylindrical cavity type microwave plasma chemical vapor deposition reactor with a ladder-shaped circumferential antenna developed for growing large area diamond films, Diam. Relat. Mater. 78 (2017) 67–72.

[54] A. Chayahara, Y. Mokuno, Y. Horino, Y. Takasu, H. Kato, H. Yoshikawa, N. Fujimori, The effect of nitrogen addition during high-rate homoepitaxial growth of diamond by microwave plasma CVD, Diam. Relat. Mater. 13 (2004) 1954–1958.

[55] Q. Liang, C. Yan, J. Lai, Y. Meng, S. Krasnicki, H. Shu, H. Mao, R.J. Hemley, Large area single-crystal diamond synthesis by 915 MHz microwave plasma-assisted chemical vapor deposition, Cryst. Growth Des. 14 (2014) 3234–3238.

[56] J. Achard, F. Silva, A. Tallaire, X. Bonnin, G. Lombardi, K. Hassouni, A. Gicquel, High quality MPACVD diamond single crystal growth: high microwave power density regime, J. Phys. D. Appl. Phys. 40 (2007) 6175.

[57] H. Yamada, A. Chayahara, Y. Mokuno, S. Shikata, Numerical microwave plasma discharge study for the growth of large single-crystal diamond, Diam. Relat. Mater. 54 (2015) 9–14.

[58] J. Lu, Y. Gu, T.A. Grotjohn, T. Schuelke, J. Asmussen, Experimentally defining the safe and efficient, high pressure microwave plasma assisted CVD operating regime for single crystal diamond synthesis, Diam. Relat. Mater. 37 (2013) 17–28.

[59] M. Muehle, J. Asmussen, M.F. Becker, T. Schuelke, Extending microwave plasma assisted CVD SCD growth to pressures of 400 Torr, Diam. Relat. Mater. 79 (2017) 150–163.

[60] P. Houston, N. Sime, Numerical modelling of MPA-CVD reactors with the discontinuous Galerkin finite element method, J. Phys. D. Appl. Phys. 50 (2017) 295202.

[61] A. Charris, S. Nad, J. Asmussen, Exploring constant substrate temperature and constant high pressure SCD growth using variable pocket holder depths, Diam. Relat. Mater. 76 (2017) 58–67.

[62] A.P. Bolshakov, V.G. Ralchenko, V.Y. Yurov, A.F. Popovich, I.A. Antonova, A.A. Khomich, E.E. Ashkinazi, S.G. Ryzhkov, A.V. Vlasov, A.V. Khomich, High-rate growth of single crystal diamond in microwave plasma in CH_4/H_2 and $CH_4/H_2/Ar$ gas mixtures in presence of intensive soot formation, Diam. Relat. Mater. 62 (2016) 49–57.

[63] H. Li, K. Yang, H.X. Liu, X.D. Zhu, Optical and mass spectroscopic properties of microwave $CH_4/H_2/Ar$ plasma for diamond deposition in a resonance cavity, Vacuum 147 (2018) 45–50.

[64] A. Tallaire, C. Rond, F. Benedic, O. Brinza, J. Achard, F. Silva, A. Gicquel, Effect of argon addition on the growth of thick single crystal diamond by high-power plasma CVD, Phys. Status Solidi A 208 (2011) 2028–2032.

[65] E.J.D. Mahoney, B.S. Truscott, M.N.R. Ashfold, Y.A. Mankelevich, Optical emission from C2–anions in microwave-activated CH_4/H_2 plasmas for chemical vapor deposition of diamond, J. Phys. Chem. A 121 (2017) 2760–2772.

[66] H. Yamada, A. Chayahara, Y. Mokuno, Effect of Ar addition on uniformity of diamond growth by using microwave plasma chemical vapor deposition, Diam. Relat. Mater. 87 (2018) 143–148.

[67] R.J. Hemley, H.-K. Mao, C. Yan, Y.K. Vohra, Apparatus and Method for Diamond Production, U.S. Patent 6,858,078, 2005.

[68] P.K. Bachmann, W. van Enckevort, Diamond deposition technologies, Diam. Relat. Mater. 1 (1992) 1021–1034.

[69] J. Asmussen, T.A. Grotjohn, T. Schuelke, M.F. Becker, M.K. Yaran, D.J. King, S. Wicklein, D.K. Reinhard, Multiple substrate microwave plasma-assisted chemical vapor deposition single crystal diamond synthesis, Appl. Phys. Lett. 93 (2008), 031502.

[70] K.W. Hemawan, C.S. Yan, Q. Liang, J. Lai, Y. Meng, S. Krasnicki, H.K. Mao, R.J. Hemley, Hot spot formation in microwave plasma CVD diamond synthesis, IEEE Trans. Plasma Sci. 39 (2011) 2790–2791.

[71] K. Hassouni, G. Lombardi, X. Duten, G. Haagelar, F. Silva, A. Gicquel, T.A. Grotjohn, M. Capitelli, J. Röpcke, Overview of the different aspects in modelling moderate pressure H_2 and H_2/CH_4 microwave discharges, Plasma Sources Sci. Technol. 15 (2006) 117.

[72] E.V. Bushuev, V.Y. Yurov, A.P. Bolshakov, V.G. Ralchenko, A.A. Khomich, I.A. Antonova, E.E. Ashkinazi, V.A. Shershulin, V.P. Pashinin, V.I. Konov, Express in situ measurement of epitaxial CVD diamond film growth kinetics, Diam. Relat. Mater. 72 (2017) 61–70.

[73] Y. Gu, J. Lu, T. Grotjohn, T. Schuelke, J. Asmussen, Microwave plasma reactor design for high pressure and high power density diamond synthesis, Diam. Relat. Mater. 24 (2012) 210–214.

[74] Q. Liang, C. Yan, Y. Meng, J. Lai, S. Krasnicki, H. Mao, R.J. Hemley, Recent advances in high-growth rate single-crystal CVD diamond, Diam. Relat. Mater. 18 (2009) 698–703.

[75] A.L. Vikharev, M.A. Lobaev, A.M. Gorbachev, D.B. Radishev, V.A. Isaev, S.A. Bogdanov, Investigation of homoepitaxial growth by microwave plasma CVD providing high growth rate and high quality of diamond simultaneously, Mater. Today Commun. 22 (2020) 100816.

[76] J.J. Su, Y.F. Li, M.H. Ding, X.L. Li, Y.Q. Liu, G. Wang, W.Z. Tang, A dome-shaped cavity type microwave plasma chemical vapor deposition reactor for diamond films deposition, Vacuum 107 (2014) 51–55, https://doi.org/10.1016/j.vacuum.2014.04.002.

[77] K.W. Hemawan, H. Gou, R.J. Hemley, Diamond synthesis at atmospheric pressure by microwave capillary plasma chemical vapor deposition, Appl. Phys. Lett. 107 (2015) 181901, https://doi.org/10.1063/1.4934751.

[78] L.C. Nistor, J. Van Landuyt, V.G. Ralchenko, E.D. Obraztsova, A.A. Smolin, Nanocrystalline diamond films: transmission electron microscopy and Raman spectroscopy characterization, Diam. Relat. Mater. 6 (1997) 159–168, https://doi.org/10.1016/S0925-9635(96)00743-1.

[79] A.L. Vikharev, A.M. Gorbachev, A.V. Kozlov, D.B. Radishev, A.B. Muchnikov, Microcrystalline diamond growth in presence of argon in millimeter-wave plasma-assisted CVD reactor, Diam. Relat. Mater. 17 (2008) 1055–1061, https://doi.org/10.1016/j.diamond.2008.01.050.

[80] A.L. Vikharev, A.M. Gorbachev, A.B. Muchnikov, D.B. Radishev, E.A. Kopelovich, M.M. Troitskiy, Investigation of the optimized parameters of microwave plasma-assisted chemical vapour deposition reactor operation in a pulsed mode, J. Phys. D. Appl. Phys. 45 (2012) 395202, https://doi.org/10.1088/0022-3727/45/39/395202.

[81] A.L. Vikharev, A.M. Gorbachev, M.A. Lobaev, D.B. Radishev, Multimode cavity type MPACVD reactor for large area diamond film deposition, Diam. Relat. Mater. 83 (2018) 8–14, https://doi.org/10.1016/j.diamond.2018.01.011.

[82] C. Jany, A. Tardieu, A. Gicquel, P. Bergonzo, F. Foulon, Influence of the growth parameters on the electrical properties of thin polycrystalline CVD diamond films, Diam. Relat. Mater. 9 (2000) 1086–1090, https://doi.org/10.1016/S0925-9635(99)00237-X.

[83] V. Ralchenko, A. Saveliev, A. Popovich, I. Vlasov, S. Voronina, E. Ashkinazi, CVD diamond coating of AlN ceramic substrates to enhance heat removal, Russ. Microelectron. 35 (2006) 205–209.

[84] V. Mortet, L. Zhang, M. Eckert, J. D'Haen, A. Soltani, M. Moreau, D. Troadec, E. Neyts, J. De Jaeger, J. Verbeeck, Grain size tuning of nanocrystalline chemical vapor deposited diamond by continuous electrical bias growth: Experimental and theoretical study, Phys. Status Solidi A 209 (2012) 1675–1682.

[85] S. Bogdanov, A. Vikharev, A. Gorbachev, A. Muchnikov, D. Radishev, N. Ovechkin, V. Parshin, Growth-rate enhancement of high-quality, low-loss CVD-produced diamond disks grown for microwave windows application, Chem. Vap. Depos. 20 (2014) 32–38.

[86] S. Yu, R. Wang, K. Zheng, J. Gao, X. Li, H. Hei, X. Liu, Z. He, Y. Shen, B. Tang, Influence of power density on high purity 63 mm diameter polycrystalline diamond deposition inside a 2.45 GHz MPCVD reactor, J. Phys. D Appl. Phys. 49 (2016) 355202.

[87] A. Popovich, V. Ralchenko, V. Balla, A. Mallik, A. Khomich, A. Bolshakov, D. Sovyk, E. Ashkinazi, V.Y. Yurov, Growth of 4″diameter polycrystalline diamond wafers with high thermal conductivity by 915 MHz microwave plasma chemical vapor deposition, Plasma Sci. Technol. 19 (2017), 035503.

[88] A. Kang, C. Liangxian, L. Jinlong, Z. Yun, Y. Xiongbo, H. Chenyi, G. Jianchao, W. Junjun, H. Lifu, L. Chengming, The effect of substrate holder size on the electric field and discharge plasma on diamond-film formation at high deposition rates during MPCVD, Plasma Sci. Technol. 19 (2017), 095505.

[89] A. Vikharev, A. Gorbachev, A. Muchnikov, D. Radishchev, Study of microwave plasma-assisted chemical vapor deposition of poly-and single-crystalline diamond films, Radiophys. Quantum Electron. 50 (2007) 913–921.
[90] T.-H. Chein, J. Wei, Y. Tzeng, Synthesis of diamond in high power-density microwave methane/hydrogen/oxygen plasmas at elevated substrate temperatures, Diam. Relat. Mater. 8 (1999) 1686–1696.
[91] K. Hemawan, T. Grotjohn, D. Reinhard, J. Asmussen, Improved microwave plasma cavity reactor for diamond synthesis at high-pressure and high power density, Diam. Relat. Mater. 19 (2010) 1446–1452.
[92] A. Bolshakov, V. Ralchenko, V. Yurov, G. Shu, E. Bushuev, A. Khomich, E. Ashkinazi, D. Sovyk, I. Antonova, S. Savin, Enhanced deposition rate of polycrystalline CVD diamond at high microwave power densities, Diam. Relat. Mater. 97 (2019) 107466.
[93] F. Köck, J. Garguilo, B. Brown, R. Nemanich, Enhanced low-temperature thermionic field emission from surface-treated N-doped diamond films, Diam. Relat. Mater. 11 (2002) 774–779.
[94] E. Gheeraert, N. Casanova, A. Tajani, A. Deneuville, E. Bustarret, J. Garrido, C. Nebel, M. Stutzmann, n-Type doping of diamond by sulfur and phosphorus, Diam. Relat. Mater. 11 (2002) 289–295.
[95] T. Grotjohn, D. Tran, M. Yaran, S. Demlow, T. Schuelke, Heavy phosphorus doping by epitaxial growth on the (111) diamond surface, Diam. Relat. Mater. 44 (2014) 129–133.
[96] H. Sachdev, R. Haubner, B. Lux, Lithium addition during CVD diamond deposition using lithium tert.-butanolat as precursor, Diam. Relat. Mater. 6 (1997) 494–500.
[97] C. Benndorf, S. Hadenfeldt, W. Luithardt, A. Zhukov, Photoelectron spectroscopic investigations and exoelectron emission of CVD diamond surfaces modified with oxygen and potassium, Diam. Relat. Mater. 5 (1996) 784–789.
[98] F. Jelezko, J. Wrachtrup, Single defect centres in diamond: a review, Phys. Status Solidi A 203 (2006) 3207–3225.
[99] R. Schirhagl, K. Chang, M. Loretz, C.L. Degen, Nitrogen-vacancy centers in diamond: nanoscale sensors for physics and biology, Annu. Rev. Phys. Chem. 65 (2014) 83–105.
[100] T. Gaebel, I. Popa, A. Gruber, M. Domhan, F. Jelezko, J. Wrachtrup, Stable single-photon source in the near infrared, New J. Phys. 6 (2004) 98.
[101] E. Neu, D. Steinmetz, J. Riedrich-Möller, S. Gsell, M. Fischer, M. Schreck, C. Becher, Single photon emission from silicon-vacancy colour centres in chemical vapour deposition nanodiamonds on iridium, New J. Phys. 13 (2011), 025012.
[102] I. Aharonovich, S. Castelletto, B.C. Johnson, J.C. McCallum, D.A. Simpson, A.D. Greentree, S. Prawer, Chromium single-photon emitters in diamond fabricated by ion implantation, Phys. Rev. B 81 (2010) 121201.
[103] I.I. Vlasov, A.S. Barnard, V.G. Ralchenko, O.I. Lebedev, M.V. Kanzyuba, A.V. Saveliev, V. I. Konov, E. Goovaerts, Nanodiamond photoemitters based on strong narrow-band luminescence from silicon-vacancy defects, Adv. Mater. 21 (2009) 808–812.
[104] S. Tamura, G. Koike, A. Komatsubara, T. Teraji, S. Onoda, L.P. McGuinness, L. Rogers, B. Naydenov, E. Wu, L. Yan, Array of bright silicon-vacancy centers in diamond fabricated by low-energy focused ion beam implantation, Appl. Phys. Express 7 (2014) 115201.
[105] D. Steinmetz, E. Neu, J. Meijer, W. Bolse, C. Becher, Single photon emitters based on Ni/Si related defects in single crystalline diamond, Appl. Phys. B 102 (2011) 451–458.
[106] E. Neu, C. Arend, E. Gross, F. Guldner, C. Hepp, D. Steinmetz, E. Zscherpel, S. Ghodbane, H. Sternschulte, D. Steinmüller-Nethl, Narrowband fluorescent nanodiamonds produced from chemical vapor deposition films, Appl. Phys. Lett. 98 (2011) 243107.

[107] A. Basov, M. Rähn, M. Pärs, I. Vlasov, I. Sildos, A. Bolshakov, V. Golubev, V. Ralchenko, Spatial localization of Si-vacancy photoluminescent centers in a thin CVD nanodiamond film, Phys. Status Solidi A 206 (2009) 2009–2011.

[108] S. Singh, V. Thomas, D. Martyshkin, V. Kozlovskaya, E. Kharlampieva, S.A. Catledge, Spatially controlled fabrication of a bright fluorescent nanodiamond-array with enhanced far-red Si-V luminescence, Nanotechnology 25 (2014), 045302.

[109] I. Aharonovich, J.C. Lee, A.P. Magyar, D.O. Bracher, E.L. Hu, Bottom-up engineering of diamond micro-and nano-structures, Laser Photonics Rev. 7 (2013) L61–L65.

[110] V. Sedov, I. Vlasov, V. Ralchenko, A. Khomich, V. Konov, A. Fabbri, G. Conte, Gas-phase growth of silicon-doped luminescent diamond films and isolated nanocrystals, Bull. Lebedev Phys. Inst. 38 (2011) 291–296.

[111] A. Dietrich, K.D. Jahnke, J.M. Binder, T. Teraji, J. Isoya, L.J. Rogers, F. Jelezko, Isotopically varying spectral features of silicon-vacancy in diamond, New J. Phys. 16 (2014) 113019.

[112] S. Grudinkin, N. Feoktistov, A. Medvedev, K. Bogdanov, A. Baranov, A.Y. Vul, V. Golubev, Luminescent isolated diamond particles with controllably embedded silicon-vacancy colour centres, J. Phys. D. Appl. Phys. 45 (2012), 062001.

[113] Y. Cui, J. Zhang, F. Sun, Z. Zhang, Si-doped diamond films prepared by chemical vapour deposition, Trans. Nonferrous Metals Soc. China 23 (2013) 2962–2970.

[114] D. Musale, S. Sainkar, S. Kshirsagar, Raman, photoluminescence and morphological studies of Si-and N-doped diamond films grown on Si (100) substrate by hot-filament chemical vapor deposition technique, Diam. Relat. Mater. 11 (2002) 75–86.

[115] V. Sedov, V. Ralchenko, A. Khomich, I. Vlasov, A. Vul, S. Savin, A. Goryachev, V. Konov, Si-doped nano-and microcrystalline diamond films with controlled bright photoluminescence of silicon-vacancy color centers, Diam. Relat. Mater. 56 (2015) 23–28.

[116] E.F. Schubert, Delta-Doping of Semiconductors, Cambridge University Press, 1996.

[117] A. Vikharev, A. Gorbachev, M. Lobaev, A. Muchnikov, D. Radishev, V. Isaev, V. Chernov, S. Bogdanov, M. Drozdov, J. Butler, Novel microwave plasma-assisted CVD reactor for diamond delta doping, Phys. Status Solidi-Rapid Res. Lett. 10 (2016) 324–327.

[118] S. Jin, T. Moustakas, Effect of nitrogen on the growth of diamond films, Appl. Phys. Lett. 65 (1994) 403–405.

[119] W. Müller-Sebert, E. Wörner, F. Fuchs, C. Wild, P. Koidl, Nitrogen induced increase of growth rate in chemical vapor deposition of diamond, Appl. Phys. Lett. 68 (1996) 759–760.

[120] T. Liu, D. Raabe, Influence of nitrogen doping on growth rate and texture evolution of chemical vapor deposition diamond films, Appl. Phys. Lett. 94 (2009), 021119.

[121] S. Dunst, H. Sternschulte, M. Schreck, Growth rate enhancement by nitrogen in diamond chemical vapor deposition—a catalytic effect, Appl. Phys. Lett. 94 (2009) 224101.

[122] G. Cao, J. Schermer, W. Van Enckevort, W. Elst, L. Giling, Growth of {100} textured diamond films by the addition of nitrogen, J. Appl. Phys. 79 (1996) 1357–1364.

[123] R. Locher, C. Wild, N. Herres, D. Behr, P. Koidl, Nitrogen stabilized <100> texture in chemical vapor deposited diamond films, Appl. Phys. Lett. 65 (1994) 34–36.

[124] Y. Avigal, O. Glozman, I. Etsion, G. Halperin, A. Hoffman, [100]-Textured diamond films for tribological applications, Diam. Relat. Mater. 6 (1997) 381–385.

[125] C. Tang, A. Fernandes, F. Costa, J. Pinto, Effect of microwave power and nitrogen addition on the formation of {100} faceted diamond from microcrystalline to nanocrystalline, Vacuum 85 (2011) 1130–1134.

[126] T. Vandevelde, T. Wu, C. Quaeyhaegens, J. Vlekken, M. D'Olieslaeger, L. Stals, Correlation between the OES plasma composition and the diamond film properties during microwave PA-CVD with nitrogen addition, Thin Solid Films 340 (1999) 159–163.

[127] D.S. Dandy, Influence of the gas phase on doping in diamond chemical vapor deposition, Thin Solid Films 381 (2001) 1–5.
[128] C.-S. Yan, Y.K. Vohra, Multiple twinning and nitrogen defect center in chemical vapor deposited homoepitaxial diamond, Diam. Relat. Mater. 8 (1999) 2022–2031.
[129] J.E. Butler, I. Oleynik, A mechanism for crystal twinning in the growth of diamond by chemical vapour deposition, Philos. Trans. R. Soc. A: Math. Phys. Eng. Sci. 366 (2008) 295–311.
[130] T. van Regemorter, K. Larsson, Effect of a NH coadsorbate on the CH_3 (or CH_2) adsorption to a surface step on diamond (100), J. Phys. Chem. C 113 (2009) 19891–19896.
[131] T.V. Regemorter, K. Larsson, Effect of coadsorbed dopants on diamond initial growth processes: CH_3 adsorption, J. Phys. Chem. A 112 (2008) 5429–5435.
[132] J. Butler, Y.A. Mankelevich, A. Cheesman, J. Ma, M. Ashfold, Understanding the chemical vapor deposition of diamond: recent progress, J. Phys. Condens. Matter 21 (2009) 364201.
[133] Z. Yiming, F. Larsson, K. Larsson, Effect of CVD diamond growth by doping with nitrogen, Theor. Chem. Accounts 133 (2014) 1432.
[134] T. Van Regemorter, K. Larsson, Effect of substitutional N on important chemical vapor deposition diamond growth steps, J. Phys. Chem. A 113 (2009) 3274–3284.
[135] B.S. Truscott, M.W. Kelly, K.J. Potter, M.N. Ashfold, Y.A. Mankelevich, Microwave plasma-activated chemical vapor deposition of nitrogen-doped diamond. II: $CH_4/N_2/H_2$ plasmas, J. Phys. Chem. A 120 (2016) 8537–8549.
[136] B.S. Truscott, M.W. Kelly, K.J. Potter, M. Johnson, M.N. Ashfold, Y.A. Mankelevich, Microwave plasma-activated chemical vapor deposition of nitrogen-doped diamond. I. N_2/H_2 and NH_3/H_2 plasmas, J. Phys. Chem. A 119 (2015) 12962–12976.
[137] J. Cui, L. Qiang, B. Zhang, X. Ling, T. Yang, J. Zhang, Mechanical and tribological properties of Ti-DLC films with different Ti content by magnetron sputtering technique, Appl. Surf. Sci. 258 (2012) 5025–5030.
[138] S. Zhang, X.L. Bui, X. Li, Thermal stability and oxidation properties of magnetron sputtered diamond-like carbon and its nanocomposite coatings, Diam. Relat. Mater. 15 (2006) 972–976.
[139] L. Qiang, B. Zhang, Y. Zhou, J. Zhang, Improving the internal stress and wear resistance of DLC film by low content Ti doping, Solid State Sci. 20 (2013) 17–22.
[140] P.V. Bharathy, D. Nataraj, P.K. Chu, H. Wang, Q. Yang, M. Kiran, J. Silvestre-Albero, D. Mangalaraj, Effect of titanium incorporation on the structural, mechanical and biocompatible properties of DLC thin films prepared by reactive-biased target ion beam deposition method, Appl. Surf. Sci. 257 (2010) 143–150.
[141] N.R. Lee, Y.S. Jun, K.I. Moon, C.S. Lee, Ti-doped hydrogenated diamond like carbon coating deposited by hybrid physical vapor deposition and plasma enhanced chemical vapor deposition, Jpn. J. Appl. Phys. 56 (2017), 035506.
[142] W. Dai, P. Ke, M.-W. Moon, K.-R. Lee, A. Wang, Investigation of the microstructure, mechanical properties and tribological behaviors of Ti-containing diamond-like carbon films fabricated by a hybrid ion beam method, Thin Solid Films 520 (2012) 6057–6063.
[143] M. Drory, J. Hutchinson, Diamond coating of titanium alloys, Science 263 (1994) 1753–1755.
[144] J.-H. Wu, M. Sanghavi, J. Sanders, A. Voevodin, J. Zabinski, D. Rigney, Sliding behavior of multifunctional composite coatings based on diamond-like carbon, Wear 255 (2003) 859–868.
[145] J.C. Sánchez-López, D. Martínez-Martínez, C. López-Cartes, A. Fernández, Tribological behaviour of titanium carbide/amorphous carbon nanocomposite coatings: from macro to the micro-scale, Surf. Coat. Technol. 202 (2008) 4011–4018.

[146] X. Liu, P. Lu, H. Wang, Y. Ren, X. Tan, S. Sun, H. Jia, Morphology and structure of Ti-doped diamond films prepared by microwave plasma chemical vapor deposition, Appl. Surf. Sci. 442 (2018) 529–536.

[147] W. Zheng, X. He, M. Wu, S. Ren, S. Cao, D. Guan, R. Liu, X. Qu, Preparation and thermal conductivities of diamond/SiC composites, Appl. Phys. A 124 (2018) 804.

[148] S. Gordeev, S. Zhukov, L. Danchukova, T. Ekstrom, Low-pressure fabrication of diamond–SiC–Si composites, Inorg. Mater. 37 (2001) 579–583.

[149] V.V. Srikanth, X. Jiang, A. Köpf, Deposition of diamond/β-SiC nanocomposite films onto a cutting tool material, Surf. Coat. Technol. 204 (2010) 2362–2367.

[150] S. Yu, Z. Chen, Y. Wang, R. Luo, T. Xu, Y. Pan, J. Liao, Fabrication of SiC/diamond composite coatings by electrophoretic deposition and chemical vapor deposition, Int. J. Appl. Ceram. Technol. 14 (2017) 644–651.

[151] A. Shul'zhenko, E. Ashkinazi, A. Sokolov, V. Gargin, V. Ral'chenko, V. Konov, L. Aleksandrova, R. Bogdanov, A. Zakora, I. Vlasov, Novel hybrid ultrahard material, J. Superhard Mater. 32 (2010) 293–300.

[152] T. Wang, H. Zhuang, X. Jiang, One step deposition of highly adhesive diamond films on cemented carbide substrates via diamond/β-SiC composite interlayers, Appl. Surf. Sci. 359 (2015) 790–796.

[153] T. Wang, S. Handschuh-Wang, Y. Yang, H. Zhuang, C. Schlemper, D. Wesner, H. Schönherr, W. Zhang, X. Jiang, Controlled surface chemistry of diamond/β-SiC composite films for preferential protein adsorption, Langmuir 30 (2014) 1089–1099.

[154] C. Zhu, J. Lang, N. Ma, Preparation of Si–diamond–SiC composites by in-situ reactive sintering and their thermal properties, Ceram. Int. 38 (2012) 6131–6136.

[155] J. Qian, G. Voronin, T. Zerda, D. He, Y. Zhao, High-pressure, high-temperature sintering of diamond–SiC composites by ball-milled diamond–Si mixtures, J. Mater. Res. 17 (2002) 2153–2160.

[156] E. Ekimov, N. Suetin, A. Popovich, V. Ralchenko, Thermal conductivity of diamond composites sintered under high pressures, Diam. Relat. Mater. 17 (2008) 838–843.

[157] B. Matthey, S. Höhn, A.-K. Wolfrum, U. Mühle, M. Motylenko, D. Rafaja, A. Michaelis, M. Herrmann, Microstructural investigation of diamond-SiC composites produced by pressureless silicon infiltration, J. Eur. Ceram. Soc. 37 (2017) 1917–1928.

[158] Z. Yang, X. He, L. Wang, R. Liu, H. Hu, L. Wang, X. Qu, Microstructure and thermal expansion behavior of diamond/SiC/(Si) composites fabricated by reactive vapor infiltration, J. Eur. Ceram. Soc. 34 (2014) 1139–1147.

[159] X. Jiang, C. Klages, Synthesis of diamond/β-SiC composite films by microwave plasma assisted chemical vapor deposition, Appl. Phys. Lett. 61 (1992) 1629–1631.

[160] H. Zhuang, L. Zhang, T. Staedler, X. Jiang, Highly selective diamond and β-SiC crystal formation at increased atomic hydrogen concentration: a route for synthesis of high-quality and patterned hybrid diamond/β-SiC composite film, Scr. Mater. 65 (2011) 548–551.

[161] X. Jiang, V.V. Srikanth, Y. Zhao, R. Zhang, Facet dependent reactivity and selective deposition of nanometer sized β-Si C on diamond surfaces, Appl. Phys. Lett. 92 (2008) 243107.

[162] H. Zhuang, X. Jiang, Growth controlling of diamond and β-SiC microcrystals in the diamond/β-SiC composite films, Surf. Coat. Technol. 249 (2014) 84–89.

[163] H. Kim, Y. Park, I. Choi, β-SiC thin film growth using microwave plasma activated CH4-SiH4 sources, Thin Solid Films 341 (1999) 42–46.

[164] V. Sedov, A. Martyanov, A. Khomich, S. Savin, V. Voronov, R. Khmelnitskiy, A. Bolshakov, V. Ralchenko, Co-deposition of diamond and β-SiC by microwave plasma CVD in H_2-CH_4-SiH_4 gas mixtures, Diam. Relat. Mater. 98 (2019) 107520.

[165] H. Masuda, T. Yanagishita, K. Yasui, K. Nishio, I. Yagi, T.N. Rao, A. Fujishima, Synthesis of well-aligned diamond nanocylinders, Adv. Mater. 13 (2001) 247–249.
[166] A. Barnard, S. Russo, I. Snook, Ab initio modeling of diamond nanowire structures, Nano Lett. 3 (2003) 1323–1328.
[167] A. Barnard, I. Snook, Phase stability of nanocarbon in one dimension: nanotubes versus diamond nanowires, J. Chem. Phys. 120 (2004) 3817–3821.
[168] L. Sun, J. Gong, D. Zhu, Z. Zhu, S. He, Diamond nanorods from carbon nanotubes, Adv. Mater. 16 (2004) 1849–1853.
[169] R. Arenal, P. Bruno, D. Miller, M. Bleuel, J. Lal, D. Gruen, Diamond nanowires and the insulator-metal transition in ultrananocrystalline diamond films, Phys. Rev. B 75 (2007) 195431.
[170] S.A. Rakha, G. Yu, J. Cao, S. He, X. Zhou, Diamond-graphite nanorods produced by microwave plasma chemical vapor deposition, Diam. Relat. Mater. 19 (2010) 284–287.
[171] S. Talapatra, J.-Y. Cheng, N. Chakrapani, S. Trasobares, A. Cao, R. Vajtai, M. Huang, P. Ajayan, Ion irradiation induced structural modifications in diamond nanoparticles, Nanotechnology 17 (2005) 305.
[172] R. Nemanich, J. Glass, G. Lucovsky, R. Shroder, Raman scattering characterization of carbon bonding in diamond and diamondlike thin films, J. Vac. Sci. Technol. A 6 (1988) 1783–1787.
[173] R. Shroder, R. Nemanich, J. Glass, Analysis of the composite structures in diamond thin films by Raman spectroscopy, Phys. Rev. B 41 (1990) 3738.
[174] R. Pfeiffer, H. Kuzmany, P. Knoll, S. Bokova, N. Salk, B. Günther, Evidence for trans-polyacetylene in nano-crystalline diamond films, Diam. Relat. Mater. 12 (2003) 268–271.
[175] H. Kuzmany, R. Pfeiffer, N. Salk, B. Günther, The mystery of the 1140 cm^{-1} Raman line in nanocrystalline diamond films, Carbon 42 (2004) 911–917.
[176] Y. Tang, Y. Li, C. Zhang, L. Zhang, L. Yang, Q. Yang, A. Hirose, Study of nanocrystalline diamond synthesis in MPCVD by bias enhanced nucleation and growth, Diam. Relat. Mater. 25 (2012) 87–91.
[177] R. Dumpala, N. Kumar, C. Kumaran, S. Dash, B. Ramamoorthy, M.R. Rao, Adhesion characteristics of nano-and micro-crystalline diamond coatings: Raman stress mapping of the scratch tracks, Diam. Relat. Mater. 44 (2014) 71–77.
[178] C. Liu, J.-H. Wang, J. Weng, Growth of micro-and nanocrystalline dual layer composite diamond films by microwave plasma CVD: influence of CO2 concentration on growth of nano-layer, J. Cryst. Growth 410 (2015) 30–34.
[179] N. Jiang, K. Eguchi, S. Noguchi, T. Inaoka, Y. Shintani, Structural characteristics and field electron emission properties of nano-diamond/carbon films, J. Cryst. Growth 236 (2002) 577–582.
[180] W. Zhu, G. Kochanski, S. Jin, Low-field electron emission from undoped nanostructured diamond, Science 282 (1998) 1471–1473.
[181] D. Zhou, D. Gruen, L. Qin, T. McCauley, A. Krauss, Control of diamond film microstructure by Ar additions to CH_4/H_2 microwave plasmas, J. Appl. Phys. 84 (1998) 1981–1989.
[182] V. Mortet, O. Elmazria, M. Nesladek, M. Assouar, G. Vanhoyland, J. D'Haen, M. D'Olieslaeger, P. Alnot, Surface acoustic wave propagation in aluminum nitride-unpolished freestanding diamond structures, Appl. Phys. Lett. 81 (2002) 1720–1722.
[183] S. Bhattacharyya, O. Auciello, J. Birrell, J. Carlisle, L. Curtiss, A. Goyette, D. Gruen, A. Krauss, J. Schlueter, A. Sumant, Synthesis and characterization of highly-conducting nitrogen-doped ultrananocrystalline diamond films, Appl. Phys. Lett. 79 (2001) 1441–1443.
[184] X. Xiao, J. Wang, C. Liu, J.A. Carlisle, B. Mech, R. Greenberg, D. Guven, R. Freda, M.S. Humayun, J. Weiland, In vitro and in vivo evaluation of ultrananocrystalline diamond for

coating of implantable retinal microchips, J. Biomed. Mater. Res. B Appl. Biomater. 77 (2006) 273–281.

[185] S. Srinivasan, J. Hiller, B. Kabius, O. Auciello, Piezoelectric/ultrananocrystalline diamond heterostructures for high-performance multifunctional micro/nanoelectromechanical systems, Appl. Phys. Lett. 90 (2007) 134101.

[186] H. Espinosa, B. Peng, B. Prorok, N. Moldovan, O. Auciello, J. Carlisle, D. Gruen, D. Mancini, Fracture strength of ultrananocrystalline diamond thin films—identification of Weibull parameters, J. Appl. Phys. 94 (2003) 6076–6084.

[187] P. Bajaj, D. Akin, A. Gupta, D. Sherman, B. Shi, O. Auciello, R. Bashir, Ultrananocrystalline diamond film as an optimal cell interface for biomedical applications, Biomed. Microdevices 9 (2007) 787–794.

[188] X. Xiao, J. Birrell, J. Gerbi, O. Auciello, J. Carlisle, Low temperature growth of ultrananocrystalline diamond, J. Appl. Phys. 96 (2004) 2232–2239.

[189] D. Zhou, T. McCauley, L. Qin, A. Krauss, D. Gruen, Synthesis of nanocrystalline diamond thin films from an Ar–CH_4 microwave plasma, J. Appl. Phys. 83 (1998) 540–543.

[190] A. Sumant, D. Grierson, J. Gerbi, J. Carlisle, O. Auciello, R. Carpick, Surface chemistry and bonding configuration of ultrananocrystalline diamond surfaces and their effects on nanotribological properties, Phys. Rev. B 76 (2007) 235429.

[191] Y. Zou, Z. Li, Y. Wu, Deposition and characterization of smooth ultra-nanocrystalline diamond film in CH_4/H_2/Ar by microwave plasma chemical vapor deposition, Vacuum 84 (2010) 1347–1352.

[192] L.-J. Chen, N.-H. Tai, C.-Y. Lee, I.-N. Lin, Effects of pretreatment processes on improving the formation of ultrananocrystalline diamond, J. Appl. Phys. 101 (2007), 064308.

[193] T. Ikeda, K. Teii, C. Casiraghi, J. Robertson, A. Ferrari, Effect of the sp 2 carbon phase on n-type conduction in nanodiamond films, J. Appl. Phys. 104 (2008), 073720.

[194] Z. Wang, Q. Wang, H. Li, J. Li, P. Xu, Q. Luo, A. Jin, H. Yang, C. Gu, The field emission properties of high aspect ratio diamond nanocone arrays fabricated by focused ion beam milling, Sci. Technol. Adv. Mater. 6 (2005) 799–803.

[195] C.-R. Lin, W.-H. Liao, D.-H. Wei, C.-K. Chang, W.-C. Fang, C.-L. Chen, C.-L. Dong, J.-L. Chen, J.-H. Guo, Improvement on the synthesis technique of ultrananocrystalline diamond films by using microwave plasma jet chemical vapor deposition, J. Cryst. Growth 326 (2011) 212–217.

[196] C. Tang, A. Fernandes, M. Granada, J. Leitao, S. Pereira, X. Jiang, J. Pinto, H. Ye, High rate growth of nanocrystalline diamond films using high microwave power and pure nitrogen/methane/hydrogen plasma, Vacuum 122 (2015) 342–346.

[197] R.N. Tiwari, L. Chang, Growth, microstructure, and field-emission properties of synthesized diamond film on adamantane-coated silicon substrate by microwave plasma chemical vapor deposition, J. Appl. Phys. 107 (2010) 103305.

Chapter 10

Future trends in microwave chemistry and biology

Aparna Das* and Bimal Krishna Banik*
Department of Mathematics and Natural Sciences, College of Sciences and Human Studies, Prince Mohammad Bin Fahd University, Al Khobar, Kingdom of Saudi Arabia
**Corresponding authors: E-mails: aparnadasam@gmail.com (Aparna Das); bimalbanik10@gmail.com, bbanik@pmu.edu.sa (Bimal Krishna Banik)*

10.1 Current trends in microwave technology

Many researchers have reported the applications of microwaves in various disciplines. For instance, microwave chemistry is applied in various fields such as pharmaceuticals, biotechnology, plastics, petroleum, and chemicals. In general, microwave-induced methods are economical, fast, high yielding, and selective in most of these cases. In the early days, household microwave ovens were extensively used in chemical laboratories for synthesis. Nowadays, dedicated instrumentation with numerous features is available. However, scientists still use kitchen microwave ovens for scientific purposes. There are serious scientific as well as practical reasons to choose dedicated instrumentation. Several scholars have made modifications in the domestic microwave ovens for different purposes. Also various commercial microwave reactors are available in the market for large-scale applications. Several modifications are made in microwave generators, high-frequency power supplies, isolators, waveguide, power monitors, and applicators to improve the performance of the device.

The feature such as the possibility of solvent-free reactions, reduction of reaction time for a chemical/pharmaceutical reaction, the possibility of parallel chemical reaction, and instantaneous and uniform heating caused the microwave technology to stay dominant over conventional heating methods. The researchers involved in drug discovery and development processes like high-speed combinatorial and medicinal chemistry mostly prefer and adopt microwave technology. The microwave technology has encouraged the scientists to initiate new unexplored areas of complex systems both in the organic and inorganic chemistry field.

Microwave-assisted catalytic and noncatalytic synthesis of *N*-heterocycles, mainly five-membered, six-membered, fused, and polycyclic *N*-heterocycles is

Microwaves in Chemistry Applications. https://doi.org/10.1016/B978-0-12-822895-1.00003-5

an important breakthrough in this field. Heterocyclic molecules are an important family of compounds because of their broad range of pharmaceutical and material science applications. Among them, nitrogen-containing heterocyclic (*N*-heterocyclic) compounds have attained the peculiar interest of researchers, owing to its various applications in different areas. Due to widespread interest in *N*-heterocycles, the synthesis of these compounds has always been among the most important research areas in synthetic chemistry. In recent years, numerous studies were exposed for the synthesis of *N*-heterocycles under different reaction conditions such as solvent-free conditions, catalytic reaction conditions, reactants immobilized on a solid support, and one-pot synthetic and microwave irradiation condition. Chemists have successfully conducted a wide range of organic reactions under microwave irradiation, such as the Ene reaction [1], Diels-Alder reaction [2], Heck reaction [3], Mannich reaction [4], Suzuki reaction [5], hydrogenation of lactam [6], esterification [7], dehydration [8], hydrolysis [9], cycloaddition [10], reduction [11], epoxidation [12], cyclization [13], and condensation [14] reactions. Studies have demonstrated the microwave-assisted synthesis of several pharmacologically significant compounds, such as pyrimidines, steroidal derivatives, thiazoles, imines, tetrazoles, triazoles, quinolines, indolizines, β-lactams, pyrroles, and quinoxalines.

Other than *N*-heterocycles, the oxygen-containing heterocyclic (*O*-heterocycles) and sulfur-containing heterocyclic (*S*-heterocycles) compounds are also important classes of compounds in chemistry due to their wide range of applications. Microwave-assisted synthesis of *O*-heterocycles, *S*-heterocycles, and related compounds was also demonstrated by several scholars.

Other than synthesis, microwave-assisted oxidation and reduction reaction is also a highly efficient and clean method. It provides a safer alternative to conventional thermal reactions. Microwave-assisted oxidation methods are fast, economical, high yielding and selective in most of the cases. A highly efficient microwave-assisted modification of conventional heating procedure for the oxidation and reduction reactions of organic compounds was described by several researchers. The advantages of this environmentally benign and safe protocol include a simple reaction setup, application of commercially available catalysts, high product yields, short reaction times, elimination of side products, and low chemical waste.

Microwave technology is important in enzymatic reactions. Microwave irradiation and enzymatic catalysis synergistically raise the reaction rate significantly. Several reports have explained the biodegradation of toxic organic pollutants using different enzymes from bacteria, fungi, and plants. Bioremediation is a cost-effective and environmentally friendly biotechnology that is controlled by microbial enzymes. The studies in this area contribute to reduce the toxicity of the pollutants and also to obtain novel useful products. Thus, microwave technology, together with enzymatic catalysis and solvent-free chemical synthesis, is a nature-friendly method with low wastage of solvent and good yield of the products. A wide range of enzymes from various microorganisms involved in the biochemical processes of various compounds assisted by microwave irradiation. Studies have showed that the activity, stability, and selectivity of the enzyme can be improved by microwave heating. However, the use of microwave irradiation in enzymatic synthesis remains still limited due to the high temperatures associated with microwave heating as enzymes are temperature-sensitive macromolecules. Several studies are going with new technology, by maintaining the temperature as low as 40°C and by precise power inputs.

Infection caused by the pathogenic Gram-positive bacteria, such as *Bacillus subtilis* and *methicillin-resistant Staphylococcus aureus*, as well as Gram-negative bacteria, including *Escherichia coli*, Salmonella, and Vibrio species, and *Pseudomonas aeruginosa*, still stays as the significant public health-threatening issues in the world [15–17]. The same for viruses and fungi, as the viruses are airborne, a good air purification system is needed at places where people gather, such as in offices, schools, hospitals, theaters, assembly halls, and restaurants. Sterilization is the process of inactivation of microorganisms. Sterilization is used essentially in several areas such as in the medical-care industry and hospital [18], in the food industry, in the material manufacturing industry. Several methods are available for the inactivation of harmful microorganisms; it can be achieved by using physical or chemical methods, such as heat, radiation, and chemical solutions or gases.

Widely used sterilization processes include ozone gas sterilization [19], heat sterilization [20], plasma sterilization [21, 22], UV light sterilization [23], and microwave (MW) sterilization [24–29]. Considering the heat sterilization, the full interior of the system should be heated during this process. Thus, the system needs much energy and time. Ozone gas sterilization requires some chemical or physical reactions, and it poses problems such as toxicity. Plasma sterilization is one of the important sterilization methods; at the same time, stabilizing the electric discharge of the airflow system in this method is difficult. UV sterilization is compact, possesses a high sterilization effect, and is very easy to use. It is widely employed in hospitals and laboratories. But UV irradiation can sterilize only the irradiated area (shadow effect). Among the sterilization methods, microwave sterilization has a lot of advantages and it is the most feasible method for several applications. In the microwave sterilization process, the material can be heated rapidly, directly, and selectively, leading to a reduction in power consumption. Besides, microwave sterilization is effective not only for the surface sterilization of material but also for the interior of the material. As microwaves are irradiated from all angles, microorganisms can be sterilized homogeneously and effectively. Owing to these advantages, microwave sterilization is widely used in many fields such as in the medical field, food industry, and material processing field. Microwave-assisted plasma sterilization methods and combined microwave systems were also shown significant results.

Microwave-enhanced chemical vapor deposition for the diamond synthesis is an emerging field. Chemical vapor deposition gives diamonds in shapes and sizes and with composition and quality control unattainable using synthesis at high pressure. Consequently, it is an enabling technology that permits the exploitation of diamond's exceptional properties in a range of technologies outside its traditional role in cutting, grinding, and dressing operations.

Over the past few decades, numerical simulation has become a promising tool to visualize and quantize the microwave heating process in several applications. Investigations based on mathematical simulations were reported for microwave heating in a solid as well as liquid samples; for instance, numerical studies were reported for heating of water, oil, food, wood, minerals, and coal. In certain studies, the microwave power absorbed was considered to decay exponentially into the sample (follows Lambert's law) [30, 31]. And, this was valid only for the large samples and high-loss dielectric materials [32]. In the case of small- or low-loss dielectric materials, the spatial variations of electromagnetic fields and microwave power absorbed within samples must be described with other mathematical expressions such as Maxwell's equations. Normally, during microwave heating of a liquid layer, the change of the microwave power level and variation of an electric conductivity changes the penetration degree and heat generation rate within the liquid layer. The rate of reflection of the microwave is related to the electric conductivity of the heating layer so that the effects of dielectric properties and the change of microwave power level must be considered. Besides, the local heating on the liquid surface by microwave induces the difference of surface tension on the surface layer and results in the flow due to the difference of surface tension. Thus, the influence of the surface tension effect must be considered during analysis. Considering all, the main factors that affect microwave heating include volume, shape, and dielectric properties of the material as well as geometric parameters and design of the microwave unit [33]. All these factors cause it difficult to properly control the heating effect to get the wanted temperature distribution in the material.

10.2 Future trends in microwave technology

The following sections describe the expecting trends in microwave technology in near future.

10.2.1 Organic and inorganic chemistry

The exploitations in microwave technology have provided for its broader usage in further areas of production and research. It is anticipated that a decade from now, microwave reactor will not be unique or uncommon in a chemistry laboratory. It is already like a hot plate or a rotary evaporator laboratory. The microwave units will be small-scale, almost unnoticeable, but it will be exploited every day for most of the chemistry executed to allow fast reactions and push the boundaries of chemistries that can open good outcomes. The microwave technology can open up new synthetic pathways as well as it can allow the usage of many environmentally friendly solvents and catalysts. This will result in cleaner products that will not need as much purification. And later on, microwaves will be simply an another tool in the standard chemistry toolkit, similar like a thin-layer chromatography.

It is expected that the development of materials, for example, semiconductor-based generators for replacing magnetrons and TWT for power supplies in large-scale and small-scale microwave reactors or in domestic cooking microwave, can increase the efficiency of the device. It also will be useful to increase the operational parameters of devices and further disseminating of effective methods for the production of valuable materials. Regarding the heating time, it will shorten heating times than those that are currently available.

The usage of automatic communication system (Internet/Bluetooth connections) in microwave reactor will make more advantages on laboratory experiments as well as on cooking. It will permit users to operate the machines from anywhere. Besides, the system will do some calculations for the user while the reaction is running.

Efficient sensors for properly using power and managing the time of chemical process will be very beneficial. Specialized sensors with the activity of detecting nonheating regions will make the microwave technology more useful. Accuracy in measuring the temperature inside the reactor is important; it is expected that replacing the standard sensor with semiconductor-based temperature sensor having high thermal stability can overcome this drawback of reactor.

Designing of new microwave reactor with stronger materials having better resistance to high temperature or materials having bigger penetration depth can increase the life time of the device.

10.2.2 Material science

Microwaves provide some distinct features and advantages in the field of materials synthesis. Because of the ability of microwaves to act directly with the sample and to turn on and off speedily, the particle growth of the corresponding materials will be much more uniform and controlled and will lead to an improvement in the desired features and property of the materials. Remarkable results have been reported in the field of material synthesis using microwave synthesizers, expecting an increased adoption rate of microwave technology in this field over the next few years.

10.2.3 Biochemistry

Another trend that is happening is the application of microwave technology in biochemistry field. Most of the biomolecules are temperature-sensitive. If it is exposed to elevated temperatures for longer periods of time, it will lose its biological activity. Also, several of the biochemical interactions are slow or not easy to occur. Studies showed that microwave technology can handle most of the challenges in this field. This technology can push biochemical interactions to take place without the high bulk temperatures that can lead to the degradation or activity loss. This shows that the great advantage of microwave energy is efficient energy transfer instead of a method to rapidly heat a solution to raised temperatures. Because of these properties, an enormous amount of acceptance in areas including peptides and proteomics is occurring.

10.2.4 Flow chemistry

Flow chemistry is an emerging field. Flow chemistry can be coupled with microwave technology. It is expected that the better performance of a reaction in a flow microwave. The efficiency of microwave reactions can be improved by using continuous-flow, stop-flow, or large batch (with large power) modes.

10.2.5 Communication

The wireless routers work in microwave frequencies with lower power compared to microwave oven. It is reported that the working of a Wi-Fi device can be interfere with other microwaves with high power (for instance, microwaves from microwave oven) that may affect the working of the Wi-fi device. To overcome this issues, the newly designed Wi-fi devices that use the 802.11ac standards send and receive data at a higher frequency of 5 GHz.

Because of the wide use of electronic equipment like cell phones, computers, communication devices, and wireless facilities, electromagnetic radiation effects in the environment increased and it can cause cancer and several other diseases. Moreover, it is important to exclude the interferences in the aircraft and other precise instruments in military. So, attenuating and minimizing electromagnetic waves are critical problems. Two-dimensional microwave-absorbing and shielding materials and structures can overcome these issues. Several studies have showed that two-dimensional materials and structures are promising for use in microwave devices. Besides, conventional materials with two-dimensional structures, other possibilities are the use of graphene-like materials, such as two-dimensional transition metal dichalcogenides and black phosphorus. These materials also have beneficial shielding and absorbing properties. In the future, these two-dimensional materials will likely play a significant role in electromagnetic wave absorption and cancellation.

10.2.6 Earth science

Researchers investigated the critical technologies for developing advanced microwave radiometers suitable for Earth science applications. An emerging technology enable microwave measurements with adequate spatial resolutions for a number of Earth science parameters, like sea ice, soil moisture, snow, precipitation, sea surface temperature, vegetation, and ocean winds. High spatial resolution microwave sensing from space with sensible swath widths and revisit times helps large real aperture radiometer systems. The aperture systems with novel lightweight compact-packaging techniques can improve the efficiency of the measurement systems. Airborne Earth Science Microwave Imaging Radiometer (AESMIR) is an example for passive microwave airborne imager. It can cover the 6–100 GHz bands that are necessary for observing key earth system elements.

10.2.7 Food industry

In the food industry, with IR scanning of every frozen or nonprepared food having a universal product code will make microwave to read the information put on the food pack and will make the reactor to use automatically the same power and time instructed in the product by the manufactures. This can also be applied in laboratory reactor using UPC codes and code reader.

10.2.8 Medicinal chemistry

Most of the currently available drugs are cytotoxic to normal as well as to diseased cells. Therefore, there is highest need for new drugs with a high degree of potency against diseased cells, low toxicity in normal cells, and unique targets of action. Physicochemical parameters play a crucial role in the prediction of medicinal activity of diverse molecules, and it aids to design more active drugs [34–46]. Microwave-assisted synthesis along with numerical analysis will help scholars to design and synthesis more potent drugs.

Sterilization is important in medical field. Sterilization methods can be improved with new microwave technology using new devices for the generation plasma in microwave-induced plasma system, for example, using semiconductor sources.

10.2.9 Space shuttles

It is expected that space shuttles could soon be launched into space by beaming power to them using microwaves. Using this technology craft can be made far lighter at launch as it is not having to carry fuel. The shuttles powered by beamed high power microwave energy with performance greatly can surpass the limit of chemical combustion rockets. So instead of chemical combustion, the designed space shuttles will be propelled by ejecting hydrogen, heated with the microwave energy. After the space shuttle reaches the destination (or orbit) and distributes the payload, it will glide back to the original position (launch pad), refuel, and will be prepared for its next journey. With this microwave technology, it will be possible to bring to market the reusable, low-cost, single-stage-to-orbit space planes.

10.2.10 Portable microwave

The emerging microwave oven technology can provide an exciting element "a portable microwave" with high efficiency. In that way, the researcher does not need to keep the microwave setup in a counter or in a fixed position.

As a first step, some companies have demonstrated mini microwaves for a car. Several peoples can benefit from it. For example, people who spend more hours in the car per week because of their work duties; who like to travel or camping in RVs; who do not have time to return for a lunch to their home; and who are staying in caravans and mobile homes can use the portable microwave oven.

The most common benefits of mini microwave ovens are as follows: These mini-microwaves are much cheaper than larger units; because of the really compact size of the mini-microwave ovens, they take only far less space; they use a ridiculously small amount of power resources (e.g., 12 V from a lighter socket in the car); despite a small size, a mini-microwave can warm/cook between 140 and 170°F (60–77°C); easy to grip handles for more convenient transportation; mini-microwave also has all basic features of a standard stove; it can heat sandwiches, pizza slices, popcorn, fast food, leftovers, liquid food, etc.; it can also heat baby bottles, food jars in few minutes; and it is very useful in the case of unexpected power outages and other emergency situations where a regular microwave oven cannot work.

Acknowledgments

AD and BKB are grateful to Prince Mohammad Bin Fahd University for support. BKB is also grateful to US NIH, US NCI, and Kleberg Foundation of Texas for financial support.

References

[1] S. Montel, D. Bouyssi, G. Balme, An efficient and general microwave-assisted copper-catalyzed conia-ene reaction of terminal and internal alkynes tethered to a wide variety of carbonucleophiles, Adv. Synth. Catal. 352 (2010) 2315–2320.

[2] M.G. Victoria, A.I. Aranda, A. Moreno, F.P. Cossio, A.D. Cozar, A. Diaz-Ortiz, A.D.L. Hoz, P. Prieto, Microwave-assisted reactions of nitroheterocycles with dienes. Diels-Alder and tandem hetero Diels-Alder/[3, 3] sigmatropic shift, Tetrahedron 65 (2009) 5328–5336.

[3] V. Declerck, J. Martinez, F. Lamaty, Microwave-assisted copper-catalyzed Heck reaction in PEG solvent, Synlett 18 (2006) 3029–3032.

[4] X. Dong, T. Liu, J. Chen, H. Ying, Y. Hu, Microwave-assisted Mannich reaction of 2-hydroxy-chalcones, Synth. Commun. 39 (2009) 733.

[5] C.N. Raut, S.M. Bharambe, Y.A. Pawar, P.P. Mahulikar, Microwave-mediated synthesis and antibacterial activity of some novel 2-(substituted biphenyl) benzimidazoles via

Suzuki-Miyaura cross coupling reaction and their N-substituted derivatives, J. Heterocyclic Chem. 48 (2011) 419–425.
[6] K. Ajay, B.K. Banik, K.J. Barakat, M.S. Manhas, Simplified rapid hydrogenation under microwave irradiation: selective transformations of B lactams, Synlett 8 (1993) 575–576.
[7] H. Shi, W. Zhu, H. Li, H. Liu, M. Zhang, Y. Yan, Z. Wang, Microwave-accelerated esterification of salicylic acid using Brönsted acidic ionic liquids as catalysts, Catal. Commun. 11 (2010) 588–591.
[8] K. Tanji, J. Koshio, O. Sugimoto, Microwave-assisted dehydration and chlorination using phosphonium salt, Synth. Commun. 35 (2005) 1983–1987.
[9] G.V. Baelen, U.W.M. Bert, Study of the microwave-assisted hydrolysis of nitriles and esters and the implementation of this system in rapid microwave-assisted Pd-catalyzed amination, Tetrahedron 64 (2008) 5604–5619.
[10] P. Appukkuttan, V.P. Mehta, E.V.V.D. Eycken, Microwave-assisted cycloaddition reactions, Chem. Soc. Rev. 39 (2010) 1467–1477.
[11] C. Schmoeger, A. Stolle, W. Bonrath, B. Ondruschka, Microwave-assisted organic reduction reactions, Curr. Org. Chem. 15 (2011) 151–167.
[12] U.R. Pillai, E. Sahle-Demessie, R.S. Varma, Microwave-expedited olefin epoxidation over hydrotalcites using hydrogen peroxide and acetonitrile, Tetrahedron Lett. 43 (2002) 2909–2911.
[13] H.M. Ibrahim, H. Behbehani, S. Makhseed, M.H. Elnagdi, Acylation of heteroaromatic amines: facile and efficient synthesis of a new class of 1,2,3-triazolo[4,5-b]pyridine and pyrazolo[4,3-b]pyridine derivatives, Molecules 16 (2011) 3723–3739.
[14] Y. Zhang, B. Zou, K. Wang, Microwave-assisted synthesis of 4-methyl-coumarin derivatives, Huaxue Shiji 32 (2010) 787–789.
[15] S.C. Clarke, R.D. Haigh, P.P.E. Freestone, P.H. Williams, Virulence of enteropathogenic Escherichia coli, a global pathogen, Clin. Microbiol. Rev. 16 (2003) 365–378.
[16] K. Hiramatsu, L. Cui, M. Kuroda, T. Ito, The emergence and evolution of methicillin-resistant Staphylococcus aureus, Trends Microbiol. 9 (2001) 486–493.
[17] D. Metry, R. Katta, New and emerging pediatric infections, Dermatol. Clin. 21 (2003) 269.
[18] A.G. Whittaker, E.M. Graham, R.L. Baxter, A.C. Jones, P.R. Richardson, G. Meek, G.A. Campbell, A. Aitken, H.C. Baxter, Plasma cleaning of dental instruments, J. Hosp. Infect. 56 (2004) 37–41.
[19] M.-W. Byun, H.-S. Yook, I.-J. Kang, C.-K. Chung, J.-H. Kwon, K.-J. Choi, Comparative effects of gamma irradiation and ozone treatment on hygienic quality of Korean red ginseng powder, Radiat. Phys. Chem. 52 (1998) 95–99.
[20] H. Honda, M. Sugiyama, A. Kakamu, A. Mizoguchi, Y. Ishikawa, T. Kobayashi, Characterization on heat sterilization of a novel autoclave with a turntable for rotation, J. Ferment. Bioeng. 81 (1996) 570–572.
[21] S. Lerouge, A.C. Fozza, M.R. Wertheimer, R. Marchand, L. Yahia, Sterilization by low-pressure plasma: the role of vacuum-ultraviolet radiation, Plasmas Polymers 5 (2000) 31–46.
[22] T.C. Montie, K. Kelly-Wintenberg, J.R. Roth, An overview of research using the one atmosphere uniform glow discharge plasma (OAUGDP) for sterilization of surfaces and materials, IEEE Trans. Plasma Sci. 28 (2000) 41–50.
[23] S. Iwaguch, K. Matsumura, Y. Tokuoka, S. Wakui, N. Kawashima, Sterilization system using microwave and UV light, Colloids Surf. B: Biointerfaces 25 (2002) 299–304.
[24] M.R. Sanborn, S.K. Wan, R. Bulard, Microwave sterilization of plastic tissue culture vessels for reuse, Appl. Environ. Microbiol. 44 (1982) 960–964.
[25] R.A. Dunsmuir, G. Gallacher, Microwave sterilization of femoral head allograft, J. Clin. Microbiol. 41 (2003) 4755–4757.

[26] F. Celandroni, I. Longo, N. Tosoratti, F. Giannessi, E. Ghelardi, S. Salvetti, A. Baggiani, S. Senesi, Effect of microwave radiation on Bacillus subtilis spores, J. Appl. Microbiol. 97 (2004) 1220–1227.

[27] D.K. Jeng, K.A. Kaczmarek, A.G. Woodworth, G. Balasky, Mechanism of microwave sterilization in the dry state, Appl. Environ. Microbiol. 53 (1987) 2133–2137.

[28] J.M. Latimer, J.M. Matsen, Microwave oven irradiation as a method for bacterial decontamination in a clinical microbiology laboratory, J. Clin. Microbiol. 6 (1977) 340–342.

[29] D.I. Martin, I. Margaritescu, E. Cirstea, I. Togoe, D. Ighigeanu, M.R. Nemtanu, C. Oproiu, N. Iacob, Application of accelerated electron beam and microwave irradiation to biological waste treatment, Vacuum 77 (2005) 501–506.

[30] C. Saltiel, A.K. Datta, Heat and mass transfer in microwave processing, in: Advances in Heat Transfer, Elsevier, 1999, pp. 1–94.

[31] K.G. Ayappa, S. Brandon, J.J. Derby, H.T. Davis, E.A. Davis, Microwave driven convection in a square cavity, AICHE J. 40 (1994) 1268–1272.

[32] Q. Zhang, T.H. Jackson, A. Ungan, Numerical modeling of microwave induced natural convection, Int. J. Heat Mass Transf. 43 (2000) 2141–2154.

[33] R.C. Anantheswaran, L. Liu, Effect of viscosity and salt concentration on microwave heating of model non-Newtonian liquid foods in a cylindrical container, J. Microw. Power Electromagn. Energy. 29 (1994) 119–126.

[34] A. Das, B.K. Banik, Dipole moment of medicinally active compounds: a sustainable approach, in: Green Approaches in Medicinal Chemistry for Sustainable Drug Design, Elsevier (UK), 2020, pp. 921–964.

[35] A. Das, B.K. Banik, Studies on dipole moment of penicillin isomers and related antibiotics, J. Indian Chem. Soc. 97 (2020) 911–915.

[36] A. Das, A.A. Alqashqari, B.K. Banik, Quantum mechanical calculations of dipole moment of diverse imines, J. Indian Chem. Soc. 97 (9b) (2020) 1563–1566.

[37] A. Das, B.K. Banik, Dipole moment studies on α-hydroxy β-lactam derivatives, J. Indian Chem. Soc. 97 (9b) (2020) 1567–1571.

[38] A. Das, B.K. Banik, Dipole Moment, Encyclopedia, MDPI, Switzerland, 2020. https://encyclopedia.pub/1604.

[39] A. Das, A.K. Bose, B.K. Banik, Synthesis of β-Lactams, Encyclopedia, MDPI, Switzerland, 2020. https://encyclopedia.pub/1575.

[40] A. Das, B.K. Banik, Thiosugars of Biological Significance, Encyclopedia, MDPI, Switzerland, 2020. https://encyclopedia.pub/1552.

[41] A. Das, B.K. Banik, Ascorbic Acid-Mediated Reactions, Encyclopedia, MDPI, Switzerland, 2020. https://encyclopedia.pub/2341.

[42] A. Das, B.K. Banik, Ascorbic Acid-Induced Reactions, Encyclopedia, MDPI, Switzerland, 2020. https://encyclopedia.pub/2300.

[43] A. Das, A.K. Bose, B.K. Banik, Stereoselective synthesis of β-lactams under diverse conditions: unprecedented observations, J. Indian Chem. Soc. 97 (2020) 917–925.

[44] A. Das, R.N. Yadav, B.K. Banik, Ascorbic acid-mediated reactions in organic synthesis, Curr. Organocatal. 7 (2020) 212–241.

[45] A. Das, B.K. Banik, Versatile thiosugars in medicinal chemistry, in: Green Approaches in Medicinal Chemistry for Sustainable Drug Design, Elsevier (UK), 2020, pp. 549–574.

[46] A. Das, B. Banik, Quantum Mechanical Studies of Physicochemical Properties on Estradiol and Isomer, Authorea Preprints, 2020.

Index

Note: Page numbers followed by *f* indicate figures, *t* indicate tables, and *s* indicate schemes.

A

Absorbing materials, 15*f*
Acceleration chamber, 30
Acetones, 133–137
Acetophenones, reduction reactions of, 224
Acetyl thiophene amino benzohydrazide (ATABH), 125–126
Acetyl thiophene benzohydrazide (ATBH), 125–126
Acid etching, 341–343, 342*f*
Acridine-acetazolamide conjugates, 184–187
Acyloins, 231–232, 234*t*
Aggregatibacter actinomycetemcomitans
  growth in microwaved and autoclaved broth, 300, 301*f*
  sterilization of, 299–300, 300*t*
Airborne Earth Science Microwave Imaging Radiometer (AESMIR), 380–381
Alcohols, 133–137
  oxidation reactions of, 202–208
Aldehydes, 222–223, 223*t*, 231, 231*s*
Alkali pretreatment combined with microwave pretreatment (APCMP), 256
Alkene groups, reduction reactions of, 233–235
(*Z*)-alkyl 3-(benzo[d]thiazol-2-yl)-4-oxo-2-(cyclohexyl/arylimino)-3,4-dihydro-2*H*-1,3-thiazine-6-carboxylates, 179–180, 181*s*, 181*t*
Alkyloxypyrrolo[1,2-a]quinoxalinones, 236, 236*s*
α-aryl malonates, 136
α, β-unsaturated carbonyl compounds, reduction reactions of, 226–227
α-chymotrypsin, 264
α-diketones, 231–232, 234*t*
2-Amino-1,3,4-thiadiazole diamides, 176–177, 177*s*
Amoxicillin, Fenton's oxidation of, 215–216
Amylopectin, 247
Amylose, 247
Amylum, 247
Analytical chemistry, 19
Aniline, 229, 230*s*
Apocynol, oxidation of, 200, 200*s*
Applicator (cavity), 33–35
Applicator part, 27–28
Arginine, 288–289
Aromatic aldehydes, 109
  bioreduction of, 228
Aromatic nitro compounds, reduction reactions of
  alumina-supported hydrazine, 230–231
  ammonium chloride and zinc dust, 228
9-Aryl-1,8-dioxo-octahyroxanthenes, microwave-induced synthesis of, 109, 109*s*
(*E*)-3-(Aryl)-1-(5-hydroxy-2*H*-chromen-6-yl) prop-2-en-1-ones, 113
2-Arylidene-2*H*-furo[2,3-f] chromen-3 (7*H*)-ones, 113, 113*s*
Arylphosphonates, 131–133
ASTEX-type reactor, 336–338, 337*f*
Astragalus polysaccharide (APS), 256–257
Atomic polarization, 13–14
Aurone, 112–113
Aza cyclo-octanes, 164
Aza-Diels-Alder reaction, 164, 164*s*
Azido reduction, 224–225

B

Bacillus subtilis, reduction kinetics, 310–313, 316*f*
Baker's yeast, 265–271
β-carboline imines, reduction reactions of, 225–226
Belt press, 333
Benzo[d]thiazole-[1,3]-thiazine hybrids, 179–180
Benzofuran-3(2H)-ones, 113, 113*s*
Benzohydrazones, 165, 165*s*, 166*t*
Benzoselenophenes, 129–130
Benzoxanthenes, 107

Biochemistry, 379–380
Biodiesel, 254, 267
Bioethanol, 271
Biomedical waste management, 286
Bioremediation, 245, 376–377
Bismuth (III) salts, 109
β-lactams, 159–163
β-lactoglobulin (β-Lg), 264
  microwave-assisted digestion of, 264
Boron doping, 349
3-Bromo-2-(1-hydroxyalkyl)benzo[b] selenophene derivatives, 130
Bunsen burner, 10–11, 10*f*
n-Butyl diphenyl methyl mercapto acetate (BDMMA), synthesis of, 247, 247*s*
n-Butyl propionate, 253
  synthesis of, 253, 253*s*

C

Carbazoles, 171–174
Carbohydrates, oxidation reactions of, 208–216
Carbon nanotubes (CNTs), 352–354
Carbonyl compounds, reduction reactions of
  glycerol, 220–222
  sodium borohydride, 231–232
  sodium borohydride supported on alumina, 228
Carboxymethylcellulose, microwave pretreatment of, 255
Cellulase, 255–259
  Cel9M of *Clostridium cellulolyticum*, 255, 256*f*
  microwave pretreatment, 255
Cellulose, microwave-assisted hydrolysis of, 257
Cetyltrimethylammonium chlorochromate (CTMACC) assisted oxidation, 213–214
Chalcones, 110, 168–170
Chemical marker method, 297
Chemical synthesis, 19
Chemical vapor deposition (CVD) method, 332–338
Chromene scaffold, 121
Chymotrypsin, 263–264
Cinnamates, 250
Cinnamic acid derivatives, 114–123, 116*t*
Cinnamyl alcohol oxidation, 206, 206*s*
Coarse-graining methodology, 92
Co-deposition, 352–354
Commercial microwave reactors, 47–55, 51–55*t*
Communication, 380
Computer vision system, 297, 297*f*
COMSOL Multiphysics, 70, 72, 87
Conducting materials, 15, 15*f*
Conjugated carbonyl compounds, 232, 234*t*
Continuous-flow microwave heating system, 289, 291*f*
Conventional heating methods, 16–17, 16–17*f*, 18–19*t*
Copper(II) 2,4-alkoxy-1,3,5-triazapentadienato complexes, 204, 205*f*
Coumarin derivatives, 114–123
Coumarin-pyrimidine amide, 146, 147*s*
Coumarin scaffold, 121
Crossed Cannizzaro reaction, 222–223
Crystal structure
  of ATP sulfurylase from *Saccharomyces cerevisiae*, 265–266, 265*f*
  of cellulase Cel9M of *Clostridium cellulolyticum*, 255, 256*f*
  of human gastric lipase, 245–246, 246*f*
  of human uropepsin, 262, 263*f*
  of *Spodoptera frugiperda* β-glycosidase, 259, 260*f*
Cubic press, 333
CVD. *See* Chemical vapor deposition (CVD) method
Cyclization reactions, 236
Cyclohexane, oxidation of, 207–208, 208*s*

D

Deep eutectic solvents (DESs), 114
Deglycosylation, 261–262
Dehydrogenation-hydrogenation, 223, 224*s*
Delta doping, 349
D-fructose, 114, 114*s*
Diamond, 329
  chemical vapor deposition (CVD) method, 332–338
  doping, 346–354
    co-deposition, 352–354
  high-pressure high temperature (HPHT) growth method, 332–333
  microwave-enhanced chemical vapor deposition, 378
  nanowires, 352–354
  natural diamonds, 330–331
  polycrystalline diamond (PCD) film synthesis, 341–345
  properties, 329–330, 331*t*
  single-crystal diamond (SCD) synthesis, 338–341

structures, 329, 330*f*
synthetic diamond, 332
ultrananocrystalline diamond (UNCD) films, 355–363
Diamond-graphite nanorods, 354
Diamond nanorods (DNRs), 352–354, 356*f*
Dielectric constants, 293, 294*f*
Dielectric heating, 293
Dielectric materials, 15
9,10-Dihydrophenanthrene, 130
2,2-Dimethylchroman-7-ol, 108
5,5-Dimethyl-1,3-cyclohexanedione, 108
Dimethylformamide (DMF), 132–133
3,6-Dinitro-9,10-dihydrophenanthrene, 130
1,8-Dioxooctahydroxanthene, 108–109, 108*s*
Diphenyl methyl mercapto acid (DPMMA), 247
Dipolar polarization, 13–14, 13*f*
Direct current (DC) arc jet method, 335, 336*f*
Disk diffusion method, 110
2,5-Disubstituted-1,3,4-oxadiazoles, 153–154, 153*s*
3,6-Di(1,3,4-oxadiazol-2-yl)-1,2,4,5-tetrazine, 175, 175*s*
3,6-Di(1,3,4-thiadiazol-2-yl)-1,2,4,5-tetrazine, 175, 176*s*
Dome-shaped cavity-type reactor, 341–343, 342*f*
Domestic microwave ovens, in laboratory, 35–47
modifications in, 35–47, 36–39*f*, 42*f*, 46*f*, 50*f*
for cycloaddition, 41, 43*f*
degradation of sodium dodecyl sulfate, 41, 44*f*
dichlorination and polychlorinated biphenyls decomposition, 41, 44*f*
Fries rearrangement, 47, 50*f*
heater/dryer system, 46, 49*f*
hydrodiffusion and gravity setup, 43–45, 48*f*
microwave melting, 42–43, 47*f*
pyrolysis of biomass, 39, 40*f*
thin-layer drying of wheat, 42, 45*f*
two reflux apparatuses, 40, 41*f*
Doping, diamond, 346–354
Dry sterilization process, in microwave N_2O plasma, 309, 309*f*

E

Earth science, 380–381
Eichhornia crassipes, 259
Electric devices, 28
Electric field, 3
Electromagnetic (EM) radiation, 3
Electromagnetic spectrum, 3–5, 4*f*, 5*t*
Electromagnetic wave, 3
Electron cyclotron resonance (ECR) microwave plasma, 307, 308*f*
Electron gun, 30
Electronic polarization, 13–14, 13*f*
Empty fruit bunch (EFB)
enzymatic saccharification, 257
lignin and holocellulose reduction, 257, 258*f*
Enzymatic saccharification
empty fruit bunch (EFB), 257
of water hyacinth, 259
Esterification
starch, 248, 250*s*
5-Ethoxymethylfurfural (EMF), 114, 114*s*
Ethyl cinnamate, transesterification reaction of, 250, 251*s*
Ethylene oxide gas (EOG) sterilization, 305–306
4-((9-Ethyl-9*H*-carbazol-3-yl)methylene) amino-1,5-dimethyl-2-phenyl-1,2-dihydro-3*H*-pyrazol-3-one, 183–184, 183*s*
Ethyl 2-methyl-4-(pyridin-2-yl)-4*H*-benzo[4,5] thiazolo[3,2-a]pyrimidine-3-carboxylate, 150, 151*s*
Ethyl 3-substituted-7-methylindolizine-1-carboxylates, 159, 159*s*
Ethyl valerate, 254

F

Faraday rotation isolators, 32
Fatty acid methyl esters (FAME), 267–268, 268*s*
Feist-Benary reaction condition, 116–118
Fenton's oxidation, 215–216
Ferrite-based isolators, 32
Field-displacement isolators, 32
Flavonolignans, 200
Flow chemistry, 380
Food industry, 381
microwave sterilization in, 288–303
Food processing applications, 19
Formic acid, 209, 212*s*
Furan derivatives, 109–114
5-(2-Furyl)-3(2*H*)-furanones, 112, 112*s*, 112*t*

G

Galacto-oligosaccharides, 259, 261, 261*s*
Galacturonic acid, 209, 210*s*
Gas-phase reactions, 333

Gastric lipase, human, 245–246, 246*f*
Glucose
  base-free oxidation of, 209, 212*s*
  microwave-assisted oxidation, 209, 210*s*, 212*s*
Glycerol, oxidation of, 202, 203*s*
3-Glycidoxypropyltrimethoxysilane, 264
Glycolic acid, 209, 211*s*
Glycosidases, 259–262
Glycosylation, 261
Glycosyltransferase trehalose-6-phosphate synthase, 266
Gram-negative bacteria, 285
Gram-positive bacteria, 285
Graphite, 333
Gyrotron, 30, 31*f*

H

Heat sterilization, 285–286, 377
Hepatic lipases, 245–246
(Hetero)aryl boronic acids, 180, 182*t*
Heterocyclic molecules. *See* *N*-heterocycles
HFCVD. *See* Hot filament chemical vapor deposition (HFCVD)
High-frequency power supplies, 32
High-pressure high temperature (HPHT) growth method, 332–333
Hormone-sensitive lipases, 245–246
Hot filament chemical vapor deposition (HFCVD), 335, 335*f*
Hot plates method, 10–11, 11*f*
Hot-spot effects, 96
HPHT method. *See* High-pressure high temperature (HPHT) growth method
Human gastric lipase, 245–246, 246*f*
Human uropepsin, crystal structure of, 262, 263*f*
Hydrazinyl thiazoles, 149–150, 150–151*s*
Hydrolysis, of triolein, 252, 252*f*
Hydrophilic chip technology, 261–262
Hydrophobic chip technology, 261–262
Hydroxycoumarin derivatives, 120–121, 120*s*
1-(5-Hydroxy-2*H*-chromen-6-yl)ethanone, 113
5-Hydroxymethyl furfural (HMF), 114
Hyperfluorescence 2-pyridone derivative (D-TPCA), 177, 177*s*

I

Imines, 151–153
5-*R*-Imino-[4-trialkylsilyl(germyl)]-1*H*-1,2,3-triazoles, 153
Indan-1-ones, oxidation of, 212–213
Indolizines, 157–159
Indoloquinolines, 157, 158*t*, 158*s*
2-[(*E*)-2-(1*H*-Indol-3-yl)vinyl]hetarenes, 169–170, 170*s*
Industrial applications, of microwave heating, 19
Infection, 377
Inorganic synthesis, 19
Insulation materials, 15, 15*f*
Interfacial polarization, 13–14, 13*f*
Ionic conduction, 13–14, 13*f*
Ionic polarization, 13–14
Ionization, 335
Ipsapirones, 174–176
Isatin-3-aminorhodanine, 175–176, 176*s*
Isatin, Wolff-Kishner reduction reaction of, 219, 220*s*
Isoamyl butyrate, 254
Isocryptolepine, tetracyclic ring systems of, 156, 157*t*
Isolators device, 32

J

Juice pasteurization, 270

K

Kernel temperature, 295, 296*f*
Ketones, 231, 233*t*
Ketorolac, 248
Klystron microwave amplifier, 28–29
Klystron microwave oscillator, 28–29
Klystron microwave tube, 28–29, 29*f*
Knoevenagel condensation, 175–176

L

Lactobacillus plantarum
  continuous microwave heating, 291, 291*f*
  conventional batch heating, 291, 291*f*
  destruction kinetics, 289
  temperature sensitivity of inactivation rates, 291, 291*f*
  time-temperature profiles, 291–292, 292*f*
Lambert's law, 61, 80–82
Lawesson's reagent (L.R.), 127
L-band, 5
Levulinic acid
  glucose oxidation to, 209, 212*s*
  reduction reactions of
    Pd/C-catalyzed reduction, 216–217
    Ru/C-catalyzed reduction, 223
Lignin, oxidation of, 199–200

Lignocellulosic materials, 257
Limulus amoebocyte lysate (LAL), 305
Lipases, 245–255
Lipoprotein lipase, 245–246
Lysine, 288–289

M

Magnetic field, 3
Magnetic silica microspheres, trypsin-immobilized, 264
Magnetron, 28, 29*f*
Maltooligosaccharyltrehalose synthase, 266
Maltooligosaccharyltrehalose trehalohydrolase, 266
Manganese(III) acetate-based oxidative cyclizations, 215
Material science, 379
Mathieu function, 65
Maxwell's equation, 61–64, 67, 80–82, 84, 87–88
Medical applications, of microwave heating, 19
Medical industry, microwave sterilization in, 286–288
Medicinal chemistry, 381
Meker burner, 10
Methicillin-resistant Staphylococcus aureus (MRSA)
  microwave-induced argon plasma system, 310, 313*f*, 315*f*
  sterilization effects of, 310
Methyl 7-amino-4-oxo-5-phenyl-2-thioxo-2, 3, 4,5-tetrahydro-1*H*-pyrano[2,3-d] pyrimidine-6-carboxylates, 146, 146*s*
Microcrystalline diamond, 347–349, 348*f*
Microwave-assisted organic synthetic method, 20
Microwave-assisted oxidation and reduction, 376
Microwave-assisted pyrolysis, 37, 39*f*
Microwave-assisted synthesis
  Cannizzaro reaction, 133–134
  Claisen-Schmidt reaction, 135
  dialkyl ω-azidoalkylphosphonates, 132, 132*s*
  Hirao reaction, 131
  N-heterocycles (*see N*-heterocycles)
  palladium-catalyzed Heck coupling reaction, 122
  Vilsmeier-Haack reaction, 122
Microwave-assisted thermal sterilization (MATS), 292–293
Microwave chemistry, 10–20
  electromagnetic spectrum, 3–5, 4*f*, 5*t*
  evolution of, 11*t*
  microwave radiation, 5–9
Microwave-convective hybrid oven, 42–43, 47*f*
Microwave device
  structure of, 27–35, 28*f*
    applicator (cavity), 33–35
    high-frequency power supplies, 32
    isolators device, 32
    microwave generator, 28–32
    power monitors, 33
    waveguide, 32, 33*f*
Microwave drying setup, 45, 49*f*
Microwave-enhanced chemical vapor deposition, 378
Microwave frequency bands, 6*t*
Microwave generation part, 27–28
Microwave generator, 28–32
Microwave heating
  continuous-flow microwave system, 61
  heating effects, 10–12, 12*f*
    advantages of, 17–19
    applications of, 19–20
    vs. conventional heating, 16–17, 16–17*f*, 18–19*t*
    principles of, 13–15
  in liquids, heat transfer modeling
    analytical model, 62, 63*f*, 67, 68*f*
    average simulated temperature, 64, 65*f*
    ballast water, 65–67
    COMSOL Multiphysics, 70, 72
    continuous-flow focused microwave system, 64–67, 64*f*
    cross-axial temperature, 67
    dielectric properties, 62, 63*f*
    distilled water, 72, 74*f*
    electromagnetic fields, 63
    evaporation process, 77
    finite difference time domain method, 63–64, 67
    fruit juices for enzyme inactivation, 73–77, 75*f*, 77*f*
    heat-resistant unwanted enzymes, inactivation of, 72
    intensification, 62
    microwave power absorption with time, deviation of, 77, 79*f*
    NaCl-water solution layer, 62
    nonuniform temperature distribution, 68–70
    numerical model, 63–64
    prepared model, 64
    radial temperature distribution, 65, 65*f*

Microwave heating *(Continued)*
rectangular waveguide, 62
resonant cavity, 64
simulated temperature differences, 68, 71*f*
single-phase microwave liquid heating, computational domain of, 77, 79*f*
superheating temperature, in organic solvents, 62
two-dimensional heat and momentum equations, 62
uniform temperature distribution, 68–70
water layer, 62
mathematical simulations, 61
plug flow approach, 61
in solids, heat transfer modeling
absorbed microwave power, 80–84, 85*f*, 88
axis-symmetric model, food system, 82, 84*f*
biomass, microwave-assisted heating, 87, 90*f*
combination-jet impingement oven, 85–86, 88*f*
complex dielectric permittivity, 86, 90*f*
computational analysis, using multiphysics software, 89–92
energy consumption, 88
finite difference method, 82
finite element method (FEM), 80–82
hot-spot effects, 96
hygroscopic porous media, microwave drying/heating of, 84
Lambert's law, 80–82
Maxwell's equations, 80, 82
microwave heating of coal, geometrical model, 93–96
NaY zeolite, dielectric properties of, 96–98, 97*f*
numerical analysis, 93
penetration depth, temperature effect, 87, 90*f*
power distributions, 82, 83*f*
SiC particles, 96
simulation domain, 80–81, 81*f*
slurry reactor, catalytic reactions, 88–89, 92
stirring speed, 92–93
temperature distribution, in wood sample along the horizontal axis, 84–85, 87*f*
transient heat equation, 80–81
zeolites, 96
surface tension effect, 61
Microwave-induced argon plasma system
Aspergillus niger, 309, 312*f*
on bacterial and fungal strains in silk fabrics, 309, 310*f*
E. coli, 305, 305*f*, 310, 314*f*
LPS inactivation, 305, 306*f*
methicillin-resistant Staphylococcus aureus (MRSA), 310, 313*f*
Pseudomonas aeruginosa, 309, 311*f*
Microwave-infrared (MW-IR) heating technology, 313–316
Microwave interactions, with living cells
destructive action, 271
nondestructive action, 271
Microwave irradiation, 301, 302*f*
accelerated removal of N-linked oligosaccharides using PNGase F, 262
cellulase production with *Trichoderma* sp., 256
diphenyl methyl mercapto butyrate synthesis, 247
galacto-oligosaccharides synthesis, 261, 261*s*
glycosidase-catalyzed reversed hydrolysis and transglycosidation, 259, 260*s*
on isoamyl butyrate ester synthesis, 254
on protease-catalyzed esterification and transesterification, 263–264
Microwave ovens, 32
Microwave photochemical reactor, 39, 40*f*
Microwave plasma chemical vapor deposition (MWCVD)
for diamond synthesis (*see* Diamond)
Microwave plasma jet chemical vapor deposition (MPJCVD), 357–360, 360*f*
Microwave plasmas, sterilization using, 304–313
Microwave radiation, 5–9, 15
advantages of, 6, 7*t*
applications of, 7–9, 7–9*t*
disadvantages of, 6, 7*t*
properties of, 6
on survival of
gram-negative bacterial cells, 288, 289*f*
gram-positive bacterial cells, 288, 289*f*
Microwave radio links, 32
Microwave sterilization, 285–286
in food industry, 288–303
in medical industry, 286–288
Microwave (MW) sterilization, 377
Microwave technology, 375–378
biochemistry, 379–380
communication, 380
earth science, 380–381
in enzymatic reactions, 376–377
flow chemistry, 380

food industry, 381
material science, 379
medicinal chemistry, 381
organic and inorganic chemistry, 378–379
portable microwave, 382
space shuttles, 381–382
Microwave-UV light sterilization system, 318, 319*f*
Microwave vacuum drying setup, 42, 45*f*
Mini microwaves, 382
Modafinil, 247
MPJCVD. *See* Microwave plasma jet chemical vapor deposition (MPJCVD)
Multicavity klystron, 28–29
Multimode microwave applicator, 34–35, 34*f*
Mycobacterium tuberculosis, 286

N

Nanocrystalline diamond (NCD), 347–349, 348*f*, 354, 358*f*, 360–361
Nanodiamond (ND) particles, 352, 353*f*
N-aroyl-*N'*-aryl thioureas, 124, 124*s*
Natural diamonds, 330–331
NCD. *See* Nanocrystalline diamond (NCD)
Nef reaction condition, 115–116
N-heterocycles, 375–376
acridine-acetazolamide conjugates, 184–187
aza cyclo-octanes, 164
β-lactams, 159–163
carbazoles, 171–174
chalcone, 168–170
imines, 151–153
indolizines, 157–159
ipsapirones, 174–176
oxadiazoles, 153–154
pyrimidines, 144–147
pyrroles, 163–164
quinolines, 156–157
quinoxalines, 170–171
Schiff bases, 165–168
steroidal derivatives, 147–149
tetrazoles, 154–156
thiazoles, 149–151, 176–184
triazoles, 154–156
NH-1,2,3-triazoloimines, 153
N-hydroxy pyrroloquinoxalinones, 236
NIRIM-type reactor, 336–338, 337*f*
4-Nitroaniline, 230, 230*s*
Nitroarenes, reduction reactions of, 218–219
Nitrobenzene, microwave-assisted reduction of, 218–219, 219*s*
Nitro compounds, reduction reactions of, 229
Nitrogen, 350
Nitrogen oxide radicals, sterilization effect of, 305–306
5-(4-Nitrophenyl)-2-phenyl-4-(phenylsulfonyl)-2,3-dihydrofuran, 215
Noncyclic ketones, isobenzofuran-1(3*H*)-ones, 111, 111*t*
Nonthermal effects (reversible), 15
Novozym 435, 248–250, 250*s*, 253–254
Numerical simulation, 378

O

Octahydroxanthenes, microwave-induced bismuth triiodide-catalyzed facile synthesis, 109
O-heterocycles, 376
Oil baths method, 10–11, 11*f*
Oligosaccharides, 259
chemical synthesis, 261
nucleotide-activated, synthesis of, 261, 262*s*
Organic and inorganic chemistry, 378–379
Organic synthesis, 19
Organosulfur compounds
discorhabdin A, 123, 124*f*
disulfiram, 123, 124*f*
penicillin, 123, 124*f*
Osthole, 118
Oxadiazoles, 153–154
Oxidation reactions
alcohols, 202–208
carbohydrates, 208–216
polyphenolic molecules, 199–202
4-Oxo-*N*-(substituted-thiazol-2-yl)-4*H*-chromene-2-carboxamides, 149, 149*s*
Oxygen-containing heterocyclic (*O*-heterocycles), microwave-assisted synthesis of
alcohols and acetones, 133–137
arylphosphonates, 131–133
benzoselenophenes, 129–130
cinnamic acid derivatives, 114–123, 116*t*
coumarin derivatives, 114–123
furan derivatives, 109–114
phenanthrene, 130
xanthene derivatives, 107–109
Ozone gas sterilization, 285–286, 377

P

Palm oil extraction process, 293–294
  pure-microwave and presteaming microwave treatment
    fruit appearance, 295, 296*f*
    kernel temperature, 295, 296*f*
    mesocarp temperature, 294–295, 295*f*
Pancreatic lipase, 245–246
Paprika, combined microwave-infrared-assisted heating of, 316–318, 317*f*
Passive microwave airborne imager, 380–381
Pepsin, 264
Peptide *N*-glycosidase F (PNGase F), 261–262
Peroxidase enzyme, 72
Peroxidative oxidation, 204–205, 206*s*
Phenanthrene, 130
1,4-Phenylenediamine, 230–231
1-Phenylethanol
  microwave-assisted oxidation of, 202–204, 204*s*
  oxidation of, 207, 208*s*
2-Phenylethanol, oxidation of, 207, 208*s*
Phthalimides, 179, 179*s*
Phytopathogenic fungi, 119
Pilot microwave sterilization system, 297, 298*f*
Planartron plasma, spore inactivation with, 313, 316*f*
Plasma sterilization, 285–286, 377
Poison plate technique, 110
Polycrystalline diamond (PCD) film synthesis, 341–345
Polyphenolic molecules, oxidation reactions of, 199–202
Polytetrafluoroethylene (PTFE), 17
Portable microwave, 382
Power monitors, 33
Pronase, 264
Propanoyl thiophene amino benzohydrazide (PTABH), 125–126
Propanoyl thiophene benzohydrazide (PTBH), 125–126
Protease, 262–265
Protein digestion method, 264
Pyranocoumarin derivatives, 121, 121*f*
Pyranoxanthenes, one-pot three-component reaction, 107–108
Pyrazolines, 168–169, 169*s*, 177, 178*s*
Pyrazolo[4′,3′:5,6] pyrido[2,3-d]pyrimidines, 144, 144*s*
Pyrimidines, 144–147
Pyrimidinones, 173–174, 173*s*
Pyrroles, 163–164

Q

Quinazolinone derivatives, 187, 187*s*, 188*t*
Quinolines, 156–157
Quinolinyl-4*H*-pyrans, 179, 180*s*
Quinoxalines, 170–171

R

Radar sets, 32
Rectangular waveguide, 62, 84–85
Reducing sugar content (RSC), 256
Reduction reactions
  acetophenones, 224
  alkene groups, 233–235
  α, β-unsaturated carbonyl compounds, 226–227
  aromatic aldehydes, bioreduction of, 228
  aromatic nitro compounds
    alumina-supported hydrazine, 230–231
    ammonium chloride and zinc dust, 228
  azido reduction, 224–225
  β-carboline imines, 225–226
  carbonyl compounds
    glycerol, 220–222
    sodium borohydride, 231–232
    sodium borohydride supported on alumina, 228
  crossed Cannizzaro reaction, 222–223
  cyclization reaction, 236
  levulinic acid
    Pd/C-catalyzed reduction, 216–217
    Ru/C-catalyzed reduction, 223
  nitroarenes, 218–219
  nitro compounds, 229
  trimethylsilyl carbonyl compounds, 219–220
  Wolff-Kishner reduction, 219
Reflex klystron, 28–30, 30*f*
Resonance-absorption-type isolator, 32, 33*f*
Resonance chamber, 30
Resonance isolators, 32
Rhizomucor miehei, 253–254
Rhodamine-derived imines, 185–187, 186*s*, 187*t*
Ring A-fused arylpyrimidines, 145, 145*s*
Ring D-fused arylpyrimidines, 145, 145*s*
Rotary system, 46*f*
Rotation of dipoles, 13–14

S

Saccharomyces cerevisiae, 265–266
  continuous microwave heating, 291, 291*f*
  conventional batch heating, 291, 291*f*

crystal structure of ATP sulfurylase from, 265–266, 265*f*
destruction kinetics, 289
microwave-assisted immobilization, 270
screening of yeast lipids in, 268–269
temperature sensitivity of inactivation rates, 291, 291*f*
time-temperature profiles, 291–292, 292*f*
Satellite communications, 32
Scanning electron microscopy (SEM), of starch samples, 248, 249*f*
SCD. *See* Single-crystal diamond (SCD) synthesis
Schiff bases, 165–168
Secondary alcohols, microwave-assisted oxidation of, 204, 205*s*
Semiconductor-based microwave generators, 31–32
Si doping, 346–347
Silibinin, microwave-assisted base-catalyzed oxidation of, 200–202, 201*s*
Single-crystal diamond (SCD) synthesis, 338–341
Single-mode microwave applicator, 34, 34*f*
Si substrate, 352, 353*f*
Solvent-free peroxidative reaction, 202–204
Space shuttles, 381–382
Specific absorption ratio (SAR), 271
Spectroscopic techniques, 5
Split-sphere (BARS) press, 333, 334*f*
Spodoptera frugiperda β-glycosidase, crystal structure of, 259, 260*f*
Starch
modification of, 247–248
oxidation of, 212
Sterilization, 285, 377, 381
of *E. coli*, 319*f*, 320
heat, 285–286
of hydrophilic contact lenses, 288
microwave, 285–286
ozone gas, 285–286
plasma, 285–286
using combined system, 313–320
using microwave plasmas, 304–313
UV light, 285–286
Steroidal derivatives, 147–149
1-(Substituted-1*H*-1,2,3-triazol-4-yl)methyl) diphenylphosphine oxides, 155–156, 155*s*
Subtilisin, 263–264
Sugars, oxidation of, 208–209
Surface polarization, 13–14
Suzuki-Miyaura cross-coupling, 180–183, 182*t*, 182–183*s*
Synthetic diamond, 332

T

Tan δ values, 14, 14*t*
Teclu burner, 10
Teflon batch-type reactor, 36, 38*f*
Terminated circulators, 32
Tert-butyl amides, 146
Tetrazoles, 154–156
Thermal effects (nonrefundable), 15
Thermomyces lanuginosus, 253–254
Thiazole fused thiosemicarbazones, 185, 186*s*
Thiazoles, 149–151, 176–184
Thioketones, 127, 128*t*
Thiophene hydrazones, 126, 126*s*
Thiosemicarbazone derivative, 172, 172*s*
Tirrill burner, 10
Titanium doping, 351, 351*f*
Transglycosidase, 266
Traveling wave tube (TWT), 30–31, 31*f*
Trehalose
biosynthesis of, 266
molecular structure of, 266, 266*f*
Trehalose synthase, 266
Triazoles, 154–156
1*H*-1,2,3-Triazoloimines, 153, 153*s*
Trichoderma viride, enhanced cellulase production of, 255–256
Tri-heterocyclic benzothiazole, 150, 151*s*
Trimethylsilyl carbonyl compounds, reduction reactions of, 219–220
Triolein, 252
hydrolysis, 252, 252*f*
Aspergillus carneus lipase, 246–247
microwave heating on lipase-catalyzed, 252–253
Two-cavity klystron tube, 28–30, 29*f*

U

Ultrananocrystalline diamond (UNCD) films, 355–363
Uropepsin, human, 262, 263*f*
UV light sterilization, 285–286, 377

V

Valeric acid, 254
Valerolactone, 216–217, 217*s*, 223, 224*s*

W

Wall heating, 16
Waste treatment, 19
Water hyacinth. *See Eichhornia crassipes*
Waveguide, 32, 33*f*
Whey protein gel (WPG), dielectric properties of, 297, 298–299*f*
Wolff-Kishner reduction reaction, 219–222, 222*s*

X

Xanthene derivatives, 107–109
Xylose oxidation, 209, 211*s*

Z

Ziehl-Neelsen (ZN) staining, 286, 287*f*

FSC
www.fsc.org
MIX
Papier | Fördert
gute Waldnutzung
FSC® C083411

Kolibri 360 GmbH
Ferdinand-Jühlke-Straße 7
99095 Erfurt, Deutschland
produktsicherheit@kolibri360.de